A Classified Inventory of Medicinal and Aromatic Plants Species of Indian Traditional Medicinal Systems

NIPA® GENX ELECTRONIC RESOURCES & SOLUTIONS P. LTD.
New Delhi-110 034

About the Author

Dr. Anurudh Kumar Singh was educated at Aligarh Muslim University (1962–1966); Patna University (1969–1972); the Max Planck Institute, Koln, Germany (1986–1987); and Cornell University, Ithaca, USA (1994–1995). He has been interested in agrobiodiversity and Medicinal and Aromatic Plants from the beginning of his career. He has made significant contributions towards the management of agrobiodiversity both nationally and internationally. More significantly, from 1978 to 1997, while at the International Crops Research Institute for the Semi-Arid Tropics (ICRISAT), and from 1998 to 2008 as a curator of the National Gene Bank and head of the Germplasm Conservation Division at the National Bureau of Plant Genetic Resources (NBPGR), New Delhi. He was responsible for the *ex-situ* conservation of crop germplasm including medicinal and aromatic plants. As member secretary of the Germplasm Registration Committee, Indian Council of Agricultural Research (ICAR), he registered several medicinal plants; and as member secretary of the Task Force for the Protection of Plant Varieties and Farmers' Rights (PPV & FR), he was involved in the identification of agrobiodiversity hotspots in India, which included medicinal plants, and production of the first inventory of cultivated species and their wild relatives. For these contributions he received the "The CGIAR Chairman's Excellence in Science Award" for "Outstanding Locally Recruited Scientist" (1997) at ICRISAT. He was awarded a fellowship of the National Academy of Sciences, India (2001), and of the National Academy of Agricultural Sciences, India (2004). He has more than 225 scientific publications to his credit, including 6 books.

Indian Traditional Medicine System - Volume 01

A Classified Inventory of Medicinal and Aromatic Plants Species of Indian Traditional Medicinal Systems

Anurudh Kumar Singh

NIPA® GENX ELECTRONIC RESOURCES & SOLUTIONS P. LTD.
New Delhi-110 034

NIPA. GENX ELECTRONIC RESOURCES & SOLUTIONS P. LTD.

101,103, Vikas Surya Plaza, CU Block
L.S.C.Market, Pitam Pura, New Delhi-110 034
Ph. +91 11 27341616, 27341717, 27341718
E-mail: newindiapublishingagency@gmail.com
www: www.nipabooks.com

For customer assistance, please contact
Phone: + 91-11-27 34 17 17
Fax: + 91-11-27 34 16 16
E-Mail: feedbacks@nipabooks.com

ISBN: 978-81-19002-50-4

Composed and Designed by NIPA®.

Acknowledgement

The author would like to acknowledge with thanks the editorial help extended by Dr. SN Nigam, former principal groundnut breeder, ICRISAT, Patancheru, Telangana. Also, acknowledgements are due for the technical help extended by various peers and colleagues in the preparation of the book, particularly Drs. JP Yadav, Dean Faculty of Science, MD University, Rohtak, Haryana and SD Purohit, Former, Director Biotechnology Centre Mohanlal Sukhadia University Udaipur, Rajasthan for perusing the part of the manuscript related to their expertise, providing comments, and sharing some information.

Thanks, are also due to NIPA Genx Electronic Resources and Solutions PVT LTD., for accepting the project proposal, and providing support for the publication of the book. In this regard, the author particularly appreciates the assistance extended by editing and composition teams in the preparation and production of the book.

Preface

The use of biological resources as medicine has been practiced in India since time immemorial, i.e., pre-*Vedic* period. This has led to a rich heritage of medical wisdom and accumulation of vast knowledge about properties of the plants, which can be used for human wellness and health care. This knowledge was further enriched with adoption and integration of exotic knowledge introduced from other cultures at different times of the history. In addition, there are plants closely associated with the traditional medicinal use by various indigenous tribes and communities. The systematic documentation of this knowledge with reasoned based application and practice led to the establishment of well recognized traditional medicinal systems, and many plants under various recognized indigenous systems of medicine, such as Ayurveda, Yoga and Naturopathy, Unani, Siddha, and Homeopathy (AYUSH) (www.indianmedicine.nic.in), and use of many more plants by tribes/communities/families based on experience of generations with casual description and/or no description. These documented and non-documented information's present a repository of information on nutritional, medicinal and other economic properties/value of many plants growing in India and used under traditional medicinal systems.

Considering that these are still being used, including along with modern systems of medicine, a formal inclusion of traditional herbal medicine in clinical practice will help to achieve the target of 'health for all'. Recognizing this, the Government of India is attempting to integrate these plants and systems in education, practice, drug manufacture, through various acts, rules, and regulations for quality control and sale. However, efforts to overcome barriers like irrational use, quality control and standardization issues, high pharmacovigilance, etc., are still lagging. Therefore, adequate knowledge about the system, proper information about the drugs and high-quality clinical trials, to ascertain their scientific effectiveness are needed for promotion of traditional medicine among common people. Mainstreaming of Indian System of Medicines along with allopathic drugs and healthy lifestyle will be helpful to provide healthcare service in the best possible way to all people not only in India but around the globe.

Identification and use of authenticated plant material is key to the success in furthering the use, development, and commercial exploitation of traditional formulations. Efforts till date in this regard have not created desired result/impact. Therefore, in the present book an attempt has been made to provide a consolidated filtered inventory of medicinal plant species as per their level of use, importance in trade, opportunities among undocumented, unexplored/underexplored in new plants, and threat to existence of many because of various reasons, particularly over-exploitation to facilitate conservation and sustained future use. Contents of book provides basic information on botanical identities, Sanskrit and other vernacular names, natural geographical distribution, ethnomedicinal properties, and associated systems. This can become the foundation for further systematic scientific research on these plants, for elaborate description, characterization of essential features, evaluation, and validation of phytochemical properties to promote greater use. This shall help in bridging the gap in information about the commonly used plant species enabling generation of information as per the present requirement of patenting [Intellectual Property Rights (IPR)] and rational use following the international standards; scientifically examine the marginally used sister plant species and/or species being used without a science base and the new potential species in search for new or alternative product or drugs formulation to overcome various wellness and health concerns and diseases. The book may become reference handbook of medicinal and aromatic plants used in Indian Traditional Medicinal Systems for teachers, researchers, students, and entrepreneurs, listing the strength and gaps for further research and development, particularly in the areas of validation of properties, identification and greater insight about principal component(s) through biochemical profiling/evaluation and the molecular characterization, to facilitate greater development of plant-based natural products to be used as nutraceuticals, therapeutics, cosmetics etc.

It is hoped that besides teachers and students the book may also help policy makers involved in the integration of Indian Traditional Medicinal Systems and indigenous traditional knowledge associated with medicinal and aromatic plants, in the medical science curriculum, research, developing new drug formulations and their clinical trials for full integration. Thereby evolving an affordable fully integrated medical systems to serve the larger human national and global populations of rural poor.

Anurudh K Singh

Contents

Abbreviations

AI	Artificial Intelligence
AICRPE	All India Coordinated Research Project on Ethnobiology
AIDS	Acquired Immune Deficiency Syndrome
AYUSH	Ayurveda, Yoga & Naturopathy, Unani, Siddha, and Homoeopathy
BSI	Botanical Survey of India
BSMs	Buyer–Seller Meets
BUSM	Boston University School of Medicine
CAGR	Compound Annual Growth Rate
CBD	Convention on Biological diversity
CDSCO	Central Drugs Standard Control Organization
CITES	Convention on International Trade in Endangered Species of Wild Fauna and Flora
CP	Contracting Parties
CSIR	Council of Scientific & Industrial Research, India
DIPP	Department of Industrial Policy and Promotion
ENVIS	Environmental Information System
EPCs	Export Promotion Councils
EPO	Exclusive Provider Organization and European Patent Office
FAO	Food and Agriculture Organization
FRLHT	Foundation for Revitalization of Local Health Traditions
GAPs	Good Agriculture Practices
GATT	The General Agreement on Tariffs and Trade
GFCPs	Good Field Collection Practices
GLPs	Good Laboratory Practices
GMPs	Good Management Practices
HIV	Human Immunodeficiency Virus
HPTLC	High-Performance Thin Layer Chromatography
ICBN	International Code of Botanical Nomenclature
ICMR	Indian Council of Medical Research
ICN	International Code of Nomenclature

IPC	Inter-Process Communication (computer science)
IPR	Intellectual Property Rights
ITK	Indigenous Technical Knowledge
ITMS	Indian Traditional Medicine System
ITPGRFA	International Treaty on Plant Genetic Resources for Food and Agriculture
IU	International Undertaking (FAO)
IUCN	International Union for Conservation of Nature
IUPGR	International Undertaking on Plant Genetic Resources
JNTBGRI	Jawaharlal Nehru Tropical Botanic Garden & Research Institute, Kerala
MADPs	Medicinal, Aromatic, and Natural Dyes Plants
MAI	Market Access Initiative
MAPs	Medicinal and Aromatic Plants
MAT	Mutually Agreed Terms
MEIS	Merchandise Exports from India Scheme
MoEF	Ministry of Environment and Forests
MoEF & CC	Ministry of Environment, Forests & Climate Change
MoU	Memorandum of Understanding
NBA	National Biodiversity Authority
NCEs	New Chemical Entities
NISR	National Institute of Sowa-Rigpa
NMITLI	New Millennium Indian Technology Leadership Initiative
NMPB	National Medicinal Plants Board
NRISR	National Research Institute for Sowa-Rigpa
NTFPs	Non-Timber Forest Products
PGR	Plant Genetic Resources
PHARNEXCIL	Pharmaceuticals Export Promotion Council
PIC	Prior Informed Consent certificate
PMR	Persistence Market Research
RBSMs	Reverse Buyer Seller Meets
SBB	State Biodiversity Board
SHEFEXIL	Shellac And Forest Products Export Promotion Council
SMTA	Standard Material Transfer Agreement
SOP	Standard Operating Procedure
TCM	Traditional Chinese Medicine
TIM	Traditional Indian Medicine
TK	Traditional Knowledge

TKDL	Traditional Knowledge Digital Library
TLC	Thin layer chromatography
TMS	Traditional Medicine System
TSM	Traditional systems of medicine
USPTO	United States Patent and Trademark Office
VCSMPP	Voluntary Certification Scheme for Medicinal Plant Produce
WHO	World Health Organization
WIPO	World Intellectual Property Organization
WTO-TRIPS	World Trade Organization-Trade-Related Aspects of Intellectual Property Rights

1

Introduction

1.1 Perspective

From the very dawn of evolution of human beings, man has been using plants as food, for wellness and as medicine to alleviate diseases and discomfort. People living in different parts of the world, both new and old, in the centres of some of the oldest civilizations, had selected organisms from time immemorial for food and medicine by a process of trial and error or even by experimentation with the biological resources, particularly among the plants growing around them. These have been referred to as ethnic food/traditional food and ethnic medicine/traditional medicine. India is one of the mega centres of biodiversity, and that of the ethnic diversity, that evolved over millennia with settlement of diverse races coming from different regions of the world, besides the indigenous people. Thereby, it inhabits some of the most diverse cultural traditions. This process has led to further evolution of some of the oldest indigenous civilizations originated in the Indian Subcontinent with amalgamation of exotic traditions and the knowledge. Integration of the knowledge associated with the use of plants with medicinal properties has enriched its great heritage of medicinal plant use, the *Ayurveda*, i.e., the knowledge of life and longevity, dating back to the pre-historic and the early *Vedic* period.

Being one of the mega centres of biodiversity, India has extraordinarily rich floristic diversity and species endemism. It contains 3 of the 34 biodiversity hotspots identified at global level namely, the Himalayas, the Eastern Indo-Myanmar region and the Western Ghats and Sri Lankan region (Conservation International, 2005). As per the recent revisit to cultivated plant species diversity (Singh, 2017), the Indian flora represents nearly 12% of the global floral diversity. The floristically rich India has about 148 endemic genera belonging to over 47 families of higher plants (Nayar, 1980, 1996). As per the latest estimates of Botanical Survey of India (BSI), India has 47,513 plants species (Singh and Dash, 2014) including species of bacteria. Of these, about 20,141 taxa are angiosperms with 17,926 species belonging to 2,991 genera and 251 families, representing approximately 7% of the described plant species in the world (Karthikeyan, 2009). About 5,725 species are endemic to India, representing 33.5% of the Indian flora (Nayar, 1996). As per the latest survey, this number has increased to 6,000 (Goyal and Arora, 2009). Of the 5,725 endemic species, Himalayas are habitat to 3,471 species followed by 2,051 species confined to peninsular India, and around 239 species to Andaman and Nicobar Islands (Nayar, 1996). However, according to a recent estimation about 4,045 taxa belonging to 975 genera in 155 families are considered as strict endemic to the present Indian political boundaries (Arisdason and Lakshminarasimhan, 2015). Nayar (1996) has identified 3 mega-centres and 25 micro-centres of endemic plants in India. The three mega-centres of endemism in India are the Western and the Eastern Himalayas, the Northeastern Indo-Myanmar region, and the Western Ghats. The Central Himalaya not only includes the fringe endemic species of Eastern and Western Himalayas but is a crucible of speciation.

The multiple waves of early migrations, over tens of millennia and the settlement of different human races from different parts of the world (primarily, East Asia, West Asia, and Africa), created the rich ethnic diversity in Indian Subcontinent. Pseudoscientific studies of the late 19th and early 20th centuries revealed predominant settlement of four human races, Caucasoid (white), Mongoloid (yellow), Negroid (black) and Australoid. The most indigenous population of the Indian Subcontinent has been assumed to be intermediate between Caucasoid and Australoid. The interactions of the ethnic diversity with natural biodiversity, besides the knowledge of other properties, resulted into generation/development of reservoir of information about the medicinal properties associated with various indigenous plants. They were used by racial communities, and tribal(s) or aboriginals inhabiting rural and remote inaccessible areas. They are often referred to as ethno- (associated with different societies and cultures), tribal, or folk traditional medicine systems. Several of the traditional forms of medicines have strong science or doctrine base that are well organized and documented, whilst many others are un-coded or unrecorded and perpetuated by word of mouth among the communities.

India has been practicing the traditional systems of medicine (TSM) of Indian origin, consisting of Ayurveda, Yoga, Naturopathy, Siddha and Sowa-Rigpa or Amchi, and exotic systems, such as, the Unani, which came from Greece and adapted in India around the eleventh century, and the Homeopathy system originated in Germany around eighteenth century and now widely practiced in India. They have been recognized as part of the national traditional medicinal systems under the Ministry of AYUSH (Ayurveda, Yoga & Naturopathy, Unani, Siddha, and Homoeopathy abbreviated as AYUSH), Government of India, established in 2014 (AYUSH: www.indianmedicine.nic.in). These systems have attained a special significance in recent years and have become important because of being well documented/coded regarding the associated knowledge about the medicinal properties of plants with a strong science and/or doctrine base. Further, these systems of medicine are predominantly dependent on plant-based drugs, although drugs of mineral and animal origin are also utilized in total health care and management.

Considering the importance about the uses of plants and animals in various human societies for various purposes including for medicine, during the 1980s, Ministry of Environment and Forests and Climate Change (MoEF & CC), Government of India launched an All India Coordinated Research Project on Ethnobiology (AICRPE). Between 1982-1998, AICRPE recorded information on traditions and knowledge systems associated with biotic and abiotic resources of the 550 tribal communities comprising over 83.3 million people belonging to diverse ethnic groups. In India, there are 550 communities of 227 ethnic groups. Under AICRPE, traditional knowledge on about 10,000 plants has been collected. This includes 8,000 wild plant species used by the various tribes for medicinal purposes. About 950 of these are found to be new claims and worthy of scientific scrutiny (Pushpangadan, *et al.*, 2018).

All over the world, the indigenous people and tribes possess a vast wealth of indigenous technological knowledge (ITK), which is unique to a given society or a culture. ITK serves as a powerful tool for bioprospecting of plant wealth for converting it into value added products ensuring health security to masses in a most befitting and sustainable manner. Indigenous communities are responsible for discovery of a range of health-giving herbal formulations including nutraceuticals and medicinal plants. These can generate considerable economic benefit for the nation and help in poverty alleviation. They can play a particularly important role in the development of the economy at national and global levels.

Thus, the knowledge associated with traditional medicinal systems and those with the traditional tribes/communities, is a treasure of accumulated knowledge and wisdom about the management and use of various plants and other materials around them for wellness and medicinal purpose. But, due to continued modernization of modern medicine system called allopathy with stronger science base, the precious traditional knowledge has been ignored and been eroding and corroding fast. It was the realization of these facts, that has motivated researchers from many countries around the world to undertake ethnobotanical investigations to recognize these alternative medicine systems and document the traditional knowledge and wisdom of the people. Their improvisation and integration may evolve more innovative complementary systems having combination of various therapies. It is now well recognized that the traditional wisdom and knowledge on utilization of the biological resources is of immense value to biodiversity planners and scientists in developing strategies for conservation, sustainable use, and generation of wealth from the bioresources. Herbal medicinal scientists consider ethnobiological/ethnobotanical knowledge system as a first effective means for identifying, as well as for locating alternative leads for drugs and pharmaceuticals, nutraceuticals, cosmetics, gums, resins etc.

At a global level, plant-based traditional medicines are the precious cultural heritage of the different societies of the world. As per the World Health Organization (WHO) (2008) estimates, they cater to the medicinal needs of 80% of the world's population. Further, the reported side effects of the modern inorganic/synthetic medicines, ease of assimilation of plant-based organic/biological medicines and their cost-effectiveness etc., have generated a renewed interest in research and development of traditional herbal medicines. Systematic scientific research efforts in this direction would help in validation of the medicinal properties of plants in detail, establishing desired science foundation and providing greater credence to the medicinal herb and the traditional systems. Such efforts may lead to protection of intellectual property rights (IPR) on indigenous traditional knowledge associated with plant properties (medicinal) as per the requirements of present scenario of IPR and facilitate their registration/ patenting. Also, help in improvisation, using modern technologies. Further, the ethnobotanical information or traditional medicinal use of plants practiced over thousands of years under the different traditional medicinal systems can provide clues for further phytopharmaceutical and therapeutic research and attract investment and entrepreneurship. Therefore, developing an inventory of plants used in different traditional medicinal systems of countries such as India can form the basis for further bioprospection in the current innovative medicinal research programs nationally and globally. Under collaborative programs such efforts can facilitate wider use towards cost-effective health services, particularly for rural poor in developing and underdeveloped countries of the world. It is more relevant for countries such as India, which has been the centre of some of the oldest civilizations of the world and home of world oldest herbal based medicinal systems (Ayurveda) with vast knowledge associated with properties of plants.

In this regard, ethnopharmacology approach involving multidisciplinary scientific exploration of biologically active agents traditionally employed, based on description, observations, and experimental investigation of indigenous drugs and their biological activities can provide a greater insight into their functioning. In its entirety, pharmacology embraces the knowledge of the history, source, chemical and physical properties, compounding, biochemical and physiological effects, mechanism of action, absorption, distribution, biotransformation, excretion, and therapeutic and other uses of drugs. A drug is broadly defined as a chemical agent that affects life processes. Therefore, briefly, the main component of ethnopharmacology may be defined as pharmacology of drugs used by ethnos (by different societies and cultures) in medicine. However, ethnopharmacology is basically about the convergence of medical ethnography and the biology of therapeutic action, i.e., a transdisciplinary exploration that spans through biological and social sciences. This suggests that ethno-pharmacologists are professionally cross trained, for example, in pharmacology and anthropology, or for that matter, ethnopharmacological research. Therefore, it is the product of collaborations among individuals whose formal training includes two or more traditional disciplines. Nevertheless, as per the requirements for explaining the action of plant-based drug, the objectives of ethnopharmacology should focus on the basic research aiming at giving rational explanation to how a traditional medicine works, and how the applied research should aim at developing a traditional medicine into a modern medicine (pharmacotherapy) or to develop its original usage by modern methods (phyto-therapy), fulfilling all the requirements as per the demands of recognition/registration/patenting as of today.

Considering the forementioned background, the present book is an effort to produce an inventory of medicinal plants used in the Indian traditional medicine systems and by the indigenous people, tribes, and communities. It has information on their vernacular name in ancient language, such as Sanskrit and language of today (including English), their distribution in the Indian Subcontinent to facilitate collection, plant parts used for obtaining the active principles and/or associated properties, contributing to use for cure of specific ailments, and the various traditional systems that are associated with the knowledge. This set of basic information shall provide clue(s) for further research and the line of investigations to be followed for their commercial exploitation on large scale, meeting the requirements for their patenting, safe use, and drug control. It may generate interest among researchers and entrepreneurs involved with developing and producing diverse plant-based products, such as fortified food, nutraceuticals, cosmetics, etc., in addition to herbal medicines/drugs.

1.1.1 The Possible Outcome/Impact

Many of the phytopharmaceuticals used as therapeutic agents have their clues from the ethnobotanical or traditional uses as medicine under various traditional medicinal systems. For this reason, collation of such information/knowledge should be the priority area for bioprospecting or systematic scientific research using recent technical advances for their biochemical profiling, identification of principle molecule(s), its/their characterization to prove the concept used, mode of action dealing with the treatment of disease and the action of remedial agents. Consequent to these, excellent leads have been discerned for many plants/parts/by-products commonly used in treatment of specific disorders. In India, this process started from middle of the 20th century resulting in publication of Wealth of India on Medicinal Plants. It mainly revolved around the plants used in Ayurveda for obtaining various drugs (Patwardhan and Mashelkar, 2009). Traditional knowledge-driven drug development can follow a reverse pharmacology path and reduce time and cost of development (Patwardhan *et al.*, 2004). In their guest editorial on the launch of the *International Journal of* Ayurvedic *Research*, Valiathan and Thatte (2010) emphasized that the use of modern science as a research tool in Ayurveda has never been more necessary, more promising, or more compelling than it is in the present and that it is time to explore.

In the case of Ayurveda, the doctrine behind it was more philosophical for a unified approach towards positive health. Much of the knowledge contained in the texts needed scientific re-interpretation in the light of present code of knowledge documentation and modern science. Since most of the drugs of Indian medicine system lack standards of identity, purity, and safety (especially compound formulations), efforts are needed to obtain/generate these details and make it available to the user. Thus, to make traditional remedies more acceptable in the modern era, it is desirable that efforts are made to validate the pharmacopoeia scientifically by incorporating scientific data as per WHO guidelines on use of herbal drugs. These efforts, besides providing a strong science base, will help revive the glory of ancient science of positive health to put forth the concepts and practices of traditional systems of medicine.

Herbs used in Ayurveda become a priority because of the system being well-organized in relation to general philosophical approaches, concepts, principles, and methods of diagnosis and treatments relating to promotion, maintenance of good health and long life, and cure or management of diseases and ailments. It prescribes drugs, diet, and other regimens that include a code of conduct conducive to the maintenance and promotion of positive healthy status of body, mind, and spirit. However, intensive efforts are needed to validate these aspects scientifically to establish medicaments of this system for global acceptance of known remedies as well as to search for further potential of Ayurvedic plants and formulations.

Further, the Traditional Knowledge Digital Library (TKDL) with vast information on plants/formulations and their traditional usage could provide initial guidance and can serve as a powerful search engine to short-list the priority target plants, principal molecules for scientific validation of therapeutic claims as well as bioprospection for potential molecules to combat hitherto incurable diseases/physiological disorders.

Claims across the ethnic groups, wherein the same plants help cure similar ailments across the communities could be another area for further research for scientific validation and drug discovery. Further, to make the traditional systems sustainable, it would be necessary to produce the target molecules in the required quantity and of desirable quality. This could be better done through organized cultivation, to ascertain the availability of desired/uniform quality and quantity of raw material for extraction of principal components. Should the commercial demand grow, their genetic enhancement for increase productivity, value addition, diversification, and development of production strategies to get the molecules of choice or their prototype, at will, through conventional and microbial "pharming" can be another area of research attention.

References

AICRPE (All India Co-ordinated Research Project on Ethnobiology), Final Technical Report, (1992-1998). Ministry of Environment and Forests (MoEF), Government of India.

Anonymous (1978) Handbook of domestic medicine and common Ayurvedic remedies. New Delhi, India: Central Council for Research in Indian Medicine and Homeopathy

Arisdason, W and Lakshminarasimhan, P. (2015) Status of plant diversity in India: An overview p 7, ENVIS Center of Floral Diversity, Hosted by Botanical Survey of India, Kolkata, India

AYUSH: www.indianmedicine.nic.in About the systems—An overview of Ayurveda, Yoga and Naturopathy, Unani, Siddha and Homeopathy. New Delhi: Department of Ayurveda, Yoga and Naturopathy, Unani, Siddha and Homoeopathy (AYUSH), Ministry and Family Welfare, Government of India.

Conservation International (2005) Biodiversity Hot Spots, 1919 M Street, NW, Suite 600, Washington, DC 20036. (202)912-1000, fax: (202)912-1030 www.conservation.org

Goyal, AK and Arora, S. (2009) Chapter 1. Overview of biodiversity status, trends and threats. In: India's Fourth National Report to the Convention on Biological Diversity pp 15-53, Ministry of Environment and Forest, Government of India, New Delhi

Karthikeyan, S. (2009) Flowering plants of India in 19th and 21st Centuries – A comparison. In: Plant and fungal biodiversity and bioprospecting (Eds: Krishnan S and Bhat DJ) pp 19-30, Goa University, Goa

Nayar, MP. (1980) Endemism and pattern of distribution of endemic genera (angiosperm). J Econ Tax Bot 1:99-110

Nayar, MP. (1996) Hot spots of endemic plants of India, Nepal and Bhutan. Tropical Botanic Garden and Research Institute, Palode, Thiruvananthapuram, Kerala, India. 253

Patwardhan, B and Mashelkar, RA. (2009) Traditional medicine-inspired approaches to drug discovery: can Ayurveda show the way forward? Drug Discovery Today 14: 804–811.

Patwardhan, B, Vaidya, ADB and Chorghade, M. (2004) Ayurveda and natural products drug discovery. Current Science 86:789-799.

Pushpangadan, P, George Varughese, Ijinu, TP and Chithra, MA. (2018) All India coordinated research project on ethnobiology and genesis of ethnopharmacology research in India including benefit sharing. Annals of Phytomedicine 7(1): 5-12, DOI: 10.21276/ap.2018.7.1.2

Singh, Anurudh K. (2017) Revisiting the Status of Cultivated Plant Species Agrobiodiversity in India: An Overview. Proc Indian Natn Sci Acad 3(1):151-174. doi: 10.16943/ptinsa/2016/v82/48406

Singh, P and Dash, SS. (2014) Plant Discoveries 2013 – New Genera, Species and New Records. Botanical Survey of India, Kolkata

The Ministry of Environment, Forest and Climate Change (MoEF & CC) (2018) The nodal agency for the planning, promotion, co-ordination and overseeing the implementation of India's environmental and forestry policies and programmes. Ministry of Environment, Indira Paryavaran Bhavan, New Delhi.

The Traditional Knowledge Digital Library (TKDL) collaborative project, initiated in India in 2001, between the Council of Scientific and Industrial Research (CSIR), Ministry of Science and Technology, and the Department of Ayurveda, Yoga and Naturopathy, Unani, Siddha and Homeopathy (AYUSH), Ministry of Health and Family Welfare, of India. It is being implemented at the CSIR 2011.

Valiathan, M and Thatte Urmila (2010) "Ayurveda: The time to experiment". International Journal of Ayurveda Research (1)1: 3. Gale Academic

WHO (World Health Organization) (2008) http://www.who.int/mediacentre/factsheets/fs134/en/ print.html

1.2 Indian Traditional Medicine Systems

The Government of India has recognized Ayurveda, Yoga and Naturopathy, Unani, Siddha, and Homeopathy (AYUSH) under the traditional medicine system (TMS). Of these, Ayurveda is the oldest one belonging to the *Vedic* period, representing the rich cultural heritage about ancient medicinal wisdom and the knowledge about the life-sciences in India. This has been practiced since time immemorial and even today it has an extensive following with strong faith and belief that has kept it in use, despite the strides made by the modern system of medicine (Allopathy) in research and development and have been recommended to be the part of integrated medicinal system at global level.

1.2.1 History

The history of Indian medicine systems started with Ayurveda. The word "Ayurveda" connotes *ayur* = life and *veda* = knowledge; its origin in India started long ago in the *Vedic* period (c3000-1700 BC). *Rigveda* and Ayurveda (2000 BC) are, perhaps, the oldest documented repository of human knowledge, representing the first mention of diseases and medicinal herbs in *Osadhi-Sukta* (Rv. 10.47, 123). Similarly, the earliest comprehensive descriptions of the drugs and diseases are found in *Atharvaveda* (1600–100 BC). In *Vedic* literature, about 200 drugs are mentioned that were in use for treatment of diseases (Singh *et al.*, 2003). In ancient (pre-historic) times, Ayurveda was the only overall health care system.

The scientific foundation of Ayurveda was laid with documentation and compilation (*Samhita*) of the knowledge generated in number of original works quoted in literature, which are not available now. For example, the work of sage Atreya, leading to *Samhitās* written by his six pupils- Agniveśa, Bhela, Jatukarpa, Parāsara, Hārita and Kaśarpani, and other great works called- *Visvamitra Samhitā, Kharnada Samhita, Kapila-Tantra, Gautama-Tantra* etc. are apparently lost. Only one of them, the *Agniveśa Samhitā*, thrice revised and re-casted, survives in skeleton and is known as the famous *Caraka Samhitā*. Presently, the three legendary and authoritative classical texts: *Charak Samhita* (500–200 BC), *Sushrata Samhita* (1000–200 BC), and *Astang Hridayam* (AD 700), collectively known as *Brihattrayi* are available. *Charak Samhita* emphasizes medicine and *Sushruta Samhita* describes surgery. Whereas the *Astang Hridayam* deals with basic principles and practices of Ayurvedic medicine, primarily based on the information contained in *Charak* and *Shushruta Samhita.* These compendia have been used for teaching of Ayurveda in India. Based on Sanskrit names, the number of plants in these compendia include 1,270 in *Sushruta*, 1,100 in *Charak*, and 1,150 in *Astang Hridayam* (Singh and Chunekar, 1972) (Fig. 1.2.1). In addition, *Bhela Samhita* enumerated medicinal plants along with a probable botanical identity. *Bhela Samhita* is one of the important classical treatises of Ayurveda mentioning 286 plants (Bhela Charya, 2000; Translation). It is quoted by many scholars from medieval periods. The *Bhela Samhita* is one of the prominent treatises of the *Samhita* period of Ayurveda (Post *Vedic* period, 100 BC to 400 BC). Bhela Charya (Bhela), the author of *Bhela Samhita,* was one of the outstanding six colleagues–disciples of Punarvasu Atreya and contemporary of Agnivesha (period 1000 BC) (Ratha *et al.*, 2018). Further elaboration of drugs (plants) and diseases is found in subsequent classical treatises like *Madhava Nidan* (AD 1200), *Sarangdhar Samhita* and *Bhava Prakash's* (AD 1600). *Nighantu* (traditional collection) mentions about 600 plants (Fig. 1.2.1) and the rest are in over 70 other lexicons (*Granths*) written between the sixth and seventh centuries. *Sarangdhar Samhita*, which dates from 1300 AD, highlights the concept of poly-herbalism in the ancient medical system (Srivastava *et al.*, 2013).

The beauty of Indian medicine systems is that they are predominantly based on Hindu ethos of serving the humanity regardless of earthly gains, for giving life even under the weight of ruins. "Ayurveda" meaning the "Science of Life", or the great world science of Medicine, is the oldest with its objectives not only limited to the protection of human life but also of the life of animals, and even of the plants. The Hindu ethos of accepting and adapting the good (knowledge/wisdom or material) from any source, irrespective of region and culture, has further enriched the Ayurveda (Singh, 2016), despite odds during different times of the history. This further contributed significantly to this great Science and Art of medicine in all the eight sections of it. It has led to the establishment of schools of different specialists with massive libraries of differentiated literature to back them. In documented history (historic period), invasion of Alexander, Scythians, and Huns, introduced many more plants with medicinal potential. Drugs such as Rhubarb, Opium, Jamaica Sarsa, etc. (Vide the writings of Bhāva Miśra), and many more became integral part of Ayurveda or Indian Medicine systems (Singh, 2016). Similarly, several exotic medicinal systems, such as Unani in ancient times, originating from Greece (460-377 BC), introduced into India by the Arabs and Persians sometime during the eighth century and Homeopathy originating from Germany in 18th century and introduced by Honigberger into India in 1839, soon becoming integral part of Indian medicine systems (Ghosh, 2018).

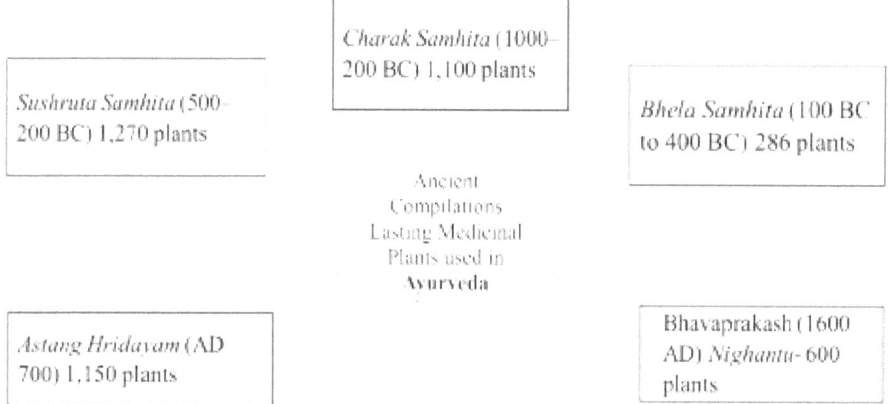

Fig. 1.2.1: Number of plants recorded under ancient compilations related to the oldest ancient traditional medicinal systems of India, the Ayurveda

Nevertheless, dark facets of the invasions led to major loss of valuable literature associated with these systems, starting from Ayurveda, the oldest Indian Medicine systems. Yasodharma Deva Vikram Aditya, the mighty destroyer of the Scythians and Huns (5th Century? CE), was able to revive only a part of the lost glory with the help of intellectual luminaries of that period. During Buddhist era, because of the difference in thoughts and the desire to establish supremacy, the decline of Ayurveda continued leading to loss of undocumented and unshared wisdom and perishing of old literature. But the worst came with the advent of Muslim invasions under which only few written manuscripts could survive. During the medieval period (Middle age), there was a negligible revival during Akbar to Shahjahan period. But the Aurungzeb regime of terror again caused ravages and loss of more literature. Consequent to these up and downs during various times of the Indian history, knowledge associated with Ayurveda and other systems suffered permanent loss of significant and well documented literature. A spurious work pass (repeat editions) for *Hārita Samhitā*. *Bhela Samhitā* has been found in crippled condition in the Tanjore State Library of the School of Surgeons headed by the Royal Master Dhanvantari (the seat created in the name of the Hindu god of medicine, an avatar of Lord Mahavishnu).

The true scientific spirit demands that we should cast off all prejudices and investigate/ undertake research with an open mind focusing on this reservoir of knowledge associated with properties of select group of plant species and of their close relatives to validate and document the available information. It is good to see that a scientific voice for such investigations has been raised and put into the action, to facilitate wider use of plant driven drugs based on indigenous knowledge of ancient civilizations by foreigners and some of our countrymen, following in their footsteps. Such initiative shall help bring

Ayurvedic and other Indian medicinal systems' knowledge into scientific domain, may be following a piecemeal approach. It can be performed either nationally and/or in collaboration with other stakeholders (Western world). It would profit Ayurveda and sister systems in getting due credit for a number of ancient innovations and discoveries about the nutritional, medicinal, aromatic, and other properties of plants. This will also stimulate further research for validation of claims, by going further deep into the genesis in search of principal components/molecules and the mechanism of action facilitating wider and more appropriate scientific and effective use. On the other hand, the western Pharmacopoeia will profit by this additional knowledge (traditional), in getting clues for further research on documented plants to develop and evolve plant-based products and motivate the western world for further investment in research and development, and entrepreneurship in plant-based products. Whereas, the Ayurvedic Pharmacopoeia will benefit and grow with greater scientific research authenticating the reported claims, fulfilling the requirement of Intellectual Property Rights (IPR) and registration of Patents, using modern technique of biochemical profiling, identification of principal molecules, their characterization and action, to claim fair and equitable benefit sharing of commercial gains accrued from their use. This would mean recognizing the contribution of indigenous tribes and communities for their technical knowledge associated with plant properties. This may help India to capture a significant component of global herbal product market, which has been presently estimated to rise more than 35 billion US $ by the end of 2021. Pharmacological and analytical laboratories working on the investigation of indigenous drugs should also work unceasingly as part of traditional Indian institutions, and for new discoveries and demonstrations on Ayurvedic lines. Consequently, the Ayurveda would be going through the process of reorientation through modern rigorous scientific scrutiny and is poised to emerge globally for much greater and effective utilization of herbal products in health care systems soon. At the present time, researchers of many disciplines are exploring Ayurveda to find remedies for lifestyle-related diseases and degenerative and psychosomatic health problems. Should we remain silent spectator to this scenario, India will lose a great economic opportunity in business of plant-based products and in strengthening the efforts towards human wellness and healthcare, particularly that of 65% rural poor. The following paragraphs discuss in brief, the traditional Indian medicinal systems.

1.2.2 The Traditional Indian Medicinal Systems

1.2.2.1 Ayurveda

As mentioned above, the scientific foundation of Ayurveda was laid during the *Samhita* (compilation) period. In ancient times, it was the only health care system that received support and patronage of both common people and the rulers. It survived owing to its strong philosophical science-based doctrine, native roots, and popularity among the people, especially those living in rural and remote areas. After independence, in recent past, it got a fillip on account of its official recognition by the Government of India. Presently, it has been upgraded to the same status as that of Allopathy in terms of official health policies, national plans, and programs.

The Ayurvedic system of medicine works through eight systems, generally referred to as *Astang Ayurveda* for management of normal health/wellness and to cure the diseases. These are *Kayachikitsa* (internal medicine treatment), *Bhootavidya* (treatment of psychological disorders), *Kaumar Bhritya* (paediatric treatment), *Rasayana* (study of geriatrics), *Vajikarana* (treatment through aphrodisiacs and eugenics), *Shalya* (surgical treatment), *Shalakya* (otorhinolaryngological and ophthalmological treatment), and *Agada Tantra* (toxicological studies) (Jaiswal and Willimas, 2017). The aim of the present book is to create an inventory of plants used in these systems as nutritive supplement and/or component in formulations of drugs/medicines used for preservation and promotion of health, and/or remedy for diseases based on the principle of Ayurvedic doctrine. Hence, it will confine to listing of the plants used in Ayurveda (predominantly *Rasayana*) to fulfil the objectives of their preservation/conservation and continued sustainable use for wellness, health care by using them or their by-product as medicine to cure diseased person, cosmetics, etc.

The concept of Ayurveda provides a unique approach of health promotion leading to a healthy, active, and long life and the preventive and curative measures for diseases (Katiyar, 2010). Conceptually, it considers "life" as the union of body, senses, mind, and soul. The total body matrix of humans comprises of three states (called *Dosa*)- *Vata*, *Pitta*, and *Kapha*, seven body tissues (*Rasa*, *Rakta*, *Mansa*, *Meda*, *Asthi*, *Majja*, and *Sukra*) and the waste products, such as faeces, urine, and sweat. A healthy person has normal state of equilibrium, revolving around balanced food, which supports growth of the body and restricts decay of body matrix, which is significantly affected by physical and other mechanisms as well as by bio fire (*Agni*). Plant-based diet (lifestyle) provides more efficient support and desired nutritive supplement to the body for appropriate growth. Organic constituents of plants, being simple and rich in antioxidants, are assimilated easily and prevent the decay of the body matrix. Therefore, food habits are the most important component of healthy life.

The human body is composed of five basic elements (*Pancha Mahabhutas*): namely, *Akasa* (vacuum-ether), *vayu* (air), *Tejas* (fire), *Jal* (water), and *Ksiti* (earth). There is a balanced condensation of these elements to suit the needs and requirements of different structures (organs) and functions of the body matrix and its parts. Sickness depends upon the absence of a balanced state of the total body matrix, including the balance between its different constituents. Both the intrinsic and extrinsic factors can cause disturbance in the natural equilibrium, giving rise to disease. Therefore, the treatment of diseases consists of restoring the balance of the disturbed body–mind matrix through regulating diet, physical activity regimen, and medicines (Kapoor, 2001; www.indianmedicine.nic.in).

For management of diseases, the two most important modes of treatments of Ayurveda are Panchakarma and Rasayana therapy, which are the basis and specialties of the system. The two are quite popular and have attracted world attention (www.indianmedicine.nic.in).

Panchakarma

Panchakarma treatment is planned for elimination of the increased and morbid *dosa* (defect/fault) (*vata, pitta,* and *kapha*) of the biological unit, the living body, responsible for its malfunctions and is known as *Sodhan Therapy* (purification therapy). It is technically termed *Pancha-Karma* (five-fold therapy). It involves internal and external purification by medically induced emesis, i.e., vomiting (*Vamana*), purgation (*Virechan*), oil enema (*Anuvasan Basti*), decoction enema therapy (*Asthapana* or *Niruha Basti*), and nasal insufflations therapy (*Sirovirechana*). Two necessary measures (including purification, food supplement, etc.) are carried out before (*Purva*) and after (*Pashchat*) performing the Panchakarma. Often, Panchakarma is induced by plant-based products.

Rasayana Therapy

The whole drug regimens of Ayurveda are divided into two distinct groups- one for maintenance and promotion of good health and longevity, and the other for treatment of diseases. The *Rasayana* group of drugs falls under the first category. The word *"Rasayana"* literally means the path that *rasa* takes (*Rasa:* plasma; *Ayana:* path). It is defined as a therapeutic measure that promotes longevity, prevents aging, provides positive health and mental faculties, increases memory; and imparts resistance and immunity against diseases (Anonymous, 1978). *Rasayanas* are broadly classified into four groups:

Kamya Rasayan- dealing with health and vigour of a healthy person (wellness).

Naimittic Rasayan- for increasing the strength of a diseased person (tonic), which may include drugs that can be used as an adjunct to the specific medical treatment for early cure.

Medhya Rasayan- for improving mental faculties.

Achara Rasayan- religious way of living with high morals and virtues following established rules of conduct, practice, usage, precept etc. (Anonymous, 1978), providing moral strength, positivity, and faith in self.

Thus, the *Rasayana* group of drugs and recipes has a multi-dimensional mode of action through nutritional dynamics of tissues. The disease-specific concepts have similar ideas of adaptogens (adoption to stress) or antistressor drugs, tonics, and "unstimmungs therapy" (German therapy, to control ailments related with mood, atmosphere, good mood, cheer, color, mood, spirits, temper) of other medical systems and traditions (Anand,1990; Singh *et al.*, 2016). After a detailed study, Wagner and Proksch (1985) opined that *Rasayana* preparation, which acts as an herbal immunostimulant and as an adaptogen, regulates the immunological and endocrine systems with relatively low doses, without damaging the auto-regulative functions of the organism.

The quintessence (perfectness) of all *Dhatus* (tissues) is known as *Ojas*. The expression of *Ojas* through *Bala* (resistance power) in turn is responsible for immunity or resistance against diseases. Ayurvedic approaches to the phenomenon of *Vyadhi-Kshamatva* (defence mechanism to resist diseases) are also fundamental and important factors to mediate through functioning of *Ojas* (AYUSH: www.indianmedicine.nic.in). The concept of modulating immune response can be equated with the *vyadhi-kshamatva* (disease resistance) and *ojas-vridhi*as (immunity enhancement) postulated in the Ayurvedic texts. Consequently, upon a range of structural and functional changes, the inevitable and progressive aging phenomenon of life takes place, and because of aging, the human system becomes prone to a variety of age-related diseases. Stress is another important factor that is directly related to body immunity. In Ayurvedic parlance, stress, adoption (choose/takeover), and immunity complexes co-exist and form the basis for better understanding of human body functions and the cause of diseases. Moreover, Rasayana drugs act inside the human body by modulating the neuro–endocrinal–immune system and have been found to be a rich source of antioxidants (Brahma and Debnath, 2003).

In recent times, extensive studies have been undertaken in the context of validating the age-old concepts from a modern point of view. Puri (1970a, 1970b, 1971, 2003) provided a detailed account of the herbs used in *Rasayana* preparations. The effects of *Rasayana* drugs on psychosomatic stress (Udupa, 1973), epilepsy (Singh and Murthy, 1989), and conclusive disorders (Dwivedi and Singh, 1992) have been studied. The role of free radicals in various diseases and 15 *Rasayana* plants with potent antioxidant activity has been reviewed for their traditional uses and mechanism of antioxidant action (Govindarajan *et al.*, 2005).

1.2.2.2 Siddha System

This is another indigenous system, which originated in South India and is associated with *Dravidian* culture. It dates back probably to the pre-*Vedic* period and thus, may be the oldest. The term "siddha" means achievement, and the person possessing knowledge of wisdom/achievements about application of medicine is considered Siddharth Purusha, i.e., "siddhas", a saint. Compilation of knowledge of such saintly figures led into the development of the Siddha system of medicine. This system of medicine is practiced predominantly in the state of Tamil Nadu and adjoining southern part of India. It is largely a system dealing with medicine/drugs concerned with the treatment of disease and the action of the remedial agents (Kumar *et al.*, 1997; Khanuja *et al.*, 2006).

The basic principles and fundamentals of this system are quite like those of *Ayurveda*, with emphasis on understanding medicine and its normal functions on living organisms (physiology) in terms of chemistry. According to this system, the human body is the replica of the universe, and the drugs are to be found in the universe irrespective of their origin. This system also considers the human body as a conglomeration of three temperaments, like Ayurveda, in *vatasu*, *pittam*, and *kapham.* The maintenance of equilibrium in humans is the key for normal or good health/wellness and its disturbance or imbalance leads to disease or sickness. The diagnosis of disease involves identification of its cause through the examination of pulse reading, examination of urine and eyes, study of voice, colour of the body, examination of the tongue, and the status of digestive system (Kumar *et al.*, 1997). The treatment is mainly directed toward restoration of equilibrium of the original stated temperament, for which one or more forms of cleansing procedure are adopted, such as vomiting, purgation, enema, and nasal drops using plant-based products. This is followed by the application of various forms of medicine facilitating restoration to normalcy.

Siddha treatment covers all types of diseases and has been effective in treating peptic ulcer, infective hepatitis, psoriasis, rheumatoid arthritis, and abnormal growth of tissues (AYUSH: www.indianmedicine.nic.in).

1.2.2.3 Sowa-Rigpa or Amchi System

Sowa-Rigpa or *Amchi* is another oldest Indian indigenous traditional medicinal system that originated in the trans-Himalayan region and is still living with well-documented medicinal practices and products, popular not only in India but also in several part of Asia. The term Sowa-Rigpa is derived from Bhoti language, which means knowledge of healing. This system further evolved and diversified in the trans-Himalayan region with the spread of Buddhism. At present, it is practiced in the Himalayan region and the regions adjacent to India, and in some of the countries with Buddhist influence, such as Bhutan, China, Mongolia, and part of southern Russia.

It draws its basic principles from Ayurveda, since more than 75 per cent texts of this system are taken from the most famous Ayurvedic treatise '*Ashtanga Hridya*'. Consequently, around 75 per cent of herbal medicines are common to the Ayurveda. The principal text, '*rGyud-bZi*' *chatush* (four) *Tantra*'- a textbook, fundamental to principles of Sowa-Rigpa, is in Sanskrit, which was later translated in Bhoti language during around 8th -12th century and further improved by Yuthok Yontan Gombo and other scholars of trans-Himalayan region as per socio-economic developments. The fundamental principles of Sowa-Rigpa are based on *Jung-wa-nga* (Pancha Mahabhutha), *Nespa-sum* (Tridosha), *Luszung-dun* (Sapta Dhatu), dos and don'ts, dietary regulations, pulse examination etc., reflecting close similarity and sisterly relationship between Sowa-Rigpa and Ayurveda besides proximity to other indigenous Indian traditional medicinal systems. Though it has some influence of Greek system as well. Medicinal treatment is composed of natural material, i.e., herbs and minerals and physical therapies to treat illness.

Government of India has recognized it under the traditional medicine systems and the alphabet 'S' under AYUSH stands for both — Siddha and Sowa-Rigpa, though the logo of the AYUSH ministry still confines to five traditional systems of medicine in India — Ayurveda, Yoga and Naturopathy, Unani, Siddha, and Homeopathy. However, the new government in India has decided to put greater emphasis on this system and plans to establish a National Research Institute for Sowa-Rigpa (NRISR) in Leh as an autonomous institute by renaming its National Institute of Sowa-Rigpa (NISR). The autonomous institute will function under the Central Council for Research in Ayurvedic Sciences, Ministry of AYUSH for its promotion in the extreme ecologies of trans-Himalayan and Himalayan region.

1.2.2.4 Unani System

It is an exotic system, which primarily originated and evolved in Greece. Its foundation was laid by Hippocrates, a Greek physician from the Age of Pericles (Classical Greece). It is based principally on experiences and documentation of the ancient cultural heritage of Egypt, Arabia, Iraq, China, Syria, and India (Kumar *et al.*, 1997). It was introduced into India during the medieval period (11th century) with invasions of the Arabs primarily to look after the suffering of moving armed forces. Following the Hindu ethos, Ayurvedic practitioners adapted the effective practices and products of this system enriching Ayurveda. This system by its own experience got amalgamated with Ayurveda and came to be known as "Unani-Tibb." During this phase, attention was paid to medicinal herbs found and used in India, and several books on the therapeutic properties of herbs were brought out by the Arab physicians, leading to development of several schools, colleges, and laboratories imparting education, training, and research, besides several hospitals and dispensaries for the treatment using this system.

Like Ayurveda, this system did suffer a setback during the British rule, causing loss of its literature and valuable information. However, Hakim Ajamal Khan, a renowned physician, initiated efforts to revive Unani system in India. After independence, like Ayurveda, this system received necessary recognition and support from the Government of India, and today, it is one of the leading traditional medicine systems practiced in India with due support of educational, research and health care institutions (AYUSH: www.indianmedicine.nic.in).

The fundamentals of Unani system are based mainly on temperament (*Mizaj*) and humour (*Akhlat*). Accordingly, specific temperament is believed to be present in every individual. Therefore, each patient is assessed based on his/her temperament. The four humours are blood, phlegm, yellow bile, and black bile, which are the fluids that the human body obtains from food consumed. Any disturbance occurring in the equilibrium of humour's is principally responsible for the ailments or diseases, and therefore, the treatment aims at restoring the equilibrium of humour's (AYUSH: www.indianmedicine.nic.in; Khanuja *et al.*, 2006; Ahmad and Akhtar, 2007).

The diagnostic process under the system involves physical examination, pulse reading, and examination of urine and stool. Considering these and the background of the patient, the cause and nature of illness is determined, and accordingly treatment is given. For therapeutics, the entire personality of the patient is considered and various types of treatments comprising regimental diets and pharmacological (drug) and surgical therapies are provided, which involve the use of naturally occurring drugs of plant, animal, and mineral origin. Since this system emphasizes the importance of temperament of the patient, it prescribes medicines suited to his/her (patient) temperament and age. This helps in accelerating the process of recovery and eliminating the risk of reaction. This system has been found suitable for chronic diseases such as skin problems, leukoderma, eczema, filaria, liver disorders, metabolic disorders, arthritis etc., and is well supported by research and development in concerned areas.

1.2.2.5 Homeopathy

This is another exotic system, which was adopted in India during colonial period and blended well because of the close similarity in philosophy and principles with those of other prevailing Indian systems of medicine (Khanuja *et al.*, 2006). Historically, the principle of Homeopathy has been known since the time of Hippocrates {around 460 – 375 BCE), the father of medicine. But it was not until the late eighteenth century that Homeopathy as it is practiced today was advanced by the great German physician Dr. Samuel Hahnemann in 1796 (Ghosh, 2018). It is based mainly on pseudoscientific principles, such as statements, beliefs, or practices that are claimed to be both scientific and factual but are incompatible with the scientific methods.

Hahnemann was appalled by the medical practice of that time and set about to develop a method of healing that would be safe, generic, and effective. He believed that human beings have a capacity for self-healing and that the symptoms of disease reflect the individual's struggle to overcome his illness.

Homeopathy signifies treatment of diseases with remedies in a simple way, usually prescribed in minute doses, which help in better diffusion of molecules of the drug in the body that can produce symptoms like those of the diseased person when taken by healthy people (*Similia Similibus Curentur*). Considering its popularity in certain parts of India (to start with Bengal), but now nationally, it has been recognized as one of the national systems of medicine. The system focuses on promotion of physical and mental values in health, disease, and treatment. Therefore, it is widely known as a holistic system.

The important demonstrable laws and principles of Homeopathy are (a) the law of similar, (b) the law of single remedies, (c) the law of direction of cure, and (d) the law of minimum dose. Every law and principle signify the use of Homeopathic medicines. A correct Homeopathic remedy or drug should be able to produce symptoms of the disease in the patient reflecting its action. A cure is considered complete, when there is full restoration of the functioning of all the vital parts of the body. A single remedy can be chosen that is like the symptom complex of the sick person. The selected remedy for a sick person should be prescribed in a minimum dose. It treats a patient but considers each patient differently from others and aims not only to alleviate the patient's present symptoms but also to possible impact(s) in long-term well-being.

It is believed that Homeopathy has definite and effective treatments for chronic diseases such as diabetes, arthritis, bronchial asthma, immunological disorders, and behavioural and mental disorders. Recently, this system has attracted debate on the cure versus molecule in placebo (medicine or procedure) and drug comparison. Therefore, extensive research is warranted to dispel doubt about the effectiveness of Homeopathic medicines for the treatment of various diseases and ailments.

1.2.2.6 Tribal and Folk Medicinal Systems

In addition to the plants of above traditional systems, there are many other plants used for their medicinal properties by Indian tribes and folks (communities), inhabiting different interior regions and rural areas of the country. However, the information/ knowledge about the use of such plants is not properly recorded or coded because of several reasons (social, traditional, confidentiality for economic gain, etc.). Considering the critical constraints of the multidimensional perspective of the complex fabric of the tribal life, culture, traditions, knowledge system, resources utilization pattern, technological capabilities, and the peculiar socio-economic problems, and to study and diagnose clearly and identify various components, the Ministry of Environment and Forests and Climate Change (MoEF & CC), Government of India sponsored a multi-institutional and multidisciplinary project. It involved about 27 centres with over 500 scientific personnel located in different institutions, spread over the length and breadth of the country. This was named as All India Coordinated Research Project on Ethnobiology (AICRPE, 1992-1998) with a multidisciplinary, multi-institutional team with action-oriented research mandate.

AICRPE during its operation (1982-1998) recorded information on the multidimensional perspectives of the life, culture, tradition, and knowledge system associated with biotic and abiotic resources of the 550 tribal communities comprising over 83.3 million people belonging to the diverse ethnic groups. In India, there are 550 communities of 227 ethnic groups. There are 116 different dialects and 227 subsidiary dialects spoken by tribal communities of India. The knowledge of these communities on the use of wild plants for food, medicine, and for meeting many other material requirements is now considered to be potential information for appropriate scientific and technical intervention for developing value added commercially marketable products. The traditional knowledge (TK) is oral, confined in traditions and not qualified for the formal IPR system. The vast amount of information collected by the AICRPE team is locked up in an unattended report for want of proper resources. Traditional knowledge on about 10,000 plants has been collected under this project (Pushpangadan *et al.*, 2018). The traditional Indian medicine systems discussed above, i.e., Ayurveda, Siddha, Unani, Amchi etc., account for use of around 2,500 plants, whereas AICRPE created a database of about 10,000 plants, which also requires further scientific validation. Out of these 8,000 wild plant species used by the tribal communities for medicinal purposes. Of these about 950 are found to be new claims and worthy of scientific scrutiny. In addition, 300 wild species used by tribal communities as piscicide (poisonous to fish) or pesticides. At least 175 of these are quite promising to be developed as safe botanicals/pesticides (Pushpangadan *et al.*, 2018).

1.2.3 Ethnobotanical and Ethnopharmacology Research in India

Traditional communities live close to nature making judicious and sustainable use of natural resources for survival and to meet the basic needs. In the process, they acquire unique knowledge by necessities, instinct, observations, trial and error, and long-time experiences. Therefore, remote regions practicing primitive ways of life have wealth of information on the properties of plants built up by the people over millennia (Schultes, 1962, 1991). Ethnobotany is a multidisciplinary science that deals with the traditional and natural relationship between human societies and plants. The study of traditional medicinal knowledge has been referred as ethnobiology (the scientific study of dynamic relationships among peoples, biota, and environments i.e., the way living things are treated or used in different human cultures) and ethnopharmacology (natural sciences research on medicinal, aromatic, and toxic plants linked with socio-cultural studies), often associated with the development of new drugs.

Recognizing the value of such wisdom, considerable ethnobotanical information has been collected in India through organized studies conducted by the late Dr E. K. Janaki Amal of the Botanical Survey of India in the later part of the twentieth century. These were further intensified by Dr S. K. Jain and other groups (Jain 1991, 2001, 2004; Ignacimuthu *et al.*, 2006). Launching of the All India Coordinated Research Project on Ethnobiology (AICRPE, 1992-1998) was also a part of these efforts.

Generation of information about traditional /knowledge led to establishment of The Traditional Knowledge Digital Library (TKDL) project in 2001 as a collaborative project between the Council of Scientific and Industrial Research (CSIR), Ministry of Science and Technology, and Department of AYUSH, Ministry of Health and Family Welfare, Government of India. By May 2010, the TKDL database had 85,500 formulations in Ayurveda, 1,20,200 formulations from Unani, and 13,470 formulations from Siddha. So far, a total of 2,20,268 medicinal formulations have been transcribed (www.csir.res.in).

The CSIR has initiated a coordinated programme for Discovery and Pre-clinical Studies of New Bioactive Molecules and Traditional Preparations. Based on the leads obtained during the last decade, 25 discovery groups in 13 institutes have been constituted. Currently, many leads are at the advanced stage in the CSIR Network Project.

To promote development in the country through science and technology, the CSIR had initiated New Millennium Indian Technology Leadership Initiative (NMITLI) Herbal Project in the year 2002. The aim of NMITLI was to place the ancient Indian system of medicine, the Ayurveda, on strong scientific base for global acceptability. The project targets some of the chronic ailments, such as diabetes mellitus, hepatitis, and arthritis. Under this project, several drugs are under clinical trials using a reverse pharmacology approach.

AYUSH, ICMR (Indian Council of Medical Research), and CSIR together are making efforts for developing formulations/plant-based drugs for identified diseases with sharing of responsibilities as per their mandate/expertise. The main aim is to develop safe, affordable, and effective standardized herbal products for the identified disease conditions for the domestic market using appropriate technologies, meeting the international standards that may facilitate both national and international patenting requirements for global acceptance. The work is in progress in 14 areas of the "Rasayan" partnership.

Recognizing the value of traditional medicinal systems and the contributions of indigenous tribes and local communities in providing knowledge about remedies to several ailments using plant-based products for centuries, in November 2002, Government of India constituted the National Medicinal Plants Board (NMPB). It is mandated with coordination of all the matters relating to medicinal plants, to support with policies and programmes for use, growth of trade, export through sustainable conservation, cultivation, research, and information documentation. The NMPB aims to augment the efforts of communities to conserve and harvest plants in a sustainable manner and to bring the benefits of this knowledge to the world with documentation and report of such knowledge through internationally recognized systems/mechanism (publications in registered journals) following international code. At the same time, the Board will ensure that communities associated with knowledge generation, cultivation, and procurement of raw material (from nature) are duly recognised for their contribution through provisions of intellectual property rights (IPR) under provisions of appropriate mechanism, to enable claim for fair and equitable benefit sharing on commercialization of the products developed using their resources and associated knowledge. The Board is a part of the Department of AYUSH of the Ministry of Health and Family Welfare (http://nmpb.nic.in).

1.2.4 Bioprospecting of Traditional and Promising Indian Medicinal Plants for Drug Formulations and Development of other Plant-based Products

The indigenous medicinal plants are not merely used in production of traditional formulations, but also serve as resource/raw material for production of nutraceuticals, cosmetics, essential oils, flavouring and colouring agents, condiments, and clinically useful prescription drugs. There are at least 121 major clinically useful prescription drugs derived from higher plants (Farnsworth and Soejarto, 1991; Kumar *et al.*, 1997; Lavania, 2005) and their number is increasing. Production of plant-based modern drugs has become an important segment of the pharmaceutical industry in the last few decades. Notable single active molecules currently in use are morphine, codeine papaverine, quinine, quinidine, hyoscine hyoscyamine, colchicine, digoxin, lanatoside, berberine, vinblastine, vincristine, reserpine, recinamine deserpidin, diosgenin, sennosides, podophyllotoxin, artemisinin, taxol, pilocarpine, and silymarin (Handa, 1992; Lavania and Lavania,1995; Kumar *et al.*, 1997; Patwardhan *et al.*, 2004). A selective list of important traditional Indian medicinal plants yielding useful drugs was compiled by Singh *et al.* (2016). It was principally drawn from the important work and compilation by Chopra *et al.* (1956); Thakur *et al.* (1989); Farnsworth and Soejarto (1991); Husain *et al.* (1992); Husain (1993); Lavania and Lavania (1994, 1995); Evans (2009); Kumar *et al.* (1997); Singh *et al.* (1999), and their own understanding. The present effort attempts to extend this list.

In the Indian traditional medicine systems, the traditions and ethnomedicines have played a prominent role for human wellness and in the development of new therapeutic agents. Many plant-based traditional remedies for wellness, health care, and for the treatment of various diseases have been recorded in the Indian traditional medicinal systems. However, their global acceptance has been limited or lacking, for the want of support from modern systematic scientific research validating their properties, identification and characterization of principal components/molecules and providing evidence towards mechanism affecting an infection and providing cure through restricting growth, etc. For this reason, most of them have obtained global recognition as food supplement (tonic) or strengthening immunity system. Approximately, 60% of the antitumor and anti-infective agents currently in use or in the stage of advanced clinical trials are of natural product origin (Cragg *et al.*, 1997). Over 5,000 species and more than 1,20,000 phytochemicals from higher plants have been examined for their therapeutic potential worldwide. During the 1991-1995 period, around 200 NCEs (new chemical entities) were introduced globally, of which 10 (2%) led to drug development and commercialization (Dev, 1997). In India, over 4,000 terrestrial plants have been studied for their biological activities and 20% of them have exhibited promising new activities (Singh *et al.*, 2016). As such, exhaustive studies on bioprospecting of plant species have been undertaken in several laboratories in India, particularly in the national laboratories of CSIR, with the hope to discover new phytopharmaceuticals and clinically validated plant-based drugs soon. In this respect, efforts on listing and dissemination of information of commonly used and selected promising Indian medicinal plants *vis-à-vis* their prospective potential, need to be intensified for further research to realize target drug molecules for treatment of specific ailments facilitating development of plant-based products and extend the search to many more. The present effort is to create of a classified inventory of wider spectrum of medicinal plants found and used in India with related issues, such as threat to their existences, cultivation for sustainable use and trade to achieve this objective.

References

Ahmad, A and Akhtar, J. (2007) Unani system of medicine. Pharmacognosy Review 1:210–214.

AICRPE (All India Co-ordinated Project on Ethnobiology), Final Technical Report, (1992-1998). Ministry of Environment and Forests (MoEF), Government of India.

Anand, N. (1990) Contribution of Ayurvedic medicines to medicinal chemistry. In: Comprehensive medicinal chemistry, vol. 1, (Ed. Kennewell PD) Oxford, England: Pergamon Press pp 113–131.

Anonymous (1978) Handbook of domestic medicine and common Ayurvedic remedies. New Delhi, India: Central Council for Research in Indian Medicine and Homeopathy.

Bhela Charya (2000) Bhela Samhita. Translated by K.H. Krishnamurthy, Edited by Sharma P.V. Chowkhamba Vishwa Bharati, Varanasi,

Brahma, SK and Debnath PK. (2003) Therapeutic importance of Rasayana drugs with special reference to their multidimensional actions. Aryavaidyan16:160–163.

Chopra, RN, Nayer, SL and Chopra IC. (1956) Glossary of Indian medicinal plants. New Delhi: Council of Scientific and Industrial Research.

Cragg, GM, Newman, DJ and Snader KM. (1997) Natural products in drug discovery and drug development. Journal of Natural Products 60:52–60.

Dev, S. (1997) Ethnotherapeutics and modern drug development: The potential of Ayurveda. Current Science 73:909–928.

Diwedi, KK and Singh RH. (1992) A clinical study of Medhya Rasayan therapy in management of convulsive disorders. Journal of Ayurveda and Siddha13:97–106.

Evans, WC. (2009) Treasa and Evans pharmacognosy, 16th ed. Philadelphia, PA: Elsevier Saunders Ltd.

Farnsworth, NR and Soejarto, DD. (1991) Global importance of medicinal plants. In: The conservation of medicinal plants, ed. O. Akerele, V. Heywood, and H. Synge, 25–63. New York: Cambridge University Press.

Ghosh, AK. (2018) History of development of Homoeopathy in India. Indian Journal of History of Science 53(1):76-83. DOI: 10.16943/ijhs/2018/v53i1/49366

Handa, SS. (1992) Medicinal plants-based drug industry and emerging plant drugs. Current Research on Medicinal Aromatic Plants 14:233–262.

Husain, A, Virmani, OP, Popli, SP, Misra, LN, Gupta, MM, Srivastava, GN, Abraham, Z and Singh, AK. (1992) Dictionary of Indian medicinal plants. Lucknow, India: Central Institute of Medicinal and Aromatic Plants.

Husain, A. (1993) Medicinal plants and their cultivation. Lucknow, India: Central Institute of Medicinal and Aromatic Plants.

Ignacimuthu, S, Ayyanar, M and Sankarasivaraman K. (2006) Ethnobotanical investigations among tribes in Madurai district of Tamil Nadu (India). Journal of Ethnobiology and Ethnomedicine 2:25. doi:10.1186/1746-4269-2-25.

Jain, SK. (1991) Dictionary of folk medicine and ethnobotany. New Delhi, India: Deep Publications.

Jain, SK. (2001) Ethnobotany in modern India. Phytomorphology Golden Jubilee Issue: Trends in Plant Sciences 51:39–54.

Jain, SK. (2004) Credibility of traditional knowledge: The criterion of multilocational and medicinal use. Indian Journal of Traditional Knowledge 3:137–153.

Jaiswal, YS and Williams LL. (2017) A glimpse of Ayurveda- The forgotten history and principles of Indian traditional medicine. J Tradit Complement Med. 7(1):50–53. doi: 10.1016/j.jtcme.2016.02.002

Kapoor, LD. (2001) Handbook of Ayurvedic medicinal plants. Boca Raton, FL: CRC Press.

Katiyar, CK. (2010) Indian initiative in the revival of traditional systems of medicine: An overview and recent leads from Ayurveda. In. Science in India: Achievements and aspirations, (Eds. Ram, HYM and Tandon PN), 243–256. New Delhi, India: Indian National Science Academy.

Khanuja SP, Singh J, Shasany AK, and Sharma A. (2006) Inventorization and strategic utilization of medicinal and aromatic plants biodiversity in Asia. In Perspectives on biodiversity and division of megadiverse countries (Eds. Verma, DD, Arora, S and Rai RK) pp 197–242. New Delhi, India: Ministry of Environment and Forests, Government of India.

Kumar, S, Singh, J, Shah, NC and Ranjan, V. (1997) Indian medicinal plants facing genetic erosion. Lucknow, India: Central Institute of Medicinal and Aromatic Plants.

Lavania, S and Lavania, UC. (1995) Sustainable progress in the realization of high value medicinal principles of plant origin. Journal of Indian Botanical Society 74A (platinum jubilee volume): 227–232.

Lavania, UC. (2005) "Pharming" plant genetic resources. Plant Genetic Resources-C 3:81–82.

Lavania, UC and Lavania, S. (1994) Plants for medicine. In: Botany in India: History and progress, (Ed. Johri BM) pp 43–61. New Delhi, India: Oxford & IBH Publishing Co.

National Medicinal Plant Board. State/UT Medicinal Plants Board (SMPBS/UTMPBS) (2014) [Last accessed on 2014 Mar 01]. Available from: http://www.nmpb.nic.in/WriteReadData/links/74009703List-of-SMPBFeb, 2014.pdf

Patwardhan, B, Vaidya, ADB and Chorghade, M. (2004) Ayurveda and natural products drug discovery. Current Science 86:789–799.

Puri, HS. (1970a) Indian medicinal plants use in elixirs and tonics. Quarterly Journal of Crude Drug Research 10:1555–1566.

Puri, HS. (1970b) Chavanprasha- An ancient Indian preparation for respiratory diseases. Indian Drugs 7:15–16.

Puri, HS. (1971) Vegetable aphrodisiacs of India. Quarterly Journal of Crude Drug Research 11:1742–1748.

Puri, HS. (2003) Rasayana- Ayurvedic herbs for longevity and rejuvenation. London: Taylor & Francis.

Pushpangadan, P, George Varughese, Ijinu, TP and Chithra MA. (2018) All India coordinated research project on ethnobiology and genesis of ethnopharmacology research in India including benefit sharing. Annals of Phytomedicine 7(1): 5-12, DOI: 10.21276/ap.2018.7.1.2

Ratha, KK, Meher, SK and Rao MM. (2018) An Enumeration and Review of Medicinal Plants Mentioned in Bhela Samhita. J Drug Res Ayurvedic Sci 3(1): 53-62.

Schultes, RE. (1962) The role of ethnobotanists in search of new medicinal plants. Lloydia 25:257–266.

Schultes, RE. (1991) The reason for ethnobotanical conservation. In: Conservation of medicinal plants (Ed. Akerele, O, Heywood, V and Synge H) pp 65–75. New York: Cambridge University Press.

Shah, NC. (2017) Some Thoughts on Hindu Medicine — An Address by Kaviraj Mahāmahopadhyāya Gananath Sen. Indian Journal of History of Science 52 (4): 445-462 DOI: 10.16943/ijhs/2017/v52i4/49266

Singh, Anurudh K. (2016) Exotic Ancient Plant Introductions: Part of Indian 'Ayurveda' Medicinal System. Plant Genetic Resources-C 14(4); 356–369. doi:10.1017/S1479262116000368.

Singh, Anurudh K. (2017) Revisiting the Status of Cultivated Plant Species Agrobiodiversity in India: An Overview. Proc Indian Natn Sci Acad. 3(1):151-174. doi: 10.16943/ptinsa/2016/v82/48406

Singh, B and Chunekar, KC. (1972) Glossary of vegetable drugs in Brihatrayi. Varanasi, India: Chaukhambah Amarbharti Prakashan.

Singh, J, Bagchi, GD and Khanuja, SPS. (2003) Manufacturing and quality control of Ayurvedic and herbal preparations. In: GMP for botanicals: Regulatory and quality issues on phytomedicine, (Ed. Mukherjee, PK and Verpoorte R) pp 201–230. New Delhi, India: Business Horizons Pharmaceutical Publishers

Singh, J, Lavania, UC and Singh S. (2016) Chapter 3. Indian Traditional and Ethno Medicines from Antiquity to Modern Drug Development. In: Genetics Resources and Chromosome Engineering of Medicinal Plants (Ed. Singh RJ) pp 53-86. USA: CRC Press

Singh, J, Sharma A, Singh SC and Kumar S. (1999) Medicinal plants for bioprospection. Lucknow, India: Central Institute of Medicinal and Aromatic Plants.

Singh, RH and Murthy, ARV. (1989) Medhya Rasayana therapy in the management of apsmaravis-à-vis epilepsies. Journal of Research and Education in Indian Medicine 8:13–16.

Srivastava, S, Lal VK and Pant KK. (2013) Polyherbal formulations based on Indian medicinal plants as antidiabetic phytotherapeutics. Phytopharmacology 2:1–15.

Thakur, RS, Puri, HS and Husain A. (1989) Major medicinal plants of India. Lucknow, India: Central Institute of Medicinal and Aromatic Plants.

The New Millennium Indian Technology Leadership Initiative (NMITLI) (2002) Public-Private Partnership (PPP) model, Council of Scientific and Industrial Research (CSIR)

Udupa, KN. (1973) Psychosomatic stress in Rasayana. Journal of Research in Indian Medicine 8:1–2.

Wagner, H and Proksch A. (1985) Immunostimulatory drugs of fungi and higher plants. In Economic and medicinal plant research (Ed. Farnsworth, N and Wagner H) pp 113–153. London: Academic Press.

2

Inventory of Commonly Used Medicinal and Aromatic Plants (MAPs) Species under Indian Traditional Medicinal Systems (ITMS)

2.1 Introduction

India is known for its richness in bio-resources including floristic diversity and the associated cultural diversity that evolved over time and holds a vast amount of indigenous traditional knowledge (ITK) assembled over millenniums. The ability to convert this knowledge into wealth of products for social good through the process of innovation based on validation of available information holds great promise. In-depth studies using the modern technologies and further examination of potential sources will determine the future of India in the 21st century, commonly referred to as the century of scientific knowledge and technological expertise/advancements that belongs to Asia/India. In this regard, five technologies namely, biochemistry, biotechnology, herbal technology, information technology, and nanotechnology are going to be the most powerful elements that are crucial for ensuring prosperity, security, and welfare of the people of the nations and the globe at large. Production of an inventory of the medicinal aromatic plants with the associated knowledge evolved and documented over millenniums under different Indian traditional medicinal systems shall be the key in promoting and strengthening their exploitation and use under various herbal medicinal systems and/or under an integrated innovative combination of these with modern medicinal systems.

A perusal of the literature about the medicinal and aromatic plants under different Indian traditional medicinal systems reflects that the medicinal herbs used can be classified into three major groups-

1. Plants commonly listed and used either directly as supplementary food or nutraceuticals, under primary healthcare, particularly for treating the common self-limiting ailments, based on the time-tested herbal remedies or in certain drug formulations to combat an ailment or pathogens.
2. Plants in wild that are taxonomically related and/or ancestral plant species of the known medicinal plant species, and other coded wild plant species reported either promising or potential medicinal plants, based on empirical observations

and/or the orthodox examination and use by the local communities and tribes, and

3. non-coded plants species visually appearing and/or reported by locals informally for medicinal properties, based on empirical, informal undocumented information, discovered during recent surveys or explorations, and deserve a scientific scrutiny.

Various publications both ancient and recent, dealing with ethnobotany have reported around 3500 species used for medicinal purposes in the Indian traditional medicinal systems recognized by Ministry of AYUSH (Ayurveda, Yoga and Naturopathy, Unani, Siddha, and Homeopathy), Government of India. For example, 1587 species in Ayurveda, 1128 species in Siddha, 503 species in Unani, 253 species in Amchi/ Sowa-Rigpa (Tibetan medicine), and 30 species in Homeopathy and modern medicine with overlapping of species across the systems (Anonymous, 2011) (Fig. 2.1.1), besides many more (4000-4500) used under folk systems and recorded in recent past. They have often been revised to about 900-2000 species in Ayurveda, 600-1300 species in Siddha, 700-1000 species in Unani, 250-500 species in Amchi/ Sowa-Rigpa (Tibetan medicine), and 30-200 species in Homeopathy and modern medicine in some later publications from time to time, documenting different number of species under different systems (Pandey *et al.*, 2013; personal communication Pushpangadan). For this reason and to provide pre-grant oppositions against the patents filed at various International Patent Offices, along with prior art evidence, AYUSH and the Council of Scientific and Industrial Research (CSIR) launched a program called Traditional Knowledge Digital Library (TKDL). Under this program a database in five international languages has been made from about 2,30,000 medicinal formulations listed in 30 million pages (www.csir.res.in › documents › tkdl). To restrict intellectual piracy and patenting of only new innovations, International Patent Classification (IPC) was established by the Strasbourg Agreement 1971. It provides for a hierarchical system of language independent symbols for the classification of patents and utility models accordingly to the different areas of technologies to which it pertains. The system is updated every five years. IPC did not recognize ITK till 2000. A five-member Task Force consisting of India, China, US, Japan, and European Union has been created to investigate the IPC number for the ITK. The need for creating new symbols of ITK or medicinal plants was established in 2002. The Government of India (GoI) has entered into an agreement with United States Patent and Trademark Office (USPTO) and Exclusive Provider Organization (EPO) for granting rights to access to the TKDL initiated in India in 2001. It is a collaboration between the CSIR, Ministry of Science and Technology, and the Department of AYUSH, Ministry of Health, GoI. The objective of the library is to protect the ancient and traditional knowledge of the country from exploitation through biopiracy and unethical patents, by documenting the available information electronically and classifying it as per inter-process communication (IPC) systems. This was mainly for the written ITK, such as Ayurveda, Siddha, Unani, Yoga, and Tibetan Systems that are in public domain. Apart from this, the non-patent database serves to foster modern research based on traditional knowledge, as it simplifies access to this vast knowledge of remedies or practices.

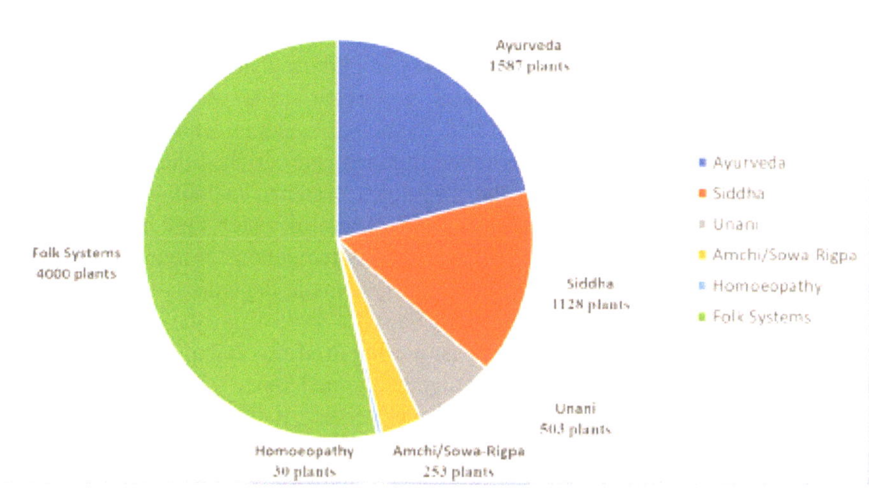

Figure 2.1.1: Number of plants recorded under different Indian traditional medicines systems. Source, Anonymous, 2011

However, there remains a vast amount of non-documented information that is available with various indigenous tribes and communities under traditional folk medicinal systems and needs attention. The Department of Industrial Policy and Promotion (DIPP), Ministry of Commerce and Industry, GoI, and World Intellectual Property Organization (WIPO) together convened a meeting on ITK in 2009. It was observed that Indigenous and Local Communities often do not have strong traditions or validated systems for ownership over knowledge that can meet the requirements of the modern forms of private ownership. For this reason, ITK often has been misappropriated and patents have been accorded many times without due recognition. This has essentially happened due to the non-availability of any database pertaining to ITK. The present inventory of medicinal aromatic plants with brief description of associated knowledge regarding the parts, properties, and the use of either whole plant or the parts must provide soft intellectual property rights (IPR) or ownership to the associated tribes/communities or at least Indian nation.

As per the recent studies and evidence collected from different lines of investigations, Indian Sub-continent is one of the oldest, richest, and most diverse cultural amalgamations, assembled over centuries, including the knowledge associated with the use of medicinal plants. The country has a great indigenous heritage of medicinal plant use, dating back to the early *Vedic* period, and the oldest human civilizations of the earth (Oak, 2018) that lists plants of economic significance. Besides, Indians have inherited traditional practices/systems of reverence towards the nature and its components including plants. Living close to nature in harmony and its sustainable use/exploitation by trial and error, empirical reasoning, and experimentation has been the part of the Hindu way of life. This led to the way for the ancient indigenous societies to evolve their own unique wealth of knowledge pertaining to conservation and sustainable use of plants, animals, and other natural resources for greater good of human beings. In addition, the Indian/*Vedic* ethos of accepting and adopting good (knowledge and/or material) from any source has resulted in introduction and adaptation of a large number of economically important plants along with associated knowledge from other indigenous cultures or civilizations across the world, bringing revolution in agriculture and traditional medicinal systems. This approach of the indigenous Indian communities has revolutionized and enriched the vast knowledge on multifarious uses of plants in the traditional systems, including the traditional Indian medicinal systems, particularly Ayurveda and of which they became integral part (Singh, 2016). The information about such indigenous resources along with associated knowledge has been preserved by systematic documentation and creating write-ups, such as *Samhita. Samhita* is a Sanskrit word from the prefix sam (सम्), 'together', and hita (हित/welfare), the past participle of the verbal root dha (धा)'put'. Therefore, *Samhita* literally means "put together, joined, union", a "collection", and "a methodically, rule-based combination of text or verses". It constitutes the oldest living part of Hindu traditions ensuring availability of such resources for future generations. However, in later period of greedy materialistic Indian societies, a large amount of the information/knowledge was informally passed on from generation to generation without proper documentation, leading to either loss or confinement of knowledge to specific clan/folk/family of medicinal practitioners.

Considering this scenario, an attempt has been made to cover commonly used medicinal and aromatic plant species known for their medicinal properties, gum, resins, dye, aroma, perfumery, etc., in this section. It includes the local seasonal vegetables, fruits, and selected medicinal herbs primarily needed/used for the nutritional balance as well as for wellness, meeting primary healthcare, particularly for treating the common self-limiting ailments. Vegetables and fruits included are those, which were once traditionally cultivated or collected from wild and used, but many might have been forgotten overtime with the western influence, such as nutritionally rich minor cereals, pulses, leafy vegetables, flowers, tubers, etc.

2.2 Nomenclature of Plants

Biological nomenclature is a tool that enables people to communicate names of plants and animals following an internationally accepted style, avoiding confusion. The terms, taxonomy and nomenclature are often confused, but have quite distinct meanings. Taxonomy is the science of classifying, describing, and characterising different groups (taxa) of living organisms. Nomenclature, on the other hand, is the science of giving distinct names to these organisms (taxonomic entities) and assist in their correct recognition/identification. Synonyms are names that have previously been applied to a taxon but are now generally superseded in the light of further investigations or the different authorities used different nomenclature for the same biological/taxonomic entity or biotypes or ecotypes.

Scientific names of plants, animals, fungi, etc. follow internationally agreed rules, which are published as "Codes of Nomenclature". Each scientific name is attached to an original 'type specimen' that can always be traced for reference. These names are essentially 'binomials' consisting of the name of a genus followed by the name of the species. This system of naming plants and animals has remained largely unchanged since Linnaeus developed it in the mid-18th century. The convention is that scientific names are written in italics with an initial upper-case letter for the genus and all lower-case letters for the species name. Technically, it should be followed by the name of the scientific authority(s), who described it.

The International Code of Nomenclature (ICN) for algae, fungi, and plants (ICN, 2011) is presently the accepted set of rules and recommendations dealing with the formal botanical names that are given to organisms, usually referred to as algae, fungi, and lower and higher plants. This code was called earlier as the International Code of Botanical Nomenclature (ICBN) but it was changed to ICN at the XVIII International Botanical Congress, held in Melbourne in July 2011, as part of the Melbourne Code, which replaced the previous Vienna Code of 2005 (Turland *et al.*, 2018). Some commonly used terms in this context are explained below in a simplified form:

Rank: the position of a taxonomic entity or its level in the taxonomic hierarchy.

Kingdom: the highest rank in the taxonomic hierarchy.

Division: the major taxonomic rank within the Plant Kingdom [Corresponding name for Phylum under the Animal Kingdom].

Class: a major taxonomic rank, between Order and Division.

Order: a taxonomic grouping of families believed to be closely related; the major taxonomic rank between Family and Class.

Family: a group of one-to-many related genera, usually clearly separable from other such groups; the major taxonomic rank between Genus and Order.

Genus: a group of related species usually clearly separable from other such groups, or a single species without close relatives; the major taxonomic rank between Species and Family.

Species: a taxon comprising one or more populations of individuals capable of interbreeding to produce fertile off-springs.

Sub-species: the main rank below species in plants, and the only formal rank below Species in animals.

Variety: a taxonomic rank below the rank of sub-species used for plants.

Taxon: a group or category, at any level, in a system for classifying plants or animals (e.g., an entity at a Species level, a Genus level, a Family level etc. - all may be called a *taxon*). 'Culton/Cultigen' has been suggested as an equivalent term for the cultivated taxon.

Type: a designated representative (voucher) for a plant or animal name (Singh *et al.*, 2013).

2.3 Cultivated and Sister or Ancestral Wild Species

The cultivated medicinal and aromatic plants owe their origin and development to human selection after identification of their properties and domestication for sustained regular use without search and extraction from the wild. This, besides sustained use, facilitates the conservation of populations in nature thereby reducing the threat to their survival and further evolution responding to ecological/climatic changes. The plants in the wild evolve subject to natural selection of variation (created by mutation and recombination through breeding). Thus, human influence plays the key role in the origin and domestication, including improvement through selection for related unique economic feature(s) that distinguishes them from their wild counterparts leading to genetic diversification, and development of best cultivation practices for sustainable availability, increased potential of the principal part and components. Wild/sister species of cultivated plant species are naturally occurring species that are part of the same larger gene pool and genetically related to each other. They may include their wild progenitors/ancestors or other taxonomically or otherwise closely related taxa that form the larger common gene pool. They are often screened and selected by the researchers/users in search of genes for higher yield of desired traits/components, and resistance to biotic and abiotic stresses. Many times, they are used as alternatives or misused as adulterants. Therefore, many times they are candidates for further investigation as alternative source or source of diversity of principal component.

2.3.1 Identity/Taxonomy Concerns

Correct identity of the medicinal and aromatic plants with distinctive features and verified taxonomic entity is key to facilitating their successful and continued utilization in herbal medicinal systems. Unfortunately, the issue of synonymy, i.e., multiple names for a single species entity [For example, *Tabernaemontana divaricata* (L.) R.Br. ex Roem. & Schult., syn. *T. coronaria* (Jacq.) Willd.; *Ervatamia* coronaria (Jacq.) Stapf, *E. divaricata* (L.) Burkill; *Hedyotis puberula* (G. Don) R.Br. ex *Arn*., syn. *Oldenlandia umbellata* L.], has been one of the critical constraints in the identification actual species recorded in the literature and used under the system(s) or by the communities. Quite often, this has led to confusion. Therefore, a systematic and detailed characterization of target/key medicinal and aromatic plant species with essential and distinctive features and with possible range of diversity is essential for continued and reproducible results/output. To reduce the possibility of misidentification and solving the problem of confusing the main species with other related or unrelated species (with wrong nomenclature), application of biosystematics (using concept of biological species) can be the initial step to ease the situation. In this regard, the application of biological species concept can play a key role, though it is under-appreciated in all biodiversity matters (Mallet, 1995), including medicinal and aromatic plant species, especially with those concerned with species diversity and synonymy, affecting their use.

Widespread acceptance of the biological species concept (Mayr, 1942) led many taxonomists in the mid-twentieth century to treat earlier names as synonyms of large polytypic species. Modern workers, perhaps increasingly motivated by phylogenetic interests, lay greater emphasis on diagnosis of taxa and less on their reproductive limits (Cracraft, 1989). Thus, a trend for re-utilization of names that were earlier treated as synonyms has been seen. Nearly 15% of the names of all valid species are today considered synonyms. Measuring biological diversity with any purpose often concerns enumerating the number of species. An understanding of the extent of species diversity is crucial to the range of studies including bioprospecting, economic exploitation, threat, ecosystem function, and conservation and use. A core aspect of this extrapolation is the number of species already described for a given medicinal plant or the taxonomic unit (genera and botanical varieties) adding confusion. For this reason, large discrepancies remain between estimates of described species of any group of considerable size of seed plants. Recent estimates of the number of described seed plants differ by 62% (Scotland and Wortley, 2003). These widely different estimates introduce a level of uncertainty in calculation of species/genetic diversity. A key factor affecting estimates of described species is that of multiple names for a single biological entity, i.e., synonymy. For any group, because of synonymy, the number of published names exceeds the actual number of described species. For seed plants, the number of published names is 1,015,000 in *Index Kewensis* but not the extent of synonymy, because many groups have not been subjected to complete (experimental) taxonomy, particularly the medicinal and aromatic plant species, which in general have received little research attention, because of their secondary value compared to food species. Govaert's (2001) estimates, therefore, imply that 58.4% of 1,015,000 binomials in *Index Kewensis* are synonyms (Scotland and Wortley, 2003), and the estimates of the species are often not based on rigorous understanding of species boundaries and pattern of synonymy. The range of synonymy rates for some sample taxa lies between 67 and 88%. The mean percentage of the names that are synonyms, counting binomials only and excluding intra-specific names, is 78%. The sample taxa cover a full range of issues relating to synonymy. For example, although *Strabilanthes* has many synonyms due to overzealous publication of new genera by Bremekamp (1944), but the closely related *Hemigraphis* despite not having taxonomic revision have a similar level of synonymy. Further, the rates of synonymy across a wide range of taxa and geography are strikingly similar (Scotland and Wortley, 2003). For large species, rich groups of plants, synonymy and number of undescribed species remain the two largest constraints in understanding the species' diversity and therefore their use. In the present effort, there has been an attempt to avoid duplicates and unaccepted or unresolved species names and include only distinct genotypes with information on phylogeny and reproductive isolation in literature. For simplicity's sake, an attempt has been made to list the synonym(s) as well along with the latest or most common taxonomic name used. It can be observed that there are many species with synonyms of doubtful identity and taxonomic status/entity. Therefore, they need further investigation to resolve the issues of doubt and poor understanding, to facilitate conservation and use of distinct and correct entities, for consistency in their effectiveness and efficacy in use. Synonyms are not only errors in applying systematic classification but are also by-products of the systematic concept used at the time of description of a species under a specific study. Future investigations, using experimental taxonomy approach and biological concept of species will certainly reveal many current names to be synonyms of others, now in circulation. Changing systematic concepts and improved technologies for detecting and defining species will continue to breathe new life into yesterday's entombed synonyms and taxonomy. This will resolve phylogenetic confusion and will also promote predictive, effective, and reproducible (consistent) result in use of medicinal and aromatic plants and facilitate conservation of distinct and exact species/ genotype avoiding sister, related or morphologically alike plant species.

Therefore, to resolve taxonomic confusions targeting correct genotype/species, studies are required to go beyond orthodox taxonomy, which deals with the convenient tabulation and grouping based only on morphological similarities and dissimilarities. This would mean a step towards experimental taxonomy. Which means going for biosystematics study regarding concerned species, fully appreciating the value of morphological differences and similarities. Biosystematics involves collection of observational and experimental data for making taxonomic decisions. It would include precise breeding (cross-compatibility) between the concerned species and other investigations, such as differences and similarities in cytology, biochemical and molecular profiling, adequate to provide precise information regarding the essential features of the species defined limits, phylogenetic relationships, variability, and dynamic taxonomic structure. It would help delimiting the natural boundaries of biotic units of the species, and would involve using following steps/methods:

1. Field study
   (i) Sampling of population systems showing variations
   (ii) Ecological observations on population systems showing variations
   (iii) Collection of specimens for herbarium preservation and of diaspores (seeds, rhizomes, tubers, etc.) for culture in experimental garden
2. Herbarium study
   (i) Recording of range of variations within species
   (ii) Recording of geographical distribution and detection of discontinuity if any (Disjunction and variance) from specimens
3. Experimental work
   (a) *Field*
      (i) Observations on variations under controlled environment
      (ii) Sympatric trial through hybridization between population systems
      (iii) Observations on vitality and vigour of hybrids
   (b) *Laboratory*
      (i) Cytological observations on chromosome number, karyomorphology, and homology of chromosomes in hybrids at meiosis
      (ii) Observations on miscellaneous aspects, viz., anatomy, embryology, palynology, etc., for relationship evaluation in case of non-hybridizing populations
      (iii) Perform molecular biosystematics experiments
4. Information documentation and analysis (Singh, 2017)

2.4 Present Inventory

The use of medicinal and aromatic plants under the auspices of various traditional Indian medicine systems in overcoming the nutritional deficiency and treatment of various human ailments is time tested and well established. However, the concerns expressed about the gaps in information in case of documented species, as per the present requirement, non-documentation of many, and their possible potential role in providing much needed economic health care to tribal, rural, and poor communities needs further research to convert these plant species into useful economical plant-based products. This will help achieving self-reliance (Aatmanibharta). Hence there is a need for a scientific revisit to traditional Indian medicine systems with focus onto the building blocks of these systems, the plant species used. In the present effort, an attempt has been made to provide a bird-eye-view about the total spectrum of medicinal and aromatic plant (biological) resources, often neglected by the modern societies/researchers of medical science. This may help generate greater interest in drawing much needed research attention from scientific communities, investments from entrepreneurs and opportunities to start-ups for their exploitation and use in the larger interest of humanity and globe.

The list of medicinal and aromatic plants commonly/popularly used in the Indian traditional systems has been organized in ascending alphabetical order of Latin name to facilitate easy search. It consists of following information on each plant species:

2.4.1 Latin Name: It presents designated binomials Latin name of the concerned species as per international code of nomenclature, including the name as per the latest taxonomic classification and assignment. Considering the problem of synonymy, as discussed earlier, in many taxa this may be followed by other binomial name available in the literature with author, to provide clarity and avoid duplication.

Threat Perception: Majority of medicinal aromatic plants are extracted from the nature (wild) and therefore, many of them are under threat proceeding towards the demise and/or in mass extinction. To reflect this and issue early warning to draw scientific attention for appropriate action to ensure protection, threat status of impacted species has been mentioned in parenthesis after the Latin name, as per the International Union of Conservation of Nature (IUCN, 1995) categories.

In addition, there are species categorized as rare. A rare species is the one that occurs in widely separated small sub-populations so that outbreeding between the sub-populations is seriously reduced or restricted, resulting inbreeding within a single population. Besides habitat loss, it is necessary to understand the biology and autecology of the endemic species to find out intrinsic causative factors leading to rarity. There are seven forms of rarity based on geographical range, habitat specificity, and local population size (Nayar, 1998). Such species require habitat protection and care and/ or conservation using various *ex situ* approaches as per the applicability.

The threatened species categorized as per the IUCN (1995) category designated in the ICUN Red List criteria include:

Extinct (EX): A taxon is Extinct when there is no reasonable doubt that the last individual has died in its natural habitat.

Extinct in The Wild (EW): A taxon is Extinct in the Wild and known only to survive in cultivation, in captivity or as a naturalized population (or populations) well outside its past range. Surveys should be conducted over a time-frame appropriate to the taxon's life cycle and life form to ascertain.

Critically Endangered (CR): A taxon is Critically Endangered when the best available evidence indicates that it is facing an extremely high risk of extinction in the wild in the immediate future (80% decline in the last ten years, 100 km^2 of area of occupancy or 10 sq. km in fragmented area; estimated 1250 mature individuals or subpopulations of not more than 50 individuals).

Endangered (EN): A taxon is Endangered when the best available evidence indicates that it is facing an extremely high risk of extinction in the wild soon (50% decline in the last 10 years; est. <5000 km^2 of the area of occupancy or 500 km^2 in fragmented areas; est. 2500 individuals or subpopulations of 250 mature individuals).

Vulnerable (VU): A taxon is Vulnerable when the best available evidence indicates that it is facing an exceedingly high risk of extinction in the wild, in the medium-term future (50% decline in the last 20 years; est. <2000 km^2 of the area of occupancy or <2000 km^2 in fragmented population est. 10,500 individuals or subpopulations of 1000 mature individuals).

Near Threatened (NT): A taxon is Near Threatened when it has been evaluated against the criteria but does not qualify for Critically Endangered, Endangered or Vulnerable now, but is close to qualifying for or is likely to qualify for a threatened category soon.

Least Concern (LC): A taxon is Least Concern when it has been evaluated against the criteria and does not qualify for Critically Endangered, Endangered, Vulnerable or Near Threatened. Widespread and abundant taxa are included in this category, which will be applicable to most listed species.

Data Deficient (DD): A taxon is data deficient when there is inadequate information to make a direct, or indirect, assessment of its risk of extinction based on its distribution and/or population status. Therefore, Data Deficient is not a category of threat but perception.

If the range of a taxon is suspected to be relatively circumscribed, and a considerable period has elapsed since the last record of the taxon, threatened status may well be justified.

Not Evaluated (NE): A taxon is Not Evaluated when it is has not yet been evaluated against the criteria as per perception.

2.4.2 Sanskrit or Ayurvedic Name(s): As the list is dominated by the plants used in the Ayurvedic system, which is the oldest ancient Indian traditional medicinal system practiced from the *Vedic* period and was further enriched with adoption of a many exotic species from other cultures/civilizations when Sanskrit was the common spoken language in India and used in documentation of information in the ancient literature. The availability of Sanskrit name reflects the use of a species from ancient times.

2.4.3 Common Names: This lists the common vernacular names used in different languages or in trade of medicinal plants to refer a herb by the ordinary people in different regions of India and other parts of the world. It may include English names as well after its adaptation. As India has significant linguistic diversity, the species dispersed and adapted across the various region of the country may have several such names from different spoken language and therefore, it is possible that many might have been left.

2.4.4 Family: Each taxon as per the International Code of Botanical Nomenclature must have been classified into or associated with taxonomic entities discussed above, including family at macro level. The family refers to a group of one-to-many related genera based on similarity of morphological characters. An attempt has been made to list the latest family associated with the concerned taxon. Updating of taxonomy of economically important plant species as per the new evidence is a continuous process. To provide the latest and most used designated family, in many cases there may be more than one family has been mentioned, for the sake of updated information and possible phylogenetic implications.

2.4.5 Distribution: Brief information about habitat and/or geographical range/extent about the occurrence of each species particularly the Phyto-geographical region or administrative state of India has been provided. At a global level, wider range of distribution, place of origin/nativity or endemism has also been enumerated. These notes have been predominantly adopted from the information given in the published floras or from the reports of agencies responsible for survey and inventory.

2.4.6 Parts and Properties Exploited: In this column, an attempt has been made to provide basic information about the parts that have been used directly or in a particular form to affect an action or as a source of the principal component that is responsible for a specific action or are used in the formulation of drug or a byproduct. This may work as a stimulus and starting point for the researchers or entrepreneurs for further investigation and their systematic use in the therapy of deficiency or treatment based on a scientific principle.

2.4.7 Associated System (s): This column lists the systems in an abbreviated form; those have been associated with the regular use of the specific plant species and were responsible for identification of medicinal potential of the concerned species and further use. It includes predominantly the systems that have been recognized by the Ministry AYUSH of Government of India. The acronym AYUSH stands for Ayurveda, Yoga and Naturopathy, Unani, Siddha, and Homeopathy, that are the six Indian systems of medicine prevalent and practiced in India and some of the neighbouring Asian countries and few in some of the developed countries. It may be observed that of many species overlap across the systems. This shows integration/adoption of these plants promoting wider use. Some plant species used in western herbal system have also found place, because of their integration into Indian systems. In addition, the plants that have been used by the local tribes, primitive communities in healing practices, based on long-time experience, ideas of body physiology and health preservation, known to selected group of indigenous people, cultures or families, though not systematically documented, but transmitted informally as open or secrete knowledge, practiced, or applied by the tribal or local culture(s) or individual families have also been included. They are commonly referred to as Folk or traditional medicine. Some western/English herbal medicinal plants have also been included, which have been integral of global indigenous systems.

The footnote at the base of this section, besides the full form of the abbreviations used, also lists the major source of information, which can be consulted for further details. In this regard referring to the set of 13 volumes, "Review on Indian Medicinal Plants", brought out by Indian Council of Medical Research (ICMR) between 2004-2012 may be worth consulting for detailed perspective about research on different medicinal and aromatic plants (Gupta and Tandon, 2004-2012).

References

Anonymous (2011) Repot of the Steering Committee on AYUSH for the 12th Five Year Plan 2012-2017. Health Division Planning Commission, Government of India. New Delhi. niti.gov.in › docs › committee › strgrp12 › st_ayush0903

Cracraft, J. (1989) Speciation and its ontology: the empirical consequences of alternative species concepts for understanding patterns and processes of differentiation. In: Speciation and its consequences (Eds. Otte D, Endler JA) pp 28–59. Sinauer, Sunderland.

Govaerts, R. (2001) How many species of seed plants are there? Taxon 50:1085–1090

Gupta, AK and Tandon Neeraj (2004-2012) Review on Indian Medicinal Plants (set of 13 Volumes), Indian Council of Medical Research, pp 9731 ISBN 9770972795006.

IUCN (The World Conservation Union) Red List Categories. Prepared by IUCN Species Survival Commission 1995 pp 8. IUCN Gland, Switzerland.

Mallet, J. (1995) A species definition for the modern synthesis. Trends Ecol Evol 10:294–299

Mayr, E. (1942) Systematics and the origin of species. Columbia University Press, New York

Nayar, MP. (1996) Hot spots of endemic plants of India, Nepal and Bhutan. pp 253. Tropical Botanic Garden and Research Institute, Palode, Thiruvananthapuram, Kerala, India.

Oak, NN. (2018) Bhishma Nirvana: An Astronomy Poison Pill pp. 298. Amazon Digital Services LIC- KDD Print US. ISBN 0983034419, 978093034414

Pandey, MM, Rastogi Subha and Rawat, AKS. (2013) Indian Traditional Ayurvedic System of Medicine and Nutritional Supplementation. Evidence-based Complementary and Alternative Medicine Vol. 13, Article ID 376327, 12 pages, 2013. https://doi.org/10.1155/2013/376327

Scotland, RW and Wortley AH. (2003) How many species of seed plants are there? Taxon 52(1):101–104

Singh, Anurudh K. (2016) Exotic Ancient Plant Introductions: Part of Indian 'Ayurveda' Medicinal System. Plant Genet Resour-C 14:356-369, doi:10.1017/S1479262116000368

Singh, Anurudh K. (2017) Wild Relatives of Cultivated Plants in India: A Reservoir of Alternative Genetic Resources and More pp 310. Springer Nature Singapore Pte Ltd. 2017 ISBN 978-981-10-5115-9

Singh, Anurudh K, Rana, RS, Bhag Mal, Singh, B and Agrawal RC, (2013) Cultivated Plants and Their Wild Relatives in India – An Inventory pp.251. Protection of Plant Varieties & Farmers' Rights Authority, NASC Complex, DPS Marg New Delhi, India.

Turland, NJ, Wiersema, JH, Barrie, FR, Greuter, W, Hawksworth, DL, Herendeen, PS, Knapp, S, Kusber, W-H, Li, D-Z, Marhold, K, May, TW, McNeill, J, Monro, AM, Prado, J, Price, MJ and Smith, GF. (Eds.) (2018) International Code of Nomenclature for Algae, Fungi, and plants (Shenzhen Code) adopted by the Nineteenth International Botanical Congress Shenzhen, China, July 2017. Regnum Vegetabile 159. Glashütten: Koeltz Botanical Books. DOI https://doi.org/10.12705/Code.2018ICN Before 2011 It was called International Code of Botanical Nomenclature (ICBN).

2.5 List of Commonly Used MAPs in ITMS

The plants used for medicinal and aromatic purposes can be classified in several ways. They can be classified based on their growth habits and breeding behaviour. As per habit they may be either a tree, shrub, climbers, herb, with annual, biennial, and perennial life cycle, propagating with the help of seed, rhizomes, tubers, etc. Similarly, the medicinal and aromatic plants (MAPs) can be classified as per their pharmacological activity due to their biologically active ingredients, commonly grouped into four groups: alkaloids, glycosides, essential oils, and other miscellaneous active substances. Classification of the MAPs based on the above criterions may help in identification and further study of the target plant species and facilitate conservation, research/evaluation, and sustainable use. Below, we present representative examples belonging to various groups-

Representative tree MAPs

S. No.	Botanical name	Common Name	Parts Used
1	*Acacia nilotica* (L.) Willd. ex Delile	Babul	Plant, pods, leaves, bark, gum
2	*Aegle marmelos* L. Corr.	Bael	Roots, leaves, fruit, bark
3	*Azadirachta indica* A.Juss	Neem	Bark leaves, flowers, seed, oil
4	*Boswellia serrata* Roxb. ex Colebr.	Indian olibanum, Salai guggul, Sallaki	Herbal extract
5	*Butea monosperma* (Lam.) Taub.	Palash	Bark, leaves, flowers, seed, gum
6	*Commiphora wightii* (Arn.) Bhandari, syn. *C. mukul* (Hook. ex Stocks) Engl.	Guggul	Resinous gum
7	*Emblica officinalis* Gaertn., syn. *Phyllanthus emblica*	Amla, Indian gooseberry	Fruit
8	*Mesua ferrea* L.	Naga	Flowers, oil.
9	*Olea europaea* Linn.	Olive	Leaves, Oil
10	*Semecarpus anacardium* L. f.	Marking nut tree, Baklava	Fruit
11	*Terminalia arjuna* (Roxb. Ex DC.) Wight & Arn.	Arjun	Bark
12	*Terminalia bellirica* (Gaertn.) Roxb.	Behera	Bark, fruit
13	*Terminalia chebula* Retz.	Harda	Fruits

Figure 2.2.1: Tree MAPs: a. Arjun; b. Neem; c. Palash; d. Guggul. Source: google.com

Representative shrub MAPs

S.No.	Botanical name	Common Name	Parts Used
1	*Artemisia nilagirica* (C.B. Clarke) Pamp.,	Davana, Thavanam	Leaves, flowering top
2	*Asparagus adscendens* Roxb.	Safed muesli	Tuberous roots
3	*Atropa acuminata* Royle *ex* Lindl., syn. *A. belladonna*	Belladonna	Leaves and roots
4	*Hippophae Rhamnoides* L.	Sea buckthorn	Fruit
5	*Lavandula bipinnata* Kuntze.	Lavender	Plant, Flowers
6	*Plumbago zeylanica* L.	Chitrak	Leaves, roots
7	*Rauvalfia serpentina* (L.) Benth. ex Kurz	Sarpa Gandha	Roots
8	*Rosmarinus officinalis* L.	*Rosemary*	Plant, leaves, essential oil
9	*Vitex negundo* L.	Negundo	Leaf, seeds

Fig. 2.2.2: Shrub MAPs: a. Sarp Gandha; b. Sea buckthorn; c. Musali; d. Chitrak.
Source: google.com

Representative Climbers (woody) MAPs

S.No.	Botanical name	Common Name	Parts Used
1	*Abrus precatorius* L.	Gun-chi, Ratti	Seed
2	*Akebia quinata* Deene	Chocolate vine	Stem, fruit
3	*Celastrus paniculatus* Wild	Jyotishmati, Malkangani	Bark, leaves, seed
4	*Gloriosa superba* L.	Kalahari	
5	*Gymnema sylvestre* R.Br. ex Schult.	Gudmar	Whole plant, leaves
6	*Piper longum* L.	Pippli	Fruits, roots
7	*Piper nigrum* L.	Kali Mirch	Fruit
8	*Tinospora cordifolia* (Willd.) Miers ex Hook.f. & Thomson	Gileo	Stem juice

Fig. 2.2.3: Climber MAPs: a. Gudmar; b. Kali Mirch; c. Gileo; d. Malkangani.
Source: google.com

Representative herb MAPs

S.No.	Botanical name	Common Name	Parts Used
1	*Centella asiatica* (L.) Urb.	Brahmi, Gotu Kola	Plant
2	*Ceropegia spiralis*	Pilachi Khantudi	Plant
3	*Datura metel* L.	Datura	Plant, leaves, flowers
4	*Mentha viridis* (L.) L., syn. *M. spicata*	Mint, Pudina	Plants, leaves
5	*Nigella sativa* L.	Kala Zira	Seed
6	*Ocimum tenuiflorum* L., syn. *O. sanctum* L.	Tulsi	Roots, leaves, seeds
7	*Papaver somniferum* L.	Afeem, Opium, Post dana	Capsule, dried latex, seed
8	*Psoralea corylifolia* L., syn. *Cullen corylifolium*	Babchi	Plant, fruit, seed,
9	*Stevia rebaudiana* Bertoni	Sweet herb	Leaves

Fig. 2.2.4: Herbal MAPs: a. Brahmi; b. Duttura; c. Tulsi; d. Opium.
Source: google.com

Representative annuals and biennials MAPs

S.No.	Botanical name	Common Name	Parts Used
1	*Apium graveolens* L*	Celery, Ajamodaa	Leaf, seed
2	*Blumea lacera* (Burm. f.)	Jangali Muli	Whole plant, leaf
3	*Carum carvi* L.*	Shah or Kala Jeera, Caraway	Leaves, seeds
4	*Cassia absus* L.*	Bankultthi	Root, leaves, seeds
5	*Celosia cristala* L.	Cock's Comb	Inflorescence
6	*Cucumis sativus* L.	Khira	Fruit, seed
7	*Cuminum cyminum* L.	Cumin, Jeera	Fruit/ Seed
8	*Euphorbia lathyrism* (L.) Fourr* syn. *Euphorbia decussata* Salisb.	Caper spurge	Seed latex
9	*Momordica charantia* L.	Bitter melon, Karela	Fruit
10	*Papaver rhoeas* L.	Red poppy, Lal post	Flowers
11	*Phyllanthus niruri* Hook, syn. *P. debilis*	Bhumi amla	Whole plant
12	*Trachyspermum ammi* (L.) Sprague	Ajwain, ajowan	Leaves, fruit, seed
13	*Trigonella foenum-graecum* L.	Fenugreek, Methi	Leaf, seed
14	*Withania somnifera* (L.) Dunal	Ashwagandha	Root

*Biennials

Fig. 2.2.5: Annual and Biennial MAPs: a. Shah or Kala Jeera; b. Karela; c. Fenugreek; d. Ashwagandha. Source: google.com

Representative perennial medicinal and aromatic grasses and herbs

S.No.	Botanical name	Common Name	Parts Used
1	*Aloe barbadensis* Mill., syn. *A. vera* L.	Gharat Kumari	Leaf juice
2	*Bacopa monnieri* L.	Brahmi	Whole plant
3	*Cissus quadrangularis* L.	Hajodi	Whole plant
4	*Cymbopogon citratus* (DC.) Stapf	Lemon grass	Leaves, essential oil
5	*Elettaria cardamomum* Maton	Cardamom, Elachi	Fruit
6	*Gentiana kurroo* Royle	Katuki	Root, whole plant
7	*Glycyrrhiza glabra* L.	Liquorice, Mulathi,	Root
8	*Thymus vulgaris* L.	Thyme	Leaves, essential oil
9	*Vetiveria zizanioides* (L.) Nash, syn. *Andropogon zizanioides*; *Chrysopogon zizanioides*	Khus grass, Vetiver	Leaves, oil

Fig. 2.2.6: Perennial plants and grasses MAPs: a. Aloe; b. Brahmi; c. Elachi; d. Khus grass. Source: google.com

Representative tubers/ rhizomes MAPs

S.No.	Botanical name	Common Name	Parts Used
1	*Allium sativum* L.	Garlic, Lahsan	Bulb, bulb juice
2	*Asparagus adscendens* Roxb	Safed musli, Pili Satavar	Tuberous root
3	*Chlorophytum borivilianum* Santapau & R.R. Fern	Safed musli	Tuberous root
4	*Curcuma amada* Roxb.	Amba haldi	Rhizomes
5	*Curcuma domestica* Valet, syn. *C. longa* L	Haldi, turmeric	Rhizomes
6	*Inula racemosa* Hook f.	Pushkarmool	Roots
7	*Manihot esculenta* Crantz.	Sakar Khand	Tubers
8	*Zingiber officinale* Roscoe	Adrak, Ginger, Sonth	Tuberous Rhizome Roots

Fig. 2.2.7: Tuberous and rhizomatous MAPs: a. Garlic; b. Safed musli; c. Turmeric; d. Ginger. Source: google.com

The List with Description of MAPs

Abelmoschus esculentus **(L.) Moench, syn.** ***Hibiscus esculentus*** **L.**

Sanskrit or Ayurvedic name(s): Bhenda

Common name (s): Bhindi

Family: Malvaceae

Distribution: Habitat to tropical, subtropical, and warm temperate regions and cultivated.

Parts and properties used: Fruit- Emollient, demulcent and diuretic, used in catarrhal infections, dysuria and gonorrhoea. Seeds- Antispasmodic, cordial, and stimulant.

Associated system (s): AFU

Abelmoschus ficulneus **(L.) Wight & Arn., syn.** ***Hibiscus ficulneus*** **L.**

Common name (s): Kattu vendai, Banbhendi, Ranbhendi, Jangli, Bhindi

Family: Malvaceae

Distribution: Tropical India.

Parts and properties used: Leaves- Used against diarrhoea. Fresh root decoction- For calcium deficiency.

Associated system (s): AF

Abelmoschus moschatus **Medik., syn.** ***Hibiscus abelmoschus*** **L.**

Sanskrit or Ayurvedic name(s): Kasturilatik, Latakasturi

Common name: Muskdana, Latakasturi

Family: Malvaceae

Distribution: Found in western Himalayas to Koraput region and cultivated.

Parts and properties used: Seed- Anti-spasmodic, diuretic stomachic, nervine, also insecticidal and aphrodisiac.

Associated system (s): AFU

Abies pindrow **(Royle ex D. Don) Royle**

Sanskrit or Ayurvedic name: Taalisha relative

Common name: Granthiparna, Talisa

Family: Pinaceae

Distribution: Found wild in west and central Himalayas and north India.

Parts and properties used: Leaf- Antibacterial. Used for respiratory ailments and anxiety.

Associated system (s): A

Abies spectabilis **(D. Don) Spach, syn.** ***Abies webbiana*** **Wall. ex D. Don**

Sanskrit or Ayurvedic name: Taalisha

Common name: Talispatra

Family: Pinaceae

Distribution: Habitat to Himalayas 1600-4000 m.

Parts and properties used: Leaves- Astringent, carminative, expectorant, stomachic and tonic. Juice used in asthma, bronchitis. Needle oil- In respiratory disorder and rheumatic pains.

Associated system (s): A

Abroma augusta **Jacq.,** ***A. augusta/augustum*** **(L.) L.f.**

Sanskrit or Ayurvedic name: Pishaacha Kaarpaasa

Common name: Ulati Kambal, Devil's cotton

Family: Malvaceae

Distribution: Found throughout hotter and humid parts of India, Punjab to Assam and peninsular India.

Parts and properties used: Root bark- Emmenagogue and uterine tonic, used in intra-uterine diseases and gynaecological disorders. Root powder- Abortifacient and anti-fertility agent. Leaves- Contain taraxerol, its acetate and lupeol are used as Homoeopathic tincture for uterine disorders.

Associated system (s): ASHSU

Abrus precatorius **L.**

Sanskrit or Ayurvedic name: Gunja

Common name: Chirmati, Chinnoti, Gundumani

Family: Leguminosae/ Fabaceae

Distribution: Found throughout the plains of India.

Parts and properties used: Plant- Uterine stimulant, abortifacient, toxic. Seed- Teratogenic. Used in tetanus, rabies and scratches, sores, wounds of dog; also, in baldness.

Associated system (s): AFHSLU

Abutilon indicum **(L.) Sweet subsp.** *guineense* **(Schumach.) Borss.**

Sanskrit or Ayurvedic name: Atibala

Common name: Tutti, Kanghi

Family: Malvaceae

Distribution: Occur throughout warm India.

Parts and properties used: Plant - Febrifuge, anthelmintic, demulcent, diuretic, anti-inflammatory. Root- Nervine tonic, given in paralysis; strangury. Also, used in gout, polyuria, haemorrhage and leucorrhoea.

Associated system (s): AFHSLU

Acacia caesia **(L.) Willd**

Sanskrit or Ayurvedic name: Khadira relative

Common name: Incha, Indu, Singapore pattai

Patai Family: Leguminosae/ Fabaceae (Mimosaceae)

Distribution: Indo-Malayan region, found wild in Kerala.

Parts and properties used: Plant extract- Treat asthma, skin, menstrual diseases, and scabies.

Associated system (s): AF

***Acacia catechu* (L. f.) Willd.**

Sanskrit or Ayurvedic name: Khadira

Common name: Katha

Family: Leguminosae/ Fabaceae (Mimosaceae)

Distribution: Occur throughout drier parts of India and cultivated.

Parts and properties used: Tannin extract- Astringent, antidiarrheal, haemostatic. Treat gingivitis leukaemia. asthma, bronchitis, and cough. Heartwood- Treat inflammations, skin diseases and urinary disorders.

Associated system (s): AFSLU

***Acacia chundra* (Roxb. ex Rottler) Willd.**

Common name: Khadir

Family: Leguminosae/ Fabaceae (Mimosaceae)

Distribution: Habitat to dry and rocky soil of Rajasthan and peninsular India.

Parts and properties used: Heartwood extract used as substitute of *A. catechu.*

Associated system (s): A

***Acacia farnesiana* (L.) Willd.**

Sanskrit or Ayurvedic name: Arimeda relative

Common name: Vilaayati Kikar

Family: Leguminosae/ Fabaceae (Mimosaceae)

Distribution: Native to West Indies; naturalized all over India and cultivated.

Parts and properties used: Bark- Astringent, demulcent, anthelmintic, antidysentery, anti-inflammatory. Used in stomatitis, ulcers, swollen gums, bronchitis, skin diseases.

Associated system (s): AFS

***Acacia leucophloea* Willd.**

Sanskrit or Ayurvedic name: Arimeda,

Common name: Godhaa-skandha, Safed Babul

Family: Leguminosae/ Fabaceae (Mimosaceae)

Distribution: Habitat to dry regions of India.

Parts and properties used: Bark- Cooling, demulcent. Used in biliousness and bronchitis. Leaves- Ant syphilitic and antibacterial. Gum- Demulcent. Seeds- Hemagglutinating.

Associated system (s): AUF

***Acacia nilotica* (L.) Willd. ex Delile subsp. *indica* (Benth.) Brenan, syn. *A. arabica* (Lam.) Willd. var. *indica* Benth.**

Sanskrit or Ayurvedic name: Babbula

Common name: Babul, Keekar

Family: Leguminosae/ Fabaceae (Mimosaceae)

Distribution: Indigenous to northern to Peninsular India and cultivated.

Parts and properties used: Plant- Anti-microbial, anti-plasmodial and antioxidant. Treat immunodeficiency, hepatitis C virus, cancer, and piles. Bark- Astringent, spasmolytic, hypoglycaemic. Used in diarrhoea and helminthiasis. Gum- Demulcent. Fruit- Used in urogenital disorder.

Associated system (s): AFS

***Acacia pennata* (L.) Willd, syn. *Acacia torta* (Roxb.) Craib.**

Sanskrit or Ayurvedic name: Khadiravallari

Common name: Lataakhadira, Aadaari, Ari.

Family: Leguminosae/ Fabaceae (Mimosaceae)

Distribution: Occur throughout India.

Parts and properties used: Bark- Ant bilious, antiasthma. Leaf- Stomachic, styptic (bleeding gum), antiseptic. Young leaf decoction- Taken in body pain, headache, and fever.

Associated system (s): AFS

***Acacia senegal* Willd.**

Sanskrit or Ayurvedic name: Shveta Babula

Common name: Babul

Family: Leguminosae/ Fabaceae (Mimosaceae)

Distribution: Native to sub-Saharan Africa, cultivated in dry western India.

Parts and properties used: Gum- Ingredient for treatment of diarrhoea, catarrh. Stembark- Anti-inflammatory, and spasmolytic. Root- Used for dysentery and urinary discharges.

Associated system (s): A

***Acacia sinuata* (Lour.) Merr., syn. *A. concinna* (Willd.) DC.**

Sanskrit or Ayurvedic name: Saptalaa

Common name: Chikakai (Shikakai)

Family: Leguminosae/ Fabaceae (Mimosaceae)

Distribution: Habitat to tropical and peninsular India.

Parts and properties used: Plant- Febrifuge, expectorant, emetic, spasmolytic, diuretic, antidiarrhea. Pod- Detergent, used to clean scalp. Ointment- Treat skin disease. Bark- Contain saponin, which is spermicidal, haemolytic, and spasmolytic, used in leprosy.

Associated system (s): AF

Acacia polycantha **Willd., syn. *A. suma* Buch. -Ham.**

Sanskrit or Ayurvedic name: Shveta Khadira

Common name: Safed Khair

Family: Leguminosae/ Fabaceae (Mimosaceae)

Distribution: Dry deciduous forests of Bengal, Bihar, and western peninsula.

Parts and properties used: Wood tannin- Powerful astringent, antidiarrheal, homeostatic; used to treat excessive mucous discharges, haemorrhages, relaxed conditions of gums, throat and mouth, stomatitis, irritable bowel.

Associated system (s): AU

Acalypha fruticosa **Forssk.**

Sanskrit or Ayurvedic name: Haritha Manjari

Common name: Chinni

Family: Euphorbiaceae

Distribution: African, found in peninsular and south India.

Parts and properties used: Leaves- Used in stomach-ache, dyspepsia, rheumatism, dermatitis, and body swellings.

Associated system (s): FS

Acalypha indica **L**

Sanskrit or Ayurvedic name: Muktavarchaa

Common name: Khokali, Arishtamanjari

Family: Euphorbiaceae

Distribution: Distributed in Old World tropics, including India.

Parts and properties used: Plant- Diuretic, anti-asthmatic/ anti-bronchitis. Leaf- Antibacterial.

Associated system (s): AFHS

Acanthospermum hispidum **DC.**

Sanskrit or Ayurvedic name: Trikantaka

Common name: Kattu nerinji

Family: Asteraceae/ Compositae

Distribution: Found as weed all over India.

Parts and properties used: Plant- Used in dermatological affections. Essential oil- Antimicrobial.

Associated system (s): A

***Acanthus ilicifolius* Linn.**

Sanskrit or Ayurvedic name: Harikusa, Krishna, Saraiyaka

Common name: Sea Holly

Family: Acanthaceae

Distribution: Habitat to tidal coastal forests of Andamans.

Parts and properties used: Decoction- Antacid, diuretic.

Associated system (s): AFS

***Achillea millefolium* L.**

Common name: Brinjasif

Family: Asteraceae/ Compositae

Distribution: Occur in cold arid zone of Ladakh and beyond. Also cultivated.

Parts and properties used: Plant- Astringent, antispasmodic, choleretic, antibacterial, diaphoretic. Given in dyspeptic ailments, fever, and cold.

Associated system (s): HSUW

***Achyranthes aspera* L.**

Sanskrit, Ayurvedic and Common name: Apamarga, Apamarga

Family: Amaranthaceae

Distribution: Found throughout tropical and sub-tropical India.

Parts and properties used: Plant- Astringent, emetic. Given in lipid disorders and obesity. Root- Blood-purifier.

Associated system (s): AFHSLU

***Achyranthes bidentata* Blume**

Sanskrit or Ayurvedic name: Shveta Apaamaarga

Common name: Apamarga

Family: Amaranthaceae

Distribution: Habitat to temperate and sub-temperate Himalayas 1200-1300 m.

Parts and properties used: Plant- An anodyne, anti-inflammatory, antirheumatic, bitter, digestive, diuretic, emmenagogue and vasodilator.

Associated system (s): ASF

***Achras zapota* Linn. *Manilkara zapota* (Linn.) P. van Royan *Manilkara achras* (Mill.)**

Common name: Sapota

Family: Sapotaceae

Distribution: Central American. Cultivated in Maharashtra, Tamil Nadu (TN) and West Bengal (WB)

Parts and properties used: Fruit- Ant bilious. Seed- Diuretic. Fruit and bark- Febrifuge.

Associated system (s): English, SU

***Aconitum balfourii* Stapf., syn. *A. atrox* (Bruchl) Mukherjee (CR - Rare)**

Sanskrit or Ayurvedic name: Vatsanaabha relative.

Common name: Vatsnabha

Family: Ranunculaceae

Distribution: Wild in alpine Himalayas between 3300-3900 m.

Parts and properties used: Root- Diaphoretic, diuretic, analgesic, febrifuge, anti-inflammatory, anti-rheumatic, anti-pyretic and vermifuge, and toxic.

Associated system (s): AF

***Aconitum chasmanthum* Stapf. (CR)**

Sanskrit or Ayurvedic name: Amrta, Vatsanabha relative.

Family: Ranunculaceae

Distribution: Wild to Chitral to Kashmir in high altitudes.

Parts and properties used: Root- Analgesic, anodyne, diaphoretic, diuretic, irritant and sedative.

Associated system (s): AFH

***Aconitum deinorrhizum* Stapf (CR)**

Sanskrit or Ayurvedic name: Vatsanaabha relative.

Common name: Vatsnabha

Family: Ranunculaceae

Distribution: Wild along alpine slopes of northwest Himalayas.

Parts and properties used: Root and leaves- Poisonous, analgesic & anti-inflammatory.

Associated system (s): AF

***Aconitum falconeri* Stapf var. *falconeri* (CR)**

Sanskrit or Ayurvedic name: Vatsanaabha relative.

Common name: Vatsnabha

Family: Ranunculaceae

Distribution: Wild in alpine and sub-alpine Himalayas, (Garhwal).

Parts and properties used: Root- Sedative, carminative, anti-inflammatory. Used in treatment of nervous and digestive systems.

Associated system (s): A

***Aconitum falconeri* Stapf var. *latilobum* Stapf**

Sanskrit or Ayurvedic name: Vatsanaabha relative.

Family: Ranunculaceae

Distribution: Endemic to Bushahr, Himachal Pradesh (HP)

Parts and properties used: Root- Sedative, carminative, anti-inflammatory. Treat nervous and digestive systems.

Associated system (s): A

***Aconitum ferox* Wall. ex Ser. (EN)**

Sanskrit or Ayurvedic name: Vatsanabha, Amrita

Common name: Meetha Telia/Vish, Vachhnag

Family: Ranunculaceae

Distribution: Occur in alpine Himalaya to Darjeeling Hills, also cultivated.

Parts and properties used: Root- Anti-pyretic, analgesic, appetizer & anti-rheumatic. Used to cure all types of fever.

Associated system (s): AFHSU

***Aconitum heterophyllum* Wall. ex Royale (EN, CR)**

Sanskrit or Ayurvedic name: Ativisha

Common name: Atis, Aconite, Ativish

Family: Ranunculaceae

Distribution: Habitat to Himalayas 2000-4000m, Ladakh. Also cultivated.

Parts and properties used: Root (tuber)- Antipyretic, Astringent. Dried tuber used in emesis and helminthiasis.

Associated system (s): AFSLU

***Aconitum palmatum* D. Don., syn. *A. bisma* (Ham.) Rap aics.**

Sanskrit or Ayurvedic name: Prattivisa

Common name: Bikhma

Family: Ranunculaceae

Distribution: Habitat to temperate to alpine Himalayas.

Parts and properties used: Root- Given in neuralgia, leprosy, fevers, cholera, and rheumatism.

Associated system (s): AF

***Acorus calamus* L., syn. *Calamus aromaticus* Garsault, (VU)**

Sanskrit, Ayurvedic and common name: Vach, Vaj, Vekhand, Sweet flag.

Family: Acoraceae/ Araceae

Distribution: Himalayan regions inhibit (4x, 3x), tropical India, Punjab to Assam, Western Ghats 3x, and cultivated too.

Parts and properties used: Rhizome- Used in gastrointestinal problems, ulcers, inflammation of the stomach, intestinal gas, and loss of appetite (anorexia). The α.- and β.-asarone, possess a wide range of pharmacological activities.

Associated system (s): AFHLSU

***Actiniopteris dichotoma* Mett. (Fern)**

Sanskrit or Ayurvedic name: Mayurasikha

Common name: Mayurpankhi

Family: Pteridaceae

Distribution: Found all over *India and* Nilgiri Hills.

Parts and properties used: Plant- Styptic, antibacterial, anti-pyretic. Used in Skin diseases.

Associated system (s): AS

***Actinodaphne hookeri* Meissn.**

Sanskrit or Ayurvedic name: Pisaa

Common name: Thali, Pisaa

Family: Lauraceae.

Distribution: Occur in Sikkim, Odisha, and Western Ghats.

Parts and properties used: Leaf's infusion- Given as urinary disinfectant, antidiabetic, spasmolytic.

Associated system (s): SF

***Adansonia digitata* L.**

Sanskrit or Ayurvedic name: Sheet-phala

Common name: Gorakimli

Family: Bombacaceae

Distribution: Native of Africa, occur in west coast of India, and cultivated too.

Parts and properties used: Plant- Refrigerant; Leaves Diaphoretic; Fruit- Antidysentery, antiseptic, antihistaminic. Source vitamin C.

Associated system (s): AFSU

***Adenanthera pavonina* L.**

Sanskrit or Ayurvedic name: Tamaraka Rakta Kanchana

Common name: Kamboji, Rakt Chandan

Family: Leguminosae/ Fabaceae (Mimosaceae)

Distribution: Distributed Assam to Andaman to Western Ghats; often cultivated.

Parts and properties used: Plant- Astringent and styptic, anti-inflammatory. Seeds- Acephalgic. Used in paralysis, pulmonary affection.

Associated system (s): AFS

***Adenia hondala* (Gaertn.) W.J. de Wilde**

Common name: Vidari

Family: Passifloraceae

Distribution: Distributed throughout Western Ghats.

Parts and properties used: Tuber- Antibacterial and antimicrobial. Used in skin disorders and hernias.

Associated system (s): AF

***Adhatoda beddomei* C.B.Clarke, syn. *Justicia beddomei* (C.B.Clarke) Bennet (CR)**

Sanskrit or Ayurvedic name: Vasa relative

Common name: Vasa

Family: Acanthaceae

Distribution: Occur wild in Kerala.

Parts and properties used: Plant and leaf- Given in fever, intrinsic haemorrhage, cough, asthma, and skin diseases.

Associated system (s): A

Adhatoda zeylanica **Medik., syn.** ***A. vasica*** **Nees;** ***Justicia adhatoda*** **L.**

Sanskrit or Ayurvedic name: Vasa

Common name: Adusa

Family: Acanthaceae

Distribution: Occur all over India and cultivated.

Parts and properties used: Leaf- Expectorant, antispasmodic, febrifuge. Used in bronchitis and dyspnoea.

Associated system (s): AFHS

Adiantum capillus-veneris **L. (Fern)**

Sanskrit or Ayurvedic name: Hansaraaja

Common name: Hansraj, Hansapadi, Parshoshan

Family: Adiantaceae

Distribution: Find in Himalayas between 1800 -2700 m.

Parts and properties used: Whole plant/leaf- Astringent, demulcent, expectorant, stimulant antitussive, emmenagogue. Used in catarrh.

Associated system (s): FU

Adiantum philippense **L. f., syn.** ***A. lunulatum*** **Burm. (Fern)**

Sanskrit or Ayurvedic name: Hamsapadi

Common name: Hansraj

Family: Adiantaceae

Distribution: Occur throughout India.

Parts and properties used: Plant- Febrifugal, antidysentery, soothing agent in erysipelas. Dried frond given in psychosis-related fear.

Associated system (s): ASF

Aegle marmelos **(L.) Correa (VU)**

Sanskrit or Ayurvedic name: Bilva

Common name: Bael, Belgiri, Bilva, Vilvam

Family: Rutaceae

Distribution: Habitat to sub-mountain and plains of India, up to Andamans and cultivated.

Parts and properties used: Fruit- stomachic, antimicrobial, digestive, astringent, spasmolytic, hypoglycaemic, used in diarrhoea. Root in dysuria; Stembark- in diabetes and lipid disorders. Also anti-arthritic. (Part of Ayurvedic *Dashamoola*)

Associated system (s): AFHSLU

***Aerva javanica* (Burm. f.), syn. *A. persica* (*Burm.f.*) Merr.**

Common name: Dholphuli

Family: Amaranthaceae

Distribution: Drier parts of tropics and subtropics, Punjab, and Peninsular India.

Parts and properties used: Plant Anti-inflammatory, diuretic, ant calculus, insecticidal. Woolly seeds- Used against rheumatism.

Associated system (s): English, FS

***Aerva lanata* (L.) Juss.**

Sanskrit or Ayurvedic name: Pattura

Common name: Cheroola

Family: Amaranthaceae

Distribution: Found in warmer India.

Parts and properties used: Plant- Diuretic and lithotripsic. Leaf- Used in hepatitis. Root- Strangury.

Associated system (s): AFS

***Aesculus hippocastanum* Linn.**

Common name: Horse Chestnut (English), Baloot (Unani), Pu. (Folk)

Family: Hippocastanaceae

Distribution: Endemic to Balkan Peninsula southeastern Europe. Introduced to India, grown as ornamental tree.

Parts and properties used: Fruit/seed- Anti-inflammatory, vasodilator, astringent (used in chronic venous insufficiency, varicose, nocturnal calves' cramps and swelling of legs.), febrifuge. Leaf- Used in whooping cough.

Associated system (s): FU

***Aesculus indica* (Wall. ex Cambess) Hook.**

Common name: Indian Horse Chesnut, Bankhor

Family: Hippocastanaceae

Distribution: Occur in Himalaya 900-3600 m and cultivated.

Parts and properties used: Fruit (seeds)- Antirheumatic, Galactogenic, anti-leucorrhoeaic

Associated system (s): F

***Aframomum melegueta* (Rosc.), syn. *Amomum melegueta* Rosc.**

Common name: Grains of Paradise

Family: Zingiberaceae

Distribution: Native to tropical Africa. Cultivated in Indian gardens.

Parts and properties used: Roots- Possess cardamom like taste and are given as a decoction in constipation; as vermifuge for tapeworms. Young leaves juice- Styptic.

Associated system (s): English, U

Aganosma dichotoma **(Roth) K. Schum.**

Sanskrit or Ayurvedic name: Madhumalati

Common name: Malti

Family: Apocynaceae.

Distribution: Northeast and Eastern Ghats; cultivated too.

Parts and properties used: Plant, leaves- Antiseptic; anodyne, anthelmintic, and emetic.

Associated system (s): A

Agaricus campestris **Linn. (mushroom)**

Sanskrit or Ayurvedic name: Chhatraka, Bhuumichhatra

Common name: Khumbi

Family: Agaricaceae

Distribution: Found in many parts of India during rainy season.

Parts and properties used: Plant- Source of protein and vitamin B complex. Plant extract- Lower blood pressure.

Associated system (s): AFSU

Agave americana **subsp.** ***americana*** **L.; syn.** ***Agave americana*** **var.** ***marginata*** **Trel.**

Sanskrit or Ayurvedic name: Ghritkumari

Common name: Aloe, American aloe

Family: Asparagaceae

Distribution: Native to America, grown all over India.

Parts and properties used: Leaf juice- Contain ten steroidal saponins. Used in warts, cancerous ulcers, tumours and as resolvent in syphilis and scrofula.

Associated system (s): A, International

Ageratum conyzoides **L.**

Sanskrit or Ayurvedic name: Dochunty

Common name: Visamustih, Goat weed, Visadodi

Family: Asteraceae/ Compositae

Distribution: Found all over India.

Parts and properties used: Leaf- Anti-inflammatory, antibacterial, antifungal, styptic.

Associated system (s): AFS

***Aglaia elaeagnoidea* (A.Juss.) Benth., syn. *A. roxburghiana* Miq. Hiern**

Sanskrit, Ayurvedic and common name: Priyangu

Family: Meliaceae

Distribution: Distributed in Western Ghats.

Parts and properties used: Fruit- Antipyretic, astringent, antidiarrheal, antidysentery, anti-inflammatory. Used in skin diseases and tumours.

Associated system (s): AF

***Aglaia odorata* Lour.**

Sanskrit or Ayurvedic name: Akaparni

Common name: Pishthparni, Pithavan

Family: Meliaceae

Distribution: Southeast Asian, cultivated in Northeast Hills

Parts and properties used: Root and leaves- Produce tonic. Dried flower- Treat mouth ulcer.

Associated system (s): F, Chinese

***Agrimonia eupatoria* auct non. L., syn. *A. pilosa* Hook.f.**

Common name: Ghaafis

Family: Rosacae

Distribution: Habitat to Himalaya between 900-3000 m.

Parts and properties used: Herb- Astringent, anti-inflammatory, hepatic, cholagogue, diuretic, haemostatic, antibacterial. Used in treatment of diarrhoea and inflammation.

Associated system (s): U

***Ailanthus excelsa* Roxb.**

Sanskrit or Ayurvedic name: Araluka

Common name: Aralu

Family: Simaroubaceae

Distribution: Occur in north and Peninsular India. Cultivated.

Parts and properties used: Bark- Astringent, febrifuge, anthelmintic, antispasmodic, expectorant. Used in asthma, dysentery, fever, Typhoid.

Associated system (s): AFS

***Ailanthus glandulosa* Desf., syn. *A. altissima* (Mill.) Swingle**

Sanskrit, Ayurvedic and common name: Aralu

Family: Simaroubaceae

Distribution: Occur in hills of north India.

Parts and properties used: Bark- Astringent, antispasmodic, parasiticidal, narcotic, cardiac depressant.

Associated system (s): A

***Ailanthus triphysa* (Dennst.) Alston**

Sanskrit or Ayurvedic name: Perumaram

Common name: Guggula Dhup

Family: Simaroubaceae

Distribution: Distributed in Western Ghats and cultivated.

Parts and properties used: Bark- dysentery and intestinal edema. Wood- Contains alkaloids and quassinoids. Used in dyspepsia, bronchitis, ophthalmia.

Associated system (s): AF

***Ajuga bracteosa* Benth.**

Sanskrit or Ayurvedic name: Neelkanthi

Common name: Nilkanti

Family: Lamiaceae/ Labiatae

Distribution: Habitat to sub-Himalayas and Gangetic plains.

Parts and properties used: Plant- Astringent, febrifugal, stimulant, aperient, diuretic. Used in gout and rheumatism. Leaves juice- Blood purifier.

Associated system (s): F

***Alangium salvifolium* (L. f.) Wang., syn. *A lamarckii* Thw.**

Sanskrit or Ayurvedic name: Ankolah

Common name: Ankodah, Dirghakilaka

Family: Alangiaceae

Distribution: Habitat to drier part of Peninsular India.

Parts and properties used: Root bark- Astringent, spasmolytic, hypotensive. Leaves- Hypoglycaemic. Fruit- Acidic, astringent, laxative.

Associated system (s): AFSL

***Albizia amara* (Roxb.) Boivin**

Sanskrit or Ayurvedic name: Shirisha

Common name: Krishnasirish

Family: Leguminosae/ Fabaceae (Mimosaceae)

Distribution: Found all over southern India.

Parts and properties used: Leaf/ flower- Anti-inflammatory, used for boils and ulcers and leaves for erysipelas. Seed- Astringent, antidiarrheal, antibacterial.

Associated system (s): AFS

***Albizia chinensis* (Osbeck) Merr., syn. *A. marginata* Merr.**

Common name: Sirish

Family: Leguminosae/ Fabaceae (Mimosaceae)

Distribution: Indo-Malayan region, occurs from Assam to Kerala.

Parts and properties used: Bark- Analgesic, anti-diarrheal, anti-inflammatory, sleep inducing. Used in scorpion/snake bite. Needs further studies.

Associated system (s): AFS

***Albizia lebbeck* (L.) Benth.**

Sanskrit or Ayurvedic name: Shirisha

Common name: Vaakaveru, Shirish

Family: Leguminosae/ Fabaceae (Mimosaceae)

Distribution: Found in many parts of India and cultivated.

Parts and properties used: Plant- Antiseptic, antibacterial, antiallergic, antidermatotic, antidysentery. Bark- used for bronchitis. Pod- Antiprotozoal. Seeds- Used in piles. Flowers- Used in cough, asthma, and pulmonary eosinophilia.

Associated system (s): AFSU

***Albizia odoratissima* (L.f.) Benth.**

Common name: Sirisa

Family: Leguminosae/ Fabaceae (Mimosaceae)

Distribution: Occur in Assam to peninsular India. Cultivated.

Parts and properties used: Have potential for free radical scavenging and antimicrobial activity.

Associated system (s): AFS

Albizia procera **(Roxb.) Benth.**

Sanskrit or Ayurvedic name: Shveta Shirisha

Common name: Safed siris

Family: Leguminosae/ Fabaceae (Mimosaceae)

Distribution: Occur wild all over India.

Parts and properties used: Decoction of bark- Used in rheumatism and haemorrhage.

Associated system (s): F

Aleurities moluccana **(Linn.) Willd., syn.** *A. triloba* **J.R. Forst. &. G. Forst**

Sanskrit or Ayurvedic name: Akshota

Common name: Jangali akharot.

Family: Euphorbiaceae.

Distribution: Native of China, grown in Kangra Valley to Assam and Bengal.

Parts and properties used: Seed oil- Purgative. Used externally in rheumatism, ulcers. Produce hair tonic.

Associated system (s): A

Alhagi pseudalhagi **(M. Bieb.) Desv., syn.** *A. camelorum* **Fisch.**

Sanskrit or Ayurvedic name: Yavasaka

Common name: Durlabha, Yavasaka

Family: Leguminosae/ Fabaceae

Distribution: Habitat to drier parts of north India.

Parts and properties used: Plant- Laxative, ant bilious, diuretic, diaphoretic, expectorant. Dried plant, used in gout and haemorrhagic. Leaves- Used in fever, headache, and rheumatism. Flowers- Blood coagulant, used for piles.

Associated system (s): ASLU

Alkanna tinctoria **(L.) Tausch.**

Common name: Ratan jot (Unani), Surulpattai, Dineshavalli.

Family: Boraginaceae

Distribution: Widely found in Europe, western Asia, and India.

Parts and properties used: Plant - Astringent, antimicrobial. Used for indolent ulcers, wounds, erysipelas, and in formulations.

Associated system (s): SU

***Allium ascalonicum* Linn.**

Sanskrit or Ayurvedic name: Grnjana

Common name: Grnjana, Shellot.

Family: Amaryllidaceae

Distribution: Introduced, cultivated.

Parts and properties used: Bulb- Anticoagulant, fibrinolytic, hypocholesterolemic.

Associated system (s): AU

***Allium cepa* L.**

Sanskrit or Ayurvedic name: Palaandu

Common name: Onion

Family: Amaryllidaceae

Distribution: Native to northern hemisphere, Central Asia. Cultivated all over India.

Parts and properties used: Bulb- Antibiotic, antibacterial, ant sclerotic, anticoagulant, anti-inflammatory, expectorant, carminative, diuretic, antidiabetic. antispasmodic, hypotensive. Stops atherosclerosis and age-related changes in the blood vessels, and loss of appetite.

Associated system (s): AFHSLU

***Allium porrum* Linn., syn. A. *ampeloprasum* L. var. *porrum* (*L.*) Gay.**

Common name: Leek, Levant garlic

Family: Amaryllidaceae

Distribution: Introduced. Cultivated

Parts and properties used: Bulb- Expectorant; used as a substitute for garlic. Medicinally used in stomach ulcer, sores, wounds, tuberculosis, reduced blood pressure and anthelmintic, reduces risk of cancer.

Associated system (s): F

***Allium sativum* L.**

Sanskrit or Ayurvedic name: Lahsuna

Common name: Lahsan, Velathulli

Family: Amaryllidaceae

Distribution: Mid Asian, extend from West Himalayas to peninsular India, and cultivated.

Parts and properties used: Bulb- Antibiotic, bacteriostatic, fungicide, anthelmintic, antithrombic, hypotensive, hypoglycaemic, hypo-cholesterol emic. Used in respiratory infections, epilepsy, and psychic disorders.

Associated system (s): AFHSLU

***Allium stracheyi* Baker (EN)**

Sanskrit or Ayurvedic name: Jambu

Common name: Faran, Jamboo and Dhungar

Family: Amaryllidaceae

Distribution: Habitat to western Himalayas. Cultivated in Uttarakhand.

Parts and properties used: Bulb- Reduce blood cholesterol. Produce tonic to digestive system. Contains Sulphur compound. Bulb juice- Moth replant.

Associated system (s): A

***Alocasia indica* (Lour.) Spach., syn. *A. macrorrhiza* (L.) G. Don**

Sanskrit or Ayurvedic name: Maankanda

Common name: Maanaka

Family: Araceae.

Distribution: Occur wild and cultivated all over India.

Parts and properties used: Root stock- Mild laxative, diuretic; used in treatment of inflammations and diseases of abdomen and spleen.

Associated system (s): AS

***Aloe barbadensis* Mill., syn. *A. vera* (L.) Burm. f.**

Sanskrit or Ayurvedic name: Kanyasara

Common name: Ghikanvar, Indian aloe, Gharat Kumari.

Family: Amaryllidaceae

Distribution: Mediterranean, habitat to coasts of western India. Cultivated all over India.

Parts and properties used: Leaf juice- Purgative, emmenagogue. Gel- Topically emollient, anti-inflammatory, antimicrobial. Used in wound, skin care, constipation, intestinal obstruction and inflammation, ulcerative colitis, appendicitis. Dry juice used in dysmenorrhoea, and liver diseases.

Associated system (s): AFHSLU

***Alpinia calcarata* (Haw.) Roscoe**

Common name: Lesser galangal, Snap Ginger, Perarathai, Chittaratha, Aratha (Malyalam)

Family: Zingiberaceae

Distribution: Habitat to tropical regions of India and cultivated.

Parts and properties used: Rhizome- Antimicrobial, anthelmintic, antinociceptive, anti-inflammatory, aphrodisiac, gastroprotective, and antidiabetic.

Associated system (s): A

***Alpinia galanga* (L.) Willd. (DD)**

Sanskrit or Ayurvedic name: Sugandhmula, Malaya, Vacha, Mahabharivacha

Common name: Kulanjan, Kulinjan, Rasnamool, Koshtakulinjan

Family: Zingiberaceae

Distribution: Found in most of India and southern Western Ghats. Cultivated.

Parts and properties used: Rhizome- Carminative, stomachic, circulatory stimulant, diaphoretic, anti-inflammatory. Used as *Rasna* (exudate) in fever, dyspepsia, rheumatism, and respiratory ailments.

Associated system (s): AFSU

***Alpinia officinarum* Hance**

Common name: Kulanjan, Khulinjan

Family: Zingiberaceae

Distribution: Chinese, grown in north India.

Parts and properties used: Rhizome- Circulatory stimulant and carminative.

Associated system (s): ASU

***Alpinia smithiae* Sabu & Mangaly**

Family: Zingiberaceae

Distribution: Native to Western Ghats. Cultivated in Uttarakhand.

Parts and properties used: Rhizome- Essential oil is extracted, which has pharmaceutical prospective.

Associated system (s): F

***Alpinia zerumbet* (Pers.) B. L. Burtt & R. M. Sm.; syn. *A. speciosa* (J.C. Wendl.) K. Suhum**

Common name: Shell ginger

Family: Zingiberaceae

Distribution: Habitat to East Himalaya and North-East India. Cultivated.

Parts and properties used: Rhizome- Substitute of *A. galanga* (Galangal)

Associated system (s): F

***Alstonia scholaris* (L.) R.Br.**

Sanskrit or Ayurvedic name: Saptaparna

Common name: Saptaparnachal, Chitvan

Family: Apocynaceae

Distribution: Found throughout moist regions of India.

Parts and properties used: Bark- Febrifuge, antiperiodic, spasmolytic, antidysentery, uterine stimulant, hypotensive. Used in internal fevers and as blood purifier.

Associated system (s): AFHSLU

***Alstonia venenata* R. Br**

Sanskrit or Ayurvedic name: Rajadana, Visaghni

Common name: Poison devil tree, Anadama

Family: Apocynaceae

Distribution: Habitat to peninsular India.

Parts and properties used: Bark, Fruit- Antiepileptic, monoamine oxidase inhibitor.

Associated system (s): AF

***Alternanthera sessilis* (L.) R.Br. ex DC.**

Sanskrit or Ayurvedic name: Matsyaki

Common name: Ponnanganni

Family: Amaranthaceae

Distribution: Occur all over hotter parts of India.

Parts and properties used: Plant- Febrifuge, galactagogue, cholagogue. Used in diseases caused by vitiated blood and skin.

Associated system (s): AFSL

***Althaea officinalis* L.**

Sanskrit or Ayurvedic name: Khatmi

Common name: Khatmi, Resha-Khatami, Shilaras, Marshmallow

Family: Malvaceae

Distribution: Native to Europe, found in Jammu and Kashmir (J & K) and HP. Cultivated.

Parts and properties used: Root and leaf- Demulcent, emollient, antitussive. Used in dry cough, inflammation of the gastric mucosa.

Associated system (s): AFSU

***Altingia excelsa* Noronha**

Sanskrit or Ayurvedic name: Shilaarasa

Common name: Silaras

Family: Altingiaceae

Distribution: Found in Assam and Arunachal Pradesh (ArP). Cultivated.

Parts and properties used: Resin- Carminative, stomachic, antiscorbutic, expectorant, antipyretic, anti-inflammatory, hepatoprotective. Used in scabies and leukoderma.

Associated system (s): AFSU

***Amaranthus caudatus* L.**

Sanskrit or Ayurvedic name: Raam-daana

Family: Amaranthaceae

Distribution: American native. Naturalized and grown in north India.

Parts and properties used: Plant- Blood purifier, diuretic. used in piles, strangury, dropsy, and anasarca.

Associated system (s): F

***Amaranthus paniculatus* L.**

Common name: Red Chaulai, Ramdana

Family: Amaranthaceae

Distribution: American native. Naturalized and grown in parts of India.

Parts and properties used: Plant- Chemo preventive, radioprotective, phytoremediative. Leaves- Astringent, used for stopping diarrhoea, bloody stools.

Associated system (s): F

***Amaranthus tricolor* L., syn. *A. gangeticus* L.; *A. mangostanus* L.**

Sanskrit or Ayurvedic name: Ramasitalika

Common name: Lal sag, Alpamarisa

Family: Amaranthaceae

Distribution: American native. Naturalized and cultivated all over India.

Parts and properties used: Plant/foliage- Astringent. Used in cough, bronchitis, consumption and externally emollient.

Associated system (s): AS

***Amberboa divaricata* Kuntze (syn. *Amberboa ramosa* (Roxb.) Jafri = *Volutarella divaricata* Benth. et Hook.f. = *Tricholepis procumbens* Wight)**

Sanskrit, Ayurvedic and common name: Brahmadandi, Baadaavard.

Family: Asteraceae/ Compositae

Distribution: Distributed from Mediterranean region to India.

Parts and properties used: Herb- Deobstruent, aperient, febrifuge nervine, Atropine antiseptic.

Associated system (s): AU

***Ambroma augusta* (L.) L. f., syn. *Abroma augusta* Jacq.**

Sanskrit or Ayurvedic name: Pisacakarpasa

Common name: Ulat-kambal, Pishacha, Kaarpasa, Pivari

Family: Sterculiaceae

Distribution: Occur throughout the hot and moist parts of India and cultivated.

Parts and properties used: Root bark- Emmenagogue, uterine tonic. Treat dysmenorrhea, amenorrhea, and gonorrhoea. Root powder- Abortifacient

Associated system (s): AFHU

***Ammannia baccifera* L.**

Sanskrit or Ayurvedic name: Agnipatri

Common name: Kurand ghas

Family: Lythraceae

Distribution: Occur as weed, all over India.

Parts and properties used: Plant- Stomachic, laxative, antirheumatic, febrifuge. Leaves- Contain lawsone. Used for ringworm, herpetic eruptions, and skin diseases

Associated system (s): AS

***Ammi majus* L.**

Common name: Ammi

Family: Apiaceae/ Umbelliferae

Distribution: Native to Nile, and Africa. Grown in Jammu and HP.

Parts and properties used: Fruit- Source of xanthotoxin, a drug to treat leukoderma. Dried powder/plant extract used topically in vitiligo.

Associated system (s): UF

***Ammi visnaga* (Linn.) Lam**

Common name: Paashaanabhedi

Family: Apiaceae/ Umbelliferae

Distribution: Mediterranean native. Grown in Jammu.

Parts and properties used: Herb- Antispasmodic, prescribed in renal colic, bronchial asthma, whooping cough.

Associated system (s): UF

***Amomum aromaticum* Roxb.**

Sanskrit or Ayurvedic name: Sthula-elaa

Common name: Bengal cardamom, Brhdela

Family: Zingiberaceae

Distribution: Occur in north Bengal and Assam.

Parts and properties used: Seed essential oil- Used like *A. subulatum* (Badi Elaichi).

Associated system (s): AS

***Amomum kepulaga* Sprague, syn. *A.compactum* Sol. ex Maton**

Common name: Elaichi

Family: Zingiberaceae

Distribution: Habitat to Southeast Asia. Cultivated in India.

Parts and properties used: Fruit (seed)- Aromatic. Used for stomachache.

Associated system (s): F

***Amomum subulatum* Roxb.**

Sanskrit or Ayurvedic name: Sthulaela

Common name: Badi Elachi

Family: Zingiberaceae

Distribution: Native to the eastern Himalaya; cultivated in Bengal, Sikkim, Assam and TN

Parts and properties used: Seed- Stomachic, antiemetic, ant bilious, astringent, alexipharmic. Used in treatment of indigestion, biliousness, abdominal pains, vomiting and liver congestion. Pericarp- Treat headache and stomatitis.

Associated system (s): AFSLU

***Amorphophallus paeoniifolius* (Dennst.) Nicolson var. *campanulatus* (Blume ex Decne.) Sivad., syn. *A. campanulatus* Blume ex Decne. (VU)**

Sanskrit or Ayurvedic name: Arsaghna, Balukand, Surana

Common name: Arsghna, Surankand, Zaminkand, Elephant foot yam or White spot giant arum

Family: Araceae

Distribution: Native to southeast Asian islands. Cultivated all over India.

Parts and properties used: Corm- Used in bronchitis, asthma, abdominal pain, emesis, dysentery, enlargement of spleen, piles, elephantiasis, rheumatic swellings, prostatic hyperplasia etc.

Associated system (s): AFSU

***Amygdalus communis* Linn.**

Sanskrit or Ayurvedic name: Vaataama

Common name: Almond, Badaam

Family: Rosaceae

Distribution: Native to the Middle East. Cultivated in Punjab and J & K.

Parts and properties used: Seed (oil)- Nutrient, nervine tonic, demulcent; oil used for skin care.

Associated system (s): AUS

***Anacardium occidentale* L.**

Common name: Kaju

Family: Anacardiaceae

Distribution: American native. Cultivated in coastal states of India.

Parts and properties used: Bark, leaves- Antimicrobial vermicide, protozoic, Cashew apple- Antiscorbutic. Seed juice- Used in mental, heart, rheumatic problems.

Associated system (s): AFHS

***Anacyclus pyrethrum* (L.) Cass.**

Sanskrit, Ayurvedic and Common name: Akarakarabha, Akarkara

Family: Asteraceae/ Compositae

Distribution: Native to Mediterranean region. Imported to India

Parts and properties used: Root, stem- Stimulant, cordial, rubefacient. Used in gargle of infusion for relaxed vulva. Roots- Used in sciatica, paralysis, hemiplegia, amenorrhoea, besides rheumatic, neuralgic affections.

Associated system (s): ASU

***Anagallis arvensis* Linn.**

Common name: Anaaghaalis (Unani), Dhabbar, Jonkmaari

Family: Primulaceae

Distribution: Occur in Himalayas, from Kashmir to hills of West Bengal, central and southern India.

Parts and properties used: Plant- Anti-inflammatory, astringent, de-obstruct ant, antifungal, nematocidal. It's toxic to leeches.

Associated system (s): English, UF

***Anamirta cocculus* (L.) Wight & Arn.**

Sanskrit or Ayurvedic name: Kaakaadan

Common name: Rakthala, Kakmari (Marathi)

Family: Menispermaceae

Distribution: Habitat to tropical Asia. Occur in Khasi Hills, Odisha, and peninsular India.

Parts and properties used: Plant- Insecticidal, antifungal. Highly valued for skin diseases; used to kill lice, etc.

Associated system (s): AFHSU

***Ananas comosus* (Stickm.) Merr.**

Sanskrit or Ayurvedic name: Anamnasam and Bahunetraphalam

Common name: Pineapple, Ananas

Family: Bromeliaceae

Distribution: Native of Brazil. Cultivated in India.

Parts and properties used: Plant- Arsenic properties. Ripe fruit juice- Antiscorbutic, diuretic, diaphoretic, aperient, refrigerant and helps in digestion of albuminous substances. Unripe fruit- Acrid, styptic, diuretic, anthelmintic, emmenagogue and abortifacient.

Associated system (s): A

***Anaphalis neelgerriana* DC.**

Sanskrit or Ayurvedic name: Vranapata

Common name: Raktaskandana

Family: Asteraceae/ Compositae

Distribution: Found in Nilgiris Hills.

Parts and properties used: Leaves- Antiseptic, applied on bruises, wounds, and cuts.

Associated system (s): AF

***Anchusa italica* Ritz., syn. *A. azurea* Mill.**

Sanskrit or Ayurvedic name: Gojihvikaa

Common name: Gojihvikaa, Gaozabaan

Family: Boraginaceae

Distribution: Found in western India. Cultivated in gardens and hills.

Parts and properties used: Herb- Stimulant, tonic, demulcent; used in bilious disorder, fever, cough, asthma, and as diuretic in bladder and kidney stones.

Associated system (s): AU

***Anchusa strigosa* Labill. (Unresolved name)**

Sanskrit or Ayurvedic name: Gojihva.

Common name: Substitute of Manu, Gaozabaan

Family: Boraginaceae

Distribution: Eastern Mediterranean regions, Imported, grown in Indian gardens.

Parts and properties used: Leaves and nutlets- Yield a substitute of Gaozaban drug.

Associated system (s): A

***Andrographis echioides* (L.) Nees**

Common name: False water willow, Ranchimani.

Family: Acanthaceae

Distribution: Habitat to warmer parts of India. Cultivated.

Parts and properties used: Plant extract, leaf- Anti-inflammatory, febrifuge, diuretic hepatoprotective.

Associated system (s): F

***Andrographis paniculata* (Burm.f.) Wall. *ex* Nees (LC)**

Sanskrit, Ayurvedic and common name: Kaalmegha, Kalmegh

Family: Acanthaceae

Distribution: Found throughout India. Cultivated.

Parts and properties used: Herb- Hepatoprotective, febrifuge, cholinergic, antispasmodic, stomachic, anthelmintic. Used as tonic in diabetes, high BP, ulcer, bronchitis, skin disorder, leprosy, flatulence, colic, influenza, dysentery, dyspepsia, and malaria.

Associated system (s): AFHSU

***Andropogon muricatus* Retz., syn. *Vetiveria zizanioides* (Linn.) Nash.**

Sanskrit or Ayurvedic name: Ushira

Common name: Vetiver, Khas

Family: Poaceae/ Gramineae

Distribution: Found all over India.

Parts and properties used: Roots- Refrigerant, febrifuge, diaphoretic, stimulant, stomachic and emmenagogue. Used in strangury, colic, flatulence, vomiting.

Associated system (s): AUS

***Anemone rivularis* Buch. -Ham. ex DC.**

Common name: Srub-ka

Family: Ranunculaceae

Distribution: Habitat to Indo-Malaya and Indo-China region. Found in Ladakh, northeast to south Indian hills.

Parts and properties used: Herb- Antiemetic and vermifuge for indigestion. Leaf juice- Applied to wounds, sores, and earaches.

Associated system (s): AL

***Anethum graveolens* L., syn. *A. sowa* Roxb. ex Flem.**

Sanskrit or Ayurvedic name: Satahva

Common name: Sowa

Family: Apiaceae/ Umbelliferae

Distribution: European, grown all over India.

Parts and properties used: Herb- Carminative, stomachic, antispasmodic, used in dyspepsia. Leaves- Used in gastrointestinal tract, kidney, urinary tract, and sleep disorder.

Associated system (s): AF

***Angelica archangelica* L.**

Sanskrit or Ayurvedic name: Chandamshukha

Common name: Chanda, Angelica

Family: Apiaceae/ Umbelliferae

Distribution: Found wild in Scandinavian countries and Russia. Grown in Kashmir.

Parts and properties used: Herb/root- Expectorant, carminative, digestant, cholagogue, antispasmodic, diaphoretic, diuretic, anti-inflammatory. Used in flatulence and peptic discomforts.

Associated system (s): AW

***Angelica glauca* Edgew. (CR)**

Sanskrit or Ayurvedic name: Chorakahkhya

Common name: Chuaraka, Choru

Family: Apiaceae/ Umbelliferae

Distribution: Endemic to Himalayas, J & K, and Chamba, HP.

Parts and properties used: Root- Stimulant, carminative, expectorant, diaphoretic. Used in constipation.

Associated system (s): AL

***Anisomeles malabarica* (L.) R.Br. ex Sims**

Sanskrit or Ayurvedic name: Sprikkaa

Common name: Karimthumpa

Family: Lamiaceae/ Labiatae

Distribution: Occur from Assam to Bihar, and TN to Western Ghats.

Parts and properties used: Plant- Antispasmodic, antipyretic, diaphoretic, antiperiodic, emmenagogue, antirheumatic. Oil- Used in rheumatic arthritis.

Associated system (s): AFS

***Annona cherimola* Mill., syn. A. *acutifolia* Saff. *ex* R. E. Fr.**

Common name: Cherimoya, Marytiphal, Hanuman phalamu

Family: Annonaceae

Distribution: Native of tropical America. Grown in northern Western Ghats.

Parts and properties used: Fruit extract- Phytochemical and pharmacological properties. Seed- Bio-insecticide and Acetogenins (polyketide)

Associated system (s): AF

***Annona reticulata* L.**

Sanskrit or Ayurvedic name: Raamphala

Common name: Aninuna, Luvuni

Family: Annonaceae

Distribution: Native to South America. Cultivated in northeast and southern hills of India.

Parts and properties used: Leaves- Insecticide, anthelmintic, styptic, used as suppurate. Dried fruit- Antidysentery anti-tumour, cardio protective. Bark- Powerful astringent. Fruit- Potent anti-tumour and cardio protective.

Associated system (s): AFS

***Annona squamosa* Linn.**

Common name: Gandagaatra

Family: Annonaceae

Distribution: Native to South America. Cultivated all over India.

Parts and properties used: Leaves- Insecticidal, treat ulcer. Seeds- Abortifacient. Root- Purgative. Fruit- Invigorating, sedative to heart, ant bilious, antiemetic, expectorant. Powder treats ulcer.

Associated system (s): ASU

***Anodendron paniculatum* A. DC.**

Common name: Sarakkodi, Kavali

Family: Apocynaceae

Distribution: Occur in Assam, Meghalaya, Western Ghats.

Parts and properties used: Whole plant- Cytotoxic, antioxidant; Bast (fibrous material) used to treat ulcer. Root- Used in vomiting and coughing

Associated system (s): F

***Anogeissus latifolia* (Roxb. ex DC.) Wall. ex Guill. & Perr.**

Sanskrit or Ayurvedic name: Shakataahya, Indravrksha

Common name: Dhawada

Family: Combretaceae

Distribution: Occur in central and south India.

Parts and properties used: Plant- Astringent, used in diarrhoea, dysentery, ulcers, piles, urinary disorders. Gum- Tonic for gestating mothers, gastrointestinal disorder, and vigour.

Associated system (s): AFSTU

***Anthemis nobilis* L.**

Common name: Babuna, Gulbabuna

Family: Asteraceae/ Compositae

Distribution: Habitat to temperate Himalayas. Cultivated.

Parts and properties used: Plant- Sedative, anticonvulsant, antispasmodic, anti-inflammatory, spasmodic, treat gastrointestinal problem and sluggish bowels.

Associated system (s): HSU

***Anthocephalus cadamba* (Roxb.) Miq.**

Sanskrit or Ayurvedic name: Kadamba

Common name: Kadam

Family: Rubiaceae

Distribution: Habitat to northeast, Eastern and Western Ghats and cultivated.

Parts and properties used: Bark- Febrifugal, antidiuretic, anthelmintic, hypoglycaemic. Dried bark is used in female genital tract and bleeding disorders. Fruit- Ant catarrhal, blood purifier, analgesic.

Associated system (s): AS

***Antiaris toxicaria* Lesch.**

Sanskrit or Ayurvedic name: Valkala vrksha

Common name: Jangali Lakuch, Jasund,

Family: Moraceae

Distribution: Distributed in Western Ghats.

Parts and properties used: Seed- Febrifuge, antidysentery. Latex- Contain Antiarin, circulatory stimulant.

Associated system (s): AS

***Aphanamixis polystachya* (Wall.) Parker, *Amoora rohituka* (Roxb.) Wight & Arn. (VU)**

Sanskrit or Ayurvedic name: Rohitaka, Daadimachhada

Common name: Rakta Rohida, Rohitak

Family: Meliaceae

Distribution: Habitat to sub-Himalayas, Western Ghats, Andamans, and cultivated.

Parts and properties used: Bark- Astringent. Used in the liver and spleen diseases, and tumours, enlarged glands. Seed oil- Used in muscular pains and rheumatism. Seed extract- Antimicrobial. Plant- Pesticidal.

Associated system (s): AHS

***Apium graveolens* L.**

Sanskrit or Ayurvedic name: Ajmodaa

Common name: Ajmoda, Celery

Family: Apiaceae/ Umbelliferae

Distribution: Native to Europe. Cultivated in hills of India.

Parts and properties used: Plant- Anti-inflammatory, diuretic, carminative, nervine, sedative, antiemetic, antispasmodic, antiseptic, emmenagogue. Seed oil- Tranquilizer, anticonvulsant, antifungal.

Associated system (s): AHSTU

***Aquilaria agallocha* Roxb., syn. *A. malaccensis* Lam (CR)**

Sanskrit or Ayurvedic name: Agaru

Common name: Agar kala, Akil

Family: Thymelaeaceae

Distribution: Habitat to South, Southeast Asia, northeast hills of India, and Assam.

Parts and properties used: Heartwood- Astringent, carminative, antiasthma, antidiarrheal/dysentery; used in gout, rheumatism, and paralysis; stimulant for sexual debility; and liniment in skin diseases. Infected wood- Aromatic (high market value).

Associated system (s): AHSTU

***Arabis tibetica* H.K.**

Common name: Perpata

Family: Brassicaceae/ Cruciferae

Distribution: Habitat to West Himalayas, including Ladakh.

Parts and properties used: Herb- Given in fever, blood disorder.

Associated system (s): Amchi

***Arctium lappa* L.**

Common name: Phaggarmul, Burdock, Jangli kuth, Pizums

Family: Asteraceae/ Compositae

Distribution: Wild in western Himalayas, Ladakh and beyond.

Parts and properties used: Plant- Hypoglycaemic. Root, Fruit extract- Diuretic, depurative, anti-fatigue, antidiabetic, anticancer, anti-inflammatory, gastroprotective, hepatoprotective, dermatological.

Associated system (s): FH

***Areca catechu* L.**

Sanskrit or Ayurvedic name: Puga

Common name: Supari

Family: Arecaceae

Distribution: Native to Southeast Asia Islands. Naturalized and cultivated in Assam and coastal regions of India.

Parts and properties used: Berry- Taeniacide, astringent, stimulant; dried- used in leucorrhoea and vaginal laxity.

Associated system (s): AFHSTU

***Argemone mexicana* L.**

Sanskrit or Ayurvedic name: Swarnaksiri

Common name: Brhami Dandi, Kusme Beeja

Family: Papaveraceae

Distribution: Native of Mexico. Naturalized throughout India.

Parts and properties used: Seed- Toxic, cause dropsy, diarrhoea. Oil and leaf juice- Treat ulcers and skin diseases.

Associated system (s): AFHSU

***Argyreia speciosa* (L. f.) Sweet, *A. nervosa* (Burm.f.) Bojer**

Sanskrit or Ayurvedic name: Antakotarapushpi, Chhagalanghhri, Vryddhadaraka, *Vriddhadara*

Common name: Samudraphal, Samundra Sokh

Family: Convolvulaceae

Distribution: Found all over India.

Parts and properties used: Root- Aphrodisiac, nervine, diuretic, antirheumatic. Seeds- Hypotensive, spasmolytic, contain hallucinogenic ergoline alkaloids. Leaves- Used in skin care, rubefacient, topically stimulant.

Associated system (s): ASTU

***Aristolochia bracteolata* Lam. (LC)**

Sanskrit or Ayurvedic name: Dhumrapatraa

Common name: Kidamari

Family: Aristolochiaceae

Distribution: Found all over India.

Parts and properties used: Plant- Oxytocic, abortifacient, emmenagogue. Leaf's/ fruit- Contain Cetyl alcohol, aristolochic acid and beta-sitosterol. Insecticidal. Seeds- Purgative.

Associated system (s): AFS

***Aristolochia indica* L.**

Sanskrit or Ayurvedic name: Isvari, Ishwar mool

Common name: Kiramar, Garudakkodi, Indian Birthwort

Family: Aristolochiaceae

Distribution: Habitat to plains and hills all over India.

Parts and properties used: Plant- Oxytocic, abortifacient, emmenagogue.

Associated system (s): AFSTU

***Aristolochia rotunda* L.**

Common name: Zarawand

Family: Aristolochiaceae

Distribution: Native to southern Europe. Imported to India.

Parts and properties used: Plant- Contains aristolochic acid, blood stimulant, heals wounds. Applied to skin complaints.

Associated system (s): FU

***Arnebia benthamii* (Wall. *ex* G. Don) I. M. Johnst. (CR)**

Common name: Gaozaban

Family: Boraginaceae

Distribution: Habitat to Alpine Himalayas 3000-3900 m.

Parts and properties used: Plant- Stimulant, cardiac tonic, expectorant, diuretic. Root- Antiseptic and antibiotic.

Associated system (s): AF

***Arnebia nobilis* Rech. f.**

Common name: Ratan Jyothi/ Ratan jot

Family: Boraginaceae

Distribution: Found in Himalayas, and north India.

Parts and properties used: Root- Food colorant, has anti-skin ageing ingredient, used in cosmetics.

Associated system (s): FU

***Artabotrys hexapetalus* (L. f.) Bhandari**

Sanskrit or Ayurvedic name: Panasagandhi

Common name: Hari-champa, Madanmast

Family: Annonaceae

Distribution: Occur in southern India, cultivated.

Parts and properties used: Fruit pericarp- Cardiac stimulant, uterine stimulant, muscle relaxant. Leaves- Have anti-fertility agents.

Associated system (s): AF

***Artemisia absinthium* L.**

Sanskrit or Ayurvedic name(s): Damanaka- bheda

Common name: Afsanteen, wormwood

Family: Asteraceae/ Compositae

Distribution: Habitat to Europe to Siberia, Kashmir between 1500-2100 m.

Parts and properties used: Plant- Choleretic, anthelmintic, stomachic, carminative, antispasmodic, anti-inflammatory, emmenagogue. Used in loss of appetite, dyspepsia, biliary dyskinesia.

Associated system (s): AHSU

Artemisia annua **L.**

Common name: Artemisia, Sweet wormwood

Family: Asteraceae/ Compositae

Distribution: Habitat to temperate north India. Cultivated.

Parts and properties used: Plant- Treat many disorders, including acanth amoebiasis and malaria.

Associated system (s): F

Artemisia brevifolia **Wall.**

Common name(s): Wormseed, Chauhar

Family: Asteraceae/ Compositae

Distribution: Found wild in western Himalayas, (Ladakh, Lahaul Spiti).

Parts and properties used: Flowering shoots and leaves- Used in antiworm, stomach disorders, artemisinin is extracted.

Associated system(s): AF

Artemisia maritima **L. ex Hook.f.**

Sanskrit or Ayurvedic name(s): Chauhaara

Common name(s): Chauhaara, Ajavayana, Chauhara, Cina

Family: Asteraceae/ Compositae

Distribution: Wild from Europe to western Himalaya, including Ladakh.

Parts and properties used: Plant- De-obstructing, stomachic, anthelmintic, antifungal, bitter tonic, and aromatic. Decoction, used in intermittent fever.

Associated system(s): AHW

Artemisia nilagirica **(C.B. Clarke) Pamp., syn.** *A vulgaris* **Linn. var.** *nilagirica* **Clarke.**

Sanskrit or Ayurvedic name(s): Damanaka

Common name(s): Thavanam

Family: Asteraceae/ Compositae

Distribution: Habitat to hilly areas of India.

Parts and properties used: Leaf- Menstrual regulator, nervine, stomachic, anthelmintic, choleretic, diaphoretic. Mainly used as emmenagogue. Infusion- Given in nervous and spasmodic affections.

Associated system(s): AHF

***Artemisia pallens* Wall. ex DC.**

Common name(s): Davna

Family: Asteraceae/ Compositae

Distribution: Habitat to Deccan Plateau and south India and cultivated.

Parts and properties used: Plant- Immuno-modulating, anthelmintic, antipyretic, antimicrobial, tonic. Used in diabetes and wounds.

Associated system(s): FS

***Artocarpus heterophyllus* Lam., syn. *A. integrifoliab* L.**

Sanskrit or Ayurvedic name(s): Kantakiphala

Common name(s): Kathal, Panasa

Family: Moraceae

Distribution: Native to south India, cultivated in hotter parts.

Parts and properties used: Latex- Bacteriolytic. Juice- Used to supress swelling. Root- Used in diarrhoea, asthma, skin diseases. Fruit unripe- Acrid; ripe- Cooling, laxative,

Associated system(s): AFS

***Artocarpus lacucha* Buch. -Ham.**

Sanskrit or Ayurvedic name(s): Granthiphala, Lakuch

Common name(s): Monkey Jack

Family: Moraceae

Distribution: Natural all over South and Southeast Asia. Cultivated too.

Parts and properties used: Bark- Heals boils, cracked skin and pimples. Seed-Purgative, hemagglutinating.

Associated system(s): ASF

***Arundo donax* L.**

Sanskrit or Ayurvedic name(s): Shuunyamadhya

Common name(s): Dhamana, Baranal, Nala

Family: Poaceae/ Gramineae

Distribution: Found in Kashmir, Assam and Nilgiris.

Parts and properties used: Rhizome- Sudorific, emollient, diuretic, anti-lactant, ant dropsical; uterine stimulant, hypotensive.

Associated system(s): AHS

***Asclepias curassavica* Linn.**

Common name(s): Blood flower, substitute of Kaakanaasikaa

Family: Asclepiadaceae.

Distribution: Native to American tropics, naturalized as ornamental in India.

Parts and properties used: Plant- Spasmogenic, cardiotonic, cytotoxic, antihemorrhagic, styptic, antibacterial. Latex- Treat warts and cancer. Roots- Astringent. Leaf juice- Antidysentery.

Associated system(s): AF

***Asparagus adscendens* Roxb.**

Sanskrit or Ayurvedic name(s): Maha Shataavari

Common name(s): Safed Musali, Pili Shatawar

Family: Liliaceae/ Asparagaceae

Distribution: Wild in western Himalaya and Punjab.

Parts and properties used: Root- Substitute for *A. officinalis* (Shatavari).

Associated system(s): ASU

***Asparagus officinalis* Linn.**

Common name(s): Shataavari

Family: Liliaceae/ Asparagaceae

Distribution: Occur from Europe to west Asia; cultivated in India.

Parts and properties used: Root- Diuretic, laxative, sedative, cardiotonic, galactagogue. Used for neuritis, rheumatism and to prevent stone formation.

Associated system(s): AU

***Asparagus racemosus* Willd.**

Sanskrit, Ayurvedic and Common name(s): Shataavari, Shatavari, Satawar

Family: Liliaceae/ Asparagaceae

Distribution: Habitat to Himalayas up to 1500 m, tropical, subtropical India, and Andamans. Cultivated

Parts and properties used: Root- Used as galactagogue, in female genitourinary tract disorder; styptic and ulcer-healing agent; intestinal disinfectant, and astringent in diarrhoea, nervine tonic, gout, and sexual debility.

Associated system(s): AFSTU

***Asparagus sarmentosus* L.**

Common name(s): Maha Satawari variety, Shakakul

Family: Liliaceae/ Asparagaceae

Distribution: Habitat to India to Bangladesh.

Parts and properties used: Root- Used like *A. racemosus* (Satawar).

Associated system(s): ASU

***Astragalus candolleanus* Royle.**

Sanskrit or Ayurvedic name(s): Rudravanti

Common name(s): Rudanti

Family: Leguminosae/ Fabaceae

Distribution: Found in western Himalayas.

Parts and properties used: Fruit- Depurative, bechic, blood purifier.

Associated system(s): A

***Atalantia monophylla* (L.) DC.**

Sanskrit or Ayurvedic name(s): Atavi-jamheera

Common name(s): Kurandubeeja

Family: Rutaceae

Distribution: Found throughout India, Assam, and Northeast.

Parts and properties used: Leaves and berry oil- Antibacterial, antifungal. Leaf decoction- Used in skin care. Fruit juice- Ant bilious.

Associated system(s): AFS

***Atropa acuminata* Royle *ex* Lindl., syn. *A. belladonna* auct. non-L. (CR)**

Sanskrit or Ayurvedic name(s): Suuchi

Common name(s): Indian Belladonna, Jharka

Family: Solanaceae

Distribution: Habitat to sub-Himalayas, Assam. Cultivated in temperate areas.

Parts and properties used: Plant- Poisonous, contains tropine, sedative, narcotic, anodyne, nervine, antispasmodic. Used in spasm and colic-like pain of gastrointestinal tract and bile ducts, muscle relaxer, anti-inflammatory, in menstrual problems, peptic ulcer, histaminic reaction and motion sickness. Valued for tropane alkaloids, gives relief in parkinsonism and neurovegetative dystonia.

Associated system(s): AFHUW

Atylosia goensis **(Dalzell) Dalzell**

Sanskrit or Ayurvedic name(s): Maashaparni

Common name(s): Katttupaayar

Family: Leguminosae/ Fabaceae

Distribution: Occur in Assam, Maharashtra, Kerala. Cultivated.

Parts and properties used: Herb- Febrifuge, ant bilious, antirheumatic.

Associated system(s): AFS

Averrhoa bilimbi **Linn.**

Common name(s): Variety of Karmaranga

Family: Averrhoaceae.

Distribution: Native to Malayan region. Cultivated in India.

Parts and properties used: Fruit syrup- Febrile excitement, haemorrhages, used in internal haemorrhoids.

Associated system(s): AUS

Averrhoa carambola **Linn.**

Sanskrit or Ayurvedic name(s): Karmaranga

Common name(s): Kamarakh

Family: Averrhoaceae.

Distribution: Native to Malayan region. Cultivated in India.

Parts and properties used: Root- Antidote to poisoning. Leaves/shoots- Treat scabies, ringworm, chickenpox. Flower- Vermicidal. Fruit- Laxative, antidysentery, anti-phlogistic, febrifuge, anti-inflammatory, antispasmodic.

Associated system(s): AFSU

Avicennia officinalis **Linn., syn.** ***A. alba*** **Blume (mangrove)**

Sanskrit or Ayurvedic name(s): Tuvara

Common name(s): Tivaria, Upattam, Indian mangrove.

Family: Acanthaceae

Distribution: Habitat to marsh and tidal creeks of Indian Ocean to the western Pacific.

Parts and properties used: Bark- Astringent. Fruit pulp- Used in skin lesions of smallpox.

Associated system(s): AFS

***Azadirachta indica* A. Juss**

Sanskrit or Ayurvedic name(s): Nimba

Common name(s): Neem

Family: Meliaceae

Distribution: Found in tropical and subtropical India. Cultivated.

Parts and properties used: Leaf, bark- Antimicrobial, Ant insecticidal, anti-inflammatory, antifertility, spermicidal, hypoglycaemic. Used for many diseases and fever. Seed oil- Contraceptive

Associated system(s): AFHSTU

***Azima tetracantha* Lam.**

Sanskrit or Ayurvedic name(s): Kundali

Common name(s): Kanta-gur-kamai, Kantangur, Kundali

Family: Salvadoraceae

Distribution: Occur in Odisha, Bengal, and peninsular India.

Parts and properties used: Root- Diuretic. Leaves- Stimulant, expectorant, antispasmodic. Used treatment of cough/ asthma. Bark- Antiperiodic, astringent, expectorant.

Associated system(s): AFS

***Bacopa monnieri* (L.) Wettst., syn. *Herpestis monnieria* (LC)**

Sanskrit or Ayurvedic name(s): Brahmi

Common name(s): Brahmi, Jal Brahmi, Nirbrahmi

Family: Plantaginaceae

Distribution: Habitat to wetlands of tropic and sub-tropical plains of India, also cultivated.

Parts and properties used: Herb- Adaptogen, astringent, diuretic, sedative, potent nervine tonic, anti-anxiety, antispasmodic. Produce brain tonic used in psychic disorders.

Associated system(s): AFHS

***Balanites aegyptiaca* (L.) Delile (LC)**

Sanskrit or Ayurvedic name(s): Angaar Vrksha

Common name(s): Hingan, Hingol, Ingua

Family: Balanitaceae

Distribution: Habitat to drier parts of India.

Parts and properties used: Seed- Expectorant, bechic. Oil- Antibacterial, antifungal. Fruit pulp- Used in cough, leukoderma, and skin diseases. Bark- Spasmolytic.

Associated system(s): AFTU

Baliospermum montanum* (Willd.) Mull. Arg., syn. *B. solanifolium [(Burm.) Suresh] (VU)

Sanskrit or Ayurvedic name(s): Danti

Common name(s): Dantimool, Nagadanti

Family: Euphorbiaceae

Distribution: Habitat to Himalayas, Northeast, Bihar, Peninsular India.

Parts and properties used: Seeds-Purgative. Leaves- Purgative, antiasthma tic. Latex- Used in body pains. Root/seed oil- Cathartic, ant dropsical, used in jaundice, etc.

Associated system(s): AFSTU

***Bambusa bambos* (L.) Voss. *B. arundinaceae* (Retz.) Willd.**

Sanskrit or Ayurvedic name(s): Vanshalochana

Common name(s): Bansalochan, Tabashir

Family: Bambuseaceae

Distribution: Occur all over India except Himalayas, often cultivated.

Parts and properties used: Shoot, leaf extracts- Aphrodisiac, cooling, used as tonic, and to treat asthma, cough, stomach, and debilitating diseases.

Associated system(s): AFU

***Bandeiraea simplicifolia* (M. Vahl ex DC.) Benth., syn. *Griffonia simplicifolia* (M. Vahl ex DC.) Baill.]**

Common name(s): Griffonia.

Family: Leguminosae/ Fabaceae (Caesalpiniaceae)

Distribution: West and central African native. Cultivated in India.

Parts and properties used: Decoction- Aphrodisia, antibiotic, remedy for diarrhoea, vomiting and stomach-ache, depression, anxiety, insomnia, fibromyalgia, and chronic headache.

Associated system(s): African traditional

***Barleria buxifolia* Linn.**

Sanskrit or Ayurvedic name(s): Iksura, Sahachara (purple)

Common name(s): Jhinti, Box leaved Barleria

Family: Acanthaceae

Distribution: Occur in peninsular India up to 1200 m.

Parts and properties used: Root, leaves- Used in cough, bronchitis, inflammations.

Associated system(s): AF

***Barleria cristata* L.**

Sanskrit or Ayurvedic name(s): Sahachara (white)

Common name(s): Jhinti, Raktapushpa, Sweta Saireyaka

Family: Acanthaceae

Distribution: Habitat to subtropical Himalayas and parts of India.

Parts and properties used: Plant extract- Spasmogenic and hypoglycaemic. Root extract- Treat anaemia. Leaves- Toothache.

Associated system(s): AST

***Barleria prionitis* L. (DD)**

Sanskrit, Ayurvedic and common name(s): Sahachara (yellow), Vajradanti

Family: Acanthaceae

Distribution: Occur throughout hotter parts of India. Horticultural potential.

Parts and properties used: Plant- Antidontalgic, treat bleeding gums. Leaf juice-Given in stomach disorders, urinary affections. Plant, root powder- Treat fever. Bark-Diaphoretic/ expectorant. Oil- Check greying of hair.

Associated system(s): AFS

***Barleria strigosa* Willd.**

Sanskrit or Ayurvedic name(s): Sahachara (blue), Nilajhint

Common name(s): Bala, Dasi, Nili

Family: Acanthaceae

Distribution: Occur in the Himalayas, Uttar Pradesh (UP), and Bengal.

Parts and properties used: Plant- Antiseptic, expectorant, gave beta-and gamma sitosterol.

Associated system(s): AST

***Barringtonia acutangula* (L.) Gaertn.**

Sanskrit or Ayurvedic name(s): Nichula,

Common name(s): Samudraphal

Family: Baringtoniaceae

Distribution: Habitat to sub-Himalayas, Assam, Madhya Pradesh (MP).

Parts and properties used: Leaf juice- Treat diarrhoea. Fruit- Acrid, anthelmintic, haemolytic, vulnerary. Given in gingivitis and goitre. Powdered seed- Expectorant. Bark- Astringent, treat diarrhoea and blennorrhoea.

Associated system(s): AFSTU

***Basella alba* L.**

Sanskrit or Ayurvedic name(s): Upodikaa

Common name(s): Pasalai, Vasalacheera pacha

Family: Basellaceae

Distribution: Native to India. Grown as pot herb.

Parts and properties used: Herb- Demulcent, diuretic, laxative. Used as cooling drug in digestive disorders.

Associated system(s): AF

***Bauhinia acuminata* Linn.**

Sanskrit or Ayurvedic name(s): Kaanchnaara

Common name(s): Kovidaara (white flower)

Family: Leguminosae/ Fabaceae (Caesalpiniaceae)

Distribution: Found in Central India.

Parts and properties used: Bark, leaves decoction- Given in biliousness, stone in bladder, venereal diseases, leprosy, and asthma.

Associated system(s): ASU

***Bauhinia purpurea* L.**

Common name(s): Sonachal

Family: Leguminosae/ Fabaceae (Caesalpiniaceae)

Distribution: Habitat to sub-Himalayas and northeast India. Cultivated.

Parts and properties used: Bark- Astringent, antidiarrheal. Buds/ flowers fried in butter, given in dysentery.

Associated system(s): AFST

***Bauhinia racemosa* Lam.**

Sanskrit or Ayurvedic name(s): Ashmantaka

Common name(s): Kachnar

Family: Leguminosae/ Fabaceae (Caesalpiniaceae)

Distribution: Habitat to sub-Himalayan tracts.

Parts and properties used: Bark- Astringent, anti-inflammatory, cholagogue. Used in skin care and ulcers. Flowers- Treat haemorrhage, piles. Seed- Antibacterial.

Associated system(s): ASU

***Bauhinia tomentosa* L.**

Sanskrit or Ayurvedic name(s): Pita-Kovidaara

Common name(s): Kachnar with yellow flower

Family: Leguminosae/ Fabaceae (Caesalpiniaceae)

Distribution: Occur in Assam, Bihar, south India.

Parts and properties used: Plant- Antidysentery. Bark- Astringent. Root bark- Vermifuge, given in liver diseases.

Associated system(s): AFS

***Bauhinia vahlii* Wight & Arn., syn. *Phanera vahlii* (Wight & Arn.) Benth.**

Sanskrit or Ayurvedic name(s): Assmantaka, Malanjhana

Common name(s): Adda, Malu

Family: Leguminosae/ Fabaceae (Caesalpiniaceae)

Distribution: Habitat to tropical India.

Parts and properties used: Stem, leaf- Antioxidant, antihyperglycemic, anti-inflammatory. Needs more investigation.

Associated system(s): AF

***Bauhinia variegata* L. (DD)**

Sanskrit, Ayurvedic and common name(s): Kaanchnaara, Kachnar

Family: Leguminosae/ Fabaceae (Caesalpiniaceae)

Distribution: Habitat to Assam, western peninsula. Often cultivated.

Parts and properties used: Buds' decoction- Given in piles. Root- Carminative. Bark- Astringent, anthelmintic. Used in scrofula and skin diseases. Also, given in lymphadenitis and goitre.

Associated system(s): AFS

***Benincasa hispida* (Thunb.) Cogn.**

Sanskrit or Ayurvedic name(s): Kusmanda

Common name(s): Ash Gourd, Kumpalanga pacha

Family: Cucurbitaceae

Distribution: Native to Indo-Malayan region. Cultivated in north India.

Parts and properties used: Fruit decoction- Laxative, diuretic, nutritious, styptic; Juice- nervous diseases.

Associated system(s): AFSTU

***Berberis aristata* DC., syn. *B. aristata* auct. Hook. f. &Thoms, *B. chitria* Lindl. (VU)**

Sanskrit or Ayurvedic name(s): Daruharidra

Common name(s): Daruhaldi, Rasnajan, Zarisuk

Family: Berberidaceae

Distribution: Habitat to western and central Himalayas. Grown in hills.

Parts and properties used: Plant extract- Cholagogue, antidiarrheal, stomachic, laxative, diaphoretic, antipyretic, antiseptic. Used in ophthalmia conjunctivitis, ulcers, swollen gums. Root, bark- Anti-inflammatory, hypoglycaemic hypotensive, ant amoebic, anticoagulant, antibacterial. Have antibacterial and anti-inflammatory properties with alkaloid Berberine (5-4.2%).

Associated system(s): AFSTU

***Berberis lycium* Royle (EN)**

Sanskrit or Ayurvedic name(s): Daaruharidraa

Common name(s): Dar-hald, Chitra

Family: Berberidaceae

Distribution: Habitat to hills of northwest India, MP and Nilgiris hills.

Parts and properties used: Plant- Used in diarrhea, intestinal colic, piles, jaundice, internal wounds, rheumatism, diabetes, ophthalmia, gingivitis, throat pain.

Associated system(s): AF

***Berberis umbellata* Wall. ex G. Don, syn. *B. aristata* Sims. (?)**

Sanskrit or Ayurvedic name(s): Daruharidra variety

Common name(s): Daruhaldi/ Rasaut

Family: Berberidaceae

Distribution: Habitat to subalpine and alpine Himalayas.

Parts and properties used: Herb- Treat eye disorders, fever, jaundice, skin diseases, etc.

Associated system(s): AF

***Berberis vulgaris* Linn.**

Sanskrit or Ayurvedic name(s): Daruharidra variety

Common name(s): Common barberry

Family: Berberidaceae

Distribution: Habitat to Northwest Himalayas.

Parts and properties used: Root and bark- Gastrointestinal, liver, gallbladder, kidney, urinary, respiratory tract. High in Berberine (80%).

Associated system(s): AF

***Bergenia ciliata* (Haw.) Sternb., syn. *B. ligulata* (Wall.) Engl. (VU)**

Sanskrit or Ayurvedic name(s): Paashaanabheda

Common name(s): Pashnabheda

Family: Saxifragaceae

Distribution: Habitat to Temperate Himalayas.

Parts and properties used: Leaf and root- Antiscorbutic, astringent, spasmolytic, antidiarrheal. Used in dysuria, spleen enlargement, pulmonary affections as a cough remedy, menorrhagia, urinary tract infections, lithiasis.

Associated system(s): AS

***Bergenia stracheyi* (Hook.f. & Thoms.) Engl.**

Common name(s): Pashnabheda, Tiang

Family: Saxifragaceae

Distribution: Habitat to cold arid zone, Ladakh and beyond.

Parts and properties used: Rhizome, leaf- Astringent, diuretic tonic, kidney disorder.

Associated system(s): A

***Beta vulgaris* Linn. subsp. *Cicla* (L.) Moq., *B. vulgaris* L.**

Sanskrit or Ayurvedic name(s): Palanki

Common name(s): Chukandar

Family: Chenopodiacae

Distribution: Mediterranean native. Cultivated in northern India.

Parts and properties used: Tuber and seed- Expectorant. Leaf- Used in burns, bruises, spleen, and liver diseases.

Associated system(s): AF

***Betula alnoides* Buch. -Ham. ex D. Don.**

Common name(s): Variety of Bhojapatra, Indian birch

Family: Chenopodiacae

Distribution: Temperate and subtropical Himalayas, Khasi Hills, Manipur.

Parts and properties used: Essential oil- Used in rheumatic ailments.

Associated system(s): A

***Betula utilis* D. Don**

Sanskrit or Ayurvedic name(s): Bhurjah, Bhojpatra

Family: Betulaceae

Distribution: Habitat to temperate Himalayas.

Parts and properties used: Resin- Laxative. Leaves- Diuretic. Bark- Used in convulsions. Oil- Astringent, antiseptic. Used to clear bacterial and inflammatory infection.

Associated system(s): AST

***Biophytum sensitivum* (L.) DC.**

Sanskrit or Ayurvedic name(s): Lajjaalu

Common name(s): Mukkutti

Family: Oxalidaceae

Distribution: Habitat to tropical India

Parts and properties used: Plant- Diuretic, anti-asthmatic/-bronchitis. Used in insomnia, convulsions, cramps, chest-complaints, inflammations, tumours, chronic skin diseases. Leaves- Diuretic, astringent, antiseptic. Applied on burns, etc.

Associated system(s): AFS

***Bixa orellana* L.**

Sanskrit or Ayurvedic name(s): Sindhuri

Common name(s): Sindhuri, Latkan dana, Annato

Family: Bixaceae

Distribution: Native of Brazil. Cultivated in MP and south India.

Parts and properties used: Plant- Astringent, ant bilious, antiemetic, blood purifier. Leaf's infusion- Given in jaundice, dysentery. Root bark- Febrifuge, antiperiodic. Fruit- Antidysentery. Seed pulp- Have Bixin, cytostatic, haemostatic, diuretic, laxative.

Associated system(s): AFHS

***Blepharis edulis* auct. non (Forssk.) Pers., syn. *B. persica* (Burm. f.) O. Kuntze.**

Sanskrit, Ayurvedic and common name(s): Utingana, Uttangan

Family: Acanthaceae

Distribution: Occur in Punjab and western Rajasthan.

Parts and properties used: Roots- Diuretic. Seeds- De-obstruent, resolvent, diuretic. Treat dysuria and impotency.

Associated system(s): AU

***Blepharis linariaefolia* Pers.; *B. linariifolia* Pers., syn. *B. sindica* T. Anders**

Sanskrit or Ayurvedic name(s): Ushtrakaandi, Uttangan variety

Common name(s): Bhangari, Unt-kantalo.

Family: Acanthaceae

Distribution: Found in northwestern India Punjab to Kutch.

Parts and properties used: Seeds boiled in milk- Invigorating tonic. Also given to cattle to increase milk.

Associated system(s): AF

***Blumea balsamifera* DC., syn. *B. densiflora* Hook. f.**

Sanskrit or Ayurvedic name(s): Kukundara

Common name(s): Kukundara, Gangaapatri

Family: Asteraceae/ Compositae

Distribution: Habitat to subtropical Himalayas.

Parts and properties used: Plant- Tranquilizer, expectorant, sudorific. Used to treat intestinal diseases and colic. Source of Ngai camphor.

Associated system(s): AU

***Blumea eriantha* DC.**

Common name(s): Kukundara variety, Kakarondaa

Family: Asteraceae/ Compositae.

Distribution: Distributed from UP to Kerala.

Parts and properties used: Herb juice- Carminative. Oil- Antimicrobial, insecticidal.

Associated system(s): AFU

***Blumea lacera* (Burm. f.)**

Sanskrit or Ayurvedic name(s): Kukundara

Common name(s): Kakranda

Family: Asteraceae/ Compositae.

Distribution: Occur throughout the plains of India.

Parts and properties used: Plant- Antipyretic. Leaf- Astringent, febrifuge, diuretic, de-obstruent, and anthelmintic. Essential oil- Antimicrobial.

Associated system(s): AFSU

***Boerhavia diffusa* L**

Sanskrit or Ayurvedic name(s): Punarnava rakta, Raktpunarnava

Common name(s): Punarnava raktha, Punarnava

Family: Nyctaginaceae

Distribution: Occur throughout India, often cultivated as green vegetable.

Parts and properties used: Plant- Antirheumatic, diuretic, hepatoprotective, anti-asthmatic and anti-bronchitis. Root- Anticonvulsant, analgesic, expectorant, CNS depressant.

Associated system(s): AFHSU

***Bombax ceiba* L., syn. *B. malabaricum* DC.; *Salmalia malabarica* (DC) Schott & Endl.**

Sanskrit or Ayurvedic name(s): Shaalmali

Common name(s): Mochras

Family: Bombacaceae

Distribution: Habitat to warmer regions of India.

Parts and properties used: Roots- Astringent, stimulant, demulcent. Used in dysentery. Fruits- Stimulant, diuretic, expectorant, used in inflammation. Bark- Used bleeding disorders and in acne vulgaris. Gum- Astringent, demulcent, styptic. Treat diarrhoea, dysentery, haemoptysis, and conception.

Associated system(s): AFS

***Borassus flabellifer* L.**

Sanskrit or Ayurvedic name(s): Tantuniyosah,Talah, Trnam-indrah, Tranam-ketuh, Tranam-raj

Common name(s): Palmyra palm, Tad

Family: Arecaceae

Distribution: Occur in coastal areas, cultivated too.

Parts and properties used: Fresh sap- Diuretic, cooling, ant phlegmatic, laxative, anti-inflammatory. Tonic for asthmatic and anaemic people. Dried male inflorescence-given in dysuria.

Associated system(s): AFSTU

***Borreria hispida* (L.) Schum. [= *Spermacoce hispida* L.], syn. *B. articularis* (Linn. f.) F. N. Williams.**

Sanskrit, Ayurvedic and Common name(s): Madana-ghanti, Thaarthaaval

Family: Rubiaceae

Distribution: Occur throughout India.

Parts and properties used: Leaf juice- Astringent in haemorrhoids. Seeds- Demulcent in diarrhoea.

Associated system(s): F

***Boswellia carterii* Bird.**

Common name(s): Sali guggul

Family: Burseraceae

Distribution: Somalian native, traded to India.

Parts and properties used: Oleo-Gum Resin- Anti-inflammatory, antimicrobial, used in wound healing.

Associated system(s): F

***Boswellia frereana* Birdw.**

Common name(s): African elemi

Family: Burseraceae

Distribution: Somalian native, cultivated in India.

Parts and properties used: Oleo-Gum Resin- Treat inflammation and arthritis.

Associated system(s): African folk system

***Boswellia serrata* Roxb. (NT, Eroded)**

Sanskrit or Ayurvedic name(s): Kunduru, Gajabhakshyaa

Common name(s): Kandur, Salai Guggul, Dupa

Family: Burseraceae

Distribution: Habitat to drier parts of peninsular India

Parts and properties used: Oleo-Gum Resin- Antiseptic, anti-inflammatory, ant atherosclerotic, emmenagogue, analgesic, sedative, hypotensive, diuretic. Oil- Used in ulcers, ringworm.

Associated system(s): ASU

Brassica alba **(L.) Rabenh**

Sanskrit or Ayurvedic name(s): Siddhaartha

Common name(s): Safed rai, Kadugu

Family: Brassicaceae/ Cruciferae

Distribution: Native to Europe. Grown in north India as cover & and green manure crop.

Parts and properties used: Seed- Stimulant to gastric mucosa, increases pancreatic secretions, emetic, diaphoretic, rubefacient, have glucosinolates.

Associated system(s): AHSU

Brassica juncea **(L.) Czern.**

Sanskrit or Ayurvedic name(s): Raajikaa

Common name(s): Kaduku, Sasuve Bili

Family: Brassicaceae/ Cruciferae

Distribution: Distributed over Central to East Asia to Europe. Cultivated in north India.

Parts and properties used: Seed- Antidysentery, stomachic, diaphoretic, anthelmintic. Increases pancreatic secretions (rai substitute). Decoction- given in indigestion, cough.

Associated system(s): AFS

Brassica napus **Linn.**

Sanskrit or Ayurvedic name(s): Krishna-Sarshapa

Common name(s): Kaali Sarson

Family: Brassicaceae/ Cruciferae

Distribution: Wild from Mediterranean to India. Grown in Gangetic plains.

Parts and properties used: Seed- Emollient, diuretic, ant catarrhal.

Associated system(s): AU

Brassica nigra **L.**

Common name(s): Black mustard, Banarasi Raai, Raajika (variety)

Family: Brassicaceae/ Cruciferae

Distribution: Native to Africa, Europe, Asia. Cultivated in north India.

Parts and properties used: Seeds- Treat coryza with thin excoriating discharge, sneezing and hacking cough and nostril blockage; powder used in headache.

Associated system(s): F (Trading potential)

Bridelia montana **(Roxb.) Willd.**

Sanskrit or Ayurvedic name(s): Ekaviraa

Common name(s): Gondni, Geia

Family: Euphorbiaceae

Distribution: Habitat to sub-Himalayas, Bihar, Odisha, Andhra Pradesh (AP)

Parts and properties used: Bark and Root- Astringent, anthelmintic. Used in treatment of bone fracture.

Associated system(s): ASF

Bridelia retusa **(Linn.) Spreng.**

Sanskrit or Ayurvedic name(s): Mahaaviraa

Common name(s): Spinus kino tree

Family: Euphorbiaceae

Distribution: Occur throughout India.

Parts and properties used: Bark- Astringent, used as liniment in rheumatism. Paste to wounds.

Associated system(s): AFS

Bryonopsis laciniosa **(Linn.) Naud.**

Sanskrit or Ayurvedic name(s): Chitraphalaa

Common name(s): Lingini, Shivalingi

Family: Cucurbitaceae

Distribution: Occur throughout India.

Parts and properties used: Seeds- Anti-inflammatory, spasmolytic and fertility promoter.

Associated system(s): ASF

Bryophyllum pinnatum **(Lam.) Kurz., syn.** ***B. calycinum*** **Salisb;** ***Kalanchoe pinnata*** **(Lam.) Pers.**

Sanskrit or Ayurvedic name(s): Priyaala

Common name(s): Parnabija, Airaavati

Family: Crassulaceae

Distribution: Occur throughout warm moist and semi-arid India.

Parts and properties used: Leaf- Disinfectant, antibacterial. Plant extract- Antifungal.

Associated system(s): AU

***Buchanania lanzan* Spreng., syn. *B. cochiinchinensis* (Lour.) (LC)**

Sanskrit or Ayurvedic name(s): Upavatth, Priyala, Prasavkh

Common name(s): Chironji

Family: Anacardiaceae

Distribution: Habitat to drier central India. Grown in agroforestry.

Parts and properties used: Kernel- Laxative, febrifuge, many medicinal uses.

Associated system(s): ASU

***Bunium persicum* (Boiss.) B. Fedts. ch, syn. *Carum bulbocastanum* W. Koch. (EN)**

Common name(s): Kala-jeera, Shah-jeera

Family: Apiaceae/ Umbelliferae

Distribution: Distributed from north Africa, southern Europe to south Asia. Grown in Himalayas to hills of south India.

Parts and properties used: Plant- Antimicrobial, antioxidant, and carminative. Used as spices.

Associated system(s): AF

***Butea monosperma* (Lam.) Taub., *B. frondosa* Willd. (EN)**

Sanskrit or Ayurvedic name(s): Palasa, Paalasha

Common name(s): Palash, Tesu phool, Lac

Family: Leguminosae/ Fabaceae

Distribution: Distributed in Peninsular India.

Parts and properties used: Bark- Astringent, styptic, anthelmintic, antifungal. Leaves- Antibacterial. Flower- Astringent, diuretic, emmenagogue, decoction given in diarrhoea and haematuria.

Associated system(s): AFSTU

***Butea superba* Willd.**

Sanskrit or Ayurvedic name(s): Lataa-Palaasha

Common name(s): Kesu phool, Tesu phool

Family: Leguminosae/ Fabaceae

Distribution: Distributed in central and southern India.

Parts and properties used: Seed- Sedative and anthelmintic. Decoction- used in piles. Oil- hypotensive. Root- Hypotensive. Sap- Used as drink. Bark- Used in tonics and elixirs.

Associated system(s): A

Caccinia crassifolia **(Vent.) C.Koch**

Common name(s): Gaozaban

Family: Boraginaceae

Distribution: Native to Baluchistan, traded to India.

Parts and properties used: Leaf- Diuretic, anti-inflammatory, demulcent. Used in strangury, asthma, and cough. Flower- Contain pyrrolizidine.

Associated system(s): U

Cadaba fruticosa **(L.) Druce, syn. *C. indica* Lam.**

Common name(s): Kodham, Pulika

Family: Capparaceae

Distribution: Found in many parts of India.

Parts and properties used: Root and leaves- DE obstruent, emmenagogue; used for uterine obstructions.

Associated system(s): FS

Caesalpinia bonduc **(L.) Roxb.**

Sanskrit or Ayurvedic name(s): Latakaranja

Common name(s): Sagargota, Gatran, Nataphal

Family: Leguminosae/ Fabaceae (Caesalpiniaceae)

Distribution: Found throughout hotter India.

Parts and properties used: Seed- Antiperiodic, antirheumatic. Roasted- antidiabetic. Leaf, bark, and seed- Febrifuge. Leaf and bark- Anthelmintic, Emmenagogue. Root- Diuretic, ant calculous.

Associated system(s): AFHS

Caesalpinia digyna **Rottler.**

Sanskrit or Ayurvedic name(s): Vaakeri

Common name(s): Teri pods, Teri Beej

Family: Leguminosae/ Fabaceae (Caesalpiniaceae)

Distribution: Occur in Bengal, Assam & Andamans.

Parts and properties used: Root- Astringent, antipyretic, treat phthisis and scrofulous affections.

Associated system(s): A

***Caesalpinia pulcherrima* (L.) Sw.**

Sanskrit or Ayurvedic name(s): Padangam, Ratnagandhi

Common name(s): Gultora, Krishnachura, Peacock flower

Family: Leguminosae/ Fabaceae (Caesalpiniaceae)

Distribution: Native to tropic, subtropic of America, grown all over India.

Parts and properties used: Leaves- Laxative, antipyretic. Leaf powder used in erysipelas. Flowers- Anthelmintic. Root decoction- Used in fevers. Bark- Emmenagogue, abortifacient.

Associated system(s): AFS

***Caesalpinia sappan* L.**

Sanskrit or Ayurvedic name(s): Pattanga

Common name(s): Pathimugam

Family: Leguminosae/ Fabaceae (Caesalpiniaceae)

Distribution: Native to tropical Asia, cultivated in Bengal and South India.

Parts and properties used: Wood decoction- Emmenagogue, antidiarrheal; used in skin diseases. Heartwood- Gives anti-inflammatory brazilin.

Associated system(s): AFSU

***Cajanus cajan* (L.) Millsp.**

Sanskrit or Ayurvedic name(s): Adhaki

Common name(s): Arhar, Tur

Family: Leguminosae/ Fabaceae

Distribution: Native, cultivated throughout India.

Parts and properties used: Leaves- Hypo-cholesterol emic. Seed- Advised to treat lipid disorders and obesity. Roots- Blood purifier.

Associated system(s): AFST

***Calamus rotang* L.**

Sanskrit or Ayurvedic name(s): Vetra, Abhrapushpa

Common name(s): Pirapan Kizhangu, Bet

Family: Arecaceae

Distribution: Occur from central to southern India.

Parts and properties used: Plant- Astringent, antidiarrheal, anti-inflammatory ant bilious, spasmolytic, strangury. Used in chronic fevers, piles, abdominal tumours, Wood- Vermifuge.

Associated system(s): AFSU

***Calendula officinalis* Linn.**

Common name(s): Thulvkka, Saamanthi

Family: Asteraceae/ Compositae

Distribution: Wild in Punjab. Cultivated throughout India.

Parts and properties used: Plant- Antiprotozoal. Flowers- Anti-inflammatory, Antiseptic (antimicrobial), stimulant, styptic, antispasmodic, emmenagogue, antihemorrhagic. Used in gastric and duodenal ulcers. Applied on cuts, bruises, burns, scalds. Essential oil- Antibacterial.

Associated system(s): SU

***Callicarpa macrophylla* Vahl.**

Sanskrit or Ayurvedic name(s): Priyangu

Common name(s): Priyangu

Family: Verbenaceae

Distribution: Habitat to sub-Himalaya tracts.

Parts and properties used: Leaves- Used hot in rheumatic pains, smoked in headache. Wood/seed paste- Used in stomatitis (Mouth/ tongue sores). Seeds and roots- Stomachic. Bark- Used in rheumatism genitourinary tract diseases. Fruit- Given in emesis and giddiness.

Associated system(s): AT

***Calophyllum apetalum* Willd. (VU)**

Common name(s): Cherupunna

Family: Clusiaceae/ Guttiferae

Distribution: Habitat to Western Ghats.

Parts and properties used: Resin- Antiphlogistic, anodyne. Seed oil- Antileprotic.

Associated system(s): FS

***Calophyllum inophyllum* L.**

Sanskrit or Ayurvedic name(s): Nagchampa, Panch Kasara, Punnaaga

Common name(s): Penaga Laut, Bintangor Laut, Punnappoovu, Sultan Champa, Tamanu

Family: Clusiaceae/ Guttiferae

Distribution: Paleo tropic. Habitat to coastal regions. Cultivated.

Parts and properties used: Seed oil- Used in scabies, skin diseases, rheumatism, venereal and genitourinary diseases. Bark juice- Purgative. Applied in orchitis, dressing of ulcers. Root bark- Antibacterial, used for indolent ulcers. Leaf- Used in vertigo, migraine, chicken pox, and skin problems.

Associated system(s): AFS

Calotropis gigantea **(L.) R.Br.**

Sanskrit or Ayurvedic name(s): Alarka, Raajaarka

Common name(s): Erukkin veru

Family: Asclepiadaceae

Distribution: Found throughout India.

Parts and properties used: Flowers- Stomachic, bechic, antiasthma-tic. Milky juice- Purgative. Roots- Used in lupus, leprosy, tuberculous, syphilitic, ulceration. Leaf juice- Poisonous, supress swellings. All parts treat bronchitis and asthma.

Associated system(s): AFHSTU

Calotropis procera **(Aiton) W.T.**

Aiton Sanskrit or Ayurvedic name(s): Alarka Surya

Common name(s): Aakh, Akada Phool

Family: Asclepiadaceae

Distribution: Habitat to arid and semi-arid zone of India.

Parts and properties used: Plant- Treat bronchial asthma. Leaves- Antimicrobial, treat dyspepsia, flatulence, constipation, and mucus in stool. Seed oil- Geriatric and tonic.

Associated system(s): AFSTU

Calycopteris floribunda **(Roxb.) Poir.**

Sanskrit or Ayurvedic name(s): Sushavi, Pullani

Common name(s): Pullaaniyila

Family: Combretaceae

Distribution: Found in Assam, Peninsular India.

Parts and properties used: Leaf- Antidysentery, treat ulcers. Fruit- Used in jaundice.

Associated system(s): AF

Canarium strictum **Roxb. (VU)**

Sanskrit or Ayurvedic name(s): Raladhupa

Common name(s): Dhoop, Raldhoop

Family: Burseraceae

Distribution: Habitat to Assam, and Western Ghats.

Parts and properties used: Resin- Used for chronic cutaneous diseases. Dammer oil- Used in rheumatism, asthma, venereal diseases.

Associated system(s): AFS

***Canna edulis* Ker Gawl.; syn. *C. indica* L.**

Sanskrit or Ayurvedic name(s): Sarvajayá, Silarumba

Common name(s): Indian shot

Family: Cannaceae

Distribution: Native to northwest and South America. Cultivated as ornamental.

Parts and properties used: Tuber extract- Used in headaches, diarrhoea, yaws, acute hepatitis, traumatic injuries, as a diuretic and against nose bleeding.

Associated system(s): A

***Cannabis sativa* L.**

Sanskrit or Ayurvedic name(s): Vijaya, Maadani, Indraasana, Trailokya-vijayaa

Common name(s): Marijuna Maya, Bhang

Family: Cannabinaceae

Distribution: Found on waste grounds, along road, water channels. Restricted cultivation all over India.

Parts and properties used: Tuber extract- Used in headaches, diarrhoea, yaws, acute hepatitis, traumatic injuries. Plant- Hallucinogenic, hypnotic, sedative, analgesic, anti-inflammatory. Treat glaucoma, while antiemetic in cancer chemotherapy. All variants produce initial excitement followed by depression. Yields 421 chemicals, such as cannabinoids, cannabispirans, and other alkaloids.

Associated system(s): AFHSTU

***Canscora decussata* Schult. & Schult.f.**

Sanskrit or Ayurvedic name(s): Daakuni.

Common name(s): Shankhuli, Nakuli, Shankahpuspi

Family: Gentianaceae

Distribution: Found throughout India.

Parts and properties used: Plant- Anticonvulsant, CNS depressant, anti-inflammatory, hepatoprotective.

Associated system(s): ASU

***Capparis aphylla* Roth., syn. *C. decidua* Edgew.**

Sanskrit or Ayurvedic name(s): Niguudhapatra

Common name(s): Apatra, Karira, Kabar, Tenti.

Family: Capparaceae

Distribution: Occur in west and south of India.

Parts and properties used: Plant- Anti-inflammatory. Fruit and seeds- Used in urinary purulent discharges & dysentery Flower, and seeds- Antimicrobial.

Associated system(s): AUSF

***Capparis divaricata* Lam.**

Common name(s): Turatti

Family: Capparaceae

Distribution: Habitat to southern Western Ghats.

Parts and properties used: Root, fruit, seeds- Antirheumatic, tonic, expectorant, antispasmodic, and analgesic.

Associated system(s): S

***Capparis moonii* Wight**

Sanskrit or Ayurvedic and Common name(s): Rudanti, Rudanti

Family: Capparaceae

Distribution: Habitat to tropical and sub-tropical India (Konkan).

Parts and properties used: Fruit extract- Used in sepsis and septic wounds. Being antimicrobial and antioxidant used in cosmetic preparations.

Associated system(s): A

***Capparis sepiaria* L.**

Sanskrit or Ayurvedic name(s): Himsraa

Common name(s): Karungurai

Family: Capparaceae

Distribution: Habitat to dry regions of India.

Parts and properties used: Plant- Antiseptic, antipyretic. Used for eczema and scabies.

Associated system(s): AFS

***Capparis spinosa* L.**

Sanskrit or Ayurvedic name(s): Himsra

Common name(s): Caper bush, Kabra, Kanther

Family: Capparaceae

Distribution: Habitat to Himalayas, north-west and peninsular India.

Parts and properties used: Plant- Anti-inflammatory, DE obstruent to liver and spleen, diuretic, anthelmintic. Bark- Given in splenic, renal, and hepatic complaints. Leaf juice and fruits- Anti-cystic, antimicrobial.

Associated system(s): AU

***Capparis zeylanica* L. (R), *C. horrida* L.**

Sanskrit or Ayurvedic name(s): Vyaghranakha

Common name(s): Sivappu Sakkarai Kizhangu

Family: Capparaceae

Distribution: Native, grown as common hedge.

Parts and properties used: Root bark- Sedative, stomachic, anticholera, diuretic febrifuge. Leaves- Used as poultice.

Associated system(s): AFS

***Capsella bursa-pastoris* (L.) Medik.**

Common name(s): Shepherds' purse

Family: Brassicaceae/ Cruciferae

Distribution: Found throughout India and cultivated.

Parts and properties used: Herb- Checks menorrhagia and haemorrhages from renal and genitourinary tract. Dried flowers and leaves- Used in lung and stomach disorders.

Associated system(s): F (has trade potential)

***Cardiospermum halicacabum* L.**

Sanskrit or Ayurvedic name(s): Karnasphota

Common name(s): Mudakkathan

Family: Sapindaceae

Distribution: Found throughout plains of India.

Parts and properties used: Herb- Used in rheumatism, lumbago, skeletal fractures, nervous diseases, erysipelas, amenorrhoea, haemorrhoids. Extract-Analgesic, anti-inflammatory and sedative on CNS, anticancer, diuretic.

Associated system(s): AFHSU

***Careya arborea* Roxb.**

Sanskrit or Ayurvedic name(s): Kumbhikah

Common name(s): Vaaikumbha, Kumbhiphoo

Family: Lecythidaceae

Distribution: Habitat to sub-Himalayan tract, Bengal, MP, and TN.

Parts and properties used: Bark- Demulcent, antipyretic and antipruritic, anthelmintic, antidiarrheal. Flower's infusion- Given in childbirth.

Associated system(s): AFS

Carica papaya **L.**

Sanskrit or Ayurvedic name(s): Erand-karkati

Common name(s): Papaya, Papita

Family: Caricaceae

Distribution: American, cultivated all over India.

Parts and properties used: Fruit- Stomachic, digestive, diuretic, carminative, galactagogue. Used in bleeding piles, haemoptysis, dysentery, and chronic diarrhoea. Latex- Yields Papin, the enzyme mixture.

Associated system(s): AFHSU

Carissa carandas **L.**

Sanskrit or Ayurvedic name(s): Karamarda, Karinkara

Common name(s): Christ's thorn, Karonda

Family: Apocynaceae

Distribution: Habitat to tropical India. Grown all over India for pickle.

Parts and properties used: Fruit- Used for acidity, flatulence, poor digestion, as a slimming diet. Bark- Treat difficult skin diseases. Root- Used in urinary disorders.

Associated system(s): AFS

Carissa opaca **Stapf. Ex Haines.**

Sanskrit or Ayurvedic name(s): Karamardikaa.

Common name(s): Jangali Karonda

Family: Apocynaceae

Distribution: Wild all-over dry India, Punjab, and J & K.

Parts and properties used: Plant- Cardiotonic. Root- Purgative.

Associated system(s): AFS

Carthamus tinctorius **L.**

Sanskrit or Ayurvedic name(s): Kusumbha,Vahinshikhaa,

Common name(s): Kusum Phool, Safflower

Family: Asteraceae/ Compositae

Distribution: Native of eastern Mediterranean (Mesopotamia). Cultivated as oilseed crop.

Parts and properties used: Flower- Stimulant, sedative, diuretic, emmenagogue. Used in fever, measles, eruptive skin. Dried flowers- Given in cardiovascular diseases, amenorrhoea, dysmenorrhoea and retention of lochia, wounds, and swelling. Oil- Prevent arteriosclerosis, coronary heart disease and kidney disorders.

Associated system(s): AFSTU

***Carum bulbocastanum* W. Koch, syn. *Bunium bulbocastanum* L.; *B. persicum* (Boiss.) (EN)**

Sanskrit or Ayurvedic name(s): Krishna jiraka

Common name(s): Jira

Family: Apiaceae/ Umbelliferae

Distribution: Native to west Europe to HP, cultivated north India and hills of south India.

Parts and properties used: Like *C. carvi* (Kala Zira).

Associated system(s): ASU

***Carum carvi* L.**

Sanskrit or Ayurvedic name(s): Krishna jiraka, Krsnajiraka

Common name(s): Shah Jeera, Kala Zira

Family: Apiaceae/ Umbelliferae

Distribution: Native to west Asia, Europe, and North Africa. Grown in Himalayas and in north Indian plains/hills.

Parts and properties used: Fruit- Carminative, antispasmodic, antimicrobial, expectorant, galactagogue, emmenagogue. Used in abdominal pain. Seed- Used in chronic fevers. Oil- Used in dyspeptic problems.

Associated system(s): AHSLU (Amchi)

***Carum carvi* var. *gracile* (Lindl) H. Wolff**

Sanskrit or Ayurvedic name(s): Krishna Jiraka

Common name(s): Caraway, Koniyot

Family: Apiaceae/ Umbelliferae

Distribution: Habitat to alpine cold arid zone of Ladakh and beyond.

Parts and properties used: Fruit- Carminative, digestive, febrifuge.

Associated system(s): Amchi

***Casearia esculenta* Roxb., syn. *C zeylanica* (Gaertn.) Thwaites**

Sanskrit or Ayurvedic name(s): Saptachakra, Svarnamulah

Common name(s): Saptarangi

Family: Flacourtiaceae

Distribution: Found in peninsular India.

Parts and properties used: Root- Antidiabetic, astringent, produce liver tonic.

Associated system(s): ASF

Casearia tomentosa **Roxb.**

Sanskrit or Ayurvedic name(s): Chilhaka

Common name(s): Chillaa, Saptrangi.

Family: Flacourtiaceae

Distribution: Habitat from Himalayas to tropical India.

Parts and properties used: Root- Hypoglycaemic. Root bark- Produce tonic for anaemia.

Associated system(s): ASF

Cassia absus **L.**

Sanskrit or Ayurvedic name(s): Chakshushyaa

Common name(s): Chaksoo

Family: Leguminosae/ Fabaceae (Caesalpiniaceae)

Distribution: On gravely hills and plains of India (Maharashtra).

Parts and properties used: Seed- Blood-purifier, astringent, diuretic stimulant. Used to treat leukoderma, ringworm, venereal ulcers, skin diseases. Yields sitosterol-beta-D-glucoside and alkaloids, chaksine, iso-chaksine. Roots- Purgative.

Associated system(s): AFSTU

Cassia alata **L., syn.** *Senna alata* **(L.) Roxb.**

Sanskrit or Ayurvedic name(s): Dadrughna

Common name(s): Dadmurdan, Dat-ka-pat

Family: Leguminosae/ Fabaceae (Caesalpiniaceae)

Distribution: Native to neotropics. Grown throughout India.

Parts and properties used: Leaf- Used in skin diseases; dried- in leprosy; decoction to ringworm. Pod- Contains Rhein (Cassic acid), which is antibacterial, emodin and aloe-emodin.

Associated system(s): AFS

Cassia angustifolia **Vahl, syn.** *C. senna* **L.;** *Senna alexandrina* **Gars. ex-Mill.**

Sanskrit or Ayurvedic name(s): Svarnapatri

Common name(s): Sona Patta, Sona Mukhi, Senna

Family: Leguminosae/ Fabaceae (Caesalpiniaceae)

Distribution: Native to Sudan. Grown in arid and semi-arid zones of India.

Parts and properties used: Herb- Purgative, used in treating biliousness, distention of stomach, vomiting and hiccups. Leaf/fruit- Laxative, used in constipation.

Associated system(s): AHSU

***Cassia auriculata* L., *Senna auriculata* (L.) Roxb.**

Sanskrit or Ayurvedic name(s): Aaavartaki

Common name(s): Avarai

Family: Leguminosae/ Fabaceae (Caesalpiniaceae)

Distribution: Habitat to and grown in arid and semi-arid zones of India.

Parts and properties used: Root- Used in skin diseases and asthma. Flower- Used in diabetes, urinary disorders, nocturnal emissions. Pod husk- Contains Nonacosane and nonacosan-6-one.

Associated system(s): AFS

***Cassia fistula* L.**

Sanskrit or Ayurvedic name(s): Aaragvadha

Common name(s): Amaltaash, Sonari Chal

Family: Leguminosae/ Fabaceae (Caesalpiniaceae)

Distribution: Native, cultivated all over India.

Parts and properties used: Flower/Pods- Purgative, febrifugal, astringent, antibailout. Pulp- Treat constipation, colic, chlorosis, and urinary disorders. Seed powder- Used in amoebiasis.

Associated system(s): AFSU

***Cassia italica* (Mill.) Lam. ex Andr., syn. *Senna italica* Mill., *C. obtusus* Roxb.**

Sanskrit or Ayurvedic name(s): Bhumyahi

Common name(s): Nila aavaarai

Family: Leguminosae/ Fabaceae (Caesalpiniaceae)

Distribution: African, habitat to, and grown in arid and semi-arid zones of India

Parts and properties used: Leaves, pods (seeds)- Purgative, decoction for stomach disorder, fever, jaundice, and venereal diseases; also, in hair treatment.

Associated system(s): F (has trade potential)

***Cassia obtusifolia* L., syn. *Senna obtusifolia* (L.) H.S. Irwin & Barneby**

Sanskrit or Ayurvedic name(s): Chakramarda,

Common name(s): Panevar, Taga

Family: Leguminosae/ Fabaceae (Caesalpiniaceae)

Distribution: Habitat to northwest sub-Himalaya and peninsular India. Cultivated.

Parts and properties used: Pod- Antidysentery, antibacterial, antifungal. Seed- Used in treatment of ringworm, skin diseases, cold cough, asthma.

Associated system(s): AF

***Cassia occidentalis* L., syn. *Senna occidentalis* (L.) Link**

Sanskrit or Ayurvedic name(s): Kaasamarda, Amimarda

Common name(s): Kasondi, Kasmardah

Family: Leguminosae/ Fabaceae (Caesalpiniaceae)

Distribution: Occur as weed throughout India.

Parts and properties used: Plant- Purgative, diuretic, febrifugal, expectorant, stomachic. Leaves- Used to treat scabies, ringworm, and skin diseases. Seeds- To treat cough and convulsions.

Associated system(s): AFS

***Cassia siamea* Lam., syn. *Senna siamea* (Lam.) H.S. Irwin & Barneby**

Common name(s): Kassod

Family: Leguminosae/ Fabaceae (Caesalpiniaceae)

Distribution: Habitat to South and Southeast Asia, also cultivated.

Parts and properties used: Plant- Tranquillizer, antipyretic, treat venereal diseases. Leaves- Treat leucorrhoea. Flowers- Antihypertensive and antipyretic.

Associated system(s): FS

***Cassia sophera* L., syn. *Senna sophera* (L.) Roxb.**

Common name(s): Kasodi, Ponthakaram

Family: Leguminosae/ Fabaceae (Caesalpiniaceae)

Distribution: Common in plains grasslands and wastelands

Parts and properties used: Leaves, seeds, bark- Cathartic for ringworm, skin diseases. Also, used in bronchitis, asthma.

Associated system(s): AFHS

***Cassia tora* L., syn. *Senna tora* (L.) Roxb.**

Sanskrit or Ayurvedic name(s): Prapunnada

Common name(s): Chakoda Beeja, Pawad, Chankuda

Family: Leguminosae/ Fabaceae (Caesalpiniaceae)

Distribution: Native of central America, adapted to India.

Parts and properties used: Leaves- Used in skin disorder. Soaked seeds- Used in treatment of spermatorrhoea. Pods- Used in dysentery.

Associated system(s): AFSTU

***Cassytha filiformis* L.**

Sanskrit or Ayurvedic name(s): Amarvalli

Common name(s): Astimu, Akash Bel

Family: Cassythaceae

Distribution: Occur in greater part of India.

Parts and properties used: Plant- Diuretic, astringent, pesticidal (used in hair wash).

Associated system(s): AS

***Casuarina equisetifolia* Linn.**

Common name(s): Jhaau, Savukku

Family: Casuarinaceae

Distribution: Invasive species. Grown in coastal areas.

Parts and properties used: Bark- Astringent, antidiarrheal. Leaf- Anti-spasmodic, used in treatment of colic.

Associated system(s): AS

***Catharanthus pusillus* G. Don., syn. *Vinca pusilla* Murr.**

Sanskrit or Ayurvedic name(s): Sangkhaphuli

Family: Apocynaceae

Distribution: Native, grown in garden all over India.

Parts and properties used: Plant- Oncolytic. Dried plant decoction- Used in treatment of lumbago. Root- Lochnericine and many alkaloids isolated. Leurosine is cytotoxic. Pusiline and pusilinine cause heart depression.

Associated system(s): AF

***Catharanthus roseus* (L.) G. Don, syn. *Vinca rosea* L.**

Sanskrit or Ayurvedic name(s): Sadaapushpaa

Common name(s): Periwinkle Sadabahar, Vinca rosea

Family: Apocynaceae

Distribution: Native to Madagascar. Naturalized and grown in garden all over India.

Parts and properties used: Root shoot extract, leaf- Cytotoxic, anti-depressant. Over 100 monomeric and bisindole alkaloids isolated. Vinblastine and vincristine are used in several diseases, including cancer.

Associated system(s): AFS

Catunaregam spinosa **(Thunb.) Tirveng., syn. *Xeromphis spinosa* (Thunb.) Keay**

Sanskrit or Ayurvedic name(s): Madana

Common name(s): Gehela, Mainphal, Maggare

Family: Rubiaceae

Distribution: Habitat to Western Ghats.

Parts and properties used: Plant- Used to treat bronchitis, asthma, and leukoderma. Bark- Treat diarrhoea, dysentery. Fruit- Help in gastrointestinal disorder.

Associated system(s): AS

Cayratia carnosa **(Wall.) Gagnep. ex Wight**

Sanskrit or Ayurvedic name(s): Gandira.

Common name(s): Amal Ved, Amal Bel, Gutt

Family: Vitaceae

Distribution: Occur throughout the warmer parts of India.

Parts and properties used: Plant- Astringent. Aerial parts- Hypothermic, CNS active.

Associated system(s): AF

Cayratia pedata **(Lam.) A. Juss. ex Gagnep. (CR)**

Sanskrit or Ayurvedic name(s): Godhaapadi.

Common name(s): Suvaha, Gummatige

Family: Vitaceae

Distribution: Found in Bihar, Bengal, Assam.

Parts and properties used: Leaves- Astringent, refrigerant. Aerial parts- Diuretic, spasmolytic.

Associated system(s): AS

Cedrus deodara **(Roxb.) G. Don**

Sanskrit or Ayurvedic name(s): Devadaru

Common name(s): Devdar

Family: Pinaceae

Distribution: Habitat to north-western Himalayas.

Parts and properties used: Bark decoction- Astringent, antidiarrheal, febrifuge. Essential oil- Antiseptic. Heartwood- In puerperal diseases

Associated system(s): ASTU

***Ceiba pentandra* (L.) Gaertn., syn. *Eriodendron pentandrum* (L.) Kurz**

Sanskrit or Ayurvedic name(s): Shveta Shaalmali

Common name(s): Kapok, Safed Semal.

Family: Bombacaceae

Distribution: Native to central and South America. Grown in west and south India around temples.

Parts and properties used: Root- Diuretic, antidiabetic, antispasmodic. Gum- Laxative, astringent, demulcent. Flower- Laxative. Young fruit- Astringent.

Associated system(s): AS

***Celastrus paniculatus* Willd. (VU)**

Sanskrit or Ayurvedic name(s): Jyotismati

Common name(s): Malkangani, Jyothismathi, Bavanthi beeja

Family: Celastraceae

Distribution: Habitat sub-Himalayan region.

Parts and properties used: Seed- Nervine, diaphoretic, febrifugal, emetic. Produce brain tonic. Seed oil- Anti-asthmatic/ anti-bronchitis/ tranquilizer, brain tonic, treat depression.

Associated system(s): ASU

***Celosia argentea* Linn.**

Sanskrit or Ayurvedic name(s): Shitivaaraka

Common name(s): Wild Cock's Comb

Family: Amaranthaceae

Distribution: Occur throughout India as weed.

Parts and properties used: Flower- Used in menorrhagia, and blood-dysentery. Seeds- Antidiarrheal, cure stomatitis.

Associated system(s): ASF

***Celosia cristaita* Linn.**

Sanskrit or Ayurvedic name(s): Jataadhaari

Common name(s): Cock's Comb

Family: Amaranthaceae

Distribution: Wild throughout India and grown as ornamental.

Parts and properties used: Flower- Used in menorrhagia, diarrhoea. Seeds- Demulcent, treat dysentery.

Associated system(s): AF

Celtis philippensis **Blanco**

Common name(s): Kakamushti substitute

Family: Ulmaceae

Distribution: Habitat to tropical India.

Parts and properties used: Roots- Astringent, treat diarrhoea. Leaf sap- Used in parasitic infections.

Associated system(s): FS

Centaurea behen **L.**

Common name(s): Behmen Safed

Family: Asteraceae/ Compositae

Distribution: Native to Iraq and Armenia and Western and Central Asia. Imported from Iran.

Parts and properties used: Root- Nervine, produce anabolic tonic for nervous system.

Associated system(s): U

Centaurea depressa **Bicb**

Common name(s): Vashakha

Family: Asteraceae/ Compositae

Distribution: Habitat to alpine Himalaya including Ladakh.

Parts and properties used: Arial part- Used in chest pain, fever, headache, and abdominal troubles.

Associated system(s): F

Centella asiatica **(L.) Urb.**

Sanskrit or Ayurvedic name(s): Mandukaparni

Common name(s): Brahmi, Brahmi booti

Family: Apiaceae/ Umbelliferae

Distribution: Habitat to wetlands of India. Cultivated

Parts and properties used: Plant- Adaptogen, nervous system relaxant, peripheral vasodilator, sedative, antibiotic, detoxifier, blood-purifier, laxative, diuretic, emmenagogue. Source of tonic. Extract- Treat stress-induced stomach and duodenal ulcers. (Used for phytoremediation too)

Associated system(s): ASH

***Centipeda orbicularis* Lour., syn. *Centipeda minima* (L.) A. Braun & Asch.**

Sanskrit, Ayurvedic and common name(s): Chhikkikaa, Chhikkini

Family: Asteraceae/ Compositae

Distribution: Habitat to damp places as weed in India.

Parts and properties used: Plant- Used in rhinitis, sinusitis, nasopharyngeal, tumours and obstructions, asthma, cold. Extract- Antitussive and expectorant.

Associated system(s): AF

***Centratherum anthelminticum* (L.) Kuntze, syn. *Vernonia anthelmintica* L.**

Sanskrit or Ayurvedic name(s): Vanyajiraka

Common name(s): Kali Zeeri

Family: Asteraceae/ Compositae

Distribution: Found throughout India.

Parts and properties used: Seed- Anthelmintic, stomachic, diuretic; skin diseases. Active principal is Delta-7-avenasterol.

Associated system(s): AHSU

***Cephaelis ipecacuanha* (Brot.) A. Rich., syn. *Psychotria (ipecacuanha* (Brot.) Standl.**

Common name(s): Ipecacuanha

Family: Rubiaceae

Distribution: Native to South America. Cultivated in Indian hills.

Parts and properties used: Root- Antiprotozoal, expectorant, diaphoretic, emetic. Mainly used in in cough and dysentery.

Associated system(s): HSU

***Ceratophyllum demersum* Linn.** (a algae)

Sanskrit or Ayurvedic name(s): Shaivaala

Common name(s): Jalnili,

Family: Caryophyllaceae

Distribution: A freshwater aquatic herb found all over India.

Parts and properties used: Herb- Purgative, ant bilious, antibacterial. Treat fever, piles, burning sensation and thirst.

Associated system(s): ASUF

***Ceropegia bulbosa* Roxb.**

Common name(s): Galot, Hedulo

Family: Asclepiadaceae

Distribution: Punjab, Thar desert, south India.

Parts and properties used: Tubers- Used in diarrhoea and dysentery. Contain alkaloid, Ceropegine, which is anti-hypertensive, diuretics, anti-allergic, tranquilizer, anti-inflammatory, anti-cancer, and analgesic.

Associated system(s): AF

***Cetraria islandica* (Linn.) Ach. (Lichen)**

Sanskrit or Ayurvedic name(s): Blcak Shaileya

Common name(s): Charela

Family: Parmeliaceae

Distribution: Habitat to hills of Himalayas.

Parts and properties used: Herb- Produce tonic given in convalescence, irritating catarrh, and bronchitis.

Associated system(s): AF

***Chenopodium album* Linn.**

Sanskrit or Ayurvedic name(s): Vaastuuka

Common name(s): Bathuaa

Family: Chenopodiaceae

Distribution: Naturalized common West Asian herb grows as weed. Grown too.

Parts and properties used: Plant- Laxative, anthelmintic, treat round-and hookworms, blood-purifier, antiscorbutic.

Associated system(s): AUS

***Chenopodium ambrosioides* Linn.**

Sanskrit or Ayurvedic name(s): Sugandh-vaastuuka

Common name(s): Khatuaa

Family: Chenopodiaceae

Distribution: Habitat to wet places of cultivated lands.

Parts and properties used: Plant- Antispasmodic, pectoral, haemostatic, emmenagogue. Given in Chorea treatment.

Associated system(s): ASF

***Chlorophytum arundinaceum* Baker**

Sanskrit or Ayurvedic name(s): Shveta-musali

Common name(s): Safed musali

Family: Asparagaceae

Distribution: Wild in Assam, Bihar, and cultivated too.

Parts and properties used: Tuber- Diuretic, nervine, produce a general tonic.

Associated system(s): AFTU

***Chlorophytum borivilianum* Santapau & R.R. Fern. (CR)**

Sanskrit or Ayurvedic name(s): Balyakanda

Common name(s): Safed musali

Family: Asparagaceae

Distribution: Habitat to peninsular India and cultivated.

Parts and properties used: Tuber- Given in natal and post-natal problems and a cure for diabetes and arthritis.

Associated system(s): AF

***Chlorophytum tuberosum* Baker**

Sanskrit, Ayurvedic and common name(s): Mushali, Musali

Family: Asparagaceae

Distribution: Wild in peninsular India, grown too.

Parts and properties used: Dried tuber- Produce tonic like that of *C. arundinaceum* and *C. borivilianum*.

Associated system(s): ASU

***Chonemorpha fragrans* (Moon) Alston, syn. *C. macrophylla* (Roxb.) G. Don. (EN)**

Sanskrit, Ayurvedic and common name(s): Muurvaa, Murva

Family: Apocynaceae

Distribution: Found throughout India.

Parts and properties used: Root, stem powder- Laxative and ant bilious.

Associated system(s): A

***Chrozophora prostrata* Dalzell & A. Gibson, syn. *C. plicata* Hook f.**

Sanskrit or Ayurvedic name(s): Suuryaavart

Common name(s): Neelkanthi

Family: Euphorbiaceae

Distribution: Found throughout India.

Parts and properties used: Root ash- Bechic, relieve cough. Leaf- Depurative. Seed- Cathartic.

Associated system(s): AF

Chrysanthemum indicum **Linn.**

Sanskrit or Ayurvedic name(s): Shatapatri

Common name(s): Pyrethrum, Guldaaudi

Family: Asteraceae/ Compositae

Distribution: Habitat to temperate zone, cultivated too.

Parts and properties used: Leaves- Given in migraine. Flower- Stomachic, aperient, anti-inflammatory.

Associated system(s): AUS

Cichorium endivia **L.**

Common name(s): Kasini, Endive

Family: Asteraceae/ Compositae

Distribution: Native to Mediterranean region, widely cultivated in north India.

Parts and properties used: Plant- Ant bilious. Root- Demulcent, febrifuge, diuretic. Used in dyspepsia, tonic to the liver and digestive system.

Associated system(s): U (has trade potential)

Cichorium intybus **L.**

Sanskrit, Ayurvedic and common name(s): Kaasani, Kasani, wild chicory

Family: Asteraceae/ Compositae

Distribution: Native to Europe, Central Asia, W. Himalaya, south India, cultivated for salad.

Parts and properties used: Herb- Diuretic, laxative, mild hepatic, cholagogue. Used in liver congestion, jaundice, rheumatism, gout, loss of appetite, dyspepsia.

Associated system(s): AHSU

Cimicifuga racemose **(Linn.) Nutt**

Common name(s): Jiuenti

Family: Ranunculaceae.

Distribution: Habitat to temperate Himalayas.

Parts and properties used: Herb- Sedative, anti-inflammatory, antitussive, diuretic, emmenagogue. Used in rheumatic nervous system, hysterical women, suffering from uterine affections.

Associated system(s): FH

***Cinchona calisaya* Wedd., syn. *C. ledgeriana* Bern. Moens ex Trimen.**

Common name(s): Cinchona

Family: Rubiaceae

Distribution: South American, introduced in 1859. Grown in Bengal, TN, and Kerala.

Parts and properties used: Bark- Source of Quinine to treat malaria.

Associated system(s): FW, has trade potential.

***Cinchona officinalis* L.**

Common name(s): Quinine

Family: Rubiaceae

Distribution: South American, cultivated in Assam, Bengal, TN.

Parts and properties used: Plant- Antimalarial, febrifuge, astringent, orexigenic, spasmolytic. Used in peptic discomforts, loss of appetite, etc. Bark- Contain Quinine, Quinidine, Cinchonine.

Associated system(s): AHU

***Cinnamomum aromaticum* Nees, syn. *C. cassia* (L.) Presl.**

Sanskrit or Ayurvedic name(s): Tvak, Swadi

Common name(s): Dalchini

Family: Lauraceae

Distribution: Native to southern China. Cultivated in southern states of India.

Parts and properties used: Bark- Antispasmodic, carminative, ant putrescent, antidiarrheal, antiemetic, antimicrobial. Used in case of flatulent dyspepsia, colic, irritable bowel, diverticulosis.

Associated system(s): ASTU

***Cinnamomum camphora* (L.) Presl**

Sanskrit or Ayurvedic name(s): Karpura

Common name(s): Kaafoor, Kapoor

Family: Lauraceae

Distribution: Native to China, cultivated in southern states of India.

Parts and properties used: Plant- yield volatile oil (camphor, safrole, linalool, eugenol, terpineol) camphor, a carminative, reflex expectorant, stimulant for heart, circulation, and respiration. Also, sedative, nervous depressant in convulsions, hysteria, epilepsy.

Associated system(s): AFHSTU

Cinnamomum macrocarpum **Hook.f. (VU)**

Common name(s): Lavang, Karuva

Family: Lauraceae

Distribution: Habitat to southern Western Ghats.

Parts and properties used: Bark, Leaf- Yield oil used in rheumatism, powder in cough (potential).

Associated system(s): AFS

Cinnamomum tamala **(Buch. - Ham.) T. Nees & Eberm. (NT)**

Sanskrit or Ayurvedic name(s): Tvakapatra, Tamalpatra

Common name(s): Tamal Patra, Tej Patra

Family: Lauraceae

Distribution: Habitat to sub-tropical Himalaya, Meghalaya, cultivated too.

Parts and properties used: Leaf- Carminative, antidiarrheal, spasmolytic, antirheumatic, hypoglycaemic. Dried given in sinusitis. Essential oil (cinnamaldehyde)- Has fungicidal properties, key use.

Associated system(s): AFSTU

Cinnamomum verum **J. Presl., syn.** ***C. zeylanicum*** **Blume**

Common name(s): Darusita, Tvak Dalchini, Qalmi-dar-chini, true cinnamon

Family: Lauraceae

Distribution: Habitat to Southwest India and Sri Lanka. Cultivated

Parts and properties used: Bark- Carminative, astringent, antispasmodic, expectorant, haemostatic, antiseptic antimicrobial. Dried used in sinusitis. Leaf- Antidiabetic, powder with honey treat cough and dysentery. Both twigs, leaves and berries crushed to make cinnamon oil.

Associated system(s): AFHLU

Cinnamomum wightii **Meisn (VU)**

Common name(s): Dalchini, Tejpatta

Family: Lauraceae

Distribution: Endemic to southern Western Ghats

Parts and properties used: Bark, leaves, oil- Treat paralytic disorders, digestion, abdominal and gynaecological disorders, cough, dysuria.

Associated system(s): S

***Cissampelos pareira* L. var. *hirsuta* (Ham. ex DC.) Forman; (= *Chondodendron tomentosum* Ruiz et Par from Peru and Brazil)**

Sanskrit or Ayurvedic name(s): Paathaa

Common name(s): Pahad mool, Laghu Patha

Family: Menispermaceae

Distribution: Habitat to tropical and subtropical peninsular India

Parts and properties used: Root- Astringent, antispasmodic, analgesic, antipyretic, diuretic, antilithic and emmenagogue. Used in gastro-intestinal disorder, asthma, bronchitis, blood-purifier. Source of Ayurvedic drug '*Patha*'.

Associated system(s): AHSTU

***Cissus quadrangularis* L.**

Sanskrit or Ayurvedic name(s): Asthisamharaka

Common name(s): Pirandai, Haatjod

Family: Vitaceae

Distribution: Habitat to open forests of tropical India and cultivated.

Parts and properties used: Plant- Anabolic and steroidal- extracted drug cure fracture. Stem- Rich Vitamin C.

Associated system(s): ASU

***Citrullus colocynthis* (L.) Schrad**

Sanskrit or Ayurvedic name(s): Indravaruni

Common name(s): Indrayan, Indravaruni, Tumba

Family: Cucurbitaceae

Distribution: Mediterranean basin to western arid zone of India, Asia.

Parts and properties used: Dried pulp of ripe fruit- Cathartic, drastic purgative, irritant, toxic. Used in cure of amenorrhea, ascites, bilious, cancer, jaundice, leukaemia, rheumatism, urogenital disorders.

Associated system(s): AFHSTU

***Citrullus lanatus* (Thunb.)**

Sanskrit or Ayurvedic name(s): Matsumara & Nakai Kalinga

Common name(s): Terbuz, Watermelon

Family: Cucurbitaceae

Distribution: Cosmopolitan. Cultivated throughout India.

Parts and properties used: Pulp- Cooling, refreshing. Source of pectin, sucrose, and carotenoids.

Associated system(s): AH

***Citrus aurantifolia* (Christm. & Panz.) Swingle, syn. *C. medica* var. *acida* Pers.**

Sanskrit or Ayurvedic name(s): Nimbuka

Common name(s): Kaagazi Nimbu

Family: Rutaceae

Distribution: Wild in warm Himalayan valleys and cultivated.

Parts and properties used: Fruit (rind)- Antiscorbutic, stomachic, appetizer, refrigerant, used in bilious. Root/leaf infusion- Given in jaundice, dysentery.

Associated system(s): AFSU

***Citrus aurantium* L.**

Common name(s): Narangi

Family: Rutaceae

Distribution: Semi-wild in northeast hills and cultivated.

Parts and properties used: Peel- Laxative, feeble stomachic, emmenagogue. Used in loss of appetite, dyspeptic ailments and as tonic. Leaves- Used in arthritis and bronchitis.

Associated system(s): FS

***Citrus* x *bergamia* Risso.**

Common name(s): Limbu chaal, Jambeeram

Family: Rutaceae

Distribution: Hybrid, cultivated.

Parts and properties used: Fruit peel- Used in skin care.

Associated system(s): ASU

***Citrus limon* (L.) Burm. f.**

Sanskrit or Ayurvedic name(s): Jambira

Common name(s): Lemon

Family: Rutaceae

Distribution: Native to southeast Asia northeast India and cultivated.

Parts and properties used: Fruit- Antiscorbutic, carminative, stomachic, antihistaminic. Used in cough, influenza, and fever. Fruit juice- Used in ringworm, erysipelas, leprosy. Leaves/stem- Antibacterial.

Associated system(s): ASU

Citrus maxima **(Burm.) Merrill.**

Sanskrit or Ayurvedic name(s): Madhukarkatikaa.

Common name(s): Chakotra, Mahaa-nibu

Family: Rutaceae

Distribution: Native to northeast India.

Parts and properties used: Fruit- Cardiotonic. Leaves, flower/rind- Sedative, given in cough, chorea, epilepsy.

Associated system(s): AFSU

Citrus medica **L.**

Sanskrit or Ayurvedic name(s): Maatulunga, Bijapura

Common name(s): Matunga, Mahanimbu, Turanj.

Family: Rutaceae

Distribution: Habitat to sub-Himalayan range.

Parts and properties used: Fruit- Antiscorbutic, refrigerant, astringent, carminative, stomachic, antibacterial. Used dyspepsia, bilious vomiting, cold and fever. Root- Anthelmintic. Flowers and buds- Astringent.

Associated system(s): AFSU

Citrus reticulata **Blanco**

Common name(s): Santra

Family: Rutaceae

Distribution: Habitat to Assam, Meghalaya, Gujarat, western and hills of India. Cultivated

Parts and properties used: Endocarp- Carminative, expectorant. Fruit- Source of vitamin C. Peel- Yield essential oil.

Associated system(s): ATU

Clausena dentata **(Willd.) Roem.**

Common name(s): Mahasindur

Family: Rutaceae

Distribution: Habitat to sub-Himalaya region, and TN.

Parts and properties used: Plant- Repellent against mosquito (Dengue). Bark- Anti-inflammatory, spasmolytic.

Associated system(s): FS

Claviceps purpurea **(Fr.) Tul (fungus)**

Sanskrit or Ayurvedic name(s): Annamaya, Sraavikaa

Common name(s): Argot

Family: Hypocreaceae

Distribution: Parasitic fungous, found particularly in rye, Nilgiris and at Chakrohi farm of J & K.

Parts and properties used: Extract- Uterine stimulant, oxytocic, abortifacient, parturient, vasoconstrictor, homeostatic. Used in obstetrics. Also, after abortion in removal of the placenta.

Associated system(s): ASU

Cleistanthus collinus **(Roxb.) Benth.**

Sanskrit or Ayurvedic name(s): Kaudigam, Nandi, Kutaja, Indrayava

Common name(s): Kutaja, Garbar

Family: Euphorbiaceae

Distribution: Found in many parts of India, Kerala, and TN.

Parts and properties used: Leaf extract- Antimicrobial. Used against skin and urinary tract infection agents. Needs more studies.

Associated system(s): AS

Clematis gouriana **Roxb. ex DC.**

Sanskrit or Ayurvedic name(s): Morata

Common name(s): Morvel, Muurvaa

Family: Ranunculaceae

Distribution: Wild in tropical forests of Ghats and all over India.

Parts and properties used: Leaf and stem- Vesicant and poisonous. Applied on bruised and skin care.

Associated system(s): AF

Clematis triloba **Heyne**

Sanskrit or Ayurvedic name(s): Laghu karni

Common name(s): Morvel

Family: Ranunculaceae

Distribution: Habitat to Deccan mountains and Konkan.

Parts and properties used: Leaf juice- Used in eye drops, purgative.

Associated system(s): AFU

***Cleome gynandra* L., syn. *Gynandropsis pentaphylla* (L.) DC.**

Sanskrit or Ayurvedic name(s): Ajagandha

Common name(s): Cat whiskers, Ridhi

Family: Cleomaceae

Distribution: Found all over the plains of India.

Parts and properties used: Crushed leaves- Used in treatment of scurvy.

Associated system(s): AFST

***Cleome viscosa* L, syn *C. icosandra* L.**

Sanskrit or Ayurvedic name(s): Tilaparni

Common name(s): Nayikadugu

Family: Cleomaceae

Distribution: Found throughout India.

Parts and properties used: Seed- Carminative, antiseptic, anthelmintic. Leaf-Sudorific. Bark- Rubefacient, vesicant. Root- Vermifuge.

Associated system(s): ASU

***Clerodendrum indicum* (L.) Kuntze**

Sanskrit or Ayurvedic name(s): Vaamana-haati

Common name(s): Bharangi

Family: Verbenaceae

Distribution: Indo-Malayan. Cultivated as ornamental.

Parts and properties used: Root- Used in asthma, cough, scrofulous affections. Leaf - Vermifuge. Resin- Antirheumatic.

Associated system(s): AS

***Clerodendru inerme* (L.) Gaertn.**

Sanskrit or Ayurvedic name(s): Putigandha, Kundali

Common name(s): Nir-Notsjil, Sangan-Kuppi,

Family: Verbenaceae

Distribution: Found throughout India, grown as hedge.

Parts and properties used: Leaf- Febrifuge, an alternative; juice used to treat fever. Plant- Used in muscular pains, skin and as substitute of quinine.

Associated system(s): AF

Clerodendrum phlomidis **L.f.**

Sanskrit or Ayurvedic name(s): Agnimantha

Common name(s): Arnimul

Family: Lamiaceae/ Labiatae

Distribution: Found throughout India.

Parts and properties used: Plant- Used in *dyspepsia*, stomach-ache, colic, cholera, dysentery, postnatal fever. Root/bark- Yield bitter tonic.

Associated system(s): ASF (Bhil)

Clerodendrum serratum **(L.) Moon**

Sanskrit, Ayurvedic and Common name(s): Bharangi

Family: Verbenaceae

Distribution: Found throughout India, Assam, Bengal.

Parts and properties used: Root- Anti-asthmatic, antihistaminic, antispasmodic, antitussive carminative. Leaf- Febrifuge.

Associated system(s): AS

Clitoria ternatea **L.**

Sanskrit or Ayurvedic name(s): Aparajitaa, Adrikarni, Girikarnika

Common name(s): Kajli

Family: Leguminosae/ Fabaceae

Distribution: Found most of India and cultivated.

Parts and properties used: Root- Cathartic like jalap. Root bark- Diuretic. Root Juice- Given in bronchitis. Dried leaf- Given in migraine, psychoneurosis, and mania.

Associated system(s): AFSTU

Coccinia grandis **(L.) Voigt, syn.** ***C. indica*** **W. & A.**

Sanskrit or Ayurvedic name(s): *Bimboshtha,* Bimbaphala, Bimbi

Common name(s): Bhrngaraj, Bimb, Kunduru, Ivy gourd, Shiv-lingi

Family: Cucurbitaceae

Distribution: Native to Tropical & Subtropical Asia, found wild on hedges. Cultivated in India.

Parts and properties used: Plant- Carminative, antipyretic, galactagogue. Root powder- Used in vomiting. Leaf juice- Antispasmodic & expectorant; applied on skin.

Associated system(s): AHSU

Cocculus hirsutus **(L.) Theob.**

Sanskrit or Ayurvedic name(s): Chhilihinta

Common name(s): Vasanvel

Family: Menispermaceae

Distribution: Found in central India and Eastern Ghats.

Parts and properties used: Root- Laxative, sudorific, alterative, antirheumatic. Leaves- Antidiabetic, antirheumatic; externally used in eczema, prurigo, and impetigo.

Associated system(s): AS

Cochlospermum religiosum **(L.) Alston, syn.** ***C. gossypium*** **DC.**

Common name(s): Katira

Family: Cochlospermaceae

Distribution: Habitat to western sub-Himalayas to peninsular India/Ghats. Cultivated

Parts and properties used: Gum- Cooling, sedative, useful in coughs, hoarse throat, diarrhoea, dysentery, scalding urine.

Associated system(s): AFSU

Cocos nucifera **L.**

Sanskrit or Ayurvedic name(s): Narikela

Common name(s): Nariyal

Family: Arecaceae

Distribution: Indo-Malayan, cultivated in coastal areas.

Parts and properties used: Flower with oil- Treat swelling. Tender fruit- Give nutritious water, stomachic, diuretic, laxative, anti-diarrheic, styptic; treat dyspepsia.

Associated system(s): AFSTU

Coffea arabica **L.**

Sanskrit or Ayurvedic name(s): Mlecchaphala

Common name(s): Coffee, Kahvaa

Family: Rubiaceae

Distribution: Native of Ethiopia, cultivated in hills of Assam, TN.

Parts and properties used: Fruit (seed)- Stimulant, nervine, and diuretic; powder treat diarrhea, and inflammation.

Associated system(s): AFSU

***Coix lacryma-jobi* L.**

Sanskrit or Ayurvedic name(s): Gavedhuka

Common name(s): Sankhlu, Dabhir

Family: Poaceae/ Gramineae

Distribution: Native and habitat to warm damp areas of southeast Asia and India.

Parts and properties used: Fruit decoction- Used in catarrhal affections and inflammation of the urinary tract. Seed- Diuretic. Root- Menstrual.

Associated system(s): AFS

***Colchicum luteum* Baker**

Sanskrit or Ayurvedic name(s): Hiranyatuttha

Common name(s): Hirantotiya, Suranjan, Tukapa

Family: Liliaceae/ Colchicaceae

Distribution: Habitat to temperate Himalaya, Ladakh

Parts and properties used: Root, fruit (seed)- non-steroidal anti-inflammatory, anti-gouts; externally used to lessen inflammation and pain.

Associated system(s): AU

***Coleus amboinicus* Lour., syn. *C. aromaticus* Benth., *Plectranthus amboinicus* Lour.**

Sanskrit or Ayurvedic name(s): Parna yavaani

Common name(s): Pattaa Ajawaayin

Family: Lamiaceae/ Labiatae

Distribution: Origin unknown, may be native to Africa, India and Indonesia.

Parts and properties used: Leaf - Used in urinary diseases, vaginal discharge, colic, and dyspepsia. Stimulates the function of liver. Also given in epilepsy and convulsive affections, asthma, etc. Bruised leaves- Applied to burns and leaf juice to chapped lips.

Associated system(s): F

***Coleus forskohlii* (Willd.) Briq., syn. *C. barbatus* Benth., *Plectranthus barbatus* Andrews**

Sanskrit or Ayurvedic name(s): Gandhira

Common name(s): Garmar

Family: Lamiaceae/ Labiatae

Distribution: Habitat to sub-tropical Himalaya. Cultivated in AP.

Parts and properties used: Root and leaf- Spasmolytic, antithrombotic, anti-inflammatory, lipolytic. Used to treat heart diseases, spasmodic and painful urination.

Source of Forskolin that reduces intraocular pressure, high BP, chest pain, respiratory disorders.

Associated system(s): AF

***Coleus zeylanicus* (Benth.) Cramer, syn. *Plectranthus vettiveroides* (K.C.Jacob) Singh & Sharma**

Sanskrit or Ayurvedic name(s): Hrivera

Common name(s): Valakah, Kuruver

Family: Lamiaceae/ Labiatae

Distribution: Endemic to south India. Cultivated in Kerala, TN.

Parts and properties used: Root- Aromatic. Used to treat leprosy, leukoderma, ulcer, vomiting, skin. Leaf juice- Cooling, carminative, used in indigestion. Source of Ayurvedic '*Hrivera*'.

Associated system(s): AS

***Colocasia esculenta* (Linn.) Schott**

Sanskrit or Ayurvedic name(s): Pindaaluka

Common name(s): Arvi, Taro

Family: Aracaeae

Distribution: Native. Grown all over India.

Parts and properties used: Petiole and corn juice- Styptic, rubefacient. Used in alopecia.

Associated system(s): AS

***Commelina benghalensis* Linn.**

Sanskrit or Ayurvedic name(s): Kanchata

Common name(s): Kanavazhai, Kenaa

Family: Commelinaceae

Distribution: Found all over moist India.

Parts and properties used: Plant- Emollient, demulcent, laxative, diuretic, antileprotic.

Associated system(s): ASF

***Commiphora myrrha* (T. Nees) Engl.**

Sanskrit or Ayurvedic name(s): Hiraabola

Common name(s): Hirabol

Family: Burseraceae

Distribution: Native to Arabian Peninsula to Africa. Grown in west/ south India.

Parts and properties used: Oleo-Gum Resin- Emmenagogue, used in indigestion, ulcers, cold, cough, asthma, arthritis, cancer, leprosy, spasms, and syphilis.

Associated system(s): ATU

***Commiphora wightii (*Arn.) Bhandari, syn. *C. mukul* (Hook. ex Stocks) Engl., *Balsamodendron mukul* Hook. ex Stocks (VU-CR)**

Sanskrit or Ayurvedic name(s): Guggulu

Common name(s): Guggul, Guggulu

Family: Burseraceae

Distribution: Native to Rajasthan and adjacent areas, also cultivated.

Parts and properties used: Oleo-Gum Resin- Used in atherosclerosis, arthritis, BP and cholesterol disorder, skin diseases, and weight loss. Ingredient of many Ayurvedic formulations.

Associated system(s): A

***Convolvulus arvensis* L.**

Sanskrit or Ayurvedic name(s): Bhadrabalaa

Common name(s): Chandvel (Prasarni)

Family: Convolvulaceae

Distribution: Found as weed throughout India.

Parts and properties used: Plant- Cooling, anticonvulsant. Root- Cathartic.

Associated system(s): AUF

***Convolvulus microphyllus Sieb.* ex Spreng., syn. *C. prostratus* Forssk.; *C. pluricaulis* Chois.**

Sanskrit, Ayurvedic and Common name(s): Sankhapuspi, Shankapushpi

Family: Convolvulaceae

Distribution: Found throughout India.

Parts and properties used: Plant- Yield brain tonic, tranquilizer, used in nervine disorders, mental aberration, internal haemorrhages and spermatorrhoea.

Associated system(s): AU

***Convolvulus scammonia* L.**

Common name(s): Saqmunia

Family: Convolvulaceae

Distribution: Mediterranean. Grown in India.

Parts and properties used: Resin of rhizome- Hydragog, cathartic, used in dropsy and anasarca.

Associated system(s): HU

Coptis teeta **Wall. (VU-EN)**

Sanskrit or Ayurvedic name(s): Mamira

Common name(s): Mamira, Titaa

Family: Ranunculaceae

Distribution: Endemic to ArP.

Parts and properties used: Rhizome- Stomachic, antiperiodic, antimicrobial. Have 9% berberine.

Associated system(s): AUF

Corallocarpus epigaeus **(Rottl. & Willd.) C.B. Clarke**

Sanskrit or Ayurvedic name(s): Shukanaasaa

Common name(s): Akasgaddah, Patalagaruda

Family: Cucurbitaceae

Distribution: Habitat to north and peninsular India.

Parts and properties used: Plant- Laxative. Tuber/root has ally of bryconin, used in dysentery and mucous enteritis.

Associated system(s): AS

Corchorus aestuans **Linn., syn. C.** *acutangulus* **Lam.;** *C. trilocularis* **L.**

Sanskrit or Ayurvedic name(s): Chunchu

Common name(s): Chinchaa

Family: Tiliaceae

Distribution: Habitat to the warm India.

Parts and properties used: Plant/seeds- Stomachic, anti-inflammatory. Used in pneumonia. Leaf extract- Used as moisturizer in cosmetics.

Associated system(s): AF

Corchorus capsularis **L., syn.** *C. fascicularis* **Lam.**

Sanskrit or Ayurvedic name(s): Kaala shaaka

Common name(s): Rajan, white jute

Family: Tiliaceae

Distribution: Habitat to all over warm India. Cultivated.

Parts and properties used: Leaves- Stomach, carminative, diuretic, antidysentery. Treat fever. Seeds- Purgative.

Associated system(s): ASF (Bhil)

***Corchorus depressus* Stocks,**

Sanskrit or Ayurvedic name(s): Bhedani

Common name(s): Bahuphali

Family: Tiliaceae

Distribution: Habitat to drier part of north India.

Parts and properties used: Herb- Treat gonorrhoea. Leaves- Treat heat stroke/ urine retention.

Associated system(s): AU

***Cordia dichotoma* G. Forst., syn. *C. myxa* Roxb. non-Linn.; *C. obliqua* Willd.**

Sanskrit or Ayurvedic name(s): Shleshmaataka

Common name(s): Lasora

Family: Cordiaceae

Distribution: Found throughout India. Often cultivated.

Parts and properties used: Fruit- Astringent, demulcent, expectorant, diuretic anthelmintic. Mucilage used in chest and urinary diseases. Also, treat malaria, intestinal disorders, conjunctivitis.

Associated system(s): ASU

***Cordia monoica* Roxb.**

Common name(s): Narivari

Family: Cordiaceae

Distribution: Dry deciduous forests of peninsular India, and Kerala.

Parts and properties used: Root- Used in vomiting & malaria. Leaves- Used in eye diseases.

Associated system(s): FS

***Cordia wallichii* G. Don, syn. *C. obliqua* Willd. var. *wallichii* (G. Don) C.B. Clarke**

Sanskrit or Ayurvedic name(s): Shleshmaataka

Common name(s): Sapistan

Family: Cordiaceae

Distribution: Habitat to peninsular India.

Parts and properties used: Fruit- Astringent, demulcent, expectorant, diuretic. Used like Lasora (*Cordia myxa*).

Associated system(s): AFS

***Cordyceps sinensis* (Berk.) Saac; mushroom (EN)**

Common name(s): Yarsa Gomba

Family: Clavicipitaceae

Distribution: Habitat to Indian Himalayas.

Parts and properties used: Mushroom- Lung and kidney tonic. Treat bronchitis, asthma, tuberculosis, respiratory diseases.

Associated system(s): F

***Coriandrum sativum* L.**

Sanskrit or Ayurvedic name(s): Dhaanyaka

Common name(s): Dhaniya, Dhana

Family: Apiaceae/ Umbelliferae

Distribution: Wild in western Asia and southern Europe. Grown all over India.

Parts and properties used: Plant- Stimulant, stomachic, carminative, antispasmodic, diuretic, hypoglycaemic, anti-inflammatory. Oil- Bactericidal, and larvicidal. Source of Coriandrin.

Associated system(s): AFSLU

***Corydalis govaniana* Wall.**

Sanskrit or Ayurvedic name(s): Bhootakeshi

Common name(s): Bhutkesi, Stong-zil

Family: Papaveraceae

Distribution: Habitat to slopes of northwest Himalayas, 3200-5600 m.

Parts and properties used: Plant- Sedative, spasmolytic, hypotensive, nervine, antiseptic. Used in cutaneous/ scrofulous affections, fever, liver complaints.

Associated system(s): AF

***Coscinium fenestratum* (Gaertn.) Coleb. (CR)**

Sanskrit or Ayurvedic name(s): Kaaliyaka

Common name(s): Maramanjal, Daruharidra

Family: Menispermaceae

Distribution: Habitat to deciduous to evergreen Forests of Western Ghats.

Parts and properties used: Root- Stomachic, diuretic, hypotensive, antidysentery, antibacterial, antifungal. Yield tonic used in dyspepsia & debility. Stem/root- Have alkaloids.

Associated system(s): AFST

***Costus speciosus* (J.Konig) Sm.**

Sanskrit or Ayurvedic name(s): Kebuka

Common name(s): Kustha, Koshtum, Kuth

Family: Costaceae

Distribution: Habitat to sub-Himalayas, northeast hills & Western Ghats.

Parts and properties used: Root- Antirheumatic, purgative, anti-asthmatic/anti-bronchitis, anthelmintic, antimicrobial. Contain saponins.

Associated system(s): AFS

***Crataegus oxyacantha* L.**

Common name(s): Hawthorn

Family: Rosaceae

Distribution: Habitat to western Himalayas, 1800-3000 m. Also cultivated.

Parts and properties used: Flower, fruit- Coronary vasodilator, antispasmodic, antihypertensive, sedative, diuretic.

Associated system(s): FH

***Crataeva religiosa* G. Forst., syn. *C. nurvula* Buch. - Ham.**

Sanskrit or Ayurvedic name(s): Varuna

Common name(s): Varun Chhal, Varunah

Family: Capparidaceae

Distribution: Habitat to semi-arid peninsular India. Cultivated.

Parts and properties used: Bark- Diuretic. Plant powder- Cholinergic. Used as anti-urolithiasis.

Associated system(s): AFSU

***Cressa cretica* Linn.**

Sanskrit or Ayurvedic name(s): Rudantikaa

Common name(s): Rudanti

Family: Convolvulaceae

Distribution: Habitat to coastal regions.

Parts and properties used: Plant- Expectorant, stomachic, ant bilious, alterative.

Associated system(s): AFS

Crinum asiaticum **L.**

Sanskrit or Ayurvedic name(s): Naagapatra

Common name(s): Chindar

Family: Amaryllidaceae

Distribution: Native to Indian Ocean islands. Cultivated as ornamental.

Parts and properties used: Bulb- Laxative, expectorant, used in biliousness and urinary infection. Leaves- Expectorant. Seed- Diuretic, emmenagogue and contain alkaloids.

Associated system(s): AS

Crinum latifolium **Linn.**

Sanskrit or Ayurvedic name(s): Sudarshana, Sukhdarshan

Family: Amaryllidaceae

Distribution: Habitat to India and Sri Lanka. Cultivated as ornamental.

Parts and properties used: Bulb- Rubefacient, antirheumatic. Used for piles. Alkaloids Crinafoline and Crinafolidine isolated. In Ayurveda it is used in fever, swelling, poisoning and skin disease.

Associated system(s): AS

Crocus sativus **L.**

Sanskrit or Ayurvedic name(s): Kumkuma

Common name(s): Kesar, Zaafraan

Family: Iridaceae

Distribution: Native of eastern Mediterranean region. Cultivated in Kashmir.

Parts and properties used: Stigma and style- Nervine, tonic, sedative, antispasmodic expectorant, stomachic, diaphoretic, emmenagogue.

Associated system(s): AHLU

Crotalaria juncea **L.**

Sanskrit or Ayurvedic name(s): Shana

Common name(s): Dhanahari, Patashan, Sunn

Family: Leguminosae/ Fabaceae

Distribution: Native of tropical and subtropical India. Cultivated all over the plains.

Parts and properties used: Leaf- Demulcent, purgative, emetic, emmenagogue, abortifacient, ant-implantation. Used in dysentery. Seed- Used in psoriasis/ impetigo.

Associated system(s): AFSU

***Crotalaria verrucosa* L.**

Sanskrit or Ayurvedic name(s): Shanpushpi

Common name(s): Banshana, San, Sanpushpi

Family: Leguminosae/ Fabaceae

Distribution: Wild in tropical India.

Parts and properties used: Leaf juice- Used in biliousness, dyspepsia, blood impurities, scabies, and impetigo. Several compounds isolated from stem oil.

Associated system(s): AFS

***Croton oblongifolius* Roxb.**

Sanskrit or Ayurvedic name(s): Naagadanti

Common name(s): Bhutankusha

Family: Euphorbiaceae

Distribution: Occur in major part of India.

Parts and properties used: Plant- Cathartic, rubefacient, irritant. Used like *C. tiglium*. Seed oil- Purgative

Associated system(s): A

***Croton tiglium* L.**

Sanskrit or Ayurvedic name(s): Jayapala

Common name(s): Jamlghota, Japala

Family: Euphorbiaceae

Distribution: Native to China. Grown in Assam, Bengal, and south India.

Parts and properties used: Plant- Cathartic, rubefacient, irritant. Used in ascites, anasarca, and enlargement of abdominal. Seed oil- Purgative.

Associated system(s): AFHSLU

***Cryptolepis buchananii* R.Br. ex Roem. & Schult.**

Sanskrit or Ayurvedic name(s): Krsnasariva

Common name(s): Medaksinghi, Dudhi, Sariva

Family: Apocynaceae

Distribution: Found throughout India.

Parts and properties used: Root, Leaves- Blood-purifier, alternative. Used in child rickets.

Associated system(s): AFS

Cucumis melo **L. var.** ***utilissimus*** **Roxb.**

Sanskrit or Ayurvedic name(s): Ervaaru

Common name(s): Kakadi beej, Magaj Kharbuja

Family: Cucurbitaceae

Distribution: Native, cultivated in northern plains.

Parts and properties used: Seed- Cooling, diuretic, used in micturition, suppression of urine, dysuria and lithiasis.

Associated system(s): ASU

Cucumis prophetarum **Linn.**

Sanskrit or Ayurvedic name(s): Indravarruni ?

Common name(s): Khat-Kachario

Family: Cucurbitaceae

Distribution: Extended from tropical Africa to western Rajasthan to Bellary, Karnataka.

Parts and properties used: Fruit pulp- Yields myriocarpin. Cause nausea, purgative; Have cucurbitacin B, C, D and Q1, and pro-pheterosterol and its acetate.

Associated system(s): AF

Cucumis sativus **L.**

Sanskrit or Ayurvedic name(s): Trapusam

Common name(s): Beej Kheera

Family: Cucurbitaceae

Distribution: Native, cultivated all over India.

Parts and properties used: Seed- Used in dysuria, irritation of the urinary tract, cystitis.

Associated system(s): AFSU

Cucumis trigonus **Roxb.**

Sanskrit or Ayurvedic name(s): Indravaaruni

Common name(s): Jangli-Indrayan

Family: Cucurbitaceae

Distribution: Wild all-over drier India.

Parts and properties used: Fruit pulp- Purgative.

Associated system(s): AFS

***Cucurbita maxima* Duch.**

Sanskrit or Ayurvedic name(s): Kuushmaandaka

Common name(s): Pumpkin

Family: Cucurbitaceae

Distribution: Native of South America. Cultivated all over India.

Parts and properties used: Fruit pulp- Sedative, emollient, and refrigerant, used as poultice. Seed- Diuretic, anthelmintic, arrest enlargement of prostate gland.

Associated system(s): ASU

***Cucurbita moschata* (Lam.) Duchesne ex Poir.**

Sanskrit or Ayurvedic name(s): Kumshmaanda variety

Common name(s): Kaddu, Kumrah

Family: Cucurbitaceae

Distribution: South American. Grown all over India.

Parts and properties used: Fruit- Anthelmintic, diuretic, used in headache, bronchitis, asthma. Dried pulp- Applied in haemoptysis.

Associated system(s): A

***Cuminum cyminum* L.**

Sanskrit or Ayurvedic name(s): Svetajiraka

Common name(s): Jeera, Shahjeera

Family: Apiaceae/ Umbelliferae

Distribution: Middle East to east India. Grown in upper Gangetic Plains.

Parts and properties used: Fruit- Carminative, antispasmodic, stimulant, diuretic, antibacterial, emmenagogue, galactagogue. Cumin oil- Antibacterial.

Associated system(s): ASU

***Cupressus sempervirens* Linn.**

Sanskrit or Ayurvedic name(s): Suraahva

Common name(s): Suram

Family: Cupressaceae.

Distribution: Native to Mediterranean region. Cultivated in northwest India.

Parts and properties used: Tincture- Vasoconstrictor, antiseptic, sedative, antispasmodic, diuretic. Used in cough, cold, etc.

Associated system(s): ASU

***Curculigo orchioides* Gaertn. (EN)**

Sanskrit or Ayurvedic name(s): Taalamuli

Common name(s): Kali musali, Musli shiya

Family: Hypoxidaceae

Distribution: Habitat to sub-tropical Himalayas and Western Ghats.

Parts and properties used: Rhizome- Nervine, apoptogenic, sedative, anticonvulsive, diuretic androgenic, anti-inflammatory. Used in jaundice. Have saponin.

Associated system(s): AFSU

***Curcuma amada* Roxb.**

Sanskrit or Ayurvedic name(s): Amra Haridraa

Common name(s): Amba haldi

Family: Zingiberaceae

Distribution: Wild in parts of India. Cultivated in Gujarat

Parts and properties used: Rhizome- Carminative, appetizer, expectorant, antipyretic, anti-inflammatory. Used to treat rheumatism.

Associated system(s): ASU

***Curcuma angustifolia* Roxb. (VU)**

Sanskrit or Ayurvedic name(s): Tvakshira

Common name(s): Tikhur

Family: Zingiberaceae

Distribution: Habitat to central Himalayas to south India. Grown marginally.

Parts and properties used: Rhizome starch- Cooling, demulcent, nutritious; used to treat asthma, bronchitis. Oil- Anti-parasitic and antimicrobial.

Associated system(s): AFSU

***Curcuma aromatica* Salisb.**

Sanskrit or Ayurvedic name(s): Karpuraa

Common name(s): Kasturi manjal, Kasturi arishna, Kapu-Kachri

Family: Zingiberaceae

Distribution: Native to eastern Himalayas/ south India. Cultivated in Bengal to Kerala.

Parts and properties used: Rhizome- Used like *C. longa*, anti-inflammatory, cholagogue, blood-purifier, regenerator of liver tissue, antiasthma anti-tumour, ant cutaneous, stomachic, carminative. Essential oil- Antimicrobial.

Associated system(s): AFSU

***Curcuma caesia* Roxb. (CR)**

Sanskrit or Ayurvedic name(s): Rajani

Common name(s): Nar-kachura, Kala-haldi

Family: Zingiberaceae

Distribution: Native of northeast of India. Grown in Bengal.

Parts and properties used: Rhizome- Carminative, Applied on bruises and sprains. Essential oil has camphor.

Associated system(s): AUS

***Curcuma longa* L., syn. *C. domestica* Valeton**

Sanskrit or Ayurvedic name(s): Haridraa, Marmarii

Common name(s): Arishna, Haldi, Kari manjal

Family: Zingiberaceae

Distribution: Native, cultivated in most of India.

Parts and properties used: Rhizome- Antiseptic applied on bruises, anti-inflammatory, blood-purifier, antioxidant, antiasthma, anti-tumour, ant cutaneous, stomachic, carminative. Essential oil- Antimicrobial.

Associated system(s): AFHSTU

***Curcuma zerumbet* Roxb., syn. *C. zeodoaria* (Christm.) Roscoe**

Sanskrit or Ayurvedic name(s): Karchuura

Common name(s): Kachur, Kachari

Family: Zingiberaceae

Distribution: Native to south and southeast Asia. Cultivated in most of India.

Parts and properties used: Stem, rhizome- Carminative, stomachic, gastrointestinal stimulant, diuretic, expectorant, demulcent, rubefacient. Used to treat flatulence, dyspepsia, also goitre.

Associated system(s): AUS

***Cuscuta reflexa* Roxb.**

Sanskrit or Ayurvedic name(s): Amarvalli

Common name(s): Aftimoon, Amar Lata Kasoos

Family: Convolvulaceae

Distribution: Parasitic climber over shrubs and trees throughout India

Parts and properties used: Stem extract, fruit- Alterative, anthelmintic, and carminative. Seed- Used in epilepsy. Have Amar Belin and Kaempferol.

Associated system(s): ASU, F (Bhil)

***Cyamopsis tetragonoloba* (Linn.) Taub.**

Sanskrit or Ayurvedic name(s): Kshudra Shimbi

Common name(s): Guar

Family: Leguminosae/ Fabaceae

Distribution: African. Grown in arid and semi-arid India.

Parts and properties used: Seed- Laxative, ant bilious. Gum- Hypoglycaemic, hypolipidemic, appetite depressor.

Associated system(s): ASU

***Cyathula prostrata* (L.) Blume**

Common name(s): Lal chirchita, Bonkhoth, Bun-shath

Family: Amaranthaceae

Distribution: Habitat to eastern and Deccan India.

Parts and properties used: Root- Rheumatic fever, dysentery, wounds, and eye trouble.

Associated system(s): AF

***Cycas circinalis* L. (CR)**

Common name(s): Madhana kama poo

Family: Cycadaceae

Distribution: Habitat to south India.

Parts and properties used: Bark, seeds- Poultice applied on sores and swellings.

Associated system(s): AFS

***Cyclea peltata* (Lam.) Hook. f. & Thomson, syn. *C. arnotii* Miers.**

Sanskrit or Ayurvedic name(s): Raaj-Paathaa

Common name(s): Paada kizhangu

Family: Menispermaceae

Distribution: Found all over east and south India.

Parts and properties used: Root- Used in smallpox, bone fractures, malarial fever, jaundice.

Associated system(s): AS

***Cydonia oblonga* Mill., syn. *C. vulgaris* Pers.; *Pyrus Cydonia* L.**

Sanskrit or Ayurvedic name(s): Amritaphala

Common name(s): Shimaimathla, Bee dana, Quince

Family: Rosaceae

Distribution: Native of Western Asia. Grown in Kashmir, Punjab & Nilgiris Hills.

Parts and properties used: Fruit pulp and seeds- Soothing and demulcent. Used in irritable bowel.

Associated system(s): AUS

***Cymbopogon citratus* (DC.) Stapf.**

Sanskrit or Ayurvedic name(s): Kattrna, Bhuutika,

Common name(s): Lemongrass Serai, Rohisha

Family: Poaceae/ Gramineae

Distribution: Native Southeast Asian islands. Grown in Punjab, Maharashtra, Gujarat, Karnataka.

Parts and properties used: Leaf- Stimulant, sudorific, antiperiodic, ant catarrhal. Essential oil- Carminative, anti-cholerine, depressant, analgesic, antipyretic, antimicrobial.

Associated system(s): A

***Cymbopogon flexuosus* (Nees ex Steud) Wats. syn. *C. travancorensis* Bor.**

Common name(s): Lemongrass

Family: Poaceae/ Gramineae

Distribution: Northeast India and southern Western Ghats. Cultivated

Parts and properties used: Arial parts & essential oil- Used in food flavouring, perfumery, treat headache, abdominal and muscle pains, stress, anxiety, irritability, insomnia, drowsiness, hair loss.

Associated system(s): F

***Cymbopogon jwarancusa* (Jones) Schult.; syn. *C. jwarancusa* (Jones) Schult. var. *assamensis* B.K. Gupta**

Sanskrit or Ayurvedic name(s): Bhuutikaa

Common name(s): Izkhar

Family: Poaceae/ Gramineae

Distribution: Habitat to northwest plains and Himalayas up to 300 m.

Parts and properties used: Grass- Blood purifier, bechic, anticholera, emmenagogue, febrifuge, antirheumatic. Essential oil- Antimicrobial.

Associated system(s): ASUL

***Cymbopogon martinii* (Roxb.) Wats.**

Sanskrit or Ayurvedic name(s): Rohisha-trn

Common name(s): Rosha grass, Lemon grass, Palm Rosa

Family: Poaceae/ Gramineae

Distribution: Native peninsular India. Grown in drier part of south India.

Parts and properties used: Leaf- Aromatic, stomachic, carminative. Essential oil- Stimulant, carminative, antimicrobial. Used in stiff joints, lumbago, skin diseases, baldness.

Associated system(s): ASL

***Cymbopogon nardus* (L.) Rendle, syn. *Andropogon nardus* L. ssp. *grandis* (Nees ex Steud.) Hack.; *A. nardus* var. *luridus* Hook.f.**

Sanskrit or Ayurvedic name(s): Jambir-tern

Common name(s): Citronella Grass, Ganjni

Family: Poaceae/ Gramineae

Distribution: Occur in Assam to Punjab, south India. Cultivated

Parts and properties used: Leaf- Stomachic, carminative, spasmolytic, mild astringent. Essential oil- Stimulant, carminative, diaphoretic, rubefacient, antiseptic, larvicidal antibacterial, antifungal.

Associated system(s): AFS

***Cymbopogon schoenanthus* Spreng.**

Sanskrit or Ayurvedic name(s): Rohisha

Common name(s): Rusaa Ghaas

Family: Poaceae/ Gramineae

Distribution: Habitat to warmer India.

Parts and properties used: Roots and rhizome- Carminative, diaphoretic, emmenagogue. Used in cold, genitourinary affections.

Associated system(s): ASU

Cymbopogon winterianus **Jowit**

Common name(s): Citronella

Family: Poaceae/ Gramineae

Distribution: Native to Southeast Asian islands. Grown northeast India.

Parts and properties used: Plant- Insect pellent. Citronella oil- Rubefacient, antispasmodic, diaphoretic, sedating, analgesic.

Associated system(s): F

Cynodon dactylon **(L.) Pers.**

Sanskrit or Ayurvedic name(s): Duurvaa

Common name(s): Durva, Dhobghas, Karuka

Family: Poaceae/ Gramineae

Distribution: Native, grown all over India.

Parts and properties used: Root- Antirheumatic, dried used in menorrhagia, metrorrhagia and burning micturition. Plant/juice- Diuretic, treat epistaxis, haematuria, inflamed tumours, wounds, skin diseases.

Associated system(s): AFHSLU

Cyperus articulatus **Linn.**

Common name(s): Kronchaadana

Family: Cyperaceae

Distribution: Native to Turkey; found in water bodies and warm regions of Bengal to Sri Lanka.

Parts and properties used: Herb- Carminative, antiemetic (useful in vomiting during pregnancy), sedative (in dyspeptic disorders).

Associated system(s): A

Cyperus esculentus **L.**

Sanskrit or Ayurvedic name(s): Chichoda

Common name(s): Musta

Family: Cyperaceae

Distribution: Occur from Punjab to Nilgiris.

Parts and properties used: Tuber- Yields a digestive tonic.

Associated system(s): AF

Cyperus rotundus **L.**

Sanskrit, Ayurvedic and Common name(s): Musta, Nagarmotha, Mustha

Family: Cyperaceae

Distribution: Found all over India.

Parts and properties used: Plant- Carminative, astringent, ant-inflammatory, hepatoprotective, diuretic nervine tonic. Treat intestinal problems. Rhizome- Used in rheumatism, inflammations, dysuria, puerperal diseases, and obesity.

Associated system(s): ASTU

Cyperus scariosus **R.Br.**

Sanskrit or Ayurvedic name(s): Bhadramustaa

Common name(s): Nagarmotha

Family: Cyperaceae

Distribution: Habitat to damp places throughout India.

Parts and properties used: Rhizome- Stomachic, cordial, antidiarrheal and diuretic. Essential oil- Hypotensive, anti-inflammatory, CNS stimulant, antimicrobial.

Associated system(s): AFSU

Dactyloctenium aegyptium **Beauv**

Sanskrit or Ayurvedic name(s): Takraa

Common name(s): Makaraa

Family: Poaceae/ Gramineae

Distribution: Found throughout plains of India.

Parts and properties used: Grass- Astringent, anthelmintic. Yields a bitter tonic.

Associated system(s): AF

Dactylorhiza hatagirea **(D. Don.) Soo. (Orchid) (Declining population; CR)**

Common name(s): Ambolakpa, Spotted heart orchid, Salam panja

Family: Orchidaceae

Distribution: Occur in alpine Himalayas, Ladakh and HP.

Parts and properties used: Root- Energetic nervine tonic. Herb- Used in dysentery, diarrhoea, fever, cough, wounds, fractures, stomach-ache, etc.

Associated system(s): AFSTU

***Dalbergia lanceolaria* L.**

Common name(s): Bithua

Family: Leguminosae/ Fabaceae

Distribution: Found throughout India.

Parts and properties used: Bark decoction- Used in dyspepsia. Oil- Antirheumatic. Leaf- Used in skin diseases.

Associated system(s): SF

***Dalbergia latifolia* Roxb**

Sanskrit or Ayurvedic name(s): Shimshapaa

Common name(s): Kala Sheesham

Family: Leguminosae/ Fabaceae

Distribution: Found in Bihar, Bengal, MP, western peninsula.

Parts and properties used: Arial part- Stimulant, appetiser, anthelmintic, spasmogenic. Used in dyspepsia, diarrhoea. Bark- Contains Hentriacontane.

Associated system(s): AFSU

***Dalbergia sissoides* Grah.**

Sanskrit or Ayurvedic name(s): Kushimshapaa

Common name(s): Sisam

Family: Leguminosae/ Fabaceae

Distribution: Found throughout south India.

Parts and properties used: Arial part- Anti-inflammatory. Root- Contains Isoflavones.

Associated system(s): ASF

***Dalbergia sissoo* DC.**

Sanskrit or Ayurvedic name(s): Shimshapaa

Common name(s): Sheesham

Family: Leguminosae/ Fabaceae

Distribution: Habitat to sub-Himalayan tract and plains. Cultivated.

Parts and properties used: Leaves- Bitter, stimulant. Wood- Anthelmintic, alterative, emetic, stomachic, antileprotic; heartwood in turbidity of the urine, calculus and lipuria. Bark- Anti-cholerine. Root- Astringent.

Associated system(s): ASLU

***Dalbergia volubilis* Roxb.**

Sanskrit or Ayurvedic name(s): Gorakhi

Common name(s): Bankharaa

Family: Leguminosae/ Fabaceae

Distribution: Occur in Himalayas, UP, and Orissa.

Parts and properties used: Leaves- Used in aphthae. Root- Used in Genitourinary tract disinfectant and scalding of urine.

Associated system(s): ASF

***Datura arborea* L., syn. *Brugmansia arborea* (L.) Steud.**

Common name(s): Tree datura, Umathi

Family: Solanaceae

Distribution: Native of America. Grown in Western Ghats.

Parts and properties used: Leaves- Anti-inflammatory, analgesic, decongestant. Used in asthma, rheumatic pain.

Associated system(s): H

***Datura innoxia* Mill.**

Sanskrit or Ayurvedic name(s): Dhuttura Kanaka or Kanakahvya

Common name(s): Datura

Family: Solanaceae

Distribution: Habitat to Western Himalayas and Maharashtra.

Parts and properties used: Plant- Source of scopolamine, used as pre-anaesthetic.

Associated system(s): AU

***Datura metel* L, syn. *D. alba* Nees, *D. fastuosa* L. and *D. meteloides* DC. ex Dunal, *D. innoxia,* Mill.**

Sanskrit or Ayurvedic name(s): Dhattuura

Common name(s): Datura

Family: Solanaceae

Distribution: American, naturalized throughout India. Cultivated.

Parts and properties used: Leaf- Antitumor, antirheumatic. Plant parts- Used in hydrophobia, epilepsy, convulsion, syphilis, smallpox, mumps, and leprosy. Plants- Used in dysuria and alopecia.

Associated system(s): ASU

***Datura stramonium* L., syn. *D. tatula* L.**

Sanskrit or Ayurvedic name(s): Krishnadhattuura

Common name(s): Datura, Umathi

Family: Solanaceae

Distribution: Invasive temperate weed of the world. Himalaya up to 2700 m, and hills of India.

Parts and properties used: Plant parts- Spasmolytic, anti-asthmatic, anticholinergic, cerebral depressant, nerve-sedative. Used in asthma, cough, tuberculosis, and bronchitis.

Associated system(s): AHU

***Daucus carota* L. var. *sativa* DC.**

Sanskrit or Ayurvedic name(s): Gaajara

Common name(s): Gaajar Beej

Family: Apiaceae/ Umbelliferae

Distribution: Native to Europe and southwest Asia. Cultivated in north India.

Parts and properties used: Root- Used in palpitation, burning micturition cough/ bronchitis. Juice- Source: of carotene. Seed- Diuretic, emmenagogue, spasmolytic.

Associated system(s): AFSU

***Decalepis hamiltonii* Wight & Arn. (EN)**

Sanskrit name: Shelupatri sariva

Common name(s): Magali

Family: Periplocaceae

Distribution: Habitat to Deccan Peninsula.

Parts and properties used: Root- Appetizer, blood purifier, bacteriostatic. Yields a tonic.

Associated system(s): SU

***Delima scandens* Burkill.**

Sanskrit or Ayurvedic name(s): Paaniya Valli

Family: Dilleniaceae.

Distribution: Occur in Assam, Bengal and Andamans forests.

Parts and properties used: Plant decoction- Used in dysentery, cough. Leaf- Treat boils. Root- Astringent. Used in burns.

Associated system(s): A

Delonix elata **(L.) Gamble =** ***D. regia*** **(Hook.) Raf.**

Common name(s): Sanchal, Gulmohar

Family: Leguminosae/ Fabaceae (Caesalpiniaceae)

Distribution: East African. Occur in Maharashtra. Grown as *D. regia.*

Parts and properties used: Bark- Antiperiodic, febrifuge. Leaf/bark extract- Anti-inflammatory.

Associated system(s): AFS

Delphinium brunonianum **Royle.**

Sanskrit or Ayurvedic name(s): Sprikkaa

Common name(s): Himalayan larkspur, Lundekaown

Family: Ranunculaceae

Distribution: Found wild in western Himalayas.

Parts and properties used: Plant- Cardiac and respiratory depressant. Insecticidal.

Associated system(s): A

Delphinium cashmerianum **Royle**

Sanskrit or Ayurvedic name(s): Tagara (?)

Common name(s): Himalayan larkspur, Lundekaown

Family: Ranunculaceae

Distribution: Found wild in western Himalayas.

Parts and properties used: Plant- Used as substitute of Tagara (*Valeriana jatamansi*) and in insecticidal pains.

Associated system(s): A

Delphinium denudatum **Wall. ex Hook. f. & Thomson (CR)**

Sanskrit or Ayurvedic name(s): Nirvishaa

Common name(s): Jadavar kath, Nirbishi

Family: Ranunculaceae

Distribution: Habitat to west Himalayas and northwest India.

Parts and properties used: Root- Astringent, vulnerary, deobstruent, alterative. Treat piles.

Associated system(s): AU

***Delphinium elatum* auct. non. Linn., syn. *D. speciosum* Janka ex Nym.; *D. vestitium* Wall. ex Royle.**

Sanskrit or Ayurvedic name(s): Nirvisha

Common name(s): Delphinium

Family: Ranunculaceae

Distribution: Habitat to Temperate Himalayas.

Parts and properties used: Plant- Cardiac and respiratory depressant, emetic, diuretic, anthelmintic. Seeds- Insecticidal

Associated system(s): A

***Dendrobium ovatum* (Willd.) Kranzl. (LC)**

Sanskrit or Ayurvedic name(s): Jivanti

Common name(s): Nagli

Family: Orchidaceae

Distribution: Habitat to Western Ghats.

Parts and properties used: Plant juice- Stomachic, carminative, antispasmodic, laxative. Yields liver tonic.

Associated system(s): AF

***Dendrophthoe falcata* (L.F) Blume**

Sanskrit or Ayurvedic name(s): Vanda

Common name(s): Bandaka

Family: Loranthaceae

Distribution: Found throughout India.

Parts and properties used: Bark- Astringent, narcotic. Used in menstrual problem.

Associated system(s): AFS

***Derris indica* (Lamk.) Bennet.**

Sanskrit or Ayurvedic name(s): Naktmaal

Common name(s): Pangu

Family: Leguminosae/ Fabaceae

Distribution: Habitat to Western Ghats.

Parts and properties used: Plant- Used for skin diseases ulcers, tumours, piles. Root juice- Sores and ulcers. Flowers- Used in diabetes.

Associated system(s): AS

Desmodium gangeticum **(L.) DC.**

Sanskrit or Ayurvedic name(s): Shaaliparni

Common name(s): Salparni, Prasniparni

Family: Leguminosae/ Fabaceae

Distribution: Habitat to Himalaya up to 1500 m and Peninsular India.

Parts and properties used: Root- Antipyretic, diuretic, astringent, ant catarrhal, laxative anthelmintic, nervine tonic. Yields an anti-inflammatory Gangetin. Leaves- Galactagogue. Bark- Cure diarrhoea haemorrhage.

Associated system(s): AFHSU (Part of Ayurvedic *Dashamoola*)

Desmodium triflorum **(L.) DC.**

Sanskrit or Ayurvedic name(s): Tripaadi

Common name(s): Hamsapapdi

Family: Leguminosae/ Fabaceae

Distribution: Found throughout India.

Parts and properties used: Leaves- Used as galactagogue and for diarrhoea. Applied on wounds. Root- Diuretic. Used to treat cough, asthma.

Associated system(s): AFS

Desmostachya bipinnata **(L.) Stapf.**

Sanskrit or Ayurvedic name(s): Kusha

Common name(s): Dharbha

Family: Poaceae/ Gramineae

Distribution: Occur throughout plains of India.

Parts and properties used: Root- Diuretic, galactagogue, astringent. Used in urinary calculi. Clumps- Used in dysentery, diarrhoea, and skin diseases.

Associated system(s): AFSU

Dianthus deltoides **L.**

Common name(s): Khenchpa

Family: Caryophyllaceae

Distribution: Native to Europe and west Asia. Grown in the Himalayas.

Parts and properties used: Whole plant- Used in stomach disorder.

Associated system(s): F

Dichrostachys cinerea **(L.) Wight & Arn.**

Sanskrit or Ayurvedic name(s): Virataru

Common name(s): Virtuli

Family: Leguminosae/ Fabaceae (Mimosaceae)

Distribution: Occur in most parts of India.

Parts and properties used: Root- Astringent, diuretic; used in renal affections, urinary calculi. Foliage- Contain tannin.

Associated system(s): AFS

Didymocarpus pedicellata **R. Br**

Sanskrit or Ayurvedic name(s): Kshudra-Paashaanabheda

Common name(s): Shilapushpi, Pathar phori, Pasanphodi

Family: Gesneriaceae

Distribution: Habitat to sub-tropical Himalaya

Parts and properties used: Leaf- Antilithic. Used in renal diseases, kidney stones, and bladder.

Associated system(s): A

Digera muricata **(Linn.) Mart.**

Sanskrit or Ayurvedic name(s): Katthinjara

Common name(s): Lahsuvaa

Family: Amaranthaceae

Distribution: Occur as weed throughout the plains

Parts and properties used: Plant- Astringent, ant bilious. Contain α - and β spina sterol Flower/seed- Diuretic.

Associated system(s): ASF

Digitalis purpurea **L.**

Sanskrit or Ayurvedic name(s): Teelpushpee, Hartpatri

Common name(s): Foxglove, Tilpushpi

Family: Plantaginaceae

Distribution: Native of Europe. Grown in hill gardens.

Parts and properties used: Plant- Has stimulatory effect upon the heart. Source of digoxin.

Associated system(s): A, English

Dillenia indica **Linn and *D. pentagyna* Roxb.**

Sanskrit or Ayurvedic name(s): Bhavya

Common name(s): Nagkesaram, Elephant Apple

Family: Dilleniaceae

Distribution: Found in Himalayas, Bengal, Bihar, Odisha and MP. Grown in plains.

Parts and properties used: Fruit- Laxative, carminative, bechic, febrifuge, antispasmodic. Bark/leaves- Astringent.

Associated system(s): AFS

Dioscorea alata **L.**

Sanskrit or Ayurvedic name(s): Kaashthaaluka

Common name(s): Sewalli kodi, Wild Yam

Family: Dioscoreaceae

Distribution: Found in Asian tropics. Grown in Assam, Bengal MP, TN.

Parts and properties used: Tuber- Cholagogue, diuretic antispasmodic, anti-inflammatory, antirheumatic.

Associated system(s): AFS

Dioscorea bulbifera **L.**

Sanskrit or Ayurvedic name(s): Vaaraahi

Common name(s): Varahi kand

Family: Dioscoreaceae

Distribution: Native to tropical India. Cultivated for bulbils.

Parts and properties used: Dried pounded tuber- Applied on swellings, boils, and ulcers. Increases sexual vigour.

Associated system(s): AFSTU

Dioscorea deltoidea **Wall. *ex* Griseb. (CR), syn. *D. deltoidea* var. *orbiculata* Prain & Burkil**

Sanskrit or Ayurvedic name(s): Vaaraahikanda

Common name(s): Shingli mingli

Family: Dioscoreaceae

Distribution: West Himalayan, being CR, brought into cultivation.

Parts and properties used: Tuber- Antipatharia. Leaf- Febrifuge. Source of Diosgenin, saponins, commercially exploited/traded.

Associated system(s): AF

***Dioscorea esculenta* Burkill., syn. *D. aculeata* Linn. *D. faciculata* Roxb., *D. spinosa* Roxb ex Wall**

Sanskrit or Ayurvedic name(s): Madhvaaluka

Common name(s): Suthani

Family: Dioscoreaceae

Distribution: Found in UP, MP, Orissa, Bengal, Assam and the Andamans.

Parts and properties used: Tubers- Starchy and free from dioscorine, contain carbohydrates albuminoids.

Associated system(s): AFS

***Dioscorea oppositifolia* L.**

Sanskrit or Ayurvedic name(s): Amlikaakanda

Common name(s): Sarpahya

Family: Dioscoreaceae

Distribution: Found south Indian hills.

Parts and properties used: Tuber- Reduce swelling.

Associated system(s): AFS

***Dioscorea pentaphylla* L.**

Sanskrit or Ayurvedic name(s): Vaaraahikanda

Common name(s): Kanta Alu

Family: Dioscoreaceae

Distribution: Found all over India.

Parts and properties used: Tuber- Reduce swelling.

Associated system(s): AF

***Diospyros ebenum* Koenig**

Sanskrit or Ayurvedic name(s): Tinduka

Common name(s): Abnus

Family: Ebenaceae

Distribution: Found in Orissa and south India.

Parts and properties used: Plant- Astringent, attenuate, lithotripsic. Heart wood- Have 2 β -naphthalhydes, 2 naphthoic acid.

Associated system(s): ASU

***Diospyros embryopteris* Pers.**

Common name(s): Tinduka variety

Family: Ebenaceae

Distribution: Found in shade throughout India.

Parts and properties used: Fruit/ bark- Astringent. Fruit infusion- Used for gargle in aphthae/ sore throat. Fruit juice- Treat wounds/ ulcers. Seed oil- Used in diarrhoea/ dysentery.

Associated system(s): ASU

***Diospyros kaki* Linn. f.**

Common name(s): Tinduka variety

Family: Ebenaceae

Distribution: Native to China. Grown in HP to Kumaon, Bengal, and Nilgiris for edible fruits.

Parts and properties used: Plant- Hypotensive, hepatoprotective, antidote to poisons and bacterial toxins. Calyx/peduncle of fruit- Used in treatment of cough and dyspnoea. Roasted seeds- Coffee substitute.

Associated system(s): A

***Diospyros lotus* L.**

Common name(s): Amlok

Family: Ebenaceae

Distribution: Native to subtropical southwest Asia. Grown in the sub-Himalayas.

Parts and properties used: Fruit- Astringent, antidiabetic, antitumor, astringent, laxative, antipyretic, used in constipation.

Associated system(s): F

***Diospyros melanoxylon* Roxb.**

Common name(s): Tinduka variety, Tendu

Family: Ebenaceae

Distribution: Habitat to north, central, and peninsular India.

Parts and properties used: Leaf- Carminative, laxative, diuretic, styptic. Bark- Astringent. Used in dyspepsia and diarrhoea.

Associated system(s): AS

Diospyros montana **Roxb. var.** ***cordifolia*** **Hiem. (EN), syn.** ***D. cordifolia*** **Roxb.**

Sanskrit or Ayurvedic name(s): Visha-tinduka,

Family: Ebenaceae

Distribution: Occur throughout greater India.

Parts and properties used: Bark extract- Anti-inflammatory, antipyretic, and analgesic. Leaves and seeds- Antibacterial. Used in fever, pneumonia, etc.

Associated system(s): AFS

Diospyros tomentosa **Roxb.**

Sanskrit or Ayurvedic name(s): Viralaa

Common name(s): Tumbi.

Family: Ebenaceae

Distribution: Habitat to sub-Himalaya and parts of India.

Parts and properties used: Plant- Astringent, anti-inflammatory, styptic. Used in treatment cough.

Associated system(s): AFS

Diplocyclos palmatus **(L.) C. Jeffrey**

Sanskrit, Ayurvedic and Common name(s): Shivlingi, Shivling beej

Family: Cucurbitaceae

Distribution: Habitat to Western Ghats.

Parts and properties used: Plant- Aphrodisiac, gynaecological, antiasthmatic, anticonvulsant, anti-venom, anti-inflammatory. Used in constipation, diarrhoea, fever. Fruit- Laxative.

Associated system(s): AFS

Dipterocarpus turbinatus **Gaertn. f., syn.** ***D. indicus*** **Bedd. (EN) and** ***D. alatus*** **Roxb.**

Sanskrit or Ayurvedic name(s): Ashwakarn, Garjana, Jaranadruma

Common name(s): Gurjan

Family: Dipterocarpaceae.

Distribution: Habitat to southern Western Ghats, Andamans and Assam.

Parts and properties used: Oleoresin, Gurjan oil- Stimulant to genitourinary system, diuretic, spasmolytic. Used in ulcers. Bark decoction- Treat rheumatism.

Associated system(s): AFS

***Dodonaea viscosa* (L.) Jacq.**

Sanskrit or Ayurvedic name(s): Raasnaa

Common name(s): Pulivailu, Bandaru

Family: Sapindaceae

Distribution: Habitat to northwest Himalayas, south Indian hills.

Parts and properties used: Leaves- Anti-inflammatory, antibacterial, febrifuge. Applied in waist pain, and gout.

Associated system(s): AS

***Dolichandrone falcata* (Wall ex DC. Seem**

Sanskrit or Ayurvedic name(s): Mesha-shringi, Vishaanikaa

Common name(s): Medsinghi

Family: Bignoniaceae

Distribution: Habitat to moist forest of central and south India.

Parts and properties used: Fruit- Carminative, anti-diabetic, urinary, skin disorders, bronchitis. Leaves- Applied on swollen glands.

Associated system(s): AS

***Dolichos biflorus* L. syn. *Macrotyloma uniflorum (*Lam.) Verdc.; *Vigna unguiculata* (L.) Walp.**

Sanskrit or Ayurvedic name(s): Kulattha

Common name(s): Kulthi, Muthira

Family: Leguminosae/ Fabaceae

Distribution: Occur in most parts of India. Cultivated.

Parts and properties used: Seed- Astringent, diuretic, treat asthma, tonic. Dry seed decoction- Given in calculus and amenorrhoea.

Associated system(s): AFSTU

***Dolichos falcatus* Seem Klein.**

Sanskrit or Ayurvedic name(s): Kulatthikaa.

Common name(s): Wild butter bean

Family: Leguminosae/ Fabaceae

Distribution: Habitat to Himalayas, western peninsula.

Parts and properties used: Root- Constipation and skin diseases. Seed decoction- Treat rheumatism.

Associated system(s): A

***Dorema ammoniacum* D. Don**

Sanskrit or Ayurvedic name(s): Uushaka

Common name(s): Ushaq

Family: Apiaceae/ Umbelliferae

Distribution: Habitat to Himalayas.

Parts and properties used: Gum- Antispasmodic, expectorant, diaphoretic, emmenagogue, used in cough, asthma, bronchitis.

Associated system(s): AHU

***Dracocephalum heterophyllum* Benth.**

Common name(s): Zinkzer

Family: Lamiaceae/ Labiatae

Distribution: Habitat to alpine Himalayas, Ladakh.

Parts and properties used: Decoction of dried flowers and leaves- Given in cold and cough.

Associated system(s): Amchi

***Dracocephalum moldavica* L**

Sanskrit or Ayurvedic name(s): Ram Tulasi

Common name(s): Zinkzer

Family: Lamiaceae/ Labiatae

Distribution: Habitat to temperate west Himalayas.

Parts and properties used: Seeds- Astringent, carminative, tonic. Used in cephalalgia, neurological disorder.

Associated system(s): AU

***Dregea volubilis* Benth**

Sanskrit or Ayurvedic name(s): Suparnikaa

Common name(s): Nakchikini

Family: Asclepiadaceae

Distribution: Occur in Assam, Bengal, Western Ghats.

Parts and properties used: Root and stalk- Emetic and expectorant, used in colds, sinusitis, and biliousness.

Associated system(s): AU

Drimia indica **(Roxb.) Jessop**

Common name(s): Jangli piaz

Family: Asparagaceae

Distribution: Habitat to western Himalayas, Western Ghats.

Parts and properties used: Bulb- Used in cardiac diseases, indigestion, dropsy, asthma, rheumatism, skin diseases, leprosy.

Associated system(s): A

Drosera peltata **Smith ex Willd. (VU)**

Sanskrit or Ayurvedic name(s): Brahma-suvarchalaa

Common name(s): Drosera

Family: Droseraceae

Distribution: Found throughout India.

Parts and properties used: Plant resin- Used in bronchitis and whooping cough. Plant- Anti-syphilitic. Applied on infectious diseases.

Associated system(s): FS

Dryobalanops aromatica **Gaertn., syn.** ***D. camphora*** **Colebr**

Sanskrit or Ayurvedic name(s): Bhimseni kapoor

Common name(s): Baraas

Family: Dipterocarpaceae

Distribution: Indo-Malayan, occur in Andaman.

Parts and properties used: Exudate- Substitute of camphor.

Associated system(s): A

Dryopteris filix-mas **(L.) Schott**

Common name(s): Shield fern, Hirvi

Family: Dryopteridaceae

Distribution: Habitat to temperate India.

Parts and properties used: Rhizome- Taeniafuge, vermifuge. Contain filicinin- Kill tapeworms.

Associated system(s): HSU

Drypetes roxburghii **(Wall.) Hurus., syn.** ***Putranjiva roxburghii*** **Wall.**

Sanskrit or Ayurvedic name(s): Apatyamjivah, Putrajivaka

Common name(s): Putrjivak

Family: Euphorbiaceae

Distribution: Found in Maharashtra, Karnataka, TN. Grown in plains.

Parts and properties used: Leaves, fruit/ stone- Given in colds and fevers, rheumatic affections.

Associated system(s): AS

***Dysoxylum malabaricum* Bedd. ex Hiern. (EN)**

Common name(s): Vellakil

Family: Meliaceae

Distribution: Habitat to southern Western Ghats.

Parts and properties used: Seed oil- Given in rheumatism and eye and ear diseases.

Associated system(s): AFS

***Ecbolium viride* (Forssk.) Alston var. *viride*, syn. *E. linneanum* Kurz**

Common name(s): Udajati, Sahacarah, Nilambari

Family: Acanthaceae

Distribution: Habitat to north-eastern peninsular India.

Parts and properties used: Plant- Used in gout and dysuria. Leaf decoction- Stricture. Roots- Used to treat jaundice, and rheumatism.

Associated system(s): ASF

***Echinops cornigerus* (L.) DC**

Common name(s): Globe thistle, Aczema

Family: Asteraceae/ Compositae

Distribution: Found in cold arid zone of Ladakh and beyond.

Parts and properties used: Dried aerial parts- Septic. Used in cure wounds.

Associated system(s): AF

***Echinops echinatus* Roxb.**

Sanskrit, Ayurvedic and Common name(s): Utkantaka, Utkanta

Family: Asteraceae/ Compositae

Distribution: Found all over India.

Parts and properties used: Plant- Alterative, diuretic, yields nerve tonic.

Associated system(s): ASU, F (Bhil)

***Eclipta prostrata* (L.) L., syn. *E. alba* Hassk.]**

Sanskrit, Ayurvedic and Common name(s): Bhrangaraja, Bhringaraj

Family: Asteraceae/ Compositae

Distribution: Found throughout India.

Parts and properties used: Plant- Deobstruent, antihepatotoxic, ant catarrhal, febrifuge, hepatoprotective. Used in hepatitis, spleen enlargements.

Associated system(s): AFS

***Ehretia laevis* Roxb. var. *aspera* (Willd.) C.B. Clarke.**

Sanskrit or Ayurvedic name(s): Charmi-vrksha

Common name(s): Addula

Family: Ehretiaceae

Distribution: Found throughout India.

Parts and properties used: Root- Used in venereal diseases.

Associated system(s): AFS

***Elaeocarpus sphaericus* (Gaertn.) K. Schum., syn. *E. ganitrus* Roxb.**

Sanskrit, Ayurvedic and Common name(s): Rudraaksa, Rudraksha

Family: Elaeocarpaceae

Distribution: Found in most of India. Occasionally cultivated.

Parts and properties used: Fruit- Thermogenic, appetizer, and sedative, used in cough, bronchitis, neuralgia, cephalalgia, anorexia, epileptic fits. Seed- Used in hypertension, insomnia, psychoneurosis, and mental diseases.

Associated system(s): AS

***Elaeocarpus serratus* Linn. (LC)**

Sanskrit or Ayurvedic name(s): Rudraaksha

Common name(s): Uttraccham

Family: Elaeocarpaceae

Distribution: Habitat to eastern Himalayas and Western Ghats

Parts and properties used: Leaf- antirheumatic. Fruit- Antidysentery. Aerial Parts- CVS and CNS active.

Associated system(s): AS

***Elaeodendron glaucum* Pers.**

Sanskrit or Ayurvedic name(s): Krishnamokshaka.

Common name(s): Jamrasi

Family: Celastraceae

Distribution: Grown overall India as ornamental.

Parts and properties used: Bark and leaves- Astringent, anti-inflammatory, emetic. Contain tannin.

Associated system(s): ASF

***Elephantopus scaber* L.**

Sanskrit or Ayurvedic name(s): Mayura-shikhaa

Common name(s): Gjihiva

Family: Asteraceae/ Compositae

Distribution: Habitat to warmer parts of India.

Parts and properties used: Plant- Astringent, cardiac tonic, diuretic, mucilaginous, emollient. Leaves- Applied on ulcer/ eczema.

Associated system(s): AFS

***Elettaria cardamomum* Maton**

Sanskrit or Ayurvedic name(s): Suksmailaa

Common name(s): Elachi Chhoti, Ilaychi

Family: Zingiberaceae

Distribution: Native of Western Ghats. Cultivated.

Parts and properties used: Fruit (seed)- Carminative antiemetic, stomachic, orexigenic, anti-ripe, anti-asthmatic. Used in dyspepsia, as cholagogue. Oil- Antispasmodic, antiseptic.

Associated system(s): ASU

***Eleusine indica* Gaertin.**

Sanskrit or Ayurvedic name(s): Nandimukha

Common name(s): Ragi

Family: Poaceae/ Gramineae

Distribution: Habitat to plains, low hills, and pasture grounds of India.

Parts and properties used: Plant- Used in biliary disorders. Decoction- Stomachic, diuretic, febrifuge, and in sprains.

Associated system(s): AF

***Elsholtzia cristata* Willd.**

Common name(s): Ban-Tulasi

Family: Lamiaceae/ Labiatae

Distribution: Habitat to Himalayas up to 3000 m.

Parts and properties used: Plant- Carminative, stomachic, astringent. Leaf - Diuretic, antipyretic.

Associated system(s): AF

***Embelia ribes* Burm.f. (Threatened - CR)**

Sanskrit or Ayurvedic name(s): Vidanga

Common name(s): Vaividang, Vidanga

Family: Myrsinaceae

Distribution: Found all over India.

Parts and properties used: Plant, Fruit- Acaricidal, astringent, anthelmintic, carminative, diuretic, anti-inflammatory, antibacterial, febrifuge. Used in chest and skin diseases. Embelin isolated from berrics.

Associated system(s): AFHSL

***Embelia tsjeriam-cottam* (Roem. & Schult.) A. DC., syn. *E. basaal* sensu (Roem. & Schult.) A. DC., non- Mez; *E. robusta* CB Clarke, non-Roxb. (VU)**

Sanskrit or Ayurvedic name(s): Vidanga

Common name(s): Vaividang

Family: Myrsinaceae

Distribution: Found in greater part of India.

Parts and properties used: Fruit- Antispasmodic, carminative, anthelmintic, antibacterial. Powder used in dysentery (Substitute for *E. ribes*)

Associated system(s): AF

***Emblica officinalis* Gaertn., syn. *Phyllanthus emblica* L.**

Sanskrit or Ayurvedic name(s): Aamalaki

Common name(s): Amla, Anwala

Family: Euphorbiaceae

Distribution: Native to Indian subcontinent. Grown all over India.

Parts and properties used: Fruit- Antianemia, anabolic, antiemetic, bechic, astringent, antihemorrhagic, antidiarrheal, diuretic, antidiabetic, carminative. Used in jaundice, etc., as antacid.

Associated system(s): ASLU

***Emilia sonchifolia* (L.) DC.**

Sanskrit or Ayurvedic name(s): Shash-shruti

Common name(s): Muyalcheviyan

Family: Asteraceae/ Compositae

Distribution: Found all over India.

Parts and properties used: Plant- Sudorific, febrifuge antiseptic. Used in infantile tympanites and bowel problems. Root- Antidiarrheal.

Associated system(s): AFS

***Enicostemma axillare* (Lam.) Raynal, syn. *E. littorale* non. Blume; *E. hyssopifolium* (Willd.) Verd.**

Sanskrit or Ayurvedic name(s): Naagjhvaa

Common name(s): Mamejava

Family: Gentianaceae

Distribution: Found throughout India.

Parts and properties used: Plant leaves extract- Carminative, blood purifier, antirheumatic, anti-inflammatory, antipsychotic, anthelmintic, cardio stimulant (substitute for *Swertia chirayita*). Used in scabies, itches, headache.

Associated system(s): AFS

***Entada pursaetha* DC., syn. *E. rheedii* Spreng.; *E. scandens* Auct. non Benth.**

Sanskrit or Ayurvedic name(s): Gila

Common name(s): Yaanai Kazharchi Kaai

Family: Leguminosae/ Fabaceae (Mimosaceae)

Distribution: Habitat to eastern Himalayas, hills, and Western Ghats.

Parts and properties used: Seed- Carminative, anodyne, spasmolytic bechic, anti-inflammatory, anthelmintic, antiperiodic. Used in liver disorders.

Associated system(s): AFS

***Ephedra gerardiana* Wall. ex JA Mey, and *E. sinica* Stapf. (EN - CR)**

Sanskrit or Ayurvedic name(s): Soma

Common name(s): Asmania, Chhapat Somalatha

Family: Ephedraceae

Distribution: Habitat to alpine arid Himalayas (Ladakh), Northeast.

Parts and properties used: Plant- Circulatory stimulant. Used in asthma, low blood pressure (BP), hepatic; yield toxic Ephedrine.

Associated system(s): AH, Amchi

***Equisetum arvense* L.**

Sanskrit or Ayurvedic name(s): Ashwa Puchha

Common name(s): Bottle brush, Horse tail

Family: Equisetaceae

Distribution: Habitat to high altitude Himalayas.

Parts and properties used: Plant- Haemostatic, haemopoietic, astringent, diuretic. Used in genitourinary affections.

Associated system(s): A

***Erigeron canadensis* Linn.**

Sanskrit or Ayurvedic name(s): Jaraayupriya

Common name(s): Canadian Fleabane

Family: Asteraceae/ Compositae

Distribution: Habitat to west Himalayas, Gangetic plains, and Western Ghats.

Parts and properties used: Plant- Astringent, haemostatic, antirheumatic, diuretic. Used in diarrhoea, kidney disorder, bronchitis, bleeding piles. Essential oil- Used in bronchial catarrh, cystitis.

Associated system(s): A

***Eriobotrya japonica* Lindl.**

Sanskrit or Ayurvedic name(s): Lottaaka

Common name(s): Loquat

Family: Rosaceae

Distribution: Native to China. Cultivated in upper Gangetic plains.

Parts and properties used: Leaves- Diabetes, mellitus and skin diseases, Sedative, antiemetic; yield lipopolysaccharides. Fruit- Sedative, antiemetic. Flower- Expectorant.

Associated system(s): ASU

***Ermania lanuginosa* (Hook. f. & Thoms) Schulz**

Common name(s): Measlo

Family: Brassicaceae/ Cruciferae

Distribution: Habitat to alpine Himalayas, including Ladakh.

Parts and properties used: Whole plant- Diuretic, purgative.

Associated system(s): Tibetan medicinal system

***Eruca sativa* Mill.**

Sanskrit or Ayurvedic name(s): Tuvarikaa

Common name(s): Rocket-Salad

Family: Brassicaceae/ Cruciferae

Distribution: Mediterranean. Grown in north and central India.

Parts and properties used: Leaf- Stimulant, stomachic, antiscorbutic, diuretic, rubefacient. Seed- Vesicant, antibacterial.

Associated system(s): ASF

***Erycibe paniculata* Roxb.**

Common name(s): Ashoka-Rohini

Family: Convolvulaceae

Distribution: Found all over India.

Parts and properties used: Bark- Anticholera. Fruit- Treat constipation.

Associated system(s): AS

***Erythrina stricta* Roxb.**

Common name(s): Mullu-murukku

Family: Leguminosae/ Fabaceae

Distribution: Found northeast, Bengal, South India.

Parts and properties used: Bark- Ant bilious, antirheumatic, febrifuge, antiasthmatic, antiepileptic, antileprotic.

Associated system(s): AS

***Erythrina suberosa* Roxb.**

Common name(s): Pangra, Vellaimuruku

Family: Leguminosae/ Fabaceae

Distribution: Indo-Malayan, occur in Kerala too.

Parts and properties used: Bark ash- Used to treat blisters and wounds (veterinary).

Associated system(s): FS

***Erythrina variegata* L., syn. *E. indica* Lam.**

Sanskrit or Ayurvedic name(s): Paaribhadra

Common name(s): Murikkila

Family: Leguminosae/ Fabaceae

Distribution: Habitat to tropical and subtropical Africa. Grown as ornamental in India.

Parts and properties used: Leaf- Cathartic, diuretic, antiseptic, anti-inflammatory. Bark- Ant bilious, anthelmintic, febrifuge, astringent, expectorant.

Associated system(s): AFSL

Erythroxylum monogynum **Roxb., syn.** ***E. indicum*** **(DC.) Bedd.**

Sanskrit or Ayurvedic name(s): Kattuchandanam

Common name(s): Gandhgiri

Family: Erythroxylaceae

Distribution: Occur in south India, up to 1000m.

Parts and properties used: Leaf - Diaphoretic, stimulant, diuretic, stomachic. Decoction- Treat malarial fever. Bark and wood- Febrifuge.

Associated system(s): AFS

Eucalyptus citriodora **Hook.**

Common name(s): Eucalyptus

Family: Myrtaceae

Distribution: Australian. Introduced and grown in north India.

Parts and properties used: Leaf (oil)- Insect repellent, antiseptic, used in cold cough.

Associated system(s): A

Eucalyptus globulus **Labill**

Sanskrit or Ayurvedic name(s): Tailaparnah

Common name(s): Eucalyptus

Family: Myrtaceae

Distribution: Australian. Introduced and grown Kerala.

Parts and properties used: Leaves- Used as an expectorant in cough, febrifuge, astringent, antiseptic and vermifuge.

Associated system(s): AFHS

Eulaliopsis binata **(Retz.) C.E. Hubb.**

Sanskrit or Ayurvedic or common name: Balvaja

Family: Poaceae/ Gramineae

Distribution: Habitat to slops of north Indian hills.

Parts and properties used: Plant- Diuretic. Used for treating lithiasis.

Associated system(s): A

***Eulophia campestris* Wall. ex Lindl.**

Sanskrit or Ayurvedic name(s): Amrita

Common name(s): Salam mishri

Family: Orchidaceae

Distribution: Found in major part of India.

Parts and properties used: Tubers- Used in stomatitis, cough, cardiac and nervine tonic.

Associated system(s): A

***Eulophia herbacea* Lindl.**

Sanskrit, Ayurvedic and Common name(s): Munjaataka, Munjataka

Family: Orchidaceae

Distribution: Habitat to Himalayas and Deccan Peninsula.

Parts and properties used: Tubers- Substitute for *Salep* (powder of orchid nutritious tuber) and for variety of diseases.

Associated system(s): AF

***Eulophia nuda* Lindl.**

Sanskrit or Ayurvedic name(s): Baalakanda

Common name(s): Ambarkand

Family: Orchidaceae

Distribution: Habitat to tropical Himalayas, Deccan Plateau and Western Ghats.

Parts and properties used: Tubers- Anti-asthmatic/anti-bronchitis, appetizer and anticancer.

Associated system(s): AF

***Euonymus tingens* Wall.**

Common name(s): Bhillotaka

Family: Celastraceae

Distribution: Habitat to tropical Himalayas 2150-3200 m.

Parts and properties used: Plant/Bark- Cholagogue, laxative, diuretic, circulatory stimulant. Used in constipation, torpidity of liver, gall bladder disorders, jaundice, and dyspepsia.

Associated system(s): AF

***Eupatorium triplinerve* Vahl.**

Sanskrit or Ayurvedic name(s): Vishalyakarani

Common name(s): Ayapani

Family: Asteraceae/ Compositae

Distribution: Found in Assam, Maharashtra. Grown in gardens.

Parts and properties used: Plant- Cardiac stimulant, laxative, emetic, expectorant, bechic, antiscorbutic, alterative. Treat ague (a fever or shivering fit). Leaf- Anti-cholerine, haemostatic.

Associated system(s): AFS

***Euphorbia antiquorum* L.**

Sanskrit or Ayurvedic name(s): Vajra-kantaka

Common name(s): Kallippalaveru

Family: Euphorbiaceae

Distribution: Habitat to warm regions of India. Cultivated as hedge.

Parts and properties used: Latex- Purgative, applied on burns. Plant- Used in dropsy, anasarca, sores, syphilis. Root- Anthelmintic. Stem- Used skin sores and scabies.

Associated system(s): AFS

***Euphorbia dracunculoides* Lamk.**

Sanskrit or Ayurvedic name(s): Charmasaahvaa

Common name(s): Titali.

Family: Euphorbiaceae

Distribution: Found throughout India.

Parts and properties used: Fruit- Removes warts topically. Plant extract- Cholinergic.

Associated system(s): AFSU

***Euphorbia hirta* L.**

Sanskrit or Ayurvedic name(s): Dugdhikaa

Common name(s): Dudhi, Dudhika

Family: Euphorbiaceae

Distribution: Habitat to warmer parts of India.

Parts and properties used: Plant- Pectoral. anti-asthmatic/ anti-bronchitis. Latex-Vermifuge. Used to treat urinogenital duct diseases.

Associated system(s): AFSU

***Euphorbia hypericifolia* auct. non. Linn., syn. *E. indica* Lam.**

Common name(s): Dugdhikaa

Family: Euphorbiaceae

Distribution: Occur in Himalaya up to 1500m and allover warm regions of India.

Parts and properties used: Plant- Used in colic, diarrhoea, and dysentery. Leaf-Astringent, antidysentery, anti-leucorrhoea.

Associated system(s): A

***Euphorbia neriifolia* L.**

Sanskrit or Ayurvedic name(s): Samant-dugdhaa

Common name(s): Pattankisend, Nanda

Family: Euphorbiaceae

Distribution: Widespread in peninsular India. Grown as fence.

Parts and properties used: Latex- Purgative, diuretic, anti-asthmatic, expectorant, rubefacient. Treat ascites, polyuria.

Associated system(s): AFSLU

***Euphorbia nivulia* Buch. -Ham**

Sanskrit or Ayurvedic name(s): Snuhi

Common name(s): Sehunda

Family: Euphorbiaceae

Distribution: Found in Maharashtra, Kerala. Planted in northcentral India.

Parts and properties used: Latex- Treat jaundice, dropsy, enlarged liver and spleen; colic; syphilis, leprosy. Applied to haemorrhoids. Coagulated latex used for bronchitis.

Associated system(s): AFS

***Euphorbia thomsoniana* Boiss.**

Sanskrit or Ayurvedic name(s): Kanchanakshiri,

Common name(s): Hiravi

Family: Euphorbiaceae

Distribution: Habitat to Kashmir 2350 m.

Parts and properties used: Root- Purgative. Latex- Used in skin eruption and diseases.

Associated system(s): AF

Euphorbia thymifolia **L.**

Sanskrit or Ayurvedic name(s): Dugdhikaa

Common name(s): Choti-Dudhi

Family: Euphorbiaceae

Distribution: Habitat to tropical plains and low hills of India.

Parts and properties used: Plant- Antispasmodic, antiasthma bronchodilator, galactagogue. Latex- Purgative. Used in ringworm, dandruff.

Associated system(s): AF (Bhil)U

Euphorbia tirucalli **L. or** ***E. Pilosa*** **L.**

Sanskrit or Ayurvedic name(s): Saptalaa

Common name(s): Modukalli

Family: Euphorbiaceae

Distribution: African, naturalized to peninsular India. Grown in gardens.

Parts and properties used: Plant/ latex- Antirheumatic, purgative, emetic, anti-asthmatic, bechic. Latex- Applied on warts.

Associated system(s): AFSU

Euphorbia tortilis **Rottler ex Ainslie (cactus)**

Common name(s): Tirugakalli, Vajratunda

Family: Euphorbiaceae

Distribution: Habitat to scrub jungle rocks of peninsular India.

Parts and properties used: Plant extract- Cytotoxic potential (Pentacyclic triterpenes) anti-HIV.

Associated system(s): AS

Euphoria longan **Steud.**

Sanskrit or Ayurvedic name(s): Aakshiki

Common name(s): Longan, Puvatti

Family: Sapindaceae

Distribution: Found Assam, Bengal, and south India.

Parts and properties used: Fruit aril- Used in insomnia, neurosis, palpitation, amnesia, anaemia.

Associated system(s): AFS

Euphrasia officinalis **L., syn.** ***E. simplex*** **D. Don.**

Common name(s): Eyebright, Kangchuk

Family: Scrophulariaceae

Distribution: Habitat to cold arid Ladakh and beyond.

Parts and properties used: Plant- Astringent, antiallergic, bechic, ant catarrhal. Used in eye ailments.

Associated system(s): FH

Euryale ferox **Salisb. ex K.D.Koenig**

Sanskrit or Ayurvedic name(s): Makhaann

Common name(s): Phoolmakhana, Gorgan Nut

Family: Nymphaeaceae

Distribution: Native to wetlands of north. Grown in middle Gangetic plains (lakes/ ponds).

Parts and properties used: Fruit (seed)- Deobstruent, nervine tonic, astringent. Used in sexual affection.

Associated system(s): ASU

Evolvulus alsinoides **(L.) L.**

Sanskrit or Ayurvedic name(s): Shankapushpi

Common name(s): Shankhavali, Shankhpush

Family: Convolvulaceae

Distribution: Found throughout India.

Parts and properties used: Plant extract- Yields brain tonic, aid in conception, astringent, antidysentery; decoction work as blood purifier. Leaves- Anti-asthmatic. Used in nervine affections.

Associated system(s): ASLU

Exacum bicolor **Roxb.**

Sanskrit or Ayurvedic name(s): Ava-chiraayata

Common name(s): Titakhana

Family: Gentianaceae

Distribution: Habitat to Upper Gangetic plains.

Parts and properties used: Plant- Stomachic, febrifuge, antifungal. Yields a tonic.

Associated system(s): AF

Excoecaria agallocha **L.**

Common name(s): Agaru, Tejbala

Family: Euphorbiaceae

Distribution: Habitat to coastal forest of India.

Parts and properties used: Latex- Antileprotic. Juice and oil- Used in rheumatism, paralysis, and leprosy.

Associated system(s): FS

***Fagonia cretica* L.**

Sanskrit or Ayurvedic name(s): Dhanvayasah

Common name(s): Dhamasa

Family: Zygophyllaceae

Distribution: Habitat to Upper Gangetic plains and peninsular India.

Parts and properties used: Plant- Astringent, antiseptic, blood purifier, febrifuge. Used in abscesses. Extract- Antiviral, ant amphetaminic, spasmogenic.

Associated system(s): ALU

***Fagopyrum esculentum* Moench**

Sanskrit, Ayurvedic and Common name(s): Kotu, Kutu

Family: Polygonaceae

Distribution: Native to Central Asia. Grown in hills of North India and the Nilgiris.

Parts and properties used: Herb/grain- Used for treating fragile capillaries, chilblains and for strengthening varicose veins and for treating high blood pressure. Rutin (flavonoid) is obtained from fresh or dried leaves and flowers.

Associated system(s): AF

***Feronia limonia* (Linn.) Swingle, syn. *Lemonia acidissima* L., *F. elephantum* Correa**

Sanskrit or Ayurvedic name(s): Kapittha, Kapityama, Dadhittha

Common name(s): Wood Apple, Kaith

Family: Rutaceae

Distribution: Native of Indo-Malay region. Cultivated throughout India.

Parts and properties used: Fruit- Antiscorbutic, carminative, stimulates the digestive system. Unripe fruit- Prescribed in sprue, malabsorption syndrome.

Associated system(s): AFSU

***Ferula foetida* (Bunge) Regel., syn. *F. assa-foetida* L.**

Sanskrit or Ayurvedic name(s): Hingu

Common name(s): Hing Heera (Baandani)

Family: Apiaceae/ Umbelliferae

Distribution: Native range from Central Asia to J & K. Cultivated.

Parts and properties used: Oleo-Gum Resin- Stimulates intestinal and respiratory tracts and nervous system. Used in digestive problems, toothache.

Associated system(s): AHSLUF (Bhil)

***Ferula jaeschkeana* Vatke (EN)**

Sanskrit or Ayurvedic name(s): Hingupatri

Common name(s): Hingupatri, Hingu, Sampharu

Family: Apiaceae/ Umbelliferae

Distribution: Native range from central Asia to Western Himalaya 2000-4000 m.

Parts and properties used: Oleo-Gum Resin- Used as abortifacient, anti-implantation.

Associated system(s): A

***Ferula narthex* Boiss.**

Sanskrit or Ayurvedic name(s): Hingu

Common name(s): Sahasravedhi

Family: Apiaceae/ Umbelliferae

Distribution: Native range extends from central Asia to Kashmir.

Parts and properties used: Gum Resin- Used as asafoetida (Hing). Oil- Bactericidal.

Associated system(s): AFHSU

***Ficus altissima* Blume.**

Sanskrit or Ayurvedic name(s): Nandi vrksha variety

Common name(s): Gadgubar

Family: Moraceae

Distribution: Native to Assam, Tripura to Malaysia.

Parts and properties used: Leaves and bark- Used in skin diseases.

Associated system(s): AF

***Ficus arnottiana* (Miq.) Miq.**

Sanskrit or Ayurvedic name(s): Nandi Vriksha

Common name(s): Kallaltholi

Family: Moraceae

Distribution: Native. Grown in Rajasthan, MP, Bihar, and western peninsula.

Parts and properties used: Leaves- Used as sterilizer after menses. Bark- Used in skin diseases.

Associated system(s): AS

***Ficus benghalensis* L.**

Sanskrit or Ayurvedic name(s): Nayagrodha jata, Nyagrodha

Common name(s): Bad, Bargad, Vadachhal, Vatathwak

Family: Moraceae

Distribution: Habitat to sub-Himalayan tract & Peninsular India. Cultivated.

Parts and properties used: Bark Infusion- Used in diabetes, dysentery, seminal weakness. Milky juice- Antirheumatic. Arial root- Used in lipid disorders.

Associated system(s): AFHSLU

***Ficus carica* L.**

Sanskrit or Ayurvedic name(s): Phalgu, Raajodumbara

Common name(s): Anjeer, Anjoora

Family: Moraceae

Distribution: Asian, cultivated since ancient times in Punjab, UP, Rajasthan and Deccan Plateau.

Parts and properties used: Fruit- Laxative. Syrup- Given in constipation. Fruit pulp- Analgesic, anti-inflammatory, treat tumours. Latex- Analgesic and toxic, treat warts, insect bites. Leaf- Used in leukoderma.

Associated system(s): ASU

***Ficus cunia* Buch. - Ham**

Sanskrit or Ayurvedic name(s): Malayu, Laakshaa-Vriksha

Common name(s): Indian Fig

Family: Moraceae

Distribution: Habitat to sub-Himalayan tract

Parts and properties used: Fruit- Spasmolytic. Bark decoction- To wash ulcers, applied to wounds and bruises. Yields- Syconium.

Associated system(s): AS

***Ficus heterophylla* Linn. f.**

Sanskrit or Ayurvedic name(s): Traayanti

Common name(s): Daantiraa

Family: Moraceae

Distribution: Habitat to all over warm India.

Parts and properties used: Fruit- Used in constipation. Leaf juice- Antidysentery. Root- Antispasmodic. Bark- Treat asthma, bronchitis.

Associated system(s): AFS

***Ficus hispida* L. f.**

Sanskrit or Ayurvedic name(s): Kaakodumbara

Common name(s): Bhui-umber, Anjir

Family: Moraceae

Distribution: Habitat to all over India outside Himalaya.

Parts and properties used: Syconium [inflorescence, (fruit)]- Galactagogue. Fruit given in jaundice, oedema, anaemia. Bark/seed- Purgative, emetic.

Associated system(s): ASU

***Ficus lacor* Buch. -Ham., syn. *F. virens* Aiton, *F. infectoria* auct. non Willd.**

Sanskrit or Ayurvedic name(s): Plaksha

Common name(s): White fig

Family: Moraceae

Distribution: Habitat to plain and lower hills of India.

Parts and properties used: Bark decoction- Used in washing ulcers and for gargle. Leaf- Estrogenic. Plant- Used in erysipelas, ulcer, epistaxis. Fruit/bark- Used in syncope, delirium, unstable mind.

Associated system(s): AS

***Ficus racemosa* L., syn. *F. glomerata* Roxb.**

Sanskrit or Ayurvedic name(s): Shitavalkah, Udumbara

Common name(s): Gular, Umbar Chal, Lakh pipal

Family: Moraceae

Distribution: Grows wild throughout India. Also, cultivated.

Parts and properties used: Plant- Astringent, antiseptic. Used in abortions, menorrhagia, leucorrhoea, urinary/ skin diseases, swellings, boils, haemorrhages. Bark- Used in lipid disorders, obesity skin diseases, inflammations, and ulcers.

Associated system(s): ALSU

***Ficus religiosa* L.**

Sanskrit or Ayurvedic name(s): Asvattha

Common name(s): Lakh pipal, Arayaalin tholi

Family: Moraceae

Distribution: Habitat to sub-Himalayan tract. Planted all over India.

Parts and properties used: Bark- Astringent, antiseptic, alterative, laxative, haemostatic, vaginal disinfectant. Leaf/twig- Applied ulcers and wounds, laxative.

Associated system(s): AFHSLU

***Ficus tsiela* Roxb. ex Buch. -Ham., syn. *F. indica; F. amplissima* Rees**

Sanskrit or Ayurvedic name(s): Plaksha, Piparee

Family: Moraceae

Distribution: Habitat to peninsular India. Cultivated in south.

Parts and properties used: Bark- Anti-diabetic, antioxidant. Leaf juice- Applied on wounds. Latex- Applied on fresh wounds.

Associated system(s): ASU

***Ficus talbotii* G. King, syn. *F. pierrei* Gagnep.**

Sanskrit or Ayurvedic name(s): Plaksha

Common name(s): Itthi (wild relative), Talbot Fig

Family: Moraceae

Distribution: Wild in peninsular India.

Parts and properties used: Bark- Antileprotic. Arial part- Diuretic, spasmolytic, CNS depressant, hypothermic.

Associated system(s): AS

***Flacourtia indica* (Burm.f.) Merr., syn. *F. sapida* Roxb.**

Sanskrit or Ayurvedic name(s): Sruvavrksa

Common name(s): Kattar, Kangu, Governer's Plum

Family: Flacourtiaceae

Distribution: Native to Africa and Asia. Cultivated in Assam, Bengal, Maharashtra

Parts and properties used: Gum- Anticholera, Applied to eczema/skin diseases. Bark- Antidysentery, Astringent, diuretic. Advised in jaundice, etc. Seed- Antirheumatic. Fruit- Stomachic.

Associated system(s): AFSL

***Flacourtia jangomas* (Lour.) Raeusch., syn. *F. cataphracta* Roxb.**

Sanskrit or Ayurvedic name(s): Praachinaamalaka, Vikaṅkata, Sruvavṛkṣa

Common name(s): Brahmi Talispatar, Pāniyāmalak, Talish pattar, Puneala plum

Family: Flacourtiaceae

Distribution: Wild in Assam, Bengal, Eastern Ghats

Parts and properties used: Leaf- Astringent, antidiarrheal, stomachic. Used in bronchitis. Fruit- Advised in liver affection.

Associated system(s): ASU

***Flemingia chappar* Buch. -Ham. *ex* Benth.**

Common name(s): Salpan

Family: Leguminosae/ Fabaceae

Distribution: Habitat to eastern Himalaya, Assam, and Bengal.

Parts and properties used: Plant- Used in epilepsy, insomnia, dysentery, stomach-ache, helminthiasis, rheumatism, ulcer, and tuberculosis. Root- Sedative and analgesic.

Associated system(s): F

***Flemingia grahamiana* Wight & Arn.**

Common name(s): Warrus/ Varus

Family: Leguminosae/ Fabaceae

Distribution: Native Africa, Arabia, Asia. Cultivated in tropical India.

Parts and properties used: Plant- Purgative. Applied externally skin diseases.

Associated system(s): F

***Flemingia macrophylla* (Willd.) Kuntze *ex* Merr.**

Common name(s): Varrus/ Varus

Family: Leguminosae/ Fabaceae

Distribution: Habitat to sub-humid to humid tropics up 2000 m.

Parts and properties used: Plant extract- Used to treat rheumatism.

Associated system(s): AF

***Flickingeria macraei* (Lindl.) Seidenf., syn. *Ephemerantha macraei* (Lindl.) Hunt & Summerh**

Common name(s): Jivanti

Family: Orchidaceae

Distribution: Habitat to Himalayas.

Parts and properties used: Leaves, flowers, pseudobulbs- Used in earache, skin infections, nervous ailments, swellings, and cholera.

Associated system(s): AF

Flickingeria nodosa **(Dalz.) Seidenf., syn.** ***Dendrobium macraei*** **auct. non-Lindl.**

Common name(s): Jivanti

Family: Orchidaceae

Distribution: Habitat to Himalayas.

Parts and properties used: Plant- Used in asthma, bronchitis, blood purifier, dermatological infections.

Associated system(s): AF

Flueggea virosa **(Roxb. ex Willd.) Royle, syn.** ***F. microcarpa*** **Blume**

Sanskrit or Ayurvedic name(s): Mishreyaa

Common name(s): Dalme, Patali

Family: Euphorbiaceae

Distribution: Habitat to tropical western India, Kerala & Pakistan.

Parts and properties used: Leaves, roots- Yield iso-coumarins, which are effective against gastric ulcers.

Associated system(s): F

Foeniculum vulgare **Gaertn.**

Sanskrit or Ayurvedic name(s): Misreya

Common name(s): Badiyan, Fennel, Saunf, Khatal

Family: Apiaceae/ Umbelliferae

Distribution: Mediterranean. Naturalized and grown as winter crop all over India.

Parts and properties used: Fruit- Carminative, stomachic, antispasmodic, emmenagogue, galactagogue, anti-inflammatory, diuretic. Advised in dyspepsia.

Associated system(s): AFSLU

Fritillaria roylei **Hook. (CR)**

Sanskrit or Ayurvedic name(s): Kshira-kakoli

Common name(s): Kakoli

Family: Liliaceae

Distribution: Habitat to western temperate Himalayas.

Parts and properties used: Bulb- Used in asthma, bronchitis, and tuberculosis.

Associated system(s): A

Fumaria indica **(Hauskn.) Pugsley, syn. *F. parviflora* Lam.**

Sanskrit or Ayurvedic name(s): *Parpat,* Parpata

Common name(s): *Pitpapra,* Shahtara

Family: Fumariaceae

Distribution: Found in high altitude regions of the Himalayas.

Parts and properties used: Plant- Detoxifying, laxative, diuretic, diaphoretic.

Associated system(s): AUS

Fumaria officinalis L

Sanskrit or Ayurvedic name(s): Parpata

Common name(s): Pittapapdo ghas, Pitpapra

Family: Fumariaceae

Distribution: Native of Eurasia. Naturalized and raised in Nilgiris Hills.

Parts and properties used: Plant- Antispasmodic, amphitheatric. Stimulant to liver and gall bladder. Used in eczema/skin diseases.

Associated system(s): AFU

Galium aparine **L.**

Common name(s): Ranche

Family: Rubiaceae

Distribution: Habitat to temperate Himalayas, including Ladakh.

Parts and properties used: Leaves/stems- Diuretic, choleretic, stomachic, alterative, detoxifier. lymphatic. Used in cleansing of enlarged lymph node and urinary disorder.

Associated system(s): English

Galium serpylloides **Royle ex Hk. F.**

Ayurvedic or Common name(s): Pimantso

Family: Rubiaceae

Distribution: Habitat to Trans-Himalayas and Ladakh.

Parts and properties used: Herb- Diuretic, purgative.

Associated system(s): A

Garcinia gummi-gutta **(L.) N. Robson, syn. *G. cambogia* (Gaertn.) Desr.] (VU)**

Sanskrit or Ayurvedic name(s): Vrkshaamla

Common name(s): Kokam, Kodampuli

Family: Clusiaceae/ Guttiferae

Distribution: Native to Indonesia. Grown in Western Ghats and Nilgiris.

Parts and properties used: Fruit rind- Used in rickets, spleen enlargement, skeletal fractures.

Associated system(s): AS

***Garcinia indica* (Dup.) Choisy (VU)**

Associated system(s): Vrkshaamla

Common name(s): Kokam, Cambogie

Family: Clusiaceae/ Guttiferae

Distribution: Found in forest lands, riversides and wastelands of Western Ghats and cultivated.

Parts and properties used: Fruit- Antiscorbutic, cholagogue, ant bilious, emollient and demulcent.

Associated system(s): AS

***Garcinia morella* (Gaertn.) Desr. (VU)**

Sanskrit or Ayurvedic name(s): Kankushtha

Common name(s): Kadukaai puli

Family: Clusiaceae/ Guttiferae

Distribution: Found in Assam, Bengal, Southern India.

Parts and properties used: Gum, resin- Hydragog, cathartic, anthelmintic. Used in droopy and amenorrhoea.

Associated system(s): AFHSLU

***Garcinia pedunculata* Roxb. ex Buch. -Ham. (EN)**

Sanskrit or Ayurvedic name(s): Amlavetasa

Common name(s): Thaikala

Family: Clusiaceae/ Guttiferae

Distribution: Found in northeast India.

Parts and properties used: Fruit- Antiscorbutic, astringent, cardiotonic, emollient. Used in anorexia, colic, dyspepsia, liver/ spleen diseases. Pericarp contains benzophenones, pedunculol, garcinol and Cambogin.

Associated system(s): AT

Garcinia xanthochymus* Hook. f. ex T. And., syn. *G. tinctoria

Common name(s): Tamaal variety

Family: Clusiaceae/ Guttiferae

Distribution: Habitat to Eastern Himalayas, and peninsular India.

Parts and properties used: Fruit- Anthelmintic, improves appetite. Also used as cardiotonic.

Associated system(s): AFS

***Gardenia gummifera* L. f. (NT)**

Sanskrit or Ayurvedic name(s): Venupatrikaa

Common name(s): Dekamalli, Dikamali

Family: Rubiaceae

Distribution: Found throughout India.

Parts and properties used: Gum- Carminative, antispasmodic, stimulant, diaphoretic, antiseptic, anthelmintic, expectorant.

Associated system(s): AFSLU

***Gardenia jasminoides* Ellis. Syn *G. florida* Linn. *G. augusta* Merrill**

Sanskrit or Ayurvedic name(s): Gandharaaja

Common name(s): Cape Jasmine, Karinga

Family: Rubiaceae

Distribution: Native to China and Japan. Cultivated in Indian gardens.

Parts and properties used: Plant- Cathartic, antiperiodic, antispasmodic, anthelmintic. Root- Antidysentery. Also used in dyspepsia and nervous disorders. Fruits- Treat gastric hyperacidity, constipation, cholestasis. Essential oil- Applied in inflammation and as a tranquilizer.

Associated system(s): AS

***Gardenia latifolia* Ait.**

Common name(s): Parpataki

Family: Rubiaceae

Distribution: Native, found throughout dry India.

Parts and properties used: Bark- Used in skin diseases.

Associated system(s): AFS

***Gardenia resinifera* Roth**

Sanskrit or Ayurvedic name(s): Naadihingu

Common name(s): Dikamali

Family: Rubiaceae

Distribution: Found in central and peninsular India.

Parts and properties used: Gum (flavonoids)- Anti-microbial, anthelmintic. Used in skin diseases.

Associated system(s): ASU

Gardenia turgida **Roxb.**

Sanskrit or Ayurvedic name(s): Mahaapindi

Common name(s): Bhaargi

Family: Rubiaceae

Distribution: Found throughout greater India.

Parts and properties used: Root- Treat indigestion in children. Fruit- Treat mammary glands. Bark- Beta-sitosterol, hederagenin.

Associated system(s): AS

Garuga pinnata **Roxb.**

Sanskrit or Ayurvedic name(s): Paaranki

Common name(s): Ghoghar, Kharpat

Family: Burseraceae

Distribution: Found throughout India.

Parts and properties used: Fruit- Stomachic. Leaf- Astringent, anti-asthmatic. Bark-Antidiabetic.

Associated system(s): AFSL

Gaultheria fragrantissima **Wall.**

Sanskrit or Ayurvedic name(s): Gandhapuura

Common name(s): Kolakkaai

Family: Ericaceae

Distribution: Habitat to central, eastern Himalayas, and Nilgiris.

Parts and properties used: Leaves- Stimulant, carminative, diuretic, antiseptic. Oil-Applied in rheumatism.

Associated system(s): AS

Gentiana kurroo **Royle (CR)**

Sanskrit or Ayurvedic name(s): Trayamana

Common name(s): Trahimaan, Kadu, Katuki

Family: Gentianaceae

Distribution: Habitat to northwest Himalayas.

Parts and properties used: Root, plant- Antimicrobial, anti-arthritic, analgesic, anti-diabetic. Cure digestive disorders.

Associated system(s): ALU

***Gentianella paludosa* (Hk.) H. Smith.**

Common name(s): Kags-ti-nagpo

Family: Gentianaceae

Distribution: Habitat to alpine Himalayas, including Ladakh.

Parts and properties used: Plant- Febrifuge. Yields a tonic.

Associated system(s): L (Tibetan)

***Geophila reniformis* D. Don, syn. *G. repens* (L.) I.M. Johnst]**

Common name(s): Kakamaci

Family: Rubiaceae

Distribution: Habitat to Assam, Western Ghats, and Andaman.

Parts and properties used: Root- Antiprotozoal, expectorant, emetic.

Associated system(s): F

***Geranium nepalense* Sweet.**

Sanskrit or Ayurvedic name(s): Bhanda

Common name(s): Ratanjot

Family: Geraniaceae

Distribution: Habitat to temperate Himalayas, and Nilgiris.

Parts and properties used: (Plant geraniin)- Astringent, styptic. Used in renal diseases, diarrhoea, bleeding.

Associated system(s): AF

***Geum urbanum* L.**

Common name(s): Wood avens

Family: Rosaceae

Distribution: Habitat to Europe to Central Asia and temperate Himalayas.

Parts and properties used: Herb and root- Astringent, styptic, stomachic, febrifuge. Rootstock used in fever, etc.

Associated system(s): F, English

***Ginkgo biloba* Linn.**

Common name(s): Ginkgo

Family: Ginkgoaceae

Distribution: Chinese. Grown in Indian garden.

Parts and properties used: Plant extract- Treat mental health conditions, Alzheimer's disease, and fatigue.

Associated system(s): Chinese

***Girardinia heterophylla* Decne., syn. var. *zeylanica* Decne., G. *diversifolia* (Link) Friis**

Sanskrit or Ayurvedic name(s): Vrishchikaa

Common name(s): Gaddanelli, Anachoriyan

Family: Urticaceae

Distribution: Habitat to temperate/sub-tropical Himalayas.

Parts and properties used: Leaves- Decoction given in fever. Applied to swollen joints and paste in headache.

Associated system(s): AF

***Givotia rottleriformis* Griff. ex Wight**

Common name(s): Polki

Family: Euphorbiaceae

Distribution: Found in Peninsular India.

Parts and properties used: Seed/Bark- Used in treatment of rheumatism, psoriasis, and dandruff.

Associated system(s): FS

***Glinus lotoides* L.**

Common name(s): Gandhi-Buti

Family: Aizoaceae

Distribution: Found throughout the plains of India.

Parts and properties used: Plant- Antidiarrheal, ant bilious, diuretic. Plant decoction- Cure piles.

Associated system(s): FS

Glinus oppositifolius **(L.) A. DC.**

Sanskrit or Ayurvedic name(s): Parpata

Common name(s): Ushnasundara

Family: Aizoaceae

Distribution: Found in Assam.

Parts and properties used: Plant- Treat inflammation, joint pains, diarrhoea, intestinal parasites, fever, skin disorders.

Associated system(s): AFS

Gloriosa superba **L. (NT)**

Sanskrit or Ayurvedic name(s): Laangali, Agnimukhi

Common name(s): Kalihari

Family: Liliaceae/ Colchicaceae

Distribution: Habitat to temperate Himalayas to tropical India. Cultivated too.

Parts and properties used: Tuberous root- Anti-inflammatory, alterative, antileprotic, anthelmintic. Fresh juice- Uterine stimulant.

Associated system(s): AFSLU

Glossocardia bosvallea **(L. f.) DC.**

Sanskrit or Ayurvedic name(s): Partpata, Pithari

Common name(s): Parpataka

Family: Asteraceae/ Compositae

Distribution: Habitat to northwest plains and Deccan peninsula.

Parts and properties used: Plant- Emmenagogue [substitute of Parpata (*Fumaria indica*), whole dried plant. Oil- Antimicrobial.

Associated system(s): AS

Glycosmis arborea **(Roxb.) DC., syn** *G. pentaphylla* **(Retz.) DC.**

Sanskrit or Ayurvedic name(s): Vana-nimbuukaa

Common name(s): Potali, Kupilah

Family: Rutaceae

Distribution: Found in Peninsular India. Cultivated.

Parts and properties used: Plant- Be chic, ant anaemic, antirheumatic. Root- Anti-inflammatory. Leaves- Given in Jaundice.

Associated system(s): AFS

***Glycyrrhiza glabra* L.**

Sanskrit or Ayurvedic name(s): Yashti, Yashtimadhu

Common name(s): Mulathi, Jeshtimadhu, Gulegafis

Family: Leguminosae/ Fabaceae

Distribution: Native to West Asia and southern Europe. Grown from J & K to south in India.

Parts and properties used: Root- Demulcent, expectorant, antiallergic, anti-inflammatory, spasmolytic. Used in cough, catarrh of respiratory tract, gastric, and duodenal ulcers.

Associated system(s): AFSLU

***Glycyrrhiza uralensis* DC.**

Common name(s): Chinese liquorice, Mulathi, Jeshtimadhu, Gulegafis

Family: Leguminosae/ Fabaceae

Distribution: Asian native. Cultivated all over India.

Parts and properties used: Root- Antimicrobial. Improve spleen function and blood circulation and reduce cough formation.

Associated system(s): F

***Gmelina arborea* Roxb.**

Sanskrit or Ayurvedic name(s): Gambhaari

Common name(s): Ghambar Chal, Shivan (Ghambari) (Chal)

Family: Lamiaceae/ Labiatae

Distribution: Found and grown throughout India.

Parts and properties used: Leaf- Demulcent, bechic. Used to remove foetid discharge of ulcer. Root- Stomachic, laxative, ant bilious, demulcent, galactagogue. Bark- Anti-cephalalgia. (Part of Ayurvedic *Dash Moola*)

Associated system(s): AFSL

***Gmelina asiatica* L., syn. *G. parviflora* Roxb.**

Sanskrit or Ayurvedic name(s): Vikarini, Gopachadra

Common name(s): Kapas Beej

Family: Verbenaceae

Distribution: Native to south India. Grown in Bengal, Maharashtra.

Parts and properties used: Root/leaf- Demulcent, alterative, blood purifier, ant catarrhal, astringent, antirheumatic.

Associated system(s): AFS

Goodyera repens **(L.) R. Br.**

Common name(s): Creeping lady's-tresses

Family: Orchidaceae

Distribution: Native to temperate Northern Hemisphere, including Himalayas, Ladakh, Northeast Hills.

Parts and properties used: Leaf infusion- Given in cold, kidney problems. Poultice applied on burns, blood purifier, syphilis.

Associated system(s): F

Gossypium arboreum **L.**

Sanskrit or Ayurvedic name(s): Kaarpaasi

Common name(s): Kapas

Family: Malvaceae

Distribution: Wild in Assam. Cultivated all over India.

Parts and properties used: Seed- Ant catarrhal, anti-gonorrhoeic, used in gleet, cystitis. Plant- Uterine stimulant.

Associated system(s): AFSU

Gossypium herbaceum **L.**

Sanskrit or Ayurvedic name(s): Kaarpaasi

Common name(s): Kapas

Family: Malvaceae

Distribution: Native to Africa and Arabia. Naturalized, and cultivated in peninsular India.

Parts and properties used: Root bark- Diuretic, oxytocic. Bark- Emmenagogue, haemostatic. Seed- Demulcent, laxative, expectorant. Oil- Used for toning breast.

Associated system(s): ASU

Gossypium hirsutum **L.**

Common name(s): Kapas

Family: Malvaceae

Distribution: Native to Central America, also found in Maharashtra. Cultivated.

Parts and properties used: Plant- Treat asthma, diarrhoea, dysentery. Boiled leaf-skin rash. Seeds- Treat swelling, ulceration of female organs, urinary diseases. Fruit (coat)- Treat fungal infection.

Associated system(s): AS, Modern

***Gouania microcarpa* DC.; *G. leptostachya* DC**

Common name(s): Sinhanbali

Family: Rhamnaceae

Distribution: Occur from Assam to Kerala.

Parts and properties used: Leaf poultice- Applied for sores. Bark- Used in washing hair, check vermin.

Associated system(s): F

***Grewia asiatica* auct.non L., syn *G. subinaequalis* DC.**

Sanskrit or Ayurvedic name(s): Parushaka

Common name(s): Phaalsaa

Family: Tiliaceae

Distribution: Found in most of Indian forest. Cultivated.

Parts and properties used: Fruit- Stomachic, astringent, cooling. Bark- Demulcent, antirheumatic.

Associated system(s): AFSLU

***Grewia hirsuta* Vahl., syn. *G. longifolia* (related spp. are *G. populifolia*, *G. tenax*)**

Sanskrit or Ayurvedic name(s): Naagabalaa

Common name(s): Kukurbicha, Asolin

Family: Tiliaceae

Distribution: Habitat to sub-Himalayan tract.

Parts and properties used: Fruit/root- Diuretic and antidiarrheal. Juice- Indigestion, spermatorrhoea.

Associated system(s): AS

***Grewia sclerophylla* Roxb. ex G. Don.**

Sanskrit or Ayurvedic name(s): Parushaka

Common name(s): Jangali Phaalsaa

Family: Tiliaceae

Distribution: Habitat to sub-Himalayan tract.

Parts and properties used: Root- Have soothing and healing properties. Plant mucilage used in irritable intestines and bladder.

Associated system(s): ASF

Grewia tiliaefolia **Vahl.**

Sanskrit or Ayurvedic name(s): Dhanurvriksha

Family: Tiliaceae

Distribution: Habitat to Upper Gangetic plain and peninsular India.

Parts and properties used: Bark- Antidysentery. Stem bark- Semen coagulant. Plant- Used in fractures.

Associated system(s): AFS

Gymnema sylvestre **R.Br. ex Schult.**

Sanskrit or Ayurvedic name(s): Meshashringi

Common name(s): Gudmar, Sirukurinjan

Family: Asclepiadaceae

Distribution: Habitat to Central and peninsular India, Eastern Ghats.

Parts and properties used: Leaf powder, plant. leaves- Anti-diabetic, diuretic, inflammatory, anti-asthmatic/anti-bronchitis. Given in gastric troubles,

Associated system(s): AFHSLU

Gymnosporia spinosa **(Forsk.) Fiori., syn.** *G. montana* **(Roth.) Benth.**

Sanskrit or Ayurvedic name(s): Vikankata

Common name(s): Baikal, Kattangi, Vikro

Family: Celastraceae

Distribution: Found throughout arid India.

Parts and properties used: Plant- Antispasmodic. Root- Used in gastroenteritis and dysentery.

Associated system(s): AFS

Gynocardia odorata **R.Br.**

Sanskrit or Ayurvedic name(s): Chaalmogra, substitute

Common name(s): Chaaval-mungari

Family: Flacourtiaceae

Distribution: Habitat to Eastern Himalayas and hills.

Parts and properties used: Seed oil- Treat psoriasis, eczema, scrofula, gout, rheumatic care.

Associated system(s): AFS

***Habenaria edgeworthii* Hook. f. ex Collett and *H. intermedia* (NT)**

Sanskrit, Ayurvedic and common name(s): Riddhi

Family: Orchidaceae

Distribution: Found in Punjab to Kumaon.

Parts and properties used: Root tuber- Yields nervine and cardiac tonic (Part of *Astavarga*).

Associated system(s): A

***Haematoxylon campechianum* Linn.**

Sanskrit or Ayurvedic name(s): Pattanga

Common name(s): Bakam-Hindi

Family: Leguminosae/ Fabaceae (Caesalpiniaceae)

Distribution: American. Grown in Indian garden.

Parts and properties used: Plant- Astringent. Used for atonic (without stress/muscle), dyspepsia, diarrhoea.

Associated system(s): AS

***Halerpestis tricuspis* (Maxtm.) Hand. -Mazz.**

Common name(s): Chu-rag-sBl-lg

Family: Ranunculaceae

Distribution: Occur Mongolia to alpine Himalayas, including Ladakh.

Parts and properties used: Plant- Given in stomach disorders.

Associated system(s): AHSL

***Hardwickia binata* Roxb.**

Sanskrit or Ayurvedic name(s): Anjana

Common name(s): Katudugu

Family: Leguminosae/ Fabaceae (Caesalpiniaceae)

Distribution: Habitat to dry forests of Deccan.

Parts and properties used: Balsam- Used in sexually transmitted diseases.

Associated system(s): AS

***Hedera helix* Linn.**

Common name(s): Maravalai

Family: Araliaceae.

Distribution: Habitat to Hills of TN. Cultivated as ornamental.

Parts and properties used: Leaves- Expectorant. Treat catarrh of respiratory passages, bronchitis. Fruit- Used in jaundice. Flower- Antidysentery.

Associated system(s): S

***Hedychium coronarium* J. Konig (NT)**

Sanskrit or Ayurvedic name(s): Shati

Common name(s): Garland flower, Ginger Lily

Family: Zingiberaceae

Distribution: Habitat to moist parts of India.

Parts and properties used: Rhizome- Anti-inflammatory, antirheumatic, febrifuge, tranquilizer.

Associated system(s): A

***Hedychium spicatum* Sm., syn. *H. acuminatum* Roscoe (VU)**

Sanskrit or Ayurvedic name(s): Shati

Common name(s): Kapoor Kachri, spiked Ginger Lily, Ban-haldi

Family: Zingiberaceae

Distribution: Habitat to central Himalayas, Assam, and hills of India. Cultivated too.

Parts and properties used: Rhizome- Carminative, analgesic, spasmolytic, hepatoprotective, anti-inflammatory, antiemetic, antidiarrheal, expectorant, antiasthma-tic, emmenagogue, hypoglycaemic, anthelmintic. Powder- Insect repellent.

Associated system(s): AFLSU

***Hedyotis corymbosa* (L.) Lam., syn. *Oldenlandia corymbosa* L.**

Sanskrit or Ayurvedic name(s): Kshetraparpata

Common name(s): Pitpapra

Family: Rubiaceae

Distribution: It's a weed of cultivated fields.

Parts and properties used: Plant- Purifies blood, improves digestion, liver stimulant.

Associated system(s): AF

***Hedyotis herbacea* L., syn. *Oldenlandia herbacea* (Willd.) Roxb.**

Common name(s): Pippapada

Family: Rubiaceae

Distribution: Found in Maharashtra, Karnataka, TN.

Parts and properties used: Plant- Useful in elephantiasis, verminosis, inflammations, asthma, bronchitis, etc.

Associated system(s): AFH

***Hedyotis puberula* (G.Don) R. Br. ex Arn., syn. *Oldenlandia umbellata* L.**

Common name(s): Chay Root, Saya Root, Chirval

Family: Rubiaceae

Distribution: Native peninsular India. Cultivated in Coromandel Coast.

Parts and properties used: Herb- Styptic, used as poultices. Leaves and roots- Used in bronchitis, asthma, tuberculosis. Also produce dye.

Associated system(s): FS

***Helianthus annuus* L.**

Sanskrit or Ayurvedic name(s): Suryamukhi

Common name(s): Sunflower

Family: Asteraceae/ Compositae

Distribution: Native to America. Cultivated.

Parts and properties used: Seed- Immunity builder; decoction given in cough/ cold, bronchial, laryngeal, pulmonary affection.

Associated system(s): AHSU

***Helicteres isora* L.**

Sanskrit or Ayurvedic name(s): Aavartani

Common name(s): Murudshing, Marodphali

Family: Sterculiaceae

Distribution: Found all over in dry forests, India.

Parts and properties used: Pods and bark- Antidiarrheal, astringent, antibilious.

Associated system(s): ASU

***Heliotropium indicum* L.**

Sanskrit or Ayurvedic name(s): Hastishundi, Shrihastini

Common name(s): Thekkada

Family: Boraginaceae

Distribution: Found throughout India.

Parts and properties used: Plant, root- Diuretic, astringent, emollient, vulnerary. Used against ulcers, wounds. Contains alkaloids.

Associated system(s): AFSU

***Helleborus niger* L.**

Sanskrit or Ayurvedic name(s): Khuraasaani Kutaki

Common name(s): Katuka Rohini, Christmas rose or black hellebore.

Family: Ranunculaceae

Distribution: European. Cultivated in gardens.

Parts and properties used: Root- Cardiac disorders, drastic, purgative, abortifacient, diuretic.

Associated system(s): AFHU

***Hemidesmus indicus* (L.) R.Br.**

Sanskrit or Ayurvedic name(s): Anandamul, Shveta Saariva

Common name(s): Anatmool, Sariwa, Sarasaparilla roots

Family: Periplocaceae

Distribution: Endemic to peninsular India, Bengal. Cultivated too.

Parts and properties used: Plant- Blood purifier, ant syphilitic, anti-leucorrhoea, Ga lactogenic, antidiarrheal, antirheumatic, febrifuge, alterative. Roots- Gonorrhoea, etc.

Associated system(s): ASLU

***Heracleum lanatum* Michx., syn. *H. candicans* Wall ex DC.; *H. nepalensis* (EN)**

Common name(s): Krandel, Rasal

Family: Apiaceae/ Umbelliferae

Distribution: Habitat to alpine west Himalayas, Ladakh. Grown.

Parts and properties used: Plant- Produce Xanthotoxin- Used to treat psoriasis, eczema, vitiligo, lymphoma due to exposer of skin to UVA/sunlight.

Associated system(s): F

***Heracleum rigens* Wall. ex DC. (DD)**

Common name(s): Cittrelam

Family: Apiaceae/ Umbelliferae

Distribution: Native to southern Western Ghats.

Parts and properties used: Plant, Fruit- Aromatic.

Associated system(s): AS

***Heterophragma roxburghii* DC., syn. *H. quadriloculare* (Roxb.) D. Schum**

Common name(s): Waarasa, Pullunga

Family: Bignoniaceae.

Distribution: Found in Gujarat, MP, Maharashtra, AP, Karnataka, TN.

Parts and properties used: Tar extracted from wood- Used in skin diseases. Leaf juice- Applied to toe sores and in chilblain.

Associated system(s): AFS

***Hibiscus rosa-sinensis* L.**

Sanskrit or Ayurvedic name(s): Japaa

Common name(s): China-rose, Gurhal, Jashwanti, Japa, Jaswand.

Family: Malvaceae

Distribution: Chinese. Grown in Indian gardens.

Parts and properties used: Flower- Used in impotency, bronchial catarrh, emmenagogue.

Associated system(s): AFSU

***Hibiscus sabdariffa* L.**

Sanskrit or Ayurvedic name(s): Ambashtthaki

Common name(s): Indian-sorrel, Lalambari,

Family: Malvaceae

Distribution: West Indian. Grown in India.

Parts and properties used: Plant- Digestive, choleretic, ant bilious, laxative, diuretic, hypotensive, antiscorbutic. Yields a cardiac and nervine tonic. Flower- Used in appetite loss, cold, respiratory/stomach catarrhs.

Associated system(s): FS

***Hibiscus surattensis* L.**

Common name(s): Ran Bhindi

Family: Malvaceae

Distribution: Pantropical, found in northeast India to Kerala.

Parts and properties used: Stem, leaf- Used in urethritis and venereal diseases. Flower- Emollient, pectoral.

Associated system(s): AF

***Hibiscus tiliaceus* L.**

Common name(s): Parutti

Family: Malvaceae

Distribution: Found in Assam, Odisha and UP.

Parts and properties used: Root- Used for gonorrhoea and herpes infection.

Associated system(s): AFS

***Hippophae rhamnoides* L.**

Common name(s): Sea Buckthorn

Family: Elaeagnaceae

Distribution: Temperate Europe, Asia northwest Himalayas 2350-5000 m. Grown in Ladakh.

Parts and properties used: Fruit- Astringent, antidiarrheal, antioxidant, multi-vitamin, carotenoids.

Associated system(s): F

***Hiptage benghalensis* Kurz.**

Sanskrit or Ayurvedic name(s): Atimukta

Common name(s): Madhav

Family: Malpighiaceae

Distribution: Found throughout the warm India.

Parts and properties used: Kernel- Advised in obesity. Leaves/bark- Used in rheumatism, asthma, and skin diseases.

Associated system(s): AS

***Holarrhena pubescens* (Buch. -Ham.) Wall. ex G.Don, syn. *H. antidysenterica* (Wall. ex DC.) Wall.**

Sanskrit or Ayurvedic name(s): Indrayava, Kutaja

Common name(s): Inderjao Kadwa, Kudachal, Kurchi.

Family: Apocynaceae

Distribution: Habitat to tropical Himalayas and forests of India.

Parts and properties used: Root/Bark- Used in amoebic dysentery; Bark- Astringent, anthelmintic, amoebicidal, diuretic. Used in colic, dyspepsia. Seed- Ant bilious.

Associated system(s): AFHSU

Holoptelea integrifolia **(Roxb.) Planch.**

Sanskrit or Ayurvedic name(s): Cirabilva

Common name(s): Aavitholi

Family: Ulmaceae

Distribution: Found throughout India.

Parts and properties used: Bark- Used in rheumatism. Paste- Scabies. Seed- Deworming of ringworm. Dried fruit- Advised in polyuria, urinary disorder.

Associated system(s): AFS

Holostemma ada-kodien **Schult., syn.** ***H. annularis*** **(Roxb.) K. Schum.;** ***H. rheedii*** **Wall. (VU)**

Sanskrit or Ayurvedic name(s): Ark-Pushpi

Common name(s): Jeevanti

Family: Asclepiadaceae

Distribution: Habitat to tropical Himalayas, western peninsula. Grown in TN.

Parts and properties used: Root (tuber)- Used in orchitis, spermatorrhoea. Also, as laxative.

Associated system(s): AS

Homonoia riparia **Lour.**

Sanskrit or Ayurvedic name(s): Substitute to Paashaana-bheda.

Common name(s): Serna, Pashan-bhedaka

Family: Euphorbiaceae

Distribution: Habitat to peninsular India

Parts and properties used: Root- Diuretic, spasmolytic, antilithic. Used in urinary discharge. Leaf/stem- Depurative.

Associated system(s): AS

Hordeum vulgare **L.**

Sanskrit or Ayurvedic name(s): Yava

Common name(s): Barley, Jaw

Family: Poaceae/ Gramineae

Distribution: Eurasian. Cultivated in north India.

Parts and properties used: Grain- Nutritive, diuretic, demulcent, expectorant, galactofuge (inhibit breastmilk). Advised in urinary disorders, muscular rigidity, sinusitis, asthma, lipid disorder and obesity..

Associated system(s): ASLU

***Houttuynia cordata* Thunb.**

Common name(s): Mosundori,

Family: Saururaceae

Distribution: Widely found in Assam and Northeast Hills. Semi-domesticated.

Parts and properties used: Herb- Aromatic, diuretic; treat pneumonia, hypertension, constipation, hyperglycemia via detoxification.

Associated system(s): F

***Hugonia mystax* L.**

Common name(s): Kamsamrah

Family: Linaceae

Distribution: Habitat to Konkan, and dry forest of TN.

Parts and properties used: Root- Anti-inflammatory, febrifuge. Disperses swellings.

Associated system(s): FS

***Humboldtia vahliana* Wight (VU -EN)**

Sanskrit or Ayurvedic name(s): Aattuvanchi,

Common name(s): Kurappunna, Korathi,

Family: Leguminosae/ Fabaceae

Distribution: Along rivers in southern Western Ghats.

Parts and properties used: Bark- Used in biliousness, leprosy, ulcers, and epilepsy.

Associated system(s): AFS

***Humulus lupulus* Linn.**

Common name(s): Hops

Family: Urticaceae

Distribution: Native to Europe, west Asia. Grown in J & K, Ladakh, HP.

Parts and properties used: Flower- Sedative, hypnotic, nervine tonic, diuretic, spasmolytic, analgesic, astringent. Used in anxiety and sleep disturbances.

Associated system(s): UF, English

***Hybanthus enneaspermus* (L.) F. Muell.**

Sanskrit or Ayurvedic name(s): Ratanpurusha

Common name(s): Padmacarini

Family: Violaceae

Distribution: Found in many parts of India.

Parts and properties used: Plant- Used in treatment of diarrhea, urinary infections, leucorrhea, dysuria, cholera inflammation.

Associated system(s): AFS

***Hydnocarpus kurzii* (King) Warb. (EN)**

Sanskrit or Ayurvedic name(s): Tuvaraka

Common name(s): Nirati, Chaulmoogra

Family: Flacourtiaceae

Distribution: Found in Assam, Tripura.

Parts and properties used: Plant- Antileprotic, derma tic, febrifuge, sedative. Used in leprosy. Source of Chaulmoogra oil.

Associated system(s): AFSU

***Hydnocarpus pentandrus* (Buch. -Ham.) Oken, (VU); syn. *H. laurifolia* (Dennst.) Sleum**

Sanskrit or Ayurvedic name(s): Katu-Kapittha

Common name(s): Marotti, Chaulmoogra

Family: Flacourtiaceae

Distribution: Found in peninsular India.

Parts and properties used: Plant- Used to treat leprosy and other skin disorders.

Associated system(s): AFS

***Hygrophila schulli* (Buch. -Ham.) M.R. & S.M. Almeida, syn. *H. auriculata* (Schumach.) Heine; *H. spinosa* T. Anderson; *Asteracantha longifolia* Nees**

Sanskrit or Ayurvedic name(s): Kokilaaksha

Common name(s): Tal makhana

Family: Acanthaceae

Distribution: Common in moist places and paddy fields.

Parts and properties used: Plant, root, fruit (seed)- Diuretic. Used in catarrh of urinary organs. Seeds- Promote sexual vigour, arrest abortion and cure diseases of vitiated blood. Source of Ayurvedic drug *Kokila Aksha* and the Unani *Tali makhana.*

Associated system(s): ASU

***Hymenodictyon excelsum* (Roxb.) Wall.**

Sanskrit or Ayurvedic name(s): Bhramar-Chhalikaa

Common name(s): Kuthan

Family: Rubiaceae

Distribution: Found in Rajasthan (Bundi, Kota, Udaipur).

Parts and properties used: Bark- Astringent, febrifuge.

Associated system(s): AF

Hyoscyamus muticus **L**

Sanskrit or Ayurvedic name(s): Paarsika-yavaani

Common name(s): Parasikaya

Family: Solanaceae

Distribution: Occur in northwest Himalaya. Grown in plains.

Parts and properties used: Plant- Sedative.

Associated system(s): AFU

Hyoscyamus niger **L.**

Sanskrit or Ayurvedic name(s): Parasika-yavani

Common name(s): Ajwain, Khursani Gaylangthang, Henbane

Family: Solanaceae

Distribution: Eurasian, does occur in temperate Himalayas, Ladakh. Grown.

Parts and properties used: Plant- Sedative. Yields a narcotic drug. Used in spasms of gastrointestinal region. Arial part- Give hyoscyamine. Treat asthma, stomach disorder, etc.

Associated system(s): AHSU

Hypecoum leptocarpum **Hook. f. et Thoms.**

Common name(s): Mirang/Parpata

Family: Papaveraceae

Distribution: Habitat to Himalayas (Ladakh, Sikkim).

Parts and properties used: Plant, juice- Antipyretic, antitussive, analgesic, and anti-inflammatory.

Associated system(s): F, Tibetan

Hypericum perforatum **L.**

Common name(s): St. John's wart

Family: Hypericaceae

Distribution: Habitat to temperate west Himalayas 2000-3000 m, including Ladakh.

Parts and properties used: Plant- Antidepressant, sedative, relaxing nervine, anti-inflammatory. Used in depression, menopausal nervousness, neuralgia menstrual cramps and rheumatism. Flower (oil)- Used in post-therapy.

Associated system(s): HU

Hyssopus officinalis **L.**

Sanskrit or Ayurvedic name(s): Dayaa-kunji

Common name(s): Zoofa, Jupha

Family: Lamiaceae/ Labiatae

Distribution: European, do occur in temperate west Himalayas. Cultivated.

Parts and properties used: Plant, fruit (seed)- Stimulant, carminative, sedative, diuretic, antispasmodic, pectoral. Used as expectorant.

Associated system(s): AFU

Ichnocarpus frutescens **(L.) W.T. Alton**

Sanskrit or Ayurvedic name(s): Gopavalli

Common name(s): Dudhi, Sariva

Family: Apocynaceae

Distribution: Occur in north India up to Sundarbans.

Parts and properties used: Root- Diuretic, demulcent, diaphoretic. Used in fevers, dyspepsia, and cutaneous affections.

Associated system(s): AFSL

Illicium griffithii **Hook. f. & Thomson**

Common name(s): Star Anise substitute

Family: Magnoliaceae

Distribution: Habitat to forest of ArP, and Himalayas.

Parts and properties used: Fruit- Stimulant and carminative.

Associated system(s): AS

Illicium verum **Hook. f.**

Common name(s): Star Anise

Family: Magnoliaceae

Distribution: Native Indo-China region. Imported

Parts and properties used: Fruit- Carminative, stimulant, diuretic. Used to treat respiratory tract and peptic discomforts, and rheumatism.

Associated system(s): SU

***Impatiens balsamina* Linn.**

Sanskrit or Ayurvedic name(s): Tarini

Common name(s): Gul-menhdi, Kasittumbai

Family: Balsaminaceae

Distribution: Native and cultivated throughout India.

Parts and properties used: Herb- Cathartic, diuretic, antirheumatic. Flowers- Used in burns.

Associated system(s): ASU

***Imperata cylindrica* (L.) Raeusch.**

Sanskrit or Ayurvedic name(s): Darbha

Common name(s): Thatch grass

Family: Poaceae/ Gramineae

Distribution: Habitat to hotter part of India.

Parts and properties used: Root- Diuretic, anti-inflammatory.

Associated system(s): AFS

***Indigofera articulata* auct. non-Gouan.**

Sanskrit or Ayurvedic name(s): Nili/Neel (sister species)

Common name(s): Surmai Nila

Family: Leguminosae/ Fabaceae

Distribution: Wild in Bihar, and peninsular India.

Parts and properties used: Root, leaf- Yields a bitter tonic. Seed- Anthelmintic.

Associated system(s): AFS

***Indigofera aspalathoides* Vahl ex DC.**

Sanskrit, Ayurvedic and Common name(s): Nili/Neel (sister species)

Family: Leguminosae/ Fabaceae

Distribution: Native, wild in plains of south India.

Parts and properties used: Plant- Antileprotic, antitumor, anti-inflammatory. Used in psoriasis and erysipelas.

Associated system(s): AFS

***Indigofera enneaphylla* Linn.**

Sanskrit or Ayurvedic name(s): Vaasukaa

Common name(s): Hanumaan-buuti

Family: Leguminosae/ Fabaceae

Distribution: Wild in Himalayas up to 1200 m.

Parts and properties used: Plant juice- Antiscorbutic, diuretic, alterative; decoction is given in epilepsy and insanity.

Associated system(s): AFS

***Indigofera oblongifolia* Forsk.**

Sanskrit or Ayurvedic name(s): Bana-Nila

Common name(s): Jhil

Family: Leguminosae/ Fabaceae

Distribution: Occur wild throughout India.

Parts and properties used: Plant- Ant syphilitic. Advised in liver and spleen enlargement.

Associated system(s): AFSU

***Indigofera pulchella* Roxb.**

Common name(s): Nirinji, (Nili- sister species)

Family: Leguminosae/ Fabaceae

Distribution: Grassland in hills of India.

Parts and properties used: Leaves and roots- Used for swelling of the stomach. Root- Used in cough; powder- in muscular pains.

Associated system(s): AS

***Indigofera tinctoria* L.**

Sanskrit or Ayurvedic name(s): Nilikaa

Common name(s): Akika, Avuri, Neel

Family: Leguminosae/ Fabaceae

Distribution: Tropical Africa to India Subcontinent to Indo China. Cultivated.

Parts and properties used: Plant- Anti-asthmatic/ anti-bronchitis, antiseptic, hepatoprotective, hypoglycaemic, nervine tonic. Treat many diseases. Advised in phobia, delusion, and mental disturbance.

Associated system(s): AHSLU

***Inula racemosa* Hook.f. (CR)**

Sanskrit or Ayurvedic name(s): Pushkaramuula

Common name(s): Pushkarmool, Manu

Family: Asteraceae/ Compositae

Distribution: Native to temperate/ alpine Himalayas 1500-4200 m. Grown.

Parts and properties used: Root- Antispasmodic, stomachic, antihistaminic, ant catarrhal. Used as expectorant.

Associated system(s): AL (Amchi)

***Ionidium suffruticosum* Ging.**

Sanskrit or Ayurvedic name(s): Amburuha

Common name(s): Ratna-Purush

Family: Violaceae.

Distribution: Habitat to warm parts of India.

Parts and properties used: Plant- Diuretic, antigonorrhoeic and demulcent. Root- Given to children in urinary/ bowel problem.

Associated system(s): AFS

***Ipomoea batatas* (Linn.) Lam.**

Sanskrit or Ayurvedic name(s): Mukhaaluka

Common name(s): Shakarkand

Family: Convolvulaceae

Distribution: Habitat to tropical America. Grown all over India.

Parts and properties used: Root- Used in strangury, urinary discharges, burning sensation. Plant- Used in fever and skin diseases.

Associated system(s): ASU

***Ipomoea mauritiana* Jacq., syn. *I. digitata* L.**

Sanskrit or Ayurvedic name(s): Kshira-vidaari

Common name(s): Palmudhukkan kizhangu

Family: Convolvulaceae

Distribution: Habitat to moist region of tropical India. Cultivated

Parts and properties used: Tuber- Cholagogue, galactagogue, alterative, demulcent, purgative. Used in tonic, *Chavanprash*.

Associated system(s): AFS

***Ipomoea nil* (L.) Roth = *I. muricata* (Linn.) Jacq., non-Cav.**

Sanskrit or Ayurvedic name(s): Antah-kotarpushpi

Common name(s): Kaladana

Family: Convolvulaceae

Distribution: Occur wild throughout India.

Parts and properties used: Plant- Purgative and blood purifier. Seed- Antifungal. Substitute of *Jalap* (*I. purga*) tuberous.

Associated system(s): ASU

***Ipomoea obscura* Ker-Gawl. = *I. marginata* (Desr.) Verdc**

Sanskrit, Ayurvedic and Common name(s): Lakshmanaa, Laksmana

Family: Convolvulaceae

Distribution: Found in Assam, Bihar, Odisha, and Maharashtra.

Parts and properties used: Plant- Dysentery, applied to sores/ pustules.

Associated system(s): AF

***Ipomoea pes-caprae* (L.) R.Br., syn. *I. biloba* Forssk.**

Sanskrit or Ayurvedic name(s): Chhagalaantri

Common name(s): Dopatilata

Family: Convolvulaceae

Distribution: Wild near sea on west coast.

Parts and properties used: Plant- Astringent, stomachic, laxative, antidiarrheal, antiemetic, analgesic. Leaf- diuretic, anti-inflammatory. Used to treat colic, prolapses rheumatism.

Associated system(s): AS

***Ipomoea petaloidea* Choisy**

Sanskrit or Ayurvedic name(s): Shyaamaa

Common name(s): Vrddhadaru

Family: Convolvulaceae

Distribution: Found throughout India.

Parts and properties used: Plant- Purgative, given in nervous diseases.

Associated system(s): ASU

***Ipomoea reptans* Poir., syn. *I. quatica* Forsk.**

Sanskrit or Ayurvedic name(s): Kalambi

Common name(s): Kalashaka

Family: Convolvulaceae

Distribution: Habitat to all over India. Grown for vegetable.

Parts and properties used: Plant- Emetic, purgative. Juice used in liver complaints.

Associated system(s): AS

***Ipomoea reniformis* Choisy**

Sanskrit or Ayurvedic name(s): Aakhuparni

Family: Convolvulaceae

Distribution: Habitat to damp places of Gangetic plains.

Parts and properties used: Plant- DE obstruent, diuretic, alterative. Used for rheumatic. affections, neuralgia, headache, skin, and urinary diseases.

Associated system(s): A

***Ipomoea sepiaria* Roxb.**

Sanskrit or Ayurvedic name(s): Banakalami

Common name(s): Thiruthaali = Lakshmana

Family: Convolvulaceae

Distribution: Found throughout India.

Parts and properties used: Plant juice- DE obstruent, diuretic, uterine tonic, hypotensive. Seeds- Cardiac depressant, hypotensive, spasmolytic. Source of Ayurvedic *Lakshmana.*

Associated system(s): AS

***Iris ensata* Thunb. = *I. germanica* L.**

Sanskrit or Ayurvedic name(s): Paarseeka Vachaa

Common name(s): Puskaramulam

Family: Iridaceae

Distribution: Habitat to temperate Himalayas, later Italian. Cultivated in Kashmir.

Parts and properties used: Plant- Used liver diseases; aerial parts contain xanthone glycosides; C-glycoside of apigenin and phenolic acids. Roots- Ceryl alcohol. Used in cosmetics.

Associated system(s): AHL

***Iris nepalensis* D. Don.**

Sanskrit or Ayurvedic name(s): Paarseeka Vachaa

Common name(s): Sosan, Shoti

Family: Iridaceae

Distribution: Habitat to temperate Himalayas and Khasi Hills.

Parts and properties used: Plant- DE obstruent, diuretic, cathartic. Used in liver diseases.

Associated system(s): AF

Ixora coccinea **L.**

Sanskrit or Ayurvedic name(s): Bandhuka

Common name(s): Thechippoovu

Family: Rubiaceae

Distribution: Native to south-western peninsular India. Grown.

Parts and properties used: Plant- Astringent, antiseptic, sedative, blood-purifier, anti- leucorrhoea, -diarrheal, -catarrhal. Root- Astringent, antiseptic.

Associated system(s): AFS

Ixora pavetta **Andr.**

Sanskrit or Ayurvedic name(s): Nevaari

Common name(s): Shulundu-kora

Family: Rubiaceae

Distribution: Found in West Bengal, Bihar, peninsular India.

Parts and properties used: Pounded flowers- Given in cough. Bark decoction- Given in anaemia, debility. Fruit/root- Female urine problems.

Associated system(s): AS

Jasminum angustifolium **(L.) Thunb.**

Sanskrit or Ayurvedic name(s): Vanamallika

Common name(s): Wild Jasmine

Family: Oleaceae

Distribution: Native to South India. Grown for offering to God.

Parts and properties used: Root- Used in deworming of ringworm. Leaf juice- Emetic, given in case of poisoning. Flower- Aromatic, gives indole.

Associated system(s): AFS

Jasminum arborescens **Roxb.**

Sanskrit or Ayurvedic name(s): Nava-mallikaa

Common name(s): Chameli

Family: Oleaceae

Distribution: Wild in sub-Himalayas, Assam, Odisha, and south India.

Parts and properties used: Leaves- Astringent and stomachic. Leaf juice- Emetic.

Associated system(s): AFLSU

***Jasminum auriculatum* Vahl.**

Sanskrit or Ayurvedic name(s): Yuuthikaa

Common name(s): Juuhi

Family: Oleaceae

Distribution: Found and grown throughout India.

Parts and properties used: Leaves- Give lupeol, its epimer, hentriacontane and *n*-triacontanol, a triterpenoid; jasminol; *d*-mannitol and volatile content jasmone. Flower- Rich in indole.

Associated system(s): AFS

***Jasminum flexile* Vahl.**

Sanskrit or Ayurvedic name(s): Maalati

Common name(s): Chameli

Family: Oleaceae

Distribution: Wild in northeast and Western Ghats.

Parts and properties used: Flower- Used like *J. officinale*, the Chameli.

Associated system(s): AFS

Jasminum heterophyllum* Roxb. non-Moench = *J. subhumile

Sanskrit or Ayurvedic name(s): Svarna-yuuthikaa

Common name(s): Juuhi, yellow variety

Family: Oleaceae

Distribution: Habitat to Nepal, Northeast of India.

Parts and properties used: Flower- Used like *J. humile*, the Juuhi.

Associated system(s): AF

***Jasminum humile* Linn.**

Sanskrit or Ayurvedic name(s): Svarna-yuuthikaa

Common name(s): Juuhi, yellow variety

Family: Oleaceae

Distribution: Native to sub-tropical Himalayas and hills of India.

Parts and properties used: Flower- Astringent. Yields cardiac tonic. Root- Deworming of ringworm. Exude- Treat fistulas.

Associated system(s): AFLS

***Jasminum multiflorum* (Burm.f.) Andr.**

Sanskrit or Ayurvedic name(s): Kunda

Common name(s): Star jasmine, Kasturi, Mogara

Family: Oleaceae

Distribution: Native to sub-tropical Himalaya to Western Ghats.

Parts and properties used: Plant- Diuretic, emetic. Boiled bark- Applied on burns.

Associated system(s): AFS

***Jasminum officinale* L. var. *grandiflorum* (L.)**

Sanskrit or Ayurvedic name(s): Kobuski, Jaati

Common name(s): Chameli, Royal jasmine

Family: Oleaceae

Distribution: Native to northwestern Himalayas. Cultivated.

Parts and properties used: Flower- Sedative, CNS depressant, astringent, anaesthetic. Used in hepatitis. Syrup- For cough. Plant- Diuretic, anthelmintic, emmenagogue. Used to treat ulcer. Oil- Used for relaxing.

Associated system(s): AFSLU

***Jasminum sambac* (L.) Aiton = *J. malabaricum* Wight.**

Sanskrit or Ayurvedic name(s): Madyantikaa

Common name(s): Bela, Mogra

Family: Oleaceae

Distribution: Native to tropical Asia. Cultivated all over India.

Parts and properties used: Root- Emmenagogue, blood purifier. Flower- Lactifuge. Source of oil. Extract- Hypotensive. Leaves- Antibacterial.

Associated system(s): AFSLU

***Jateorhiza palmata* Miers., syn. *J. calumba* Miers.**

Sanskrit or Ayurvedic name(s): Kalambaka

Common name(s): Kolumbu

Family: Menispermaceae

Distribution: Indigenous to south-east tropical Africa. Imported to India.

Parts and properties used: Root- Bitter, carminative, gastric tonic, ant flatulent, hypotensive, orexigenic, uterine stimulant, sedative. Used in anorexia, poor digestion, hypochlorhydria, amoebic dysentery and menstrual disorders. Antifungal, appetizer.

Associated system(s): AS

***Jatropha curcas* L.**

Sanskrit or Ayurvedic name(s): Vyaaghrairanda

Common name(s): Physic Nut

Family: Euphorbiaceae

Distribution: Tropical American. Grown in parts of India.

Parts and properties used: Seed- Toxic. Seed oil- Purgative. Plant- Treat scabies, ringworm, eczema, warts, syphilis.

Associated system(s): AFHSU

***Jatropha glandulifera* Roxb.**

Sanskrit or Ayurvedic name(s): Vyaaghrairanda

Common name(s): Jangali-erandi

Family: Euphorbiaceae

Distribution: Native, from south India to Bengal to Myanmar.

Parts and properties used: Root/seed oil- Purgative Oil- Antirheumatic, ant paralytic.

Associated system(s): AFS

***Jatropha gossypifolia* L**

Sanskrit or Ayurvedic name(s): Rakta-Vyaaghrairanda

Common name(s): Laal Bagharenda

Family: Euphorbiaceae

Distribution: Tropical American. Grown in parts of India.

Parts and properties used: Leaf/seed- Purgative, emetic; Oil- Used in treatment paralytic affection and skin care. Leaf- Antidermatotic. Bark- Emmenagogue.

Associated system(s): AFS

***Juglans regia* L.**

Sanskrit or Ayurvedic name(s): Aksoda

Common name(s): Akhrot, Walnut

Family: Juglandaceae

Distribution: Habitat to Balkans to Himalayas. Grown in Kashmir.

Parts and properties used: Leaves/bark- Alterative, laxative, antiseptic, anti-inflammatory, ant scrofula, detergent. Leaf infusion- Used in herps superficial inflammation of the skin and excessive perspiration.

Associated system(s): AFHSLU

Juniperus communis **L., syn.** ***J. effuses*** **L.**

Sanskrit or Ayurvedic name(s): Hapushaa

Common name(s): Hauber, Sukpa

Family: Cupressaceae

Distribution: Habitat to northern temperate hemisphere to Khasi Hills. Cultivated.

Parts and properties used: Berries- Diuretic, urinary antiseptic, carminative, digestive, sudorific, anti-inflammatory. Used in cystitis, dyspepsia.

Associated system(s): AHSLU

Juniperus macropoda **Boiss.**

Common name(s): Hapusa, Sukpa

Family: Cupressaceae

Distribution: Habitat to temperate Himalayas.

Parts and properties used: Mature Berries- Used like *J. communis*. Purgative.

Associated system(s): A

Jurinea macrocephala **DC., syn.** ***J. dolomiaea*** **Boiss. (NT)**

Sanskrit or Ayurvedic name(s): Jaatukanda

Common name(s): Dhoop, Guggul

Family: Asteraceae/ Compositae

Distribution: Habitat to Himalayas, including Ladakh.

Parts and properties used: Root- Used as incense. Stimulant, given in colic and fever.

Associated system(s): AF

Jussiaea tenella **Burm. f.**

Sanskrit or Ayurvedic name(s): Jala-lavanga

Family: Onagraceae

Distribution: Habitat to swamps of Bihar, Orissa, south India.

Parts and properties used: Root infusion- Used in syphilis. Poultice- To cure pimple.

Associated system(s): A

Justicia gendarussa **Burm. f., syn.** ***Gendarussa vulgaris*** **Nees**

Sanskrit or Ayurvedic name(s): Indrani, Krishna Vaasa

Common name(s): Nilanirgundi

Family: Acanthaceae

Distribution: Endemic to greater part of India. Cultivated.

Parts and properties used: Plant- Febrifuge, diaphoretic, emetic, emmenagogue. Leaf infusion- Treat cephalalgia, hemiplegia and facial paralysis.

Associated system(s): ASU

***Justicia procumbens* Linn.**

Common name(s): Parpata substitute

Family: Acanthaceae

Distribution: Habitat to Western Ghats.

Parts and properties used: Plant contains, naphthofuranones, justicidin A, B, C, D, G and H, and dyphylline. Treats osteoporosis.

Associated system(s): A

***Kaempferia galanga* L. (CR)**

Sanskrit or Ayurvedic name(s): Sugandha-vachaa

Common name(s): Galanga, Kachoram, Kapoor Kachri

Family: Zingiberaceae

Distribution: Native to South and Southeast Asia. Grown all over India.

Parts and properties used: Tuber- Stimulant, carminative, Diuretic. Used as expectorant, for cough, bronchitis, and asthma.

Associated system(s): AFS

***Kaempferia rotunda* L.**

Sanskrit or Ayurvedic name(s): Bhuumi-champaka

Common name(s): Bhuichampak, Karunkuvalai

Family: Zingiberaceae

Distribution: Native to Indochina. Cultivated.

Parts and properties used: Tuber- Antitumoral. Used in swellings, mumps, wounds.

Associated system(s): AFS

***Kalanchoe integra* (Medic.) Kuntze**

Sanskrit or Ayurvedic name(s): Parnabija

Common name(s): Rungru

Family: Crassulaceae

Distribution: Habitat to tropical Himalayas, Kashmir to Bhutan.

Parts and properties used: Plant- Hypotensive, antiarrhythmic.

Associated system(s): AFS

***Kalanchoe laciniata* (L.) DC.**

Sanskrit or Ayurvedic name(s): Parnabija

Common name(s): Hemsagarar

Family: Crassulaceae

Distribution: Found in Bengal, Deccan, Maharashtra.

Parts and properties used: Plant- Used in fever, dyspepsia, skin allergy, asthma, bronchitis. Leaves- Astringent, antiseptic, astringent. Source of lotion used insect bites.

Associated system(s): AS

***Kedrostis rostrata* (Rottler) Cogn., syn. *K. foetidissima* (Jacq.) Cogn.**

Common name(s): Arunkovai

Family: Cucurbitaceae

Distribution: Habitat to peninsular India.

Parts and properties used: Root/fruit- Demulcent. Used in asthma and piles.

Associated system(s): FS

***Kingiodendron pinnatum* (Roxb. ex DC.) Harms. (EN)**

Sanskrit or Ayurvedic name(s): Samparni

Common name(s): Kodapalai

Family: Leguminosae/ Fabaceae (Caesalpiniaceae)

Distribution: Found in hills of south India.

Parts and properties used: Oleo-gum-resin- Used in catarrh, and sours.

Associated system(s): AS

***Kyllinga triceps* Roxb.**

Sanskrit or Ayurvedic name(s): Nirvishaa

Common name(s): Nirbisi, Mustu

Family: Cypraceae

Distribution: Found in north-west, and south India.

Parts and properties used: Root- Febrifuge, antidermatotic. antidiabetic.

Associated system(s): AF

***Lactuca serriola* Linn.**

Sanskrit or Ayurvedic name(s): Vanya-Kaahuu

Common name(s): Salad

Family: Asteraceae/ Compositae

Distribution: Habitat to western Himalaya 1800-3300 m.

Parts and properties used: Plant- Diuretic, antiseptic, diaphoretic, expectorant. Seed- Demulcent; powder- used in cough.

Associated system(s): ALSU

***Lagenaria siceraria* (Molina) Standley**

Sanskrit or Ayurvedic name(s): Tumbini

Common name(s): Bottle-gourd

Family: Cucurbitaceae

Distribution: Native to tropical Africa, neutralized in Asia. Cultivated all over India.

Parts and properties used: Fruit pulp- Purgative, emetic. Leaf- Used in jaundice.

Associated system(s): AFSLU

***Lagerstroemia speciosa* (L.) Pers., syn. *L. flos-reginae* Retz.**

Sanskrit or Ayurvedic name(s): Jarula, Kramuka

Common name(s): Jarul

Family: Lythraceae

Distribution: Habitat to tropical India.

Parts and properties used: Root- Astringent, stimulant, febrifuge. Leaves- Purgative, diuretic, DE obstruent. Bark infusion- Given in diarrhoea. Seed- Narcotic.

Associated system(s): AFS

***Lagerstroemia parviflora* Roxb.**

Sanskrit or Ayurvedic name(s): Siddhaka

Common name(s): Lendia

Family: Lythraceae

Distribution: Found throughout India.

Parts and properties used: Plant- Astringent, fungitoxic.

Associated system(s): AF

Lallemantia royleana **(Wall. ex Benth) Benth.**

Common name(s): Tukhme-Balanga

Family: Lamiaceae/ Labiatae

Distribution: Occur from Europe to Asia. Grown in plains of Punjab and Kumaon Hills.

Parts and properties used: Seed- Cooling, diuretic, sedative; used in urinary troubles & cough; poultice- to abscesses, boils, and inflammations. Essential oil- Antibacterial.

Associated system(s): FU

Lamium album **Linn.**

Common name(s): White Dead Nettle

Family: Lamiaceae/ Labiatae

Distribution: Habitat to Western Himalayas.

Parts and properties used: Plant- Haemostatic, astringent, diuretic, anti-inflammatory, antispasmodic, expectorant. Used in respiratory catarrh, leucorrhoea.

Associated system(s): H

Lannea coromandelica **(Houtt.) Merr.**

Sanskrit or Ayurvedic name(s): Jingini

Common name(s): Kalasan, Moya

Family: Anacardiaceae

Distribution: Habitat to deciduous forests of India. Grown all over.

Parts and properties used: Bark- Stimulant and astringent; used in gout. Boiled leaves- Applied in sprains, and bruises. Gum- Given in asthma.

Associated system(s): AF

Lantana camara **L.**

Sanskrit or Ayurvedic name(s): Chaturaangi

Common name(s): Lantana

Family: Verbenaceae

Distribution: Grown as hedge in India.

Parts and properties used: Plant- Antirheumatic, Antimalarial. Leaf- Applied on cuts.

Associated system(s): AFS

***Launaea pinnatifida* Cass.**

Sanskrit or Ayurvedic name(s): Gojihvaa

Common name(s): Vana-gobhi

Family: Asteraceae/ Compositae

Distribution: Habitat to sandy coast of India.

Parts and properties used: Plant- Galactagogue, soporific, diuretic, aperient.

Associated system(s): AF

***Laurus nobilis* L.**

Common name(s): Hub-ul-ghar

Family: Lauraceae

Distribution: Mediterranean, cultivated throughout India.

Parts and properties used: Leaves- Gastric tonic, cholagogue, diaphoretic, antiseptic, antifungal. Oil- Rheumatism, hair

Associated system(s): FU

***Lavandula bipinnata* Kuntze.**

Sanskrit or Ayurvedic name(s): Shankhapushpi

Common name(s): Wild Lavender

Family: Lamiaceae/ Labiatae

Distribution: Habitat to many parts of India.

Parts and properties used: Plant- Used as a substitute for Shankhpushpi (*Convolvulus pluricaulas*).

Associated system(s): A

***Lavandula stoechas* L.**

Common name(s): Ustakuthus, Spanish lavender or French lavender

Family: Lamiaceae/ Labiatae

Distribution: Native to circa-Mediterranean countries. Imported in India.

Parts and properties used: Flower- Ant depressive, sedative, anticonvulsant, carminative, antispasmodic, antibacterial, antiseptic; used in depression. Oil- Rubefacient, anti-microbial.

Associated system(s): UF

***Lawsonia inermis* L.**

Sanskrit or Ayurvedic name(s): Madayanti

Common name(s): Henna, Mehndi

Family: Lythraceae

Distribution: Native to Iran to western India. Grown in semi-arid regions of India.

Parts and properties used: Leaf- Astringent, antihemorrhagic, antispasmodic, oxytocic, ant fertility, antimicrobial. Used in skin care, dysuria, jaundice, bleeding, etc.

Associated system(s): ASLU

***Leea aequata* Linn.**

Sanskrit or Ayurvedic name(s): Kaakajanghaa

Common name(s): Surapadi

Family: Vitaceae

Distribution: Habitat to northeast, and peninsular India.

Parts and properties used: Stem and root- Anthelmintic. Used in indigestion, jaundice, malaria. Oil- Astringent, Schroeder (cutter)

Associated system(s): AF

***Leea indica* Merrill.**

Sanskrit or Ayurvedic name(s): Karkatajihvaa

Common name(s): Karkani

Family: Vitaceae

Distribution: Habitat to tropical, subtropical forests of India.

Parts and properties used: Root- Antidiarrheal, antidysentery, antispasmodic, cooling, sudorific. Leaves- Digestive. Yields an ointment for vertigo.

Associated system(s): AFS

***Leea macrophylla* Roxb.**

Sanskrit or Ayurvedic name(s): Hastikanda

Common name(s): Hatikana

Family: Leeaceae

Distribution: Found all over hot India.

Parts and properties used: Plant- Astringent, anodyne, styptic, antiseptic. Tubers- Astringent, mucilaginous; applied to wounds, treat worms.

Associated system(s): AF

Leonotis nepetifolia **R. Br.**

Sanskrit or Ayurvedic name(s): Granthiparni

Common name(s): Hajurchei

Family: Lamiaceae/ Labiatae

Distribution: Found all over warm India.

Parts and properties used: Leaves- Spasmolytic. Flower ash- Treat burns, scalds, skin diseases and ringworm. Root- Advised in cough, bronchitis, and dyspnoea.

Associated system(s): AF

Lepidium sativum **L.**

Sanskrit or Ayurvedic name(s): Chandrashuura

Common name(s): Alivirai, Kurassaani, Halam, Chaunsar

Family: Brassicaceae/ Cruciferae

Distribution: Native to Egypt and west Asia. Cultivated all over India.

Parts and properties used: Plant- Anti-inflammatory, antirheumatic, anti-asthmatic/ anti-bronchitis. Seeds- Diuretic, emmenagogue lacteous. Powder advised in gout.

Associated system(s): ASU

Leptadenia reticulata **(Retz.) Wight & Arn.**

Sanskrit or Ayurvedic name(s): Jivanti

Common name(s): Dodishaak

Family: Ascelpiadaceae

Distribution: Habitat to sub-Himalayas to Deccan Peninsula.

Parts and properties used: Plant- Stimulant, restorative and improves eyesight. Leaves and root- Skin diseases.

Associated system(s): AFSL

Leucas aspera **(Willd.) Link**

Sanskrit or Ayurvedic name(s): Dronapushpi

Common name(s): Dronpushpi, Dharanpushpi

Family: Lamiaceae/ Labiatae

Distribution: Found all over India in cultivated fields.

Parts and properties used: Plant- Carminative, antihistaminic, antipyretic, febrifuge, antiseptic. Used in jaundice, anorexia.

Associated system(s): AFHS

Leucas cephalotes **(Roth.) Spreng.**

Sanskrit or Ayurvedic name(s): Dronpushpi

Common name(s): Guumaa, Tumbai

Family: Lamiaceae/ Labiatae

Distribution: Occur as weed all over India, up to 1800 m, Himalayas.

Parts and properties used: Dried plant- Stimulant, diaphoretic, antiseptic, insecticidal. Used in jaundice, asthma, etc. Flower syrup- Coughs and colds.

Associated system(s): AFLS

Lilium polyphyllum **D. Don (CR)**

Sanskrit or Ayurvedic and common name(s): Kaakoli

Family: Liliaceae

Distribution: Habitat to Uttarakhand and HP.

Parts and properties used: Root tuber- Roasted. Tonic in emaciation (Part of *Astavarga*).

Associated system(s): A

Limonia crenulata **Roxb., syn.** ***Naringi crenulata*** **(Roxb.) Nicolson**

Sanskrit or Ayurvedic name(s): Kakoli Bilvaparni

Common name(s): Ran-limbu

Family: Rutaceae

Distribution: Indo-Malayan. Grown throughout India.

Parts and properties used: Plant- Anti-inflammatory. Dried fruit- Stomachic; used in pestilent fevers, and antidote to poisons. Root- Purgative, sudorific.

Associated system(s): AFS

Linum usitatissimum **L.**

Sanskrit or Ayurvedic name(s): Atasi

Common name(s): Alsi

Family: Linaceae

Distribution: Central Asian. Grown in north-west-central India.

Parts and properties used: Seed- Demulcent, emollient, laxative, antilipidemic, pectoral antitussive. Used in constipation. Flower- Give nervine /cardiac tonic.

Associated system(s): AFHSLU

Lipasis rostrata **Rehd.**

Sanskrit or Ayurvedic name(s): Jivaka-Rshabhaka

Family: Orchidaceae

Distribution: Habitat to temperate regions of India.

Parts and properties used: Plant- Give age-sustaining, invigorating tonics.

Associated system(s): A

Lippia nodiflora **Rich.**

Sanskrit or Ayurvedic name(s): Jalapippali

Common name(s): Paduthalai

Family: Verbenaceae

Distribution: Found throughout wet parts of India.

Parts and properties used: Plant- Cooling, febrifuge, diuretic. Poultice used for boils.

Associated system(s): A

Liquidambar formosana **Hance.**

Common name(s): Silhak variety (Silaaras variety)

Family: Altingiaceae

Distribution: East Asian, introduced to Lalbagh Bangalore

Parts and properties used: Gum- Anti-inflammatory, stimulating expectorant, antiparasitic, antiseptic, anti-microbial. Bark- Used in skin diseases.

Associated system(s): ASU

Liquidambar orientalis **Mill.**

Sanskrit or Ayurvedic name(s): Turushka, Silhaka

Common name(s): Sweetgum or Turkish sweetgum

Family: Altingiaceae

Distribution: Mediterranean. Introduced to hills of India.

Parts and properties used: Resin/Gum- Used in skin diseases like fungi, scabies, gastric problems; asthma and bronchitis. Also, cancer.

Associated system(s): ASU

Litsea glutinosa **(Lour.) C.B. Rob., syn.** ***L. chinensis*** **Lam.**

Sanskrit or Ayurvedic name(s): Medaasakah

Common name(s): Menda-Lakadi, Naramamidi, Maidachal

Family: Lauraceae

Distribution: Found in north and south India.

Parts and properties used: Leaf- Antispasmodic, emollient. Bark- Demulcent, emollient, astringent, ant diarrheal, anodyne Berry oil and essential oil- Anti-rheumatic, anti-microbial.

Associated system(s): ASU

Litsea monopetala **(Roxb.) Pers.**

Sanskrit or Ayurvedic name(s): Maidaa-lakdi

Common name(s): Maidalagadil

Family: Lauraceae

Distribution: Habitat to Assam, Eastern Himalayas, TN.

Parts and properties used: Bark- Stimulant, astringent, spasmolytic, stomachic, antidiarrheal, contain alkaloids.

Associated system(s): AS

Lobelia inflata **Linn.**

Common name(s): Devanala

Family: Lobeliaceae

Distribution: Imported from US

Parts and properties used: Leaves- Anti-asthmatic, antispasmodic, bronchodilator, expectorant, and relaxant. Used in asthma, bronchitis.

Associated system(s): A, English

Lodoicea maldivica **(J.F. Gmel.) Pers. ex H.Wendl.**

Sanskrit or Ayurvedic name(s): Aklari, Samudra naarikela

Common name(s): Dariai narial, double coconut

Family: Arecaceae

Distribution: A cultivated palm.

Parts and properties used: Fruit water and kernel- Antacid and ant bilious. Advised in gastroenteritis.

Associated system(s): ASU

Lolium temulentum **Linn.**

Sanskrit or Ayurvedic name(s): Mochani

Common name(s): Visha-ghaasa

Family: Poaceae/ Gramineae

Distribution: Habitat to western Himalayas.

Parts and properties used: Herb- Sedative.

Associated system(s): AF

Loranthus falcatus **Linn. f.**

Sanskrit or Ayurvedic name(s): Bandaaka

Common name(s): Baandaa.

Family: Loranthaceae

Distribution: Found throughout India, as a parasite.

Parts and properties used: Bark- Stringent, and narcotic. Young shoots have tannin

Associated system(s): AF

Luffa acutangula **(L.) Roxb.**

Sanskrit or Ayurvedic name(s): Katukoshataki

Common name(s): Torai, Ridged gourd

Family: Cucurbitaceae

Distribution: Cosmopolitan. Cultivated throughout India

Parts and properties used: Plant- Purgative, diuretic. Used in oedema, splenic enlargement, cough asthma. Seeds- Emetic, expectorant.

Associated system(s): AHSLU

Luffa aegyptiaca **Mill., syn. *L. cylendrica* (L.) MJ Roem.**

Sanskrit or Ayurvedic name(s): Dhaamaargava

Common name(s): Ghia-tori

Family: Cucurbitaceae

Distribution: Native. Cultivated throughout India.

Parts and properties used: Plant- Used in pharyngitis, rhinitis, mastitis, oedema, swellings, burns. Leaves and juice- Treat bronchitis, amenorrhoea respectively.

Associated system(s): AFU

Luffa echinata **Roxb.**

Sanskrit or Ayurvedic name(s): Devadaali

Common name(s): Panibira, Bandaal

Family: Cucurbitaceae

Distribution: Wild to Northwest India.

Parts and properties used: Fruit- Purgative. Infusion- Given in biliary and intestinal colic. Have flavonoids.

Associated system(s): AFHSL

Luvunga scandens **(Roxb.) Buch-Ham. ex Wight & Arn.; (CR) and *L. eleutherandra* Dalz (unresolved)**

Sanskrit or Ayurvedic name(s): Lavanga-lataa

Common name(s): Sugandh-kokila

Family: Rutaceae

Distribution: Habitat to Bengal, Assam, and Khasi Hills.

Parts and properties used: Berries- Have Coumarin. Essential oil- Antifungal.

Associated system(s): AF

Lycopodium clavatum **L.**

Common name(s): Club Moss, Bendarali

Family: Lycopodiaceae

Distribution: Found in northeast, and Western Ghats

Parts and properties used: Plant- Sedative, antispasmodic, diuretic. Used in urinary disorder.

Associated system(s): FH

Lycopus europaeus **Linn.**

Common name(s): Gipsywort, Gandam-gundu

Family: Labiatae

Distribution: Habitat to Western Himalayas.

Parts and properties used: Plant- Cardioactive, diuretic, vasoconstrictor, sedative/ narcotic, antihemorrhagic, thyro-static, used in thyroid hyperfunction.

Associated system(s): F

Lygodium flexuosum **(L.) Sw. (Fern)**

Common name(s): Tsjeru-Valli-Panna

Family: Lygodiaceae

Distribution: Found throughout India.

Parts and properties used: Plant- Expectorant; decoction given in gastric attack. Roots- Rheumatism, sprains, cut wound, eczema, scabies.

Associated system(s): F

Madhuca indica **J.F. Gmel., syn.** ***Bassia latifolia*** **Roxb.**

Sanskrit or Ayurvedic name(s): Madhuuka

Common name(s): Mahuaa

Family: Sapotaceae

Distribution: Habitat to dry deciduous, offorests of Western and Eastern Ghats. Cultivated too.

Parts and properties used: Bark/seed oil/gum- Antirheumatic. Flower- Stimulant, demulcent, laxative, anthelmintic, bechic. Used in asthma and phthisis. Seed oil- Galactogenic, emetic; used in pneumonia, skin diseases. Bark- Astringent, emollient. Treat tonsillitis, gum, diabetes, ulcers.

Associated system(s): AFSL

Madhuca longifolia **(Koen) Mach.** ***M. indica*** **JF Gmel. (EN)**

Sanskrit or Ayurvedic name(s): Madhuulaka

Common name(s): Mahua phool

Family: Sapotaceae

Distribution: Found in central and north Indian plains & forests. Grown in UP, MP, Bihar, etc.

Parts and properties used: Properties same as *M. indica* Flowers- Used in renal diseases. Fruits- Used in rheumatism, cough, asthma, and consumption. Seed oil- used in rheumatism.

Associated system(s): AFS

Maerua arenaria **Hook. f. &Thoms.**

Sanskrit or Ayurvedic name(s): Madhusravaa

Common name(s): Murhari.

Family: Capparidaceae.

Distribution: Found over most of India.

Parts and properties used: Root- Used in piles, alterative (restorer) in fevers. As tonic in muscular debility.

Associated system(s): AFS

Maharanga emodi **(Wall.) DC., syn.** ***Onosma emodi*** **(Wall.) DC.**

Sanskrit or Ayurvedic name(s): Raktadala

Common name(s): Maharangi,

Family: Boraginaceae

Distribution: Native to Himalayas in Bhutan 3500-4000m, Assam to peninsular India.

Parts and properties used: Herb- Cooling, laxative, anthelmintic, used in diseases related to blood, bronchitis, abdominal pain, etc. Root- Treat wounds, pain, fevers, infectious diseases, hair dye. Flower- Stimulant/ cardiac tonic.

Associated system(s): F

***Malaxis acuminata* D. Don., syn. *Microstylis wallichii* L. (NT)**

Sanskrit or Ayurvedic name(s): Jivakah, Rishbhak

Common name(s): Jeevak, Jivak

Family: Orchidaceae

Distribution: Habitat to Himalayas between 1800- 2300 m altitude.

Parts and properties used: Plant- Aphrodisiac, antioxidant with cooling action, cures bleeding diathesis, fever, phthisis, and burning sensation. Produce rejuvenating tonic. Ingredient of *Astavarga*, under *Jivaniya Varga* of *Charaka Samhita.* Rhizome- Used in skin diseases, piles, bronchitis, tonic.

Associated system(s): A

***Malaxis muscifera* (Lindl.) Kuntze, syn. *Microstylis musifera* Ridley (CR)**

Sanskrit or Ayurvedic name(s): Jivaka

Common name(s): Rsabhakah, substitute of *Pueraria tuberosa*

Family: Orchidaceae

Distribution: Habitat to Himalayas 1500-2800 m.

Parts and properties used: Tuber- Rejuvenating tonic. Cure sterility, seminal weakness, dysentery, tonic. Part of *Astavarga* and *Chyavanprash.*

Associated system(s): A

***Mallotus philippensis* (Lam.) Muell. -Arg. (R)**

Sanskrit or Ayurvedic name(s): Kampillaka

Common name(s): Kameela

Family: Euphorbiaceae

Distribution: Found throughout tropical India.

Parts and properties used: Gland/ fruit hair- Purgative, styptic anthelmintic, treat tapeworm. Fruit- Hypoglycaemic, spasmolytic, antibacterial, anti-cancer (?).

Associated system(s): AFHSU

Malus pumila **Mill., syn -** ***M. domestica*** **Borkh.,** ***M. sylvestris*** **Hort. non-Mill.** ***Pyrus malus*** **Linn.**

Sanskrit or Ayurvedic name(s): Sinchitikaa

Common name(s): Seb, Sev

Family: Rosaceae

Distribution: Native to Europe and West Asia. Grown in Kashmir, HP, Kumaon, Assam, Nilgiris.

Parts and properties used: Bark- Anthelmintic, refrigerant, hypnotic, given in intermittent, remittent, and bilious fevers. Leaves- Inhibit the growth of gram-positive, gram-negative bacteria.

Associated system(s): AF

Malva rotundifolia **Linn.**

Sanskrit or Ayurvedic name(s): Suvarchalaa

Common name(s): Khubhaazi

Family: Malvaceae

Distribution: Habitat to hills and plains of north India.

Parts and properties used: Leaves- Demulcent, emollient, used in sugar, stomach problem. Seed- Demulcent. Treat bronchitis, cough, bladder inflammation.

Associated system(s): AU

Malva sylvestris **L.**

Sanskrit or Ayurvedic name(s): Suvarchalaa

Common name(s): Khawaji, Khubaji

Family: Malvaceae

Distribution: Native to Europe to Himalayas. Grown in hills of south India.

Parts and properties used: Fruit- Emollient, laxative, pectoral, antitussive, antibacterial. Used in irritative cough, mucosa, throat problems.

Associated system(s): AFU

Malva verticillata **L.**

Common name(s): Sucheli, Chinese mallow or cluster mallow

Family: Malvaceae

Distribution: Eurasian, Habitat to Himalaya, including Ladakh.

Parts and properties used: Seed- Used in renal disorders, the retention of fluids, frequent thirst, and diarrhea. Yields a liver tonic.

Associated system(s): L, Tibetan

***Mammea suriga* (Buch. -Ham. ex Roxb.) Kosterm.**

Common name(s): Nagakesr substitute

Family: Clusiaceae/ Guttiferae

Distribution: Habitat to Western Ghats.

Parts and properties used: Flower bud- Stimulant, astringent, carminative. Used in dyspepsia.

Associated system(s): AF

***Manihot esculenta* Crantz.**

Common name(s): Tapioca, Maravalli kizhangu

Family: Euphorbiaceae

Distribution: Native of Brazil. Grown southern states of India.

Parts and properties used: Tuber- Food. Used to treat hypertension, headache, and other pains, irritable bowel syndrome and fever, particularly bitter varieties.

Associated system(s): FS

***Mangifera indica* L.**

Sanskrit or Ayurvedic name(s): Amra

Common name(s): Aam, Aamba

Family: Anacardiaceae

Distribution: Native. Cultivated all over India.

Parts and properties used: Leaves- Anti-inflammatory, antibacterial, chlorotic, diuretic. Fruit, unripe- Antiscorbutic, astringent; ripe- invigorating and refrigerant. Kernel- Anti-inflammatory, antimicrobial, anthelmintic, antispasmodic, antiscorbutic; dried used in diarrhoea and dysentery.

Associated system(s): AFHSLU

***Manilkara kauki* (L.) Dubard.**

Sanskrit or Ayurvedic name(s): Khirni

Common name(s): Palai.

Family: Sapotaceae

Distribution: Habitat to tropical Asia. Grown in gardens.

Parts and properties used: Root/ bark- Astringent. Given in diarrhoea. Fruit juice- cooling. Seed- Febrifuge, anthelmintic, antileprotic.

Associated system(s): AS

***Maranta arundinacea* L.**

Common name(s): Arrowroot, Citalapattiri

Family: Marantaceae

Distribution: Native to Americas rainforest. Grown in Kerala.

Parts and properties used: Rhizome- Nutritive, demulcent. Source of neutral starch.

Associated system(s): AFS

***Marsdenia roylei* Wight**

Sanskrit or Ayurvedic name(s): Muurvaa

Common name(s): Murva

Family: Apocynaceae

Distribution: Habitat to Himalayas, and Assam Hills.

Parts and properties used: Root- Purgative.

Associated system(s): AF

***Marsdenia sylvestris* (Retz.) P.I. Forst.; syn. *Gymnema sylvestre* (Retz.) Schult.**

Common name(s): Gurmari, Gymnema cow plant

Family: Apocynaceae

Distribution: African. Grown in Southern and Central India.

Parts and properties used: Leaf tea- Reduces blood sugar, stimulates the circulatory system, increases urine secretion.

Associated system(s): F

***Marsdenia tenacissima* (Roxb.) Moon**

Sanskrit or Ayurvedic name(s): Muurvaa

Common name(s): Nishod sufed

Family: Apocynaceae

Distribution: Occur in Himalayas, Assam, and peninsular India.

Parts and properties used: Root- Purgative, antispasmodic, mild CNS depressant. Used in colic.

Associated system(s): AF

***Marsilea quadrifolia* L.**

Ayurvedic and common name(s): Cupatiya

Family: Marsileaceae

Distribution: European. Grows in moist places of India.

Parts and properties used: Plant- Anti-inflammatory, diuretic, depurative, febrifuge, and refrigerant.

Associated system(s): AFS

Marsilea minuta **Linn.**

Sanskrit or Ayurvedic name(s): Sunishannaka,

Family: Marsileaceae

Distribution: Found throughout India in marsh lands.

Parts and properties used: Plant- Sedative. Used in insomnia, epilepsy, and epilepsy. Active principle- marsilin (1-triacontanol cerotate). Arrest greying of hair.

Associated system(s): A

Martynia annua **L.**

Sanskrit or Ayurvedic name(s): Kaakanasika

Common name(s): Kaknasa, Hathajor

Family: Pedaliaceae/ Martyniaceae

Distribution: Native to America, naturalized in India.

Parts and properties used: Leaf- Used in epilepsy. Fruit- Anti-inflammatory.

Associated system(s): AFS

Matricaria chamomilla **L.**

Common name(s): Chamomile, Baabunaa

Family: Asteraceae/ Compositae

Distribution: Native to Europe. Grown in J & K, HP.

Parts and properties used: Plant- Sedative, antiseptic. Used in inflammatory diseases, stomach care, externally skin, bowel, sleep.

Associated system(s): HF

Meconopsis aculeata **Royle (CR)**

Sanskrit or Ayurvedic name(s): Akshatsermum

Common name(s): Himalaya Blue Poppy, Gul-e-Nilam, Kunda

Family: Papaveraceae

Distribution: Endemic to western Himalayas 3300-4500 m. Cultivated.

Parts and properties used: Plant- Diuretic, blood purifier. Root- Narcotic. Used in ulcer, liver and lung disorders. Tibetan use for inflammation from fractures and pain of legs and ribs. *Jwaraghna* (antipyretic; reduce fever) and heat troubles.

Associated system(s): AL (Amchi)

***Medicago sativa* L.**

Sanskrit or Ayurvedic name(s): Vilaayatigawuth

Common name(s): Alfalfa

Family: Leguminosae/ Fabaceae

Distribution: Native of South-central Asia and peninsular India. Grown.

Parts and properties used: Leaves- Anticholesterolemic. Used to prevent high BP and night blindness.

Associated system(s): AHUF (Bhil)

***Melia azedarach* L.**

Sanskrit or Ayurvedic name(s): Mahaanimba

Common name(s): Bakayan phal

Family: Meliaceae

Distribution: Habitat to sub-Himalaya, grown all over India.

Parts and properties used: Leaf- Anthelmintic, antilithic, diuretic; Leaf/flower- Febrifuge, sedative, emmenagogue. Dried bark- Used in turbid urine, skin diseases, nausea. Seed oil- Antirheumatic, insecticidal.

Associated system(s): AFSLU

***Melilotus indica* (Linn.) All. *M. parviflora* Desf.**

Sanskrit or Ayurvedic name(s): Vana-methikaa

Common name(s): Senji

Family: Leguminosae/ Fabaceae (Papilionaceae)

Distribution: Native to Eurasia. Cultivated in north India as fodder.

Parts and properties used: Plant- Astringent, discutient, emollient, poultice. Gave coumarins.

Associated system(s): AFU

***Melothria maderaspatana* (L.) Cogn.**

Sanskrit or Ayurvedic name(s): Ahilekhana

Common name(s): Agmaki

Family: Cucurbitaceae

Distribution: Occur throughout India.

Parts and properties used: Tender shoots- Aperient, diuretic, stomachic; decoction used in biliousness and flatulence.

Associated system(s): AFS

***Melocanna bambusoides* Trin.**

Common name(s): Bansalochan, bamboo variety

Family: Poaceae/ Gramineae

Distribution: Found in tropical India.

Parts and properties used: Clum- Tabasheer, siliceous concretion (solid mass) of culms- Yield tonic. Used to treat respiratory diseases.

Associated system(s): F

***Memecylon edule* Roxb., syn. *M. umbellatum* Burm. f.**

Sanskrit or Ayurvedic name(s): Anjani

Common name(s): Yaalki

Family: Melastomataceae.

Distribution: Found in Odisha, Assam, and Western peninsula.

Parts and properties used: Fruit and leaf- Astringent. Leaf lotion- Anti-leucorrhoea, spasmolytic, hypoglycaemic.

Associated system(s): AFS

***Mentha aquatica* L.**

Common name(s): Water-mint

Family: Lamiaceae/ Labiatae

Distribution: Sub-cosmopolitan. Cultivated in Indian garden.

Parts and properties used: Leaf- Stimulant, astringent. Used in diarrhoea and dysmenorrhoea.

Associated system(s): F

***Mentha arvensis* L.**

Common name(s): Pudina, Podina pati

Family: Lamiaceae/ Labiatae

Distribution: Sub-cosmopolitan. Cultivated in J & K.

Parts and properties used: Plant- Carminative, cholagogue, expectorant, antimicrobial. Mint oil- Used in gastrointestinal problems and catarrhs.

Associated system(s): FSU

Mentha longifolia (L.) L.

Common name(s): Horse mint, Phuloling (wild mint)

Family: Lamiaceae/ Labiatae

Distribution: European. Cultivated in J & K and several states.

Parts and properties used: Leaf and flower top- Antiseptic and carminative. Used for headache and stomach disorder.

Associated system(s): U

Mentha piperita **L.**

Common name(s): Menthol, Peppermint

Family: Lamiaceae/ Labiatae

Distribution: European. Cultivated in J & K, Punjab, Maharashtra.

Parts and properties used: Leaf- Gastrointestinal tract septic. Oil- Digestive, carminative, chlorotic, antiseptic antispasmodic, diuretic, antiviral antiemetic, diaphoretic. Used in mixtures.

Associated system(s): AHU

Mentha viridis **(L.) L., syn.** ***M. spicata*** **Linn. emend. Nathh.**

Sanskrit or Ayurvedic name(s): Pudinah

Common name(s): Pudino

Family: Lamiaceae/ Labiatae

Distribution: Sub-cosmopolitan. Cultivated in Punjab, UP, Maharashtra.

Parts and properties used: Plant- Carminative, stimulant, antispasmodic, antiemetic, diaphoretic, antiseptic. Given in tracheobronchitis & hypertension.

Associated system(s): AFU

Merremia emarginata **(Burm. f.) Hallier f.**

Ayurvedic and Common name(s): Underkarni, Mooshkarni

Family: Convolvulaceae

Distribution: Habitat to tropics, Kerala, TN.

Parts and properties used: Leaves- Diuretic, alterative. Used in rheumatism, neuralgia, cough, headache, fever due to liver and kidney diseases.

Associated system(s): AS

Merremia hederacea **(Burm. f.) Hallier f.**

Ayurvedic and Common name(s): Kudici-Valli

Family: Convolvulaceae

Distribution: Found in Assam, Odisha, Meghalaya.

Parts and properties used: Plant- Antimicrobial. Treat cold, febrile disease, sunstroke, oliguria, tonsil inflammation, laryngitis. Seeds- Used in fevers, sore throats, hematuria, conjunctivitis & boils.

Associated system(s): AFS

***Merremia tridentata* (L.) Hallier f.**

Sanskrit, Ayurvedic and Common name(s): Prasaarini, Prasarani

Family: Convolvulaceae

Distribution: Habitat to tropics. Grown in upper Gangetic Plains.

Parts and properties used: Plant- Laxative, astringent, anti-inflammatory. Used in piles, rheumatic affection.

Associated system(s): AFS

***Mesua ferrea* L., syn. *M. nagassarium* (Burm. f.) Kosterm.]**

Sanskrit or Ayurvedic name(s): Nagakeshara

Common name(s): Nagakesari, Sirunagappo

Family: Clusiaceae/ Guttiferae

Distribution: Habitat to Eastern Himalayas, Northeast region, and Western Ghats.

Parts and properties used: Flower bud- Antidysentery. Flower- Astringent, haemostatic, anti-inflammatory, stomachic. Used in cough, piles, metrorrhagia. Dry stamens- Used in gout, urinary and haemorrhagic disorders.

Associated system(s): AFHSLU

***Meyna laxiflora* Robyns**

Sanskrit or Ayurvedic name(s): Pinditaka

Common name(s): Aaliv, Helu

Family: Rubiaceae

Distribution: Found in West Bengal, Bihar, Odisha.

Parts and properties used: Fruit- Cholagogue; decoction used in biliary complaints, hepatic congestion; dried- as narcotic.

Associated system(s): AFS

***Michelia champaca* L. (VU)**

Sanskrit, Ayurvedic and common name(s): Champaka, Champaka, Champa

Family: Magnoliaceae

Distribution: Habitat to Eastern Himalaya, hills of South India. Cultivated.

Parts and properties used: Root- Purgative & emmenagogue. Bark- Stimulant, diuretic, febrifuge. Flower- Carminative, anti-spasmodic, demulcent, antiemetic, diuretic, antipyretic. Fruit- Given in dyspepsia, renal diseases.

Associated system(s): AFST

Miliusa velutina **Hook. f. &Thoms.**

Sanskrit or Ayurvedic name(s): Rshiyaproktaa

Common name(s): Gandha-Palaasa

Family: Annonaceae.

Distribution: Habitat to sub-Himalayas, Indian peninsula, and east coast.

Parts and properties used: Bark- Used in treatment of gout.

Associated system(s): AF

Mimosa pudica **L.**

Sanskrit or Ayurvedic name(s): Anjalikarik, Ajàlikalika, Namaskâri Lajjaalu

Common name(s): Lajwanti, Lajjalu

Family: Leguminosae/ Fabaceae (Mimosaceae)

Distribution: Habitat to tropical and subtropical India.

Parts and properties used: Leaf- Astringent, alterative, antiseptic, styptic, blood purifier, treat several diseases. Root- Used in gravel and urinary complaints, decoction in asthma.

Associated system(s): AFSL

Mimusops elengi **L.**

Sanskrit, Ayurvedic and common name(s): Bakula, Bakul

Family: Sapotaceae

Distribution: Found in tropical forests of South Asia. Cultivated in most of India.

Parts and properties used: Flower, fruit, bark- Astringent, antiulcer. Fruit pulp- Cure chronic dysentery. Leaves- Contain sterols. Seeds- Purgative, spermicidal. Essential oil- Mycotoxin. Used in dental ailments, dysentery and to promote fertility.

Associated system(s): AFSLU

Mirabilis jalapa **L.**

Sanskrit or Ayurvedic name(s): Sandhyarága, Krishnakeli, Trisandhi

Common name(s): Gul-abash. four o'clock flower

Family: Nyctaginaceae

Distribution: Habitat to tropical South America & Himalayas, Bengal Northeast India. Grown as ornamental.

Parts and properties used: Plant- Antitumor, virus-inhibitory. Leaf- To treat uterine discharge, as poultice for boils. Root- Purgative, spasmolytic.

Associated system(s): AFSU

Mitragyna parvifolia **(Roxb.) Korth.**

Sanskrit or Ayurvedic name(s): Giri-kadamba

Common name(s): Kadamb

Family: Rubiaceae

Distribution: Native, found all over India.

Parts and properties used: Bark- Relieve muscular pain. Bark/root- Febrifuge, anti-spasmodic.

Associated system(s): AFS

Mollugo cerviana **Ser.**

Sanskrit or Ayurvedic name(s): Grishma-Sundara (Parpata Substitute)

Family: Molluginaceae

Distribution: Occur in most parts of India.

Parts and properties used: Plant- Stomachic, aperient, febrifuge, antiseptic, blood purifier, emmenagogue. Flower/stem- Diaphoretic. Root- Used in rheumatism/ gout.

Associated system(s): AFS

Mollugo pentaphylla **L. and *M. spergula* Linn**

Sanskrit or Ayurvedic name(s): Grishma-sundara

Common name(s): Turapoondu

Family: Molluginaceae

Distribution: Found in greater part of India.

Parts and properties used: Plant- Stomachic, aperient, antiseptic. Promotes digestion, and cure eye sore.

Associated system(s): FS

Momordica balsamina **Linn.**

Common name(s): Balsam Apple, Jangali Karela

Family: Cucurbitaceae

Distribution: Wild in northwest India, Andhra Pradesh

Parts and properties used: Plant- Contains a ribosome inactivating protein, momordin II; Esters exhibited antihypertensive, analgesic, and antibacterial. Fruit- Treat burns.

Associated system(s): AF

Momordica charantia **L.**

Sanskrit or Ayurvedic name(s): Karavallaka

Common name(s): Bitter melon, Karela

Family: Cucurbitaceae

Distribution: Native. Widely cultivated all over India.

Parts and properties used: Fruit- Antirheumatic, stomachic, laxative, ant bilious, emetic. Used in respiratory and skin diseases, intestinal worms, gout, etc. Seed/fruit- Cure diabetes. Root- Astringent, to haemorrhoids.

Associated system(s): AFSLU

Momordica cochinchinensis **Spreng.**

Sanskrit or Ayurvedic name(s): Karkataka

Common name(s): Kakrol

Family: Cucurbitaceae

Distribution: Southeast Asian, cultivated all over India.

Parts and properties used: Leaf/fruit- Used in lumbago, ulceration, bone fracture. Seed- Be chic, aperient, emmenagogue, anti-inflammatory, de-obstruent. Used in liver and spleen obstructions. Root- Have Saponin.

Associated system(s): AF

Momordica dioica **Roxb. *ex* Willd.**

Sanskrit or Ayurvedic name(s): Karkotikaa

Common name(s): Jangli-Kareli

Family: Cucurbitaceae

Distribution: Native to India. Cultivated all over India.

Parts and properties used: Tuberous root- Febrifuge, antiseptic, anthelmintic, spermicidal. Used in piles, urinary affections. Root extract- Anti-allergy.

Associated system(s): AFS

Momordica tuberosa **(Roxb.) Cogn., syn. *M. tuberosa*** **(Roxb.) Cogn.**

Sanskrit or Ayurvedic name(s): Kaarali-Kanda

Common name(s): Kakrol

Family: Cucurbitaceae

Distribution: Wild to peninsular India on bushes along water course.

Parts and properties used: Tuberous root-Emmenagogue, abortifacient; acrid; contain bitter glycoside. Treat diabetes.

Associated system(s): AFS

Monochoria vaginalis **(Burm. f.) Presl.**

Sanskrit or Ayurvedic name(s): Indivara

Common name(s): Nukha, Kolai

Family: Pontederiaceae

Distribution: Habitat to moist places of India.

Parts and properties used: Leaf juice- Used in cough. Root- Stomach and liver disorder.

Associated system(s): AFS

Morchella esculenta **(L.) Pers. and** ***M. conica*** **Pers. (Edible mushroom)**

Common name(s): Guchhi

Family: Morchellaceae

Distribution: Habitat to humid soil of forest and gardens in J & K, wild medicinal. Cultivated.

Parts and properties used: Fruiting body- Flavoring agent. Used in indigestion, phlegm, breathing shortness.

Associated system(s): F

Morinda citrifolia **L.**

Sanskrit or Ayurvedic name(s): Ashyuka

Common name(s): Noni, Nunaa

Family: Rubiaceae

Distribution: Habitat to Indian coasts, Andaman Nicobar. Grown marginally.

Parts and properties used: Fruit juice- Emmenagogue, anti-leucorrhoea, anti-dysentery, ant catarrhal. Used in asthma.

Associated system(s): AFS

Morinda pubescens **J.E. Sm.,** ***M. coreia*** **Buch. -Ham.;** ***M. tinctoria*** **Roxb. var.** ***tomentosa*** **(Heyne ex Roth) Hook.f.**

Ayurvedic and Common name(s): Manjanatthi

Family: Rubiaceae

Distribution: Wild in Eastern Ghats.

Parts and properties used: Root, stem bark- Used in rheumatism.

Associated system(s): AFS

***Moringa concanensis* Nimmo ex Dalzell & A. Gibson (VU)**

Sanskrit or Ayurvedic name(s): Shigru (wild relative)

Common name(s): Kadvo saragvo

Family: Moringaceae

Distribution: Wild in southern peninsular India.

Parts and properties used: Plant- Reduce cholesterol. Leaves- Aphrodisiac. Used in fatigue, eye care. Flower- Cure thyroid problem. Bark- Reduce pain.

Associated system(s): ASL

***Moringa oleifera* Lam., syn. *M. pterygosperma* Gaertn. (NE)**

Sanskrit or Ayurvedic name(s): Shigru

Common name(s): Sahenjana chaal, Drumstick

Family: Moringaceae

Distribution: Native and widely cultivated all over India.

Parts and properties used: Pods- Antipyretic, anthelmintic. Advised in diabetes. Flower- Cholagogue, stimulant, diuretic. Root juice- Diuretic, anti-epileptic. Yields cardiac tonic. Used in nervous debility, asthma, inflamed liver and spleen, and calculus affection.

Associated system(s): AFSLU

***Morus alba* L.**

Sanskrit or Ayurvedic name(s): Tutam

Common name(s): Shahtoot, Toot

Family: Moraceae

Distribution: Native to Indochina. Cultivated in northern India.

Parts and properties used: Fruit- Cooling, laxative. Treat sore throat, dyspepsia, and melancholia. Leaf- Anti-inflammatory, emollient, diaphoretic, used gargle. Leaves/root bark- Diuretic, expectorant, hypotensive. Bark extract- Hypoglycaemic.

Associated system(s): AFLU

***Morus nigra* Linn.**

Common name(s): Black Mulberry, Tuut Siyaah

Family: Moraceae

Distribution: Native to southwestern Asia. Grown in Kashmir & Darjeeling.

Parts and properties used: Berries/root bark- Laxative. Treat respiratory catarrh. Berries- Refrigerant. Given during convalescence.

Associated system(s): U

Mucuna monosperma **DC.**

Sanskrit or Ayurvedic name(s): Dadhipushpi Kaakaandolaa.

Common name(s): Kaagadolia

Family: Leguminosae/ Fabaceae

Distribution: Indomalaya region, northeast India, Andamans.

Parts and properties used: Seeds- Sedative, restorative, expectorant. Used in treatment of coughs, asthma.

Associated system(s): AFS

Mucuna pruriens **(L.) DC., syn.** ***M. prurita*** **Hook.;** ***M. pruriens*** **var.** ***utilis*** **W.;** ***M. deeringiana*** **(Bort) Merr.**

Sanskrit or Ayurvedic name(s): Aatmagupta, Kapikachchhu, Vanari

Common name(s):): Cowach, Konchh, Poonaikkaali, Bengal bean, buffalo bean, velvet bean.

Family: Leguminosae/ Fabaceae

Distribution: American extending to tropical Africa and Asia, including India. Cultivated.

Parts and properties used: Seed- Nervine tonic, stimulant. Treat impotency, spermatorrhoea, urinary troubles and leucorrhoea. Leaf- Applied on ulcers.

Associated system(s): AFSLU

Murdannia nudiflora **(L.) Brenan.all. ex Wight.**

Sanskrit or Ayurvedic name(s): Koshapushpi

Common name(s): Kanshura

Family: Commelinaceae

Distribution: Found all over moist and marshy places of India.

Parts and properties used: Plant- Used in treatment of burns, boils, and sores.

Associated system(s): AF

Murraya koenigii **(L.) Spreng.**

Sanskrit or Ayurvedic name(s): Surabhini-nimba

Common name(s): Kariveppila

Family: Rutaceae

Distribution: Native and cultivated in parts of India.

Parts and properties used: Leaf- Aromatic, stomachic, anti-protozoal, spasmolytic, anti-dysenteric. Externally, used in skin eruptions. Rich in carbazole alkaloids.

Associated system(s): AFSU

Musa sapientum **L., syn.** ***M. paradisiaca*** **Linn.**

Sanskrit or Ayurvedic name(s): Sakrtphala

Common name(s): Kadali, Kela

Family: Musaceae

Distribution: Grown in tropical and subtropical regions of India.

Parts and properties used: Fruit- Laxative, reduce diarrhoea and dysentery, heal intestinal lesions in ulcerative colitis. Unripe- Cure diabetes. Fresh rhizome- Advised in dysuria, polyuria, menstrual disorders. Flower- Used in asthma, bleeding, vaginal discharges, leucorrhoea.

Associated system(s): ASU

Mussaenda frondosa **L.**

Sanskrit or Ayurvedic name(s): Shrivati.

Common name(s): Sribati

Family: Rubiaceae

Distribution: Habitat to tropical Himalayas to peninsular India, Andamans.

Parts and properties used: Flower- Diuretic, anti-asthmatic, antiperiodic. Leaf/ flower- Treat ulcers. Root- Treat white leprosy.

Associated system(s): AFS

Myrica esculenta **Buch. -Ham. ex D. Don, syn.** ***M. nagi*** **auct. non Thunb.**

Sanskrit or Ayurvedic name(s): Katphala

Common name(s): Katphala, Kaiphal

Family: Myricaceae

Distribution: Habitat to sub-tropical Himalayas 900-2100 m.

Parts and properties used: Bark- Carminative, antiseptic. Used in cough, asthma. Fruit- Pectoral, sedative. Fruit wax- Used to treats ulcer.

Associated system(s): AFSLU

Myristica dactyloides **auct. non Gaertn., syn.** ***M. malabarica*** **(VU)**

Sanskrit or Ayurvedic name(s): Paashikaa, Raamapatri

Common name(s): Jaiphal, Javatri, Nutmeg

Family: Myristicaceae

Distribution: Endemic to south India and Sri Lanka.

Parts and properties used: Fruit (Aril/ Mace, Kernel)- Used in treatment of cough, bronchitis, fever, inflammation of joints, skin, wounds, sleeplessness, indigestion, liver disorders and deworming.

Associated system(s): AS

***Myristica fragrans* Houtt.**

Sanskrit or Ayurvedic name(s): Jaatiphala

Common name(s): Jatipatre, Mace

Family: Myristicaceae

Distribution: Indonesian. Cultivated in Bengal, South India.

Parts and properties used: Fruit- Carminative, spasmolytic, antiemetic, orexigenic; topically anti-inflammatory. Mace (dried cover of seed)- Stimulant carminative. Hexane extract- Has myristicin, an anti-inflammatory principle, and Licarin-B, *Dehydro-di-isoeugenol,* which exhibit CNS depressant.

Associated system(s): AFHSLU

***Myristica malabarica* Lam. (VU)**

Sanskrit or Ayurvedic name(s): Paashikaa

Common name(s): Rampathiri

Family: Myristicaceae

Distribution: Habitat to Western Ghats.

Parts and properties used: Plant- Topically stimulant. Applied on indolent ulcers. Fruit rind- Yields diarylnonanoids and a lignin, malabaricanol.

Associated system(s): AFS

***Myrsine africana* L.**

Sanskrit or Ayurvedic name(s): Vaayavidanga

Common name(s): Chapra, Vidanga

Family: Myrsinaceae

Distribution: Habitat to outer Himalayas, Khasi Hills 300-2700 m.

Parts and properties used: Fruit- Anthelmintic, anti-spasmodic, purgative. Externally used in treatment ringworm, skin affection.

Associated system(s): AU

***Naravelia zeylanica* (Linn.) DC.**

Sanskrit or Ayurvedic name(s): Dhanavalli

Common name(s): Vathomkolli

Family: Ranunculaceae.

Distribution: Habitat to tropical forests of India.

Parts and properties used: Plant- Astringent, anti-inflammatory, vulnerary, anthelmintic. Used for colic, inflammations, rheumatic pain, ulcers, intestinal worms, leprosy, and skin diseases.

Associated system(s): AS

Nardostachys grandiflora **Wall. ex DC., syn.** ***N. jatamansi*** **(D. Don). (EN-CR)**

Sanskrit or Ayurvedic name(s): Jataamansi

Common name(s): Balchad, Jatamansi

Family: Valerianaceae

Distribution: Habitat to alpine Himalayas, Northeast region.

Parts and properties used: Rhizome- Sedative, hypotensive. Used in epilepsy, hysteria, convulsive affections, palpitation of heart and intestinal colic. Root powder/ decoction- Used in treatment High BP.

Associated system(s): ASU

Nasturtium officinale **R. Br.**

Sanskrit or Ayurvedic name(s): *Kumbhayoni*

Common name(s): Chhooch, Manju Lata, Watercress, *Simrayo or Sim-Saag*

Family: Brassicaceae/ Cruciferae

Distribution: Habitat to north hemisphere. Cultivated.

Parts and properties used: Leaves- Antiscorbutic, expectorant, diuretic, detoxifying. Used in catarrh of respiratory tract.

Associated system(s): F

Nelumbo nucifera **Gaertn.**

Sanskrit or Ayurvedic name(s): Ambuj, Kamala

Common name(s): Kamal phool.

Family: Nelumbonaceae

Distribution: Native to warmer parts of India.

Parts and properties used: Filament- Astringent and haemostatic. Used in treatment of bleeding piles and menorrhagia. Flower decoction- Cholera, fever, palpitation of heart. Root- Astringent, diuretic, antiemetic. Used in dysentery, dyspepsia, piles, skin affections and as anticoagulant.

Associated system(s): AFSLU

Nepeta cataria **L.**

Common name(s): Badranjboya

Family: Lamiaceae/ Labiatae

Distribution: Habitat to trans-Himalaya, including Ladakh. Cultivated in gardens.

Parts and properties used: Leaves and flowers- Sedative, carminative, antispasmodic, antidiarrheal, diaphoretic, febrifuge. Used in restlessness, convulsions, colic, influenza. Plant- Insect repellent.

Associated system(s): FU

Nepeta hindostana **(B. Heyne ex Roth) Haines**

Common name(s): Badranjboya

Family: Lamiaceae/ Labiatae

Distribution: Habitat to Himalayas to south India up to 2400 m.

Parts and properties used: Alcoholic extract of plant yields triterpenoid aldehyde, nepheline. Used like *Nepeta cataria.*

Associated system(s): FU

Neptunia oleracea **Lour. fl. Cochin**

Sanskrit or Ayurvedic name(s): Lajjaalu

Common name(s): Paani-lajak

Family: Leguminosae/ Fabaceae

Distribution: Pantropical, found throughout India in water bodies.

Parts and properties used: Plant, extract- Astringent, refrigerant, toxic, and hepatoprotective.

Associated system(s): AFS

Nerium oleander **L., syn. *N. indicum* Mill.; *N. odoratum* Soland**

Sanskrit or Ayurvedic name(s): Karavira

Common name(s): Oleander, Kaner

Family: Apocynaceae

Distribution: Native from Mediterranean to India. Grown in Indian gardens.

Parts and properties used: Root- Resolvent, attenuate; paste applied on haemorrhoids, ulcerations, leprosy; oil to skin diseases. Leaves- Cardioactive, diuretic, toxic. anti-inflammatory, antifungal, insecticidal.

Associated system(s): AFHS

Nervilia aragoana **Gaudich. (EN)**

Sanskrit or Ayurvedic name(s): Padmachaarini

Common name(s): Sthalapadma

Family: Orchidaceae

Distribution: Habitat to tropical Himalayas 1200-1500 m.

Parts and properties used: Plant- Astringent, diuretic. Used in asthma, vomiting, diarrhea.

Associated system(s): A

Nervilia plicata **(Andrews) Schltr.**

Ayurvedic and Common name(s): Padmacarini

Family: Orchidaceae

Distribution: Found in Kerala.

Parts and properties used: Plant- Antimicrobial.

Associated system(s): AF

Nicotiana tabacum **L.**

Sanskrit or Ayurvedic name(s): Taamraparna

Common name(s): Tambaku

Family: Solanaceae

Distribution: Native to Caribbean. Grown in many parts India.

Parts and properties used: Leaf decoction- Applied to relax muscle in dislocation, strangulated hernia and orchitis. Source of Nicotine.

Associated system(s): AFHSU

Nigella sativa **L.**

Sanskrit or Ayurvedic name(s): Upakuncikaa

Common name(s): Jeera (Black), Kalongi

Family: Ranunculaceae

Distribution: Occur from Europe to west Asia. Grown in Punjab, Bihar, Bengal and Assam.

Parts and properties used: Seed- Stimulant, carminative, diuretic, lactiferous, emmenagogue. Used in puerperal fever.

Associated system(s): AFSU

*Nilgirianthus ciliatus (*Nees) **Bremek., syn.** *Strobilanthes ciliatus* **Nees (EN)**

Sanskrit or Ayurvedic name(s): Sahachara

Common name(s): Kurinji, Karvi

Family: Acanthaceae

Distribution: Habitat to evergreen forests of Western Ghats.

Parts and properties used: Plant- Used in neurological disorders, sciatica, glandular swellings, oedema.

Associated system(s): AS

***Notonia grandiflora* DC.**

Common name(s): Mosakathu-thalai

Family: Asteraceae/ Compositae

Distribution: Habitat to Deccan Hills and Western Ghats.

Parts and properties used: Plant- Aperient. Used to treat pimples.

Associated system(s): SF

***Nyctanthes arbor-tristis* L.**

Sanskrit, Ayurvedic and common name(s): Paarijaata and Paruatak pan

Family: Oleaceae/ Nyctanthaceae

Distribution: Native to Assam, Bengal. Cultivated.

Parts and properties used: Leaves- Cholagogue, febrifuge, anti-inflammatory, antispasmodic, hypotensive, respiratory stimulant. Used in rheumatism.

Associated system(s): AHSU

***Nymphaea alba* L.**

Sanskrit or Ayurvedic name(s): Kumuda

Common name(s): Neel kamal, Neelofar

Family: Nymphaeaceae

Distribution: Naturally found in J & K. Cultivated.

Parts and properties used: Flower and rhizome- Demulcent, mild sedative, antiseptic, spasmolytic, antimicrobial. Used as infusion in diarrhoea. Seed- Used in diabetes.

Associated system(s): ASLU

***Nymphaea nouchali* Burm.f., syn. *N. rubra* Roxb. ex Salisb**

Sanskrit or Ayurvedic name(s): Kumuda

Common name(s): Niloth phal

Family: Nymphaeaceae

Distribution: Habitat to the warmer region of India. Cultivated.

Parts and properties used: Flower- Astringent, cardiac tonic. Used in palpitation of heart. Rhizome- Used in dysentery and dyspepsia.

Associated system(s): AFS

***Nymphaea stellata* Willd.**

Sanskrit or Ayurvedic name(s): Utpala

Common name(s): Nilofar (blue)

Family: Nymphaeaceae

Distribution: Found in ponds of India.

Parts and properties used: Flower- Used like *N. alba*.

Associated system(s): ASLU

***Ochna jabotapita* Linn., syn. *O. squarrosa* Linn.**

Ayurvedic and Common name(s): Kanaka Champaa

Family: Ochnaceae

Distribution: Habitat to Assam, Bihar, Odisha, Deccan Peninsula. Cultivated.

Parts and properties used: Bark- Digestive tonic. Root decoction- Used in asthma, tuberculosis, menstrual disorders. Boiled leaves- Used as emollient cataplasm. Poultice in lumbago.

Associated system(s): FS

***Ochrocarpus longifolius* Bentb. & Hook. f.**

Sanskrit or Ayurvedic name(s): Surapunnaaga

Common name(s): Surangi

Family: Clusiaceae/ Guttiferae

Distribution: Habitat to evergreen forests of Western Ghats.

Parts and properties used: Flower bud- Stomachic, analgesic, antibacterial. Used for gastritis, haemorrhoids, blood diseases, leprosy.

Associated system(s): AFS

***Ocimum americanum* L., syn. *O. canum* Sims**

Common name(s): Kali Tulsi, Bantulsi

Family: Lamiaceae/ Labiatae

Distribution: Habitat to plains and lower hills of India.

Parts and properties used: Plant- Stimulant, carminative, diaphoretic. Leaf- Be chic, febrifuge; treat cold, bronchitis, catarrh, skin diseases. Essential oil- Antifungal. Seed- Hypoglycaemic.

Associated system(s): AFHS

***Ocimum basilicum* L.**

Sanskrit or Ayurvedic name(s): Ajgandhikaa

Common name(s): Sweet basil, Kali Tulsi, Ban Tulsi

Family: Lamiaceae/ Labiatae

Distribution: Pantropic, occur in Ladakh to northeast India. Grown all over.

Parts and properties used: Flower- Stimulant, carminative, antispasmodic, diuretic, demulcent. Seed- Antidysentery. Essential oil/juice- Anti-microbial.

Associated system(s): AFS

***Ocimum gratissimum* L.**

Sanskrit or Ayurvedic name(s): Bilvapami, Vriddha, Tulasi

Common name(s): Ram Tulasi

Family: Lamiaceae/ Labiatae

Distribution: Pantropic, found in Western Ghats. Cultivated all over India.

Parts and properties used: Plant- Neurological, rheumatic affections in seminal weakness and child aphthae. Seed- Used in neuralgia and cephalalgia. Essential oil-Anti-microbial.

Associated system(s): AFHS

***Ocimum kilimandscharicum* Gurke**

Sanskrit, Ayurvedic and common name(s): Karpura Tulasi, Karpoora Tulasi

Family: Lamiaceae/ Labiatae

Distribution: Habitat to Karnataka. Grown in parts of India.

Parts and properties used: Plant- Spasmolytic, antibacterial. Essential oil + Camphor-Insecticidal.

Associated system(s): A

***Ocimum tenuiflorum* L., syn. *O. sanctum* L.**

Sanskrit or Ayurvedic name(s): Tulasi, Sulabhaa

Common name(s): Tulasi

Family: Lamiaceae/ Labiatae

Distribution: Native to Himalayas up to 1800 m, and rest of India. Cultivated.

Parts and properties used: Leaf- Carminative, stomachic, antispasmodic, antiasthma tic, antirheumatic, expectorant, hepatoprotective, antipyretic and diaphoretic; advised in rhinitis and influenza. Seed- Treat genitourinary diseases. Root- Anti-malarial. Essential oil- Antimicrobial.

Associated system(s): AFHS

***Olax scandens* Roxb.**

Sanskrit or Ayurvedic name(s): Dheniaani

Common name(s): Rimil-beeri

Family: Olacaceae

Distribution: Habitat to sub-Himalayas and most of India.

Parts and properties used: Bark- Used in anaemia, diabetes, and fever.

Associated system(s): AFS

***Olea europaea* Linn.**

Common name(s): Olive

Family: Oleaceae

Distribution: Mediterranean. Cultivated in the West Himalayas.

Parts and properties used: Leaves/bark- Febrifugal, diuretic, antihypertensive. Oil- Used in cholangitis, flatulence, meteorism.

Associated system(s): U

***Onosma bracteatum* Wall.**

Sanskrit or Ayurvedic name(s): Gojihvaa

Common name(s): Gule-Gazbaan

Family: Boraginaceae

Distribution: Occur from J & K to Kumaon.

Parts and properties used: Plant- Astringent, diuretic cardiac tonic. Treat cold cough, bronchitis, fever depression, jaundice.

Associated system(s): ASU

***Onosma echioides* L., syn. *O. hispidum* Wall. ex G. Don**

Common name(s): Ratanjot, Gajowan

Family: Boraginaceae

Distribution: European. Occur in the western Himalayas 1000-1500 m.

Parts and properties used: Plant- Astringent and styptic. Root- Applied on bruised/ eruption. Locals used to relieve restlessness. Flowers- Stimulant, cardiac, tonic.

Associated system(s): UF

***Operculina turpethum* (L.) J. Silva Manso, syn. *Merremia turpethum* (L.) Shah & Bhat (NT)**

Sanskrit or Ayurvedic name(s): Trivrtaa

Common name(s): Nishoth

Family: Convolvulaceae

Distribution: Found throughout India.

Parts and properties used: Root- Anti-inflammatory, purgative. Used in rheumatic/ paralytic affections. Active principle is glycosidic resin.

Associated system(s): AFHSLU

Ophiorrhiza mungos **Linn.**

Sanskrit or Ayurvedic name(s): Sarpaakshi

Common name(s): Sarahati

Family: Rubiaceae

Distribution: Habitat to Khasi Hills, Western Ghats, and Andaman.

Parts and properties used: Root- Yields a bitter tonic. Leaf- Used in dressing of ulcers.

Associated system(s): AFS

Opuntia dillenii **(Ker-Gawl.) Haw and** ***O. vulgaris*** **Mill. (Cactus)**

Sanskrit or Ayurvedic name(s): Naagaphani

Common name(s): Nagathal

Family: Cactaceae

Distribution: Found throughout India.

Parts and properties used: Leaves- Used as poultice on inflammation. Baked fruit- Given in cough. Flower- Tincher in homoeopathy.

Associated system(s): AHSU

Orchis latifolia **L.**

Sanskrit or Ayurvedic name(s): Munjaataka, Salabmisri

Common name(s): Salam Misri

Family: Orchidaceae

Distribution: Habitat to western Himalayas 3-4000 m. Cultivated.

Parts and properties used: Plant- Aphrodisiac. Yields a nervine tonic. Tuber- Nutritive, demulcent, restorative. Given for convalescent.

Associated system(s): ASU

Origanum majorana **L., syn.** ***Majorana hortensis*** **Moench.**

Sanskrit or Ayurvedic name(s): Sukhaatmaka

Common name(s): Gandhira, Sweet Marjoram

Family: Lamiaceae/ Labiatae

Distribution: Native of Europe. Cultivated in India.

Parts and properties used: Plant- Emmenagogue, expectorant, antispasmodic, carminative. Leaves/ seed- Astringent, antispasmodic.

Associated system(s): AHSU

***Origanum vulgare* L.**

Common name(s): Ban Tulsi, Maruaa

Family: Lamiaceae/ Labiatae

Distribution: Habitat to high altitudes in Himalayas, including Ladakh.

Parts and properties used: Leaves/stem- Antiseptic, antispasmodic, carminative, cholagogue, diaphoretic, emmenagogue, expectorant, stimulant, stomachic. Volatile oil- Antioxidant, antimicrobial.

Associated system(s): AFU

***Oroxylum indicum* (L.) Benth. ex Kurz. (VU)**

Sanskrit or Ayurvedic name(s): Shyonaaka

Common name(s): Tetuchaal, Arlu

Family: Bignoniaceae

Distribution: Found throughout India.

Parts and properties used: Tender fruit- Carminative, stomachic, spasmolytic. ark- Antirheumatic, diuretic. Seed- Purgative. Root, bark- Antirheumatic, antidiarrheal (Part of *Dash moola*).

Associated system(s): AFSL

***Orthosiphon glabratus* Benth., syn. *O. tomentosus* Benth. var. *glabratus* Hook. f.**

Sanskrit, Ayurvedic and Common name(s): Prataanikaa, Pratanika

Family: Lamiaceae/ Labiatae

Distribution: Habitat to Odisha, Gujarat, south India up to 1000m in hills.

Parts and properties used: Plant decoction- Treat diarrhoea. Leaves- Applied on cuts and wounds.

Associated system(s): AF

***Oryza sativa* L.**

Sanskrit or Ayurvedic name(s): Shaali, Tandula

Common name(s): Aval, Akki, Rice, Thavidu

Family: Poaceae/ Gramineae

Distribution: Native. Cultivated all over India.

Parts and properties used: Grains- Diuretic, reduce lactation. Dried root- Given in dysuria and lactic disorders.

Associated system(s): ASLU

***Osmanthus fragrans* Lour.**

Sanskrit or Ayurvedic name(s): Vasuka

Family: Obleaceae

Distribution: Occur in Himalayas from Kumaon to Sikkim.

Parts and properties used: Herb- Diuretic, genitourinary tract disinfectant. Flower- Antiseptic, insecticidal.

Associated system(s): AF

***Ougeinia dalbergioides* Benth., syn. *O, oojeinensis* (Roxb.) Hochr.**

Sanskrit or Ayurvedic name(s): Tinishaa

Common name(s): Saanan

Family: Leguminosae/ Fabaceae

Distribution: Habitat to outer Himalayas 1500 m to Deccan peninsula.

Parts and properties used: Bark- Febrifuge, antidiarrheal, spasmolytic. Leaves- Contain flavonoids.

Associated system(s): AFS

***Oxalis corniculata* L. and *O. acetosella* Linn.**

Sanskrit or Ayurvedic name(s): Chaangeri

Common name(s): Puliyaarila

Family: Oxalidaceae

Distribution: Found in peninsular India.

Parts and properties used: Plant- Diuretic, refringent, used in tympanites, dyspepsia, biliousness, dysentery, urinary infection.

Associated system(s): AFSU

***Oxystelma secamone* (Linn.) Karst.**

Sanskrit or Ayurvedic name(s): Dugdhikaa

Common name(s): Dugdhani

Family: Asclepiadaceae

Distribution: Habitat to plains and lower hills of India.

Parts and properties used: Herb- Antiseptic, depurative, galactagogue; decoction used for gargle. Latex- Vulnerary (wound healing).

Associated system(s): AF

Paederia foetida **L., syn.** ***P. scandens*** **(Lour.) Merill., syn.** ***P. tomentosa*** **Blume**

Sanskrit or Ayurvedic name(s): Talanili, Gandhaprasaarini.

Common name(s): Gandha, Prasaarani

Family: Rubiaceae

Distribution: Temperate forests of Bihar to Eastern Himalayas and Northeast Hills.

Parts and properties used: Root- Anti-inflammatory. Leaf juice- Carminative, anti-inflammatory, astringent, spasmolytic, antidiarrheal, diuretic, antilithic. Used in diarrhoea.

Associated system(s): AFS

Paeonia emodi **Wall. ex Royle (VU)**

Common name(s): Ud saleev

Family: Paeoniaceae

Distribution: Habitat to north-western Himalaya 2-3000 m.

Parts and properties used: Herb- CNS depressant, diuretic, hypothermic, anti-inflammatory. Root- Used in nervous affection, uterine diseases, and as a blood purifier. Flower- Antidiarrheal.

Associated system(s): FU

Panax ginseng **C.A. Mey (CR)**

Common name(s): Chinese Ginseng

Family: Araliaceae

Distribution: Chinese, grows naturally and cultivated for root.

Parts and properties used: Root extract- Has old-age revitalizer compound, used as a tonic. It has significant trade value.

Associated system(s): Chinese Medicinal system (Pharmacopoeia)

Panax pseudoginseng **Wall. (VU-CR)**

Sanskrit or Ayurvedic name(s): Lakshmanaa

Common name(s): Indian Ginseng (equal to Chinese)

Family: Araliaceae

Distribution: Habitat to Western Himalayas.

Parts and properties used: Root extract- Adaptogen, digestive, relaxant, helps old-age revitalization. Antiarrhythmic.

Associated system(s): AFH

***Pandanus fascicularis* Lam., syn. *P. utilis* Bory**

Sanskrit or Ayurvedic name(s): Ketaki

Common name(s): Kewada, Ketaki

Family: Pandanaceae

Distribution: Widely distributed in eastern coastal regions of Eastern Ghats, Andaman, and cultivated.

Parts and properties used: Flower- Carminative, stomachic antiseptic, treat ulcer, and skin diseases. Root- Osteoarthritis, leucorrhoea, amenorrhoea; decoction used in abdominal inflammation and venereal diseases. Leaves- Scabies, leprosy smallpox.

Associated system(s): ASU

***Pandanus odoratissimus* Linn.**

Sanskrit or Ayurvedic name(s): Ketaki

Common name(s): Kevda

Family: Pandanaceae

Distribution: Distributed in eastern coast and UP. Cultivated in south Odisha.

Parts and properties used: Plant parts- Used in headache, epilepsy, rheumatism spasm, ulcers, colic, hepatitis, leukoderma, smallpox, syphilis, and cancer.

Associated system(s): A

***Papaver rhoeas* L.**

Sanskrit or Ayurvedic name(s): Rakta Posta

Common name(s): Lal post

Family: Papaveraceae

Distribution: Habitat to northern hemisphere, J & K, and plains of north India. Grown in gardens.

Parts and properties used: Capsule latex- Narcotic. Patel- Expectorant, antitussive, sudorific. Used as sedative and for respiratory tract diseases.

Associated system(s): AFS

***Papaver somniferum* L.**

Sanskrit or Ayurvedic name(s): Ahiphenam, Aphukam Khaskhasa

Common name(s): Postdana, Khaskhas

Family: Papaveraceae

Distribution: Habitat to northern hemisphere. Widely grown, including in India.

Parts and properties used: Capsule latex- Narcotic, sedative, hypnotic, analgesic, sudorific, anodyne, antispasmodic. Seed- Nutritive, demulcent, emollient, spasmolytic. Source of opium, morphine heroin.

Associated system(s): AFHSU

***Paris polyphylla* Sm. (VU)**

Sanskrit or Ayurvedic name(s): Haimavati

Common name(s): Daiswa paris, Svetavaca

Family: Melanthiaceae

Distribution: Habitat to temperate Himalayas.

Parts and properties used: Plant- Sedative, analgesic, haemostatic, anthelmintic. Rhizome- Has glucoside, *Alpha-paristyphnin*, exhibiting a depressant action.

Associated system(s): AF

***Parmelia perlata* (Huds.) Ach., syn. *Parmotrema perlatum* (Hudson) M. Choisy. (lichens)**

Sanskrit or Ayurvedic name(s): Saileya

Common name(s): Charelaa, Stone flower, Dagad Phool

Family: Parmeliaceae

Distribution: Globally distributed. Grown in the Himalayas.

Parts and properties used: Fruiting thallus- Astringent, resolvent, diuretic. Applied for pain in renal and lumbar regions.

Associated system(s): ASLU

***Paspalum scrobiculatum* Linn., syn. *Panicum frumentaceum* Rottb.**

Sanskrit or Ayurvedic name(s): Kodrava, Korduusha

Common name(s): Kodo millet

Family: Poaceae/ Gramineae

Distribution: Cultivated throughout India.

Parts and properties used: Plant- Used as a CNS drug for treating schizophrenia and psychoses. Grains- High in polyphenols, and an antioxidant, used in diabetes.

Associated system(s): F

***Passiflora foetida* L.**

Common name(s): Foul Passiflora

Family: Passifloraceae

Distribution: American. Grown tropical India for fruit.

Parts and properties used: Plant extract- Antibacterial, analgesic, anti-diarrheal.

Associated system(s): S

***Paullinia asiatica* Linn.**

Sanskrit or Ayurvedic name(s): Kaanchana

Common name(s): Jangali Kaalimirchi

Family: Rutaceae

Distribution: Found in Konkan Maharashtra.

Parts and properties used: Root bark- Febrifuge, diaphoretic, antiperiodic. Infusion is used in constitutional debility and convalescence.

Associated system(s): AFS

***Pavetta indica* Linn.**

Sanskrit or Ayurvedic name(s): Kathachampaa

Common name(s): Paapadi

Family: Rubiaceae

Distribution: Habitat to greater part of India.

Parts and properties used: Root- Aperient. Given in visceral obstructions, renal dropsy, and ascites.

Associated system(s): AFS

***Pavonia odorata* Willd.**

Sanskrit or Ayurvedic name(s): Vaalaka

Common name(s): Moramasi, Sugandhabala

Family: Malvaceae

Distribution: Found in northwest India, Bengal, and Konkan.

Parts and properties used: Plant- Anti-inflammatory, spasmolytic used in rheumatism. Root- Stomachic, astringent, demulcent. Used in dysentery, ulcers& bleeding.

Associated system(s): ASL

***Pavonia zeylanica* (L.) Cav.**

Sanskrit or Ayurvedic name(s): Bala

Common name(s): White leadwort, Sittanmutti

Family: Malvaceae

Distribution: Habitat to most parts of India.

Parts and properties used: Plant- Purgative, febrifuge, anthelmintic, anticancer, antimicrobial. Contains several bioactive compounds, plumbagin. Stem- Gives saponin.

Associated system(s): AFS

***Pedalium murex* L.**

Sanskrit or Ayurvedic name(s): Brihatgokshura

Common name(s): Gokhru Bada

Family: Pedaliaceae

Distribution: Distributed throughout India.

Parts and properties used: Plant- Diuretic. Root- Ant bilious. Fruit- Yields digestive tonics. Used in ulcers, fevers, wounds, men impotency. Have diosgenin and vanillin, source of steroidal contraceptive.

Associated system(s): ASU

***Peganum harmala* L.**

Sanskrit or Ayurvedic name(s): Haramal

Common name(s): Harmal, Sepan

Family: Zygophyllaceae

Distribution: Occur in Himalayas Ladakh, allover north India, and drier region of Deccan.

Parts and properties used: Plant- Emmenagogue, abortifacient. Seeds- Narcotic, hypnotic, antispasmodic, anodyne, emetic. Given in asthma, colic neuralgia, dysmenorrhoea hysteria. Produce drug *Harmala*.

Associated system(s): ASU

***Pelargonium graveolens* L. Heritt.**

Common name(s): Scented geranium, Rose geranium

Family: Geraniaceae

Distribution: Introduced from South Africa. Grown in Uttarakhand, Karnataka, and TN hills.

Parts and properties used: Flower- Yields Geraniol oil- Antitumor, also, *in vitro* antifungal activity observed.

Associated system(s): South African Medicinal system

***Pentapetes phoenicea* L.**

Sanskrit or Ayurvedic name(s): Bandhujiva

Common name(s): Dopohoria, Pushparakta

Family: Sterculiaceae

Distribution: Found in Assam, Odisha. Grown in gardens of warmer regions.

Parts and properties used: Capsule- Mucilaginous, treats bowel problems. Decoction- Emollient. Root- Ant bilious, ant phlegmonous, febrifuge.

Associated system(s): AFS

***Pentatropis capensis* (Linn. f.) Bullock, syn. *P.microphylla* W. & A.**

Sanskrit or Ayurvedic name(s): Kaakanaasaa, Sunasika

Common name(s): Ambarvel

Family: Asclepiadaceae

Distribution: Habitat to dry areas and forests of Bengal and peninsular India.

Parts and properties used: Plant- Antifungal, antiseptic, keratolytic. Used in various skin conditions.

Associated system(s): AF

Pergularia daemia* (Forssk.) Choiv., syn. *P. extensa

Sanskrit or Ayurvedic name(s): Uttamaarani

Common name(s): Atrilal

Family: Asclepiadaceae

Distribution: Habitat to warmer parts of India.

Parts and properties used: Plant- Uterine stimulant expectorant, emetic. Used in asthma, bronchitis. Leaves- Used in amenorrhoea, dysmenorrhoea. Applied to boils.

Associated system(s): AFS

***Peristrophe bicalyculata* (Retz.) Nees**

Sanskrit or Ayurvedic name(s): Kakajangha, Nadikanta

Common name(s): Atrilal

Family: Acanthaceae

Distribution: Occur throughout India.

Parts and properties used: Plant- Febrifuge. Essential oil- Inhibit growth of tuberculosis strains.

Associated system(s): AF

***Petroselinum crispum* (Mill.) Airy-Shaw., *Apium crispum* Mill.**

Common name(s): Parsley, Fitraasaaliyum

Family: Apiaceae/ Umbelliferae

Distribution: European. Grown all over India.

Parts and properties used: Herb- Diuretic, antispasmodic, uterine, tonic, emmenagogue, sedative, carminative, expectorant, aperient, and antiseptic. Used to clear urinary tract.

Associated system(s): U, English

***Phaseolus adenanthus* G.F.W. Mey.**

Sanskrit or Ayurvedic name(s): Aranya-mudga

Common name(s): Kattupayaru

Family: Leguminosae/ Fabaceae

Distribution: Found and grown all over Indian plains.

Parts and properties used: Decoction- Used in bowel and stricture. Root- Helps stop salivation.

Associated system(s): AS

***Phaseolus lunatus* Linn., syn. *P. inamoenus* L.**

Sanskrit or Ayurvedic name(s): Shimbi

Common name(s): Lima Bean

Family: Leguminosae/ Fabaceae

Distribution: American. Grown in India.

Parts and properties used: Seeds- Astringent. Treat fever. Shoots, pods- Contain alkaloids.

Associated system(s): AU

***Phaseolus trilobus* sensu Ait. & auct., syn. *Vigna trilobata* L.**

Sanskrit or Ayurvedic name(s): Mudgaparni

Common name(s): Jangali Moong

Family: Leguminosae/ Fabaceae

Distribution: Occur throughout India.

Parts and properties used: Plant- Febrifuge. Leaves- Sedative, ant bilious. Decoction- Given in fever.

Associated system(s): AFS

Phlogacanthus thyrsiflorus **Nees, syn.** ***Justicia thyrsiflora*** **Roxb.**

Common name(s): Titaphul

Family: Acanthaceae

Distribution: Wild edible plant of Assam.

Parts and properties used: Herb- Antimicrobial antidiabetic, anti-inflammatory, ant cancerous, hypolipidemic and hepatoprotective. Used like *Adhatoda vasica* in cough.

Associated system(s): F

Phoenix dactylifera **L.**

Sanskrit, Ayurvedic and Common name(s): Kharjura, Khajur

Family: Arecaceae

Distribution: Native of Fertile Crescent. Neutralized in India. Grown in Punjab, Rajasthan.

Parts and properties used: Fruit pulp- Antitussive, diuretic, expectorant, demulcent, laxative, nutritive. Gum- Used in diarrhoea and genitourinary diseases.

Associated system(s): ASU

Phoenix loureirii **Kunth, syn.** ***P. humilis*** **Royle;** ***P. pusilla*** **Gaertn.**

Sanskrit name(s): Parushaka

Common name(s): Sitreechu

Family: Arecaceae

Distribution: Occur in north India up to Assam.

Parts and properties used: Fruit- Restorative, analgesic yields a tonic. Gum- Used diarrhoea and genitourinary diseases.

Associated system(s): S

Phoenix paludosa **Roxb.**

Sanskrit, Ayurvedic and common name(s): Hintala, Hital

Family: Arecaceae

Distribution: Habitat to coastal swamps of Bengal, Odisha and Andaman.

Parts and properties used: Fruit- Antiphlogistic, cooling. Used in flatulence.

Associated system(s): AF

Phoenix sylvestris **(L.) Roxb.**

Sanskrit, Ayurvedic and Common name(s): Kharjuri, Desi Khajur

Family: Arecaceae

Distribution: Occur from Himalaya to MP, AP, Western Ghats. Cultivated.

Parts and properties used: Fruit- Restorative, gastric stimulant. Root- Cure nervous debility. Sap- Yields toddy drink, jaggery.

Associated system(s): AFST

Phragmites karka **Trin. ex Steud., syn.** ***P. roxburghii*** **(Kunth) Steud.**

Sanskrit or Ayurvedic name(s): Nala

Common name(s): Narakul

Family: Poaceae/ Gramineae.

Distribution: Found throughout India.

Parts and properties used: Stem and rhizome- Diuretic, diaphoretic. Used in insect bite.

Associated system(s): AFS

Phyla nodiflora **(L.) Greene, syn.** ***Lippia nodiflora*** **Rich.**

Sanskrit or Ayurvedic name(s): Jalapippali

Common name(s): Chota-okra, Jalpippali

Family: Verbenaceae

Distribution: Found throughout India near water bodies.

Parts and properties used: Plant juice/paste- Spasmolytic, diuretic, febrifuge.

Associated system(s): AF (Bhil)S

Phyllanthus amarus **Schumach. & Thonn., syn.** ***P. fraternus*** **Webst.**

Sanskrit or Ayurvedic name(s): Taamalaki

Common name(s): Bhumiamla

Family: Euphorbiaceae

Distribution: Occur as weed all over warm areas of India. Cultivated too.

Parts and properties used: Plant- Diuretic, DE obstruent, styptic, astringent, anti-inflammatory. Used in jaundice to skin eruption.

Associated system(s): AFS

Phyllanthus acidus **(L.) Skeels, syn.** ***P. distichus*** **Muell. -Arg.**

Sanskrit or Ayurvedic name(s): Lavali-phala

Common name(s): Aranelli

Family: Euphorbiaceae

Distribution: Occur in Brahmaputra Valley, Grown in Indian gardens.

Parts and properties used: Fruit- Astringent, tonic. Help in biliousness, constipation, vomiting, bronchitis. Root/seed- Cathartic.

Associated system(s): AS

***Phyllanthus indofischeri* Bennet.**

Common name(s): Konda Usirika

Family: Euphorbiaceae

Distribution: Occur in Eastern Ghats.

Parts and properties used: Fruit- High in vitamin C. Used in dyspepsia, jaundice.

Associated system(s): AF

***Phyllanthus maderaspatensis* L.**

Common name(s): Kanocha

Family: Euphorbiaceae

Distribution: Habitat to all over dry India.

Parts and properties used: Leaf infusion- Used in headache. Seeds- Carminative, diuretic.

Associated system(s): SU

***Phyllanthus niruri* Hook, syn. *P. debilis* Klein ex Willd.; *P. fraternus* G.L. Webster**

Sanskrit, Ayurvedic and name(s): Bhuumyaamalaki, Bhumyamalaki

Family: Phyllanthaceae/ Euphorbiaceae

Distribution: American. Naturalized to parts of India.

Parts and properties used: Herb- Antispasmodic, antipyretic, diuretic, anti-viral bacterial. Treat genitourinary system, stomach, liver, kidney, spleen.

Associated system(s): AS

***Phyllanthus reticulatus* Poir., syn. *Kirganelia reticulata* (Poir.) Baill.**

Common name(s): Buinowla

Family: Euphorbiaceae

Distribution: Occur in northeast, and peninsular India.

Parts and properties used: Dried bark leaves decoction- Diuretic, alterative, cooling, and boost fertility. Treat smallpox.

Associated system(s): AFS

***Phyllanthus urinaria* L.**

Sanskrit or Ayurvedic name(s): Bhuumyaamataki

Common name(s): Lal-Bhui-anvalah

Family: Euphorbiaceae

Distribution: Occur all over plains of India.

Parts and properties used: Plant- Used as (substitute) *P. amarus.*

Associated system(s): AFS

***Phyllanthus virgatus* G. Forst., syn. *P. simplex* Retz.**

Sanskrit or Ayurvedic name(s): Bhumyaamalaki

Common name(s): Niruri

Family: Euphorbiaceae

Distribution: Occur throughout India.

Parts and properties used: Plant- Antiseptic. Leaves, flower, fruit- Used in gonorrhoea.

Associated system(s): AF

***Physalis alkekengi* L.**

Sanskrit or Ayurvedic name(s): Raajaputrikaa

Common name(s): Shoklo, Winter cherry

Family: Solanaceae

Distribution: European. Grown in Himalayas, Ladakh and Indian gardens.

Parts and properties used: Berries- Diuretic, antitussive, oxytocic, analgesic, febrifuge. Used in urinary disorder.

Associated system(s): AU

***Physalis angulata* L.**

Common name(s): Budda Kodisha

Family: Solanaceae

Distribution: American. Grown in Indian gardens.

Parts and properties used: Plant- Diuretic. Leaf paste- Relaxant, applied on wounds.

Associated system(s): F

***Physalis minima* L.**

Sanskrit or Ayurvedic name(s): Tankaari

Common name(s): Lal tang, Bajar Bang

Family: Solanaceae

Distribution: Occur as weed throughout India up to 2300 m.

Parts and properties used: Berries- Diuretic, aperient, alterative. Used in gout, urinary diseases.

Associated system(s): ASL

***Physalis peruviana* Linn.**

Sanskrit or Ayurvedic name(s): Parpoti variety

Common name(s): Rasbhari, Mako

Family: Solanaceae

Distribution: American, grown in Indian gardens.

Parts and properties used: Plant- Diuretic. Leaf- Anthelmintic, infusion used in abdominal disorders. Fruits- Source of carotene and ascorbic acid, edible.

Associated system(s): AFS

***Physochlaina praealta* (Decne.) Miers (VU)**

Common name(s): Bharangi, Kadavi, Langthang

Family: Solanaceae

Distribution: Habitat to Himalayas 3300-4600 m. (Ladakh)

Parts and properties used: Plant- Anthelmintic, narcotic. Source of atropine.

Associated system(s): F

Picea smithiana* Boiss., and *P. abies

Common name(s): Roi, Rhai, Spruce

Family: Pinaceae

Distribution: Native to Himalayas 1500-3500 m. Grown.

Parts and properties used: Essential oil- Antiseptic, used in bath salts, catarrhal illness, rheumatic and neuralgic pains.

Associated system(s): English, F

***Picrasma quassioides* (D. Don) Benn.**

Sanskrit or Ayurvedic name(s): Bhurangi

Common name(s): Kutaki

Family: Scrophulariaceae

Distribution: Habitat to Garhwal, HP.

Parts and properties used: Wood- Stomachic, amoebicidal, anthelmintic and insect repellent. Yields tonic.

Associated system(s): AF

***Picrorhiza kurroa* Royle *ex* Benth (Near EX)**

Sanskrit or Ayurvedic name(s): Katukaa

Common name(s): Buinowla, Kutki, Kaor

Family: Scrophulariaceae

Distribution: Habitat to alpine Himalayas, Ladakh to Sikkim. Harvested from wild.

Parts and properties used: Root- Appetizer, liver protectant yields tonic. Used in jaundice, intermittent fever, dyspnoea, and skin diseases. Yields Kutkin.

Associated system(s): ASLU

***Pimpinella anisum* L.**

Sanskrit or Ayurvedic name(s): Anisuna

Common name(s): Ajamoda, Badiyan Khathayi

Family: Apiaceae/ Umbelliferae

Distribution: Mediterranean. Cultivated in north and northeast India.

Parts and properties used: Fruit- Carminative, diuretic, anticholera, antispasmodic, expectorant. Used for flatulence, coughs, bronchitis, dyspeptic problem, and catarrhs.

Associated system(s): AHSU

***Pinus gerardiana* Wall. (Rare)**

Sanskrit or Ayurvedic name(s): Nikochaka

Common name(s): Chilgozaa.

Family: Pinaceae

Distribution: Habitat to northwest Himalayas.

Parts and properties used: Kernel- Stimulant, carminative, expectorant. Seed oil- Treat wounds and ulcers.

Associated system(s): AU

***Pinus roxburghii* Sarg., and *P. excelsa* Wall. ex D. Don.; *P. wallichiana* A.B. Jackson.**

Sanskrit or Ayurvedic name(s): Sarala

Common name(s): Gandabiroja

Family: Pinaceae

Distribution: Habitat to Himalayas.

Parts and properties used: Needle, oil- Decongestant, expectorant, antiseptic. Used in cough cold, rheumatism and stiff muscle.

Associated system(s): AHSLU

***Piper betle* L.**

Sanskrit or Ayurvedic name(s): Nagavalli, Taambula

Common name(s): Betle, Paan

Family: Piperaceae

Distribution: Native of South, Southeast Asia. Grown in Assam, Northeast, Bengal, Bihar, UP, Kerala. Karnataka,

Parts and properties used: Leaf- Carminative, stimulant, astringent, antiseptic, laxative expectorant. Yields tonic. Essential oil- Antispasmodic. Used in catarrhs.

Associated system(s): AFSLU

***Piper chaba* Hunter, syn. *P. retrofractum* Vahl.**

Sanskrit or Ayurvedic name(s): Cavya

Common name(s): Kabab chini, Chavya, Gajpippal

Family: Piperaceae

Distribution: Native, cultivated Northeast India

Parts and properties used: Fruit- Stimulant, carminative. Used in haemorrhoidal affections; as tonic and in spleen, abdomen diseases, like *P. longum* and *P. nigrum.*

Associated system(s): ALS

***Piper cubeba* L. f.**

Sanskrit or Ayurvedic name(s): Kankola

Common name(s): Tailed Pepper

Family: Piperaceae

Distribution: Native, and cultivated in Assam, Karnataka.

Parts and properties used: Fruit- Carminative, diuretic, expectorant. Used in bronchial, urinary, amoebiasis, dysuria. Oil- Antibacterial. Used in genitourinary diseases and cystitis.

Associated system(s): AFHSLU

***Piper longum* L. (NT)**

Sanskrit or Ayurvedic name(s): Pippalimula, Pippali

Common name(s): Pipal, Thippili, Pippali

Family: Piperaceae

Distribution: Indo-Malayan. Found in warmer India, from Himalaya to Western Ghats. Grown.

Parts and properties used: Fruit- Used in diseases of respiratory tract. As sedative, as cholagogue, as emmenagogue, as digestive, carminative, tonic, and haematinic. Yields antipyretic, hypotensive, analeptic, stimulant CNS, Piperine.

Associated system(s): AFSLU

***Piper nigrum* L. (VU)**

Sanskrit or Ayurvedic name(s): Maricha

Common name(s): Pipal Gol, Kalimirch

Family: Piperaceae

Distribution: Native to northeast India and southern Western Ghats. Cultivated.

Parts and properties used: Stem/fruit- Stimulant, carminative, diuretic, anticholera, sialagogue, be chic, antiasthma tic. Used in fevers, indigestion, gastro-intestinal, coughs, breathing and heart problems.

Associated system(s): AHSLU

***Pisonia grandis* R. Br.**

Common name(s): Chandi keerai

Family: Nyctaginaceae

Distribution: Widely distributed all over India. Cultivated near coasts.

Parts and properties used: Fresh leaf- Diuretic. Used in inflammations. Root-Purgative.

Associated system(s): S

***Pistacia integerrima* Stew. ex Brand., syn. *P. chinensis* Bunge var. *integerrima* Brandis**

Sanskrit, Ayurvedic and common name(s): Karkatashringi, Kakarsinghi

Family: Pistaciaceae

Distribution: Habitat to Himalayas.

Parts and properties used: Galls- Astringent, expectorant, antiasthma tic, antidysentery, styptic. Used in cough, bronchitis, and dyspnoea.

Associated system(s): ASU

***Pistacia lentiscus* L.**

Sanskrit, Ayurvedic and Common name(s): Rumimastagi, Mastangi

Family: Pistaciaceae

Distribution: Mediterranean. Resin imported.

Parts and properties used: Gum- Carminative, diuretic, stimulant, astringent.

Associated system(s): SU

***Pistacia vera* L.**

Sanskrit or Ayurvedic name(s): Mukuulaka

Common name(s): Magaj Pista Irani

Family: Pistaciaceae

Distribution: Central Asian. Cultivated in north India.

Parts and properties used: Kernel- Cardiac. Yields brain tonic. Flower- Treat leucorrhoea. Husk- Against dysentery.

Associated system(s): AU

***Pistia stratiotes* L.**

Sanskrit, Ayurvedic and common name(s): Jalakumbhi, Jalkumbhi

Family: Araceae

Distribution: Habitat to sub-tropical India.

Parts and properties used: Plant/ root- Diuretic. Leaves- Antitussive, demulcent, antidysentery.

Associated system(s): AS

***Pithecellobium dulce* Benth.**

Common name(s): Karapilly, Dakhini babool.

Family: Leguminosae/ Fabaceae (Mimosaceae).

Distribution: Native to the Pacific coasts. Cultivated over plains of India.

Parts and properties used: Bark- Astringent, febrifuge, anti-dysenteric, spasmolytic. Seeds- Anti-inflammatory.

Associated system(s): FS

***Pittosporum floribundum* Wight & Arn., syn. *P. napaulense* (DC.) Rehder & EH Wilson *P. wightii* AK Mukherjee**

Common name(s): Kattu, Sampangi, Vekhali

Family: Pittosporaceae

Distribution: Habitat to sub-tropical Himalayas to Western Ghats

Parts and properties used: Bark- Narcotic, anti-inflammatory, antispasmodic. Treat chronic bronchitis. Himalayan form yield oil.

Associated system(s): FS

***Plantago afra* L.**

Common name(s): Kala Isabgol

Family: Plantaginaceae

Distribution: Cosmopolitan naturalized to Eastern Ghats. Sown in hills.

Parts and properties used: Seed husk- Laxative.

Associated system(s): UF

***Plantago amplexicaulis* Cav.**

Common name(s): Brown isabgol

Family: Plantaginaceae

Distribution: Pantropic to dry sand and deserts. Grown in Gujarat, Rajasthan, MP.

Parts and properties used: Seed and husk- Astringent, demulcent. Like *P. ovata*

Associated system(s): UF

***Plantago lanceolata* L.**

Ayurvedic and Common name(s): Baltanga, English plantain

Family: Plantaginaceae

Distribution: Occur in Western Himalayas.

Parts and properties used: Leaf & root- Astringent, bechic, antiasthma tic, diuretic, anti-inflammatory, hypothermic. Seed- Cathartic, haemostatic. Used for catarrhs.

Associated system(s): AU

***Plantago major* L.**

Sanskrit or Ayurvedic name(s): Ashvagola

Common name(s): Lahuriya, Ripple grass

Family: Plantaginaceae

Distribution: Habitat to alpine Himalayas 600-3500 m, including Ladakh.

Parts and properties used: Plant- Haemostatic, antibacterial antihistaminic. Leaves-Astringent, diuretic, vulnerary, febrifuge. Used in dysentery.

Associated system(s): AHSU

***Plantago ovata* Forssk., syn. *P. orbignyana* Steinh. ex Decne.**

Common name(s): Isabgol

Family: Plantaginaceae

Distribution: Native to west and south Asia. Grown in MP, Rajasthan, Gujarat.

Parts and properties used: Seed + husk- Laxative, diuretic, demulcent, bechic. Cholinergic. Used in chronic constipation and irritable bowel.

Associated system(s): AFSU

***Plectranthus amboinicus* (Lour.) Spreng., syn. *Coleus amboinicus* Lour.**

Ayurvedic Common name(s): Patharcur

Family: Lamiaceae/ Labiatae

Distribution: Native to Assam. Cultivated.

Parts and properties used: Leaves- Antioxidant Anti-microbial, and anti-epileptic. Herb- Diuretic, aromatic.

Associated system(s): A

***Plectranthus hadiensis* var. *tomentosus* (Benth. ex E.Mey.) Codd; syn. *Coleus zeylanicus* Benth.**

Common name(s): Vick's plant

Family: Lamiaceae/ Labiatae

Distribution: Native to South Africa, naturalized, grown in south India.

Parts and properties used: Root- Used as enema.

Associated system(s): AF

***Plectranthus rotundifolius* (Poir.) Spreng.; syn. *Coleus parviflorus* Benth; *C. tuberosus* Blume; *C. rotundifolius* (Poir.) A.Chev. & Perrot.; *Solenostemon rotundifolius* (Poir)**

Common name(s): Chinese Potato

Family: Lamiaceae/ Labiatae

Distribution: Native of tropical Africa. Cultivated in south India.

Parts and properties used: Leaves- Used in dysentery. Plant- Used in blood in urine/ eye disorder.

Associated system(s): AF

***Plesmonium margaritiferum* Schott.**

Sanskrit or Ayurvedic name(s): Vajrakanda

Common name(s): Kharhar

Family: Araceae

Distribution: Found in Bengal, Jharkhand to TN.

Parts and properties used: Tuber- Anti-inflammatory. Seed- Applied on bruises.

Associated system(s): AF

***Pleurospermum angelicoides* (DC.) C.B. Clarke -**

Common name(s): Choru, Gandhrayan (Uttarakhand)

Family: Apiaceae/ Umbelliferae

Distribution: Habitat to Himalayas

Parts and properties used: Root- Used in cardiac complaints.

Associated system(s): AF

***Pluchea lanceolata* (DC.) Oliv. & Hiern**

Sanskrit, Ayurvedic and Common name(s): Raasnaa, Rasna

Family: Asteraceae/ Compositae

Distribution: Occur in northwest Indian plains.

Parts and properties used: Arial part- Muscle relaxant. Stem- Anti-inflammatory.

Associated system(s): AF

***Plumbago auriculata* Lam., *P. capensis* Thunb.**

Sanskrit or Ayurvedic name(s): Nilachitraka

Common name(s): Nila Chitrak

Family: Plumbaginaceae

Distribution: Native to South Africa. Grown in Gardens.

Parts and properties used: Herb/ roots- Treat warts, broken bones, and wounds. Snuff of headaches. Used like *P. zeylanica*.

Associated system(s): South African

***Plumbago indica* L**

Sanskrit or Ayurvedic name(s): Rakta-chitraka

Common name(s): Chitrak

Family: Plumbaginaceae

Distribution: Found in Sikkim, Khasi Hills. Grown in gardens.

Parts and properties used: Root- Used like *P. zeylanica* against anaemia, rheumatic pain, dysmenorrhea, leprosy, ulcers, and elimination of intestinal parasites.

Associated system(s): AS

***Plumbago zeylanica* L.**

Sanskrit or Ayurvedic name(s): Chitraka

Common name(s): Chitrak, Chitrakmool, Sheetra

Family: Plumbaginaceae

Distribution: Distributed in several regions and peninsular India. Grown in gardens.

Parts and properties used: Root- Stimulate and normalize digestion- Used in malabsorption syndrome, piles, inflammation, etc. Root paste- Applied on abscesses, leprosy.

Associated system(s): AFSU

***Plumeria alba* Linn., and *P. rubra* L.**

Sanskrit or Ayurvedic name(s): Kshira Champaka

Family: Apocynaceae

Distribution: Native to Central America. Grown in gardens.

Parts and properties used: Root bark- For blennorrhagia (mucous discharge). Bark-latex- For herpes, scabies syphilitic ulcers.

Associated system(s): AS

***Podophyllum hexandrum* Royle, syn. *P. emodi* Wall. ex Honig. (EN-CR)**

Sanskrit or Ayurvedic name(s): Giriparpata

Common name(s): Bankakri, Himalayan mayapple, Venivel, Patval

Family: Berberidaceae

Distribution: Grows across Himalayas 1800-4000 m. Harvested from nature, attempts to grow.

Parts and properties used: Rhizomes/roots- Antineoplastic, anti-cancer, radioprotection, anti-gynaecological. Used as topically as an ointment.

Associated system(s): AF

***Pogostemon cablin* (Blanco) Benth.**

Sanskrit or Ayurvedic name(s): Paachi

Common name(s): Patchouli

Family: Lamiaceae/ Labiatae

Distribution: Native to Philippines. Widely grown in India.

Parts and properties used: Plant- Insecticidal. Leaf infusion- Scented. Used in treatment menstrual troubles.

Associated system(s): AF

***Pogostemon heyneanus* Benth., syn. *P. patchoulis* Hook. f. non-Pelletier.**

Sanskrit or Ayurvedic name(s): Pancholi

Common name(s): Patchouli

Family: Lamiaceae/ Labiatae

Distribution: Indo-Malayan region, Western Ghats hills. Cultivated.

Parts and properties used: Plant- Insecticidal. Leaf- Bechic, antiasthma tic. Poultice used in boil & headache.

Associated system(s): AFS

***Polianthes tuberosa* Linn.**

Sanskrit or Ayurvedic name(s): Rajanigandhaa

Common name(s): Gulcheri

Family: Amaryllidaceae

Distribution: Mexican. Grown in Indian gardens.

Parts and properties used: Flowers and bulbs- Diuretic. Bulb rubbed to control skin eruptions.

Associated system(s): AFS

***Polyalthia longifolia* (Sonn.) Thwaites**

Sanskrit or Ayurvedic name(s): Devadaari, Putrajiva

Common name(s): Ashok's substitute

Family: Annonaceae

Distribution: Indian native, widely grown in gardens.

Parts and properties used: Bark- (adulterant of *Saraca asoca*)- Febrifuge, causes cardiac depression. Leaf- Fungitoxic.

Associated system(s): AFS

***Polycarpaea corymbosa* Lamm.**

Sanskrit or Ayurvedic name(s): Parpata

Common name(s): Bhisatta

Family: Caryophyllaceae

Distribution: Found in most parts of India.

Parts and properties used: Leaves- Anti-inflammatory. Used as poultice and in jaundice.

Associated system(s): AFS

Polygonatum cirrhifolium **(Wall.) Royle (EN)**

Sanskrit, Ayurvedic and Common name(s): Maha-medaa, Meda-m-meda

Family: Asparagaceae, Liliaceae

Distribution: Habitat to northern Himalayas, 1500-3300 m.

Parts and properties used: Root- Used in kidney and hip pain, swelling and fluids in bone joints. Yields tonic to rejuvenate nervous system. (Part of *Astavarga*)

Associated system(s): AF

Polygonatum multiflorum **(L.) All.**

Sanskrit or Ayurvedic name(s): Maha-medaa

Common name(s): Solomnons seal

Family: Liliaceae/ Asparagaceae

Distribution: Habitat to Himalayas from Kashmir to Manipur.

Parts and properties used: Root- Anti-inflammatory astringent. Infusion- Nervine, demulcent, emetic. Given in diosgenin.

Associated system(s): A

Polygonatum verticillatum **(L.) All. (NT-EN)**

Sanskrit or Ayurvedic name(s): Medaa

Common name(s): Meda-m-meda

Family: Liliaceae/ Asparagaceae

Distribution: Habitat to temperate, from Himalayas to Manipur.

Parts and properties used: Root- Cardiotonic, antiperiodic, demulcent, diuretic, aphrodisiac, sedative, antitumor, hypoglycemic. Tibetan treat emaciation, senility, pulmonary affections. Have diosgenin. (Part of *Astavarga*).

Associated system(s): AL

Polygonum aviculare **L.**

Common name(s): Anjabar, Kesri Machoti,

Family: Polygonaceae

Distribution: North American. Naturalized from Kashmir to Kumaon.

Parts and properties used: Stem, fruit (seed)- Astringent and diuretic. Used in bleeding and catarrhs.

Associated system(s): AFU

***Polygonum glabrum* Willd.**

Sanskrit or Ayurvedic name(s): Rakta-rohidaa

Common name(s): Raktharohitha, Attalaree

Family: Polygonaceae

Distribution: Wild throughout India up to 1900m in marshy lands.

Parts and properties used: Plant juice and rootstock- Used in pain, jaundice, piles, pneumonia, burn, wound, etc.

Associated system(s): AFS

***Polygonum hydropiper* L.**

Common name(s): Paakur-muula

Family: Polygonaceae

Distribution: Found throughout wet regions of India. Cultivated.

Parts and properties used: Herb, leaves- Haemostatic, anti-inflammatory, astringent, carminative, diaphoretic, diuretic, emmenagogue, stomachic, styptic.

Associated system(s): F

***Polygonum plebeium* R.Br. (var. *sindica* rare)**

Sanskrit or Ayurvedic name(s): Sarpaakshi (?)

Common name(s): Muniyaaraa

Family: Polygonaceae

Distribution: Pantropical. Occur all over warm India.

Parts and properties used: Herb- Ga lactogenic, antidiarrheal. Powdered given in pneumonia.

Associated system(s): AF

***Polygonum viviparum* L.**

Common name(s): Unjwar, Anjbar

Family: Polygonaceae

Distribution: Widely found in Indian Himalayas 3300-4800 m.

Parts and properties used: Rootstock- Astringent, antiseptic, antidiarrheal, antiperiodic. Used in haemoptysis. Decoction for gargle.

Associated system(s): FU

***Polypodium vulgare* L. (fern)**

Common name(s): Bisphaiz

Family: Polypodiaceae

Distribution: Habitat to temperate world, including Kashmir.

Parts and properties used: Rhizome- Expectorant, laxative, cholagogue, alterative. Used in bronchitis, catarrh, appetite loss.

Associated system(s): AUF

***Pongamia pinnata* (L.) Pierre, syn. *P. glabra* Vent. ; *Derris indica* (Lam.) Bennett**

Sanskrit or Ayurvedic name(s): Karanja

Common name(s): Honge beej, Karanj, Pongum seed

Family: Leguminosae/ Fabaceae

Distribution: Found all over India on riverbanks. Cultivated.

Parts and properties used: Plant bark- Given in piles. Oil- Applied in scabies, herpes, leukoderma, skin diseases, on chest in pneumonia and cold.

Associated system(s): ASU

***Portulaca oleracea* L.**

Sanskrit or Ayurvedic name(s): Brihat Lonikaa, Kozuppa

Common name(s): Kulfaa.

Family: Portulacaceae

Distribution: Distributed worldwide. Grown all over India as vegetable.

Parts and properties used: Herb- Refrigerant, spasmodic, diuretic, antiscorbutic. Used in diseases of liver, spleen, kidney, and bladder. Seeds- Diuretic, antidysentery.

Associated system(s): AFSU

***Portulaca quadrifida* Linn.**

Sanskrit or Ayurvedic name(s): Laghu-lonikaa

Family: Portulacaceae

Distribution: Habitat to warmer India as weed.

Parts and properties used: Herb- Used in asthma, cough, urinary discharges, ulcers, and inflammations. Used like Kulfaa, *P. oleracea*.

Associated system(s): AS

***Potentilla anserina* L.**

Common name(s): Silver weed, Troma

Family: Rosaceae

Distribution: Habitat to western Himalayas 2100-4800 m, including Ladakh.

Parts and properties used: Flower and leaf- Astringent, anti-inflammatory, antispasmodic, haemostatic. Used in diarrhoea, leucorrhoea, dysmenorrhoeal disorders.

Associated system(s): F

***Potentilla arbuscula* D. Don.**

Sanskrit or Ayurvedic name(s): Bajradanti

Family: Rosaceae

Distribution: Habitat to temperate Himalayas 1200-4350 m.

Parts and properties used: Rootstock- Antidiarrheal, reinforce gum and tooth in tooth powders.

Associated system(s): A

***Pothos scandens* L.**

Common name(s): Bendarli

Family: Araceae

Distribution: West Indian. Grown ornamental.

Parts and properties used: Leaf extract- Anti-inflammatory, antiseptic, antimicrobial.

Associated system(s): West Indian F

***Prangos pabularia* Lindl.**

Sanskrit or Ayurvedic name(s): Avipriya

Common name(s): Prangos, plans

Family: Apiaceae/ Umbelliferae

Distribution: Habitat to alpine and sub-alpine Himalayas.

Parts and properties used: Root and fruit- Diuretic, emmenagogue. Used in treatment for leucoplakia, digestive disorder, healing scar and stopping bleeding. Infusion (drink)- Stomach disorder. Source of coumarin.

Associated system(s): AF

***Premna herbacea* Roxb.**

Common name(s): Siru Thekku, Bharangi.

Family: Lamiaceae/ Labiatae

Distribution: Habitat to sub-tropical Himalayas, Assam to south India.

Parts and properties used: Root/leaves- Given in asthma, rheumatism.

Associated system(s): SF

***Premna corymbosa* Rottl. & Willd.**

Sanskrit or Ayurvedic name(s): Agnimantha, Shriparni, Arani, Nadey

Common name(s): Arni, Munnai

Family: Verbenaceae

Distribution: Occur in Assam, Khasi Hills, Andaman.

Parts and properties used: Plant- Carminative, galactagogue. Tender plant- used in neuralgia rheumatism. Roots- Used in asthma, bronchitis, as an expectorant in cold, catarrh and fever. (Part of *dashamoola*).

Associated system(s): AFS

***Premna serratifolia* L. [= *P. integrifolia* L.] *P. obtusifolia* R. Br.**

Sanskrit or Ayurvedic name(s): Agnimantha

Common name(s): Headache tree Arin, Arnimool

Family: Verbenaceae

Distribution: Indo-Malayan. Occur in plains, deciduous forests of Assam to Western Ghats.

Parts and properties used: Root wood- Acteoside isolated. It is a glucoside derivative. Root bark- Contain cytotoxic and antioxidant diterpene. Part of Ayurvedic *dashamoola.* Widely used in various ailments.

Associated system(s): AFS

***Primula vulgaris* Huds.**

Common name(s): Nakhud

Family: Primulaceae

Distribution: Habitat to sub-Himalayan region.

Parts and properties used: Plant- Anti-inflammatory, vulnerary, vermifuge, emetic. Used externally.

Associated system(s): FU

***Prosopis cineraria* (L.) Druce, syn. *P. spicigera* Linn.**

Sanskrit or Ayurvedic name(s): Shami

Common name(s): Jhand, Chonkar

Family: Leguminosae/ Fabaceae (Mimosaceae))

Distribution: Habitat to northwest arid and semi-arid region. Planted in agroforestry.

Parts and properties used: Pod- Astringent, pectoral, demulcent. Bark- Anti-inflammatory, antirheumatic.

Associated system(s): AFS

***Prunus armeniaca* L.**

Sanskrit or Ayurvedic name(s): Urumaana

Common name(s): Apricot, Khuubaani

Family: Rosaceae

Distribution: Habitat to northwest Himalayas 3000 m. Cultivated

Parts and properties used: Kernel powder- Antitussive, and antiasthma-tic.

Associated system(s): AUFL

***Prunus avium* (L.) L.**

Sanskrit or Ayurvedic name(s): Elavalukam

Common name(s): Sweet cherry, Aileya

Family: Rosaceae

Distribution: Eurasian. Grown in Kashmir.

Parts and properties used: Fruit stalk- Diuretic, anti-inflammatory, astringent. Used in oedema, urinary inflammation.

Associated system(s): AF

***Prunus cerasoides* D. Don.**

Common name(s): Himalayan cherry, sour cherry

Family: Rosaceae

Distribution: Habitat to Himalayas to Southwest China up to 1200-2400 m.

Parts and properties used: Plant- Treat skin diseases, improving complexion and used as uterine tonic.

Associated system(s): A

***Prunus domestica* L.**

Sanskrit or Ayurvedic name(s): Aaruka, Aaluubukhaaraa

Common name(s): Aaluuchaa

Family: Rosaceae

Distribution: European. Grown in J & K, HP, Kumaon.

Parts and properties used: Fruit- Febrifuge, laxative, stomachic, nutritive. Improves haemoglobin. Root- Astringent.

Associated system(s): ASU

***Prunus dulcis* (Mill.) D.A. Webb, syn. *P. amygdalus* Batsch.**

Sanskrit or Ayurvedic name(s): Vaataama

Common name(s): Baadaam

Family: Rosaceae

Distribution: Native to Iran and around. Cultivated in Kashmir.

Parts and properties used: Kernel- Nutritious, demulcent, nervine tonic, stimulant, help in heart problems. Oil- Demulcent, slightly laxative.

Associated system(s): ASU

***Prunus mahaleb* L**

Sanskrit or Ayurvedic name(s): Gandha-priyangu

Common name(s): Mahaleb, Priyangu

Family: Rosaceae

Distribution: Native to southern Europe to central Asia. Grown in India.

Parts and properties used: Kernel paste- Applied to treat freckles and blemishes. Has coumarin.

Associated system(s): AU

***Prunus persica* Batsch., syn. *Amygdalus persica* Linn.**

Sanskrit or Ayurvedic name(s): Aaluka,

Common name(s): Peach

Family: Rosaceae

Distribution: Chinese. Grown in J & K, HP, Kumaon.

Parts and properties used: Leaves and bark- Expectorant, sedative, stomachic, demulcent, antiscorbutic, diuretic. Leaf tea- treat morning sickness, dry and hard cough. Fresh leaves- Anthelmintic. Leaf powder- Styptic (Applied externally). Fruit- Stomachic, antiscorbutic, tranquillizer, expectorant, diuretic, antipyretic.

Associated system(s): A

***Pseudarthria viscida* (L.) Wight & Arn. (NT)**

Sanskrit or Ayurvedic name(s): Sanaparni

Common name(s): Moovila

Family: Leguminosae/ Fabaceae

Distribution: Found in many parts of India.

Parts and properties used: Root- Astringent, febrifuge, antirheumatic. Decoction/ powder- Used to treat biliousness and diarrhoea. Also, asthma, nervous dysfunction.

Associated system(s): AFS

***Psidium guajava* L**

Sanskrit or Ayurvedic name(s): Peruka

Common name(s): Gujava, Jaamphal, Amrood

Family: Myrtaceae

Distribution: Central and South American. Grown in most of India.

Parts and properties used: Unripe fruit- Antidiarrheal. Leaves- Used in dysentery, diabetes, cough and cold.

Associated system(s): ASU

***Psoralea corylifolia* L., syn. *Cullen corylifolium* (L.) Medik.**

Sanskrit or Ayurvedic name(s): Baakuchi

Common name(s): Baabchi, Bavanchi, Bu Gu Zhi

Family: Leguminosae/ Fabaceae

Distribution: Habitat to hotter/dry regions of peninsular India. Cultivated marginally.

Parts and properties used: Plant- Produce oleoresin to treat kidney disorders. Yields tonic for impotency, lumbago, premature ejaculation, etc. Seed- Treat leprosy, leukoderma, vitiligo, psoriasis, inflammation.

Associated system(s): AHSUL

***Pterocarpus marsupium* Roxb. (EN)**

Sanskrit or Ayurvedic name(s): Asana, Bijaka

Common name(s): Damul-akhwain, Vijaysar

Family: Leguminosae/ Fabaceae

Distribution: Habitat to tropical region of hilly India.

Parts and properties used: Heartwood- Used in anaemia, worm infestation, skin diseases, urinary and lipid disorders, and obesity. Stem bark- Used in diabetes. Bark-Kino (gum)- Astringent, anti-haemorrhagic, antidiarrheal. Flower- Febrifuge. Leaves-Applied in skin diseases.

Associated system(s): AHSLU

***Pterocarpus santalinus* L. f. (EN)**

Sanskrit or Ayurvedic name(s): Raktachandana

Common name(s): Lal Chandan, Sandal Surkh

Family: Leguminosae/ Fabaceae

Distribution: Endemic to AP, TN, Karnataka. Cultivated.

Parts and properties used: Heartwood- Antbilious, anti-inflammatory, hypoglycaemic, astringent, diaphoretic, febrifuge. Wood-paste- Used in headache.

Associated system(s): ASLU

Pterospermum acerifolium **Willd.**

Sanskrit or Ayurvedic name(s): Muchukunda

Common name(s): Vennangu

Family: Sterculiaceae

Distribution: Found in many parts of India.

Parts and properties used: Flower- Anti-inflammatory, styptic.

Associated system(s): ASU

Pterospermum canescens **Roxb., syn.** ***P. suberifolium*** **Lam. non-Roxb.**

Sanskrit or Ayurvedic name(s): Muchukunda

Common name(s): Sempulavu

Family: Sterculiaceae

Distribution: Occur in south India. Grown in Bengal.

Parts and properties used: Flower- Yields Anodyne (painkilling drug). Paste/leaves- Used in migraine, and headache.

Associated system(s): AS

Pueraria lobata **(Willd.)**

Sanskrit or Ayurvedic name(s): Ohwi, Vidaari (?)

Common name(s): Mudgaparni, Surpaparni

Family: Leguminosae/ Fabaceae

Distribution: Habitat to Eastern Himalaya, Assam, Khasi Hills.

Parts and properties used: Root- Antipyretic, anti-inflammatory, spasmolytic. Flower- Hepatoprotective.

Associated system(s): A

Pueraria tuberosa **(Roxb. ex Willd.) DC. (NT - VU)**

Sanskrit or Ayurvedic name(s): Vidaari, Vidarikanda

Common name(s): Vidhari kand, Vidari, Patal Kumbha

Family: Leguminosae/ Fabaceae

Distribution: Found in northwest, and central India.

Parts and properties used: Tuber- Diuretic, cardiac tonic, galactagogue, help fertility control. Root- Used as demulcent, refrigerant, and cataplasm.

Associated system(s): AFL

***Punica granatum* L. var. *granatum* and *P. granatum* L. var. *nana* Pers.**

Sanskrit or Ayurvedic name(s): Daadima, Raktapushpa

Common name(s): Dadam chal, Anaar

Family: Punicaceae

Distribution: West Asian. Grown arid and semi-arid India.

Parts and properties used: Fruit Rind- Astringent, stomachic, digestive. Treat diarrhoea, colitis, dysentery, uterine disorders, and dyspepsia. Leaves- Stomatitis.

Associated system(s): AHSLU

***Putranjiva roxburghii* Wall.**

Sanskrit or Ayurvedic name(s): Putranjiva

Common name(s): Jiyaapotaa, Child-life tree

Family: Euphoriaceae

Distribution: Native, grown throughout tropical India.

Parts and properties used: Fruit powder- Treat cough, cold and sprue. Rosaries hard stones- Protect children from infections.

Associated system(s): AFS

***Pupalia lappacea* (L.) Juss., syn. *Achyranthes lappacea* L.**

Common name(s): Chirchitta

Family: Amaranthaceae

Distribution: Found in Assam and many parts of India.

Parts and properties used: Herb- Treat bone fracture, wounds, boils, cough, toothache, fever, malaria.

Associated system(s): AF

***Pygmaeopemna herbacea* (Roxb.) Mold.**

Sanskrit or Ayurvedic name(s): Bhumi-jambu

Common name(s): Siru Thekku, Bhaarangi

Family: Verbenaceae

Distribution: Habitat to Eastern and southern Western Ghats.

Parts and properties used: Rootstock- Anti-asthmatic, anti-rheumatic. Leaf- Bechic, febrifuge, anti-rheumatic.

Associated system(s): AS

Pyrus communis **Linn.**

Common name(s): Bagu-goshaa

Family: Rosaceae

Distribution: European. Grown in J & K, HP, Punjab.

Parts and properties used: Plant/fruit- Anti-inflammatory, sedative, antipyretic, analgesic, hypolipidemic, anti-aging, and spasmolytic. Source of pectin, help maintain acid balance in the body. Advised in diabetes.

Associated system(s): F

Quercus ilex **Linn.**

Sanskrit or Ayurvedic name(s): Mayaaphala

Common name(s): Holly oak

Family: Fagaceae

Distribution: Habitat to Himalayas 900-2600 m, J & K.

Parts and properties used: Leaves- Antioxidant. Galls- Have 41% tannin.

Associated system(s): A

Quassia indica **(Gaertn.) Noot.**

Common name(s): Nibam, Lokhandi

Family: Simaroubaceae.

Distribution: Habitat to west coast along back waters.

Parts and properties used: Bark- Febrifuge. Leaf juice- Applied to skin diseases. Infusion- Given as an emmenagogue. Seed- Emetic, purgative; used in bilious fevers; seed oil- used in rheumatism.

Associated system(s): FS

Quercus infectoria **G. Olivier**

Sanskrit or Ayurvedic name(s): Maajuphalaka

Common name(s): Majuphal, Oak

Family: Fagaceae

Distribution: Native of Greece. Grown in the Himalayas.

Parts and properties used: Herb- Astringent. Bark/fruit- Used in eczema and impetigo. Galls- Used in gums and oral cavity. Added in breast and vaginal firming creams.

Associated system(s): AFSU

Quisqualis indica **Linn.**

Common name(s): Rangoon-ki-Bel, Laal-chameli

Family: Combretaceae

Distribution: Native to tropical Asia. Grown in Indian gardens.

Parts and properties used: Fruit and seed- Anthelmintic. Seeds- Soporific, given in diarrhoea. Soaked in oil applied to skin diseases.

Associated system(s): AFS

Randia dumetorum **Poir.**

Sanskrit or Ayurvedic name(s): Madana, Vishapushpaka

Common name(s): Mainphal

Family: Rubiaceae

Distribution: Habitat to foothills, scrub jungles, dry deciduous forests of northeast, Travancore, and Andaman.

Parts and properties used: Fruit- Nervine, calmative, emetic, antispasmodic, anthelmintic, abortifacient. Dried fruit- Advised in chlorosis, common cold, rhinitis and skin.

Associated system(s): AFSU

Randia uliginosa **DC.**

Sanskrit or Ayurvedic name(s): Pindaalu

Common name(s): Mainphal, Wagatta

Family: Rubiaceae

Distribution: Habitat to open areas of forests of Assam, Sikkim, and rest of India.

Parts and properties used: Unripe fruit- Astringent. Root- Diuretic; used for biliousness, diarrhoea, and dysentery.

Associated system(s): AFS

Ranunculus sceleratus **Linn.**

Sanskrit or Ayurvedic name(s): Sukaandaka

Common name(s): Jal-dhaniyaa

Family: Ranunculaceae

Distribution: Found in warm valleys of Himalayas, and northern plains.

Parts and properties used: Fresh plant- Acrid, rubefacient, vesicant and toxic. Dried used in skin diseases.

Associated system(s): AFH

***Raphanus sativus* Linn.**

Sanskrit or Ayurvedic name(s): Muulakapotikaa

Common name(s): Muuli, Raddish

Family: Brassicaceae/ Cruciferae.

Distribution: India is secondary center of origin. Grown allover India.

Parts and properties used: Root preparations are used in liver, gallbladder, and urinary complaints. Given in peptic disorders, and catarrhs.

Associated system(s): ASU

***Rauvolfia serpentina* (L.) Benth. ex Kurz (Rare- CR)**

Sanskrit or Ayurvedic name(s): Sarpagandha

Common name(s): Sarpagandha, Pagal Buti

Family: Apocynaceae

Distribution: Habitat to sub-Himalayan and oriental region of Indian. Subcontinent. Cultivated.

Parts and properties used: Root decoction- Increase uterine contractions and expulsion of foetus in difficult cases. Root extract- Induces sedation, bradycardia, hypotension. Key application- Hypertension, High BP, anxiety insomnia, fever. constipation, joint, liver, and mental disorders.

Associated system(s): AFHSLU

***Reinwardtia indica* Dumort**

Sanskrit or Ayurvedic name(s): Balbasant Baasanti

Common name(s): Abai, Yellow Flax, Pyoli

Family: Linaceae.

Distribution: Habitat to grasslands of Indochina, J & K, Sikkim. Grown in gardens.

Parts and properties used: Branch/leaves- Used in treatment of paralysis. Crushed leaves- Applied on maggot's wounds. Have trade potential.

Associated system(s): AF

***Reissantia indica* Halle.**

Common name(s): Odangod, Kazurati

Family: Celastraceae

Distribution: Habitat to north-east India.

Parts and properties used: Rootbark- Cure respiratory troubles. Stem- Febrifuge. Leaf powder- Applied to sore and wounds.

Associated system(s): FS

Rhamnus wightii **Wight & Arn.**

Sanskrit or Ayurvedic name(s): Rakta-Rohidaa

Common name(s): Kokkuvalli, Peyipoola

Family: Rhamnaceae

Distribution: South Asian hills and peninsular India.

Parts and properties used: Bark/fruit- Astringent, Deobstruent and tonic. Leaves- Give phenol, etc.

Associated system(s): AFS

Rhaphidophora pertusa **(Roxb.) Schott, syn.** ***R. laciniata*** **(Burm. f.) Merr.;** ***Pothos pertusus*** **Roxb (VU)**

Common name(s): Ganeshkanda

Family: Araceae

Distribution: Occur in Coromandel Coast, Deccan Peninsula, and Western Ghats. Cultivated.

Parts and properties used: Plant- Antidote to poisonous bites, anti-inflammatory and analgesic.

Associated system(s): AFS

Rheum australe **D. Don,** ***R. emodi*** **Wall. ex Meissn.] (VU)**

Sanskrit or Ayurvedic name(s): Amlaparni, Revandachini

Common name(s): Revan chini

Family: Polygonaceae

Distribution: Habitat to sub-alpine Himalayas 3300-5200m. Grown, traded.

Parts and properties used: Rhizome- Anticholesterolemic, antiseptic, antispasmodic, antitumor, aperient, astringent, diuretic, laxative. Yields a tonic.

Associated system(s): ASU

Rheum moorcroftianum **Royle**

Ayurvedic name(s): Amlaparni

Common name(s): Revand-chini, Indian rhubarb

Family: Polygonaceae

Distribution: Recorded in sub-alpine Himalayas 3000-5200m.

Parts and properties used: Rhizome- Used like *R. emodi*. Can be a substitute.

Associated system(s): A

Rheum palmatum **L.**

Common name(s): Revand-chini, Chinese rhubarb

Family: Polygonaceae

Distribution: Habitat to East Himalayas. Grown in gardens.

Parts and properties used: Rhizome- Cathartic, poultice for fevers & edema. Best type rhubarb

Associated system(s): ASU

Rheum spiciforme **Royle**

Ayurvedic name(s): Amlaparni,

Common name(s): Rhubarb, Revalchini

Family: Polygonaceae

Distribution: Habitat to Ladakh, Kumaon and Sikkim 2700-4800 m.

Parts and properties used: Rhizome- Purgative + rhubarb property. Source of anthraquinone.

Associated system(s): A

Rheum webbianum **Royle**

Common name(s): Rhubarb, Revalchini

Family: Polygonaceae

Distribution: Habitat to west Himalayas 3000-5000 m.

Parts and properties used: Rhizome- Purgative, antiseptic, antispasmodic, muscle relaxant + rhubarb property.

Associated system(s): AU

Rhinacanthus nasutus **(L.) Kurz.**

Sanskrit or Ayurvedic name(s): Yuuthikaparni. Paalaka-Juuhi.

Common name(s): Juyiparni, Vitamallikai, Nagamalli

Family: Acanthaceae

Distribution: Occur in most parts of India, Dehradun to TN. Grown in gardens.

Parts and properties used: Root, leaf, seed- Antidiabetic, antiseptic, antiparasitic. Paste- Applied to skin diseases, eczema, and control ringworm.

Associated system(s): ASU

Rhodiola rosea **L.**

Common name(s): Sanjeevani, solo, arctic root (trade name)

Family: Crassulaceae

Distribution: Habitat to arctic Europe, North America. Recently found in high-altitude arctic Himalayas.

Parts and properties used: Herb/root- Well known for adaptogen properties (reducing fatigue and exhaustion in prolonged stressful situations). Traded with the name of arctic root.

Associated system(s): European, F

Rhododendron anthopogon **D. Don. (VU)**

Common name(s): Talisapatra

Family: Ericaceae

Distribution: Habitat to alpine Himalayas 3000-5000 m.

Parts and properties used: Leaves- Stimulant, antitussive, diaphoretic, digestive and have quercetin, etc. Yields incense.

Associated system(s): F

Rhus parviflora **Roxb**

Sanskrit or Ayurvedic name(s): Tintindeeka.

Common name(s): Raitung

Family: Anacardiaceae

Distribution: Habitat to dry hot slopes of Himalayas.

Parts and properties used: Fruit juice- Vermifuge. Leaves- Flavonoids.

Associated system(s): AFU

Rhus succedanea **L.**

Sanskrit, Ayurvedic and common name(s): Karkatashringee, Karkataka shringi

Family: Anacardiaceae

Distribution: Habitat to temperate Himalayas 600-2500 m.

Parts and properties used: Lump on branches- Astringent, expectorant. Given in diarrhoea, dysentery, and vomiting. Fruit- Expectorant. Pith- Have polyphenols.

Associated system(s): ASLU

Rhynchosia beddomei **Baker (VU)**

Common name(s): Adavi Kandi

Family: Leguminosae/ Fabaceae

Distribution: Endemic to Eastern Ghats, AP.

Parts and properties used: Leaves- Abortifacient. Used in rheumatic pains, wounds, etc.

Associated system(s): F

***Ricinus communis* L.**

Sanskrit or Ayurvedic name(s): Eranda, Gandharva hasta, Raktaerund,

Common name(s): Shuklaerund, Arand

Family: Euphorbiaceae

Distribution: Native and cultivated in arid and semi-arid regions of India.

Parts and properties used: Leaf and seed oil- Purgative. Used in dermatosis and eczema, anti-asthmatic/anti-bronchitis, poultice. Root decoction- Advised in rheumatism, urinary bladder pain, lumbago, abdomen diseases and inflammations.

Associated system(s): AFHSLU

***Rivea hypocrateriformis* Choisy.**

Sanskrit, Ayurvedic and common name(s): Phanji

Family: Convolvulaceae

Distribution: Found throughout India.

Parts and properties used: Root- Yields a tonic, given after childbirth. Leaves- Astringent, used in haemorrhagic diarrhoea, and dysentery.

Associated system(s): AFS

***Rivea ornate* (Roxb.) Choisy.**

Common name(s): Phaang (Phanji variety)

Family: Convolvulaceae

Distribution: Found throughout India.

Parts and properties used: Plant juice- Used topically in haemorrhagic diseases and piles.

Associated system(s): AFS

***Rorippa indica* (L.) Hiern, syn. *R. dufia* Hara.**

Common name(s): Chamsuru, Kattukadugu

Family: Brassicaceae/ Cruciferae

Distribution: Found throughout India.

Parts and properties used: Plant- Antiscorbutic, stimulant, diuretic. Seeds- Laxative, given in asthma.

Associated system(s): FSU

***Rosa alba* L. (hybrid parentage)**

Sanskrit or Ayurvedic name(s): Shveta Taruni,

Common name(s): Gulseoti

Family: Rosaceae

Distribution: European. Naturalized, and cultivated in Indian gardens.

Parts and properties used: Plant extract- Liver disorders. Flower- Cardiac tonic for heart palpitation, febrifuge. Petal- Laxative.

Associated system(s): AU

***Rosa bourboniana* Desportes., syn. *R. chinensis* Jacq. *R. indica* L.**

Sanskrit or Ayurvedic name(s): Taruni, Taruni-Kantaka

Common name(s): Desi Gulaab

Family: Rosaceae

Distribution: Occur in Bengal. Grown Indian gardens.

Parts and properties used: Fruit- Applied on wounds, sprains, ulcers.

Associated system(s): AS

***Rosa centifolia* L.**

Sanskrit or Ayurvedic name(s): Satapatrikaa

Common name(s): Gulab

Family: Rosaceae

Distribution: European. Grown in Assam, Bihar, UP.

Parts and properties used: Flower decoction- Anti-inflammatory, aphrodisiac. Used in treatment of mouth, pharynx inflammation and intestine ulcers.

Associated system(s): AFSU

***Rosa damascena* Mill.**

Sanskrit or Ayurvedic name(s): Taruni

Common name(s): Gulab, Rose flowers

Family: Rosaceae

Distribution: Habitat to and cultivated in Indian gardens.

Parts and properties used: Flower buds- Astringent, expectorant, laxative. Used in abdominal and chest pain, digestive, inflammation, yield scent and essential oil.

Associated system(s): AFSHU

***Rosa moschata* Hook. f. non-Mill. Nec Herrm.**

Sanskrit or Ayurvedic name(s): Kubjaka

Common name(s): Kujai

Family: Rosaceae

Distribution: Habitat to central, and western Himalayas.

Parts and properties used: Plant- Used in bilious affections, skin irritation, and eye diseases. Flower- Produce rose water.

Associated system(s): AF

Rosa multiflora* Thunb., syn. *R. polyantha

Sanskrit or Ayurvedic name(s): Rakta-Taruni

Common name(s): Gulaab.

Family: Rosaceae

Distribution: Native to northeast India. Grown in Kulu (HP).

Parts and properties used: Fruit- Antiseptic. Applied to wounds, injuries, sprains ulcers.

Associated system(s): AF

***Rosa rubra* Blackw., syn. *R. gallica* L.**

Sanskrit or Ayurvedic name(s): Rakta-Taruni

Common name(s): Gulaab.

Family: Rosaceae

Distribution: European. Grown in Indian gardens.

Parts and properties used: Dried petals- Astringent. Yields a tonic. Oil and rose water- Used in bronchial asthma, skin irritation.

Associated system(s): A

***Rosa webbiana* Wall., and *R. macrophylla* Lindl.**

Sanskrit or Ayurvedic name(s): Ladakhi-Sevati

Common name(s): Siah

Family: Rosaceae

Distribution: Wild to dry-inner Himalayas 1500- 4000 m.

Parts and properties used: Fruit- Rich in vitamin C. Control Hepatitis, and cancer.

Associated system(s): AF

***Roscoea purpurea* Sm., *R. procera* Wall. (NT)**

Sanskrit, Ayurvedic and common name(s): Kshira-Kaakoli (Kshirkakoli)

Family: Zingiberaceae

Distribution: Habitat to Himalayas 1500-2500 m and Khasi Hills.

Parts and properties used: Tuberous root- Antidiabetic. Given in diarrhoea, hypertension, inflammation, and as revitalizing tonic (Part of *Astavarga*)

Associated system(s): AF

Rosmarinus officinalis **L.**

Common name(s): Rosemary

Family: Lamiaceae/ Labiatae

Distribution: Native of Mediterranean. Cultivated in Nigiri Hills.

Parts and properties used: Leaf, flower essential oil- Anti-inflammatory, astringent, antiseptic, stomachic, carminative; used in muscle pain, memory, immunity, circulatory system, and hair care. Leaf- Used in dyspeptic and rheumatic complaints.

Associated system(s): FH

Rotula aquatica **Lour.**

Common name(s): Pasanbheda

Family: Lythraceae

Distribution: Native to many parts of tropical India.

Parts and properties used: Plant- Aromatic. Root- Diuretic, astringent. Used in kidney and bladder stones, dysuria, coughs, heart, and blood disorders.

Associated system(s): SF

Rourea santaloides **Wight & Arn., syn.** ***R. minor*** **(Gaertn.) Alston**

Sanskrit or Ayurvedic name(s): Vridha

Common name(s): Vaakeri, Varadara

Family: Connaraceae

Distribution: Habitat to Assam, Northeast Hills, Bengal, southern Western Ghats.

Parts and properties used: Plant- Antiseptic, antitubercular. Root and twigs- Yields uterine tonic, used in rheumatism, pulmonary disorder, scurvy, and diabetes.

Associated system(s): AF

Rubia cordifolia **L., syn.** ***R. munjesta*** **Roxb.;** ***R. manjith*** **Roxb. (VU)**

Sanskrit or Ayurvedic name(s): Manjishthaa

Common name(s): Maddar, Manjitti, Majith, b-Tsod

Family: Rubiaceae

Distribution: Found all over India up to 3700 m, including Ladakh.

Parts and properties used: Root decoction or dried twigs- Anthelmintic, blood purifier, given in skin, urinogenital disorders; ulcers, dysentery, inflammations, piles, erysipelas, and rheumatism.

Associated system(s): AFSLU

***Rubia tinctorum* Linn.**

Common name(s): European Madder

Family: Rubiaceae

Distribution: Native of southern Europe. Also, found in Kashmir.

Parts and properties used: Root- Contains anthraquinone, and their glycosides, used for menstrual, urinary, and liver disorders.

Associated system(s): English

***Rumex acetosa* L.**

Sanskrit or Ayurvedic name(s): Chukram

Common name(s): Garden Sorrel, Hammaaz-Barri

Family: Polygonaceae

Distribution: Native to Eurasia, Himalayas (J & K, Kumaon). Grown.

Parts and properties used: Herb- Laxative, diuretic, refrigerant antiscorbutic. Seeds- Astringent.

Associated system(s): AU

***Rumex acetosella* Linn.**

Sanskrit or Ayurvedic name(s): Chukrikaa

Common name(s): Hammaaz

Family: Polygonaceae

Distribution: Wild worldwide, including East Himalayas, Sikkim, Nilgiris.

Parts and properties used: Herb- Diuretic, diaphoretic, antiscorbutic, refrigerant. Fresh used in urinary, kidney disorder.

Associated system(s): AU

***Rumex crispus* Linn. (R)**

Sanskrit or Ayurvedic name(s): Chukrikaa, Shatvedhani.

Common name(s): Yellow Dock

Family: Polygonaceae

Distribution: European. Grown in Mount Abu.

Parts and properties used: Root- Laxative, astringent. Used in rheumatism, bilious disorders, in piles and haemorrhagic affections.

Associated system(s): A

Rumex maritimus **Linn.**

Sanskrit or Ayurvedic name(s): Kunanjara

Common name(s): Jangali Paalak

Family: Polygonaceae

Distribution: From Canada to temperate Himalayas, to Western Ghats Hills.

Parts and properties used: Leaves- Cathartic, applied on burns. Seeds- Used of sex-tonics formulations.

Associated system(s): AU

Rumex vesicarius **Linn.**

Sanskrit or Ayurvedic name(s): Chukra, Chakravarti.

Common name(s): Shakkankeerai

Family: Polygonaceae

Distribution: African. Grown all over India.

Parts and properties used: Plant- Astringent antiscorbutic, stomachic, diuretic. Used in lymphatic, glandular, renal disorder, asthma, etc. Seeds- Antidysentery.

Associated system(s): ASU

Rungia pectinata **(L.) Nees.**

Sanskrit or Ayurvedic name(s): Parpata

Common name(s): Punakapundu

Family: Acanthaceae

Distribution: Found throughout India.

Parts and properties used: Leaf juice- Aperient, febrifuge, refrigerant. Root-Febrifuge.

Associated system(s): AS

Rungia repens **Nees.**

Sanskrit or Ayurvedic name(s): Parpata (subs.)

Common name(s): Kharmor

Family: Acanthaceae

Distribution: Found throughout moist India (Assam - UP).

Parts and properties used: Herb- Vermifuge, diuretic. Dried and pulverized herb used for cough, fever, and fresh on bruised.

Associated system(s): AFS

Ruta graveolens **L.**

Common name(s): Sadab

Family: Rutaceae

Distribution: European. Grown all over India.

Parts and properties used: Herb- Stimulating, antispasmodic, stomachic, irritant, abortifacient, used as emmenagogue, and in cough and colic. Leaves- Treat atonic amenorrhoea, menorrhoea, and colic.

Associated system(s): AFHSU

Saccharum munja **Roxb.**

Sanskrit or Ayurvedic name(s): Bhadramuja

Common name(s): Sarpata

Family: Poaceae/ Gramineae

Distribution: Habitat to plains and low hills allover India.

Parts and properties used: Plant- Refrigerant, used in burning sensation, thirst, dyscrasia, erysipelas and urinary complaints. Root- Used in dysuria, giddiness, and vertigo.

Associated system(s): ASL

Saccharum officinarum **L.**

Sanskrit or Ayurvedic name(s): Iksu

Common name(s): Karumbu

Family: Poaceae/ Gramineae

Distribution: Native. Cultivated in many parts of India.

Parts and properties used: Cane juice- Restorative, cooling, laxative, demulcent, diuretic, antiseptic. Advised in debility, haemorrhagic diseases, and anuria.

Associated system(s): AFHLSU

Saccharum spontaneum **L.**

Sanskrit, Ayurvedic and common name(s): Kaasha, Kusha

Family: Poaceae/ Gramineae

Distribution: Native to all over India.

Parts and properties used: Plant- Astringent, galactagogue, refrigerant, diuretic.

Advised in burning sensation, dysuria, dyscrasia, dysentery, piles. Root- Diuretic, galactagogue. Advised in calculus, dysuria, and haemorrhagic diseases.

Associated system(s): AFSLU

Saccolabium papillosum **Lindl.**

Sanskrit or Ayurvedic name(s): Naakuli (substitute), Vrkshaadani (variety)

Common name(s): Raasanaa, Indian birthwort

Family: Orchidaceae

Distribution: Habitat to outer range of Himalaya.

Parts and properties used: Roots- Used for rheumatism.

Associated system(s): AF

Salacia chinensis **L.**

Sanskrit or Ayurvedic name(s): Saptachakraa

Common name(s): Courondi

Family: Celastraceae

Distribution: Habitat to throughout India.

Parts and properties used: Root- Used in diabetes, genitourinary, venereal diseases. obesity.

Associated system(s): AFS

Salacia oblonga **Wall. ex Wight & Arn., syn.** ***S. pomifera*** **Wall. (EN)**

Common name(s): Chundan, Kadalainjil, Ponkorandi,

Family: Celastraceae

Distribution: Found in rain forest of Western Ghats.

Parts and properties used: Root bark- Used in diarrhea, fever, arthritis, gonorrhea, skin diseases. Have trade potential.

Associated system(s): AFS

Salix alba **Linn.**

Sanskrit or Ayurvedic name(s): Jalavetasa.

Common name(s): Vivir

Family: Salicaceae

Distribution: Native northwest Himalayas 1800 m. Cultivated.

Parts and properties used: Plant- Analgesic, anti-inflammatory, febrifuge. Advised in fever, headache, rheumatism.

Associated system(s): AUF

***Salix babylonica* Linn.**

Common name(s): Aatru Paalai, Giur

Family: Salicaceae

Distribution: Habitat to north India along water course.

Parts and properties used: Leaves/bark- Astringent, antipyretic. Used in fevers. Bark- Anthelmintic. Biological activity- Antiviral, CNS active, hypothermic.

Associated system(s): FS

***Salix caprea* Linn.**

Sanskrit or Ayurvedic name(s): Vetasa

Common name(s): Bed Mushk

Family: Salicaceae

Distribution: Europe to J & K, Punjab, HP, UP and grown.

Parts and properties used: Flower water- Cordial, stimulant. Externally applied to headache. Stem/leaves- Astringent, febrifuge.

Associated system(s): AU

***Salix tetrasperma* Roxb.**

Sanskrit or Ayurvedic name(s): Jalavetasa

Common name(s): Indian willow, Vaanira

Family: Salicaceae

Distribution: Habitat to northeast India, Western Ghats.

Parts and properties used: Dried leaves- Anti-inflammatory, given in swelling, rheumatism fever, piles. Bark- Febrifuge.

Associated system(s): AFS

***Salsola kali* Linn.**

Sanskrit or Ayurvedic name(s): Sarjikaa

Common name(s): Barilla

Family: Chenopodiaceae

Distribution: Habitat to northwest Himalayas.

Parts and properties used: Plant- Cathartic. Juice and seed-vessels- Diuretic. Ash used in Unani medicine.

Associated system(s): AFU

***Salvadora oleoides* Decne.**

Sanskrit, Ayurvedic and common name(s): Piluh (var.), Pilu, Kalawa

Family: Salvadoraceae

Distribution: Habitat to western India arid zone. Planted.

Parts and properties used: Leaf- Bechic. Bark- Vesicant. Fruit- Febrifuge. Used in enlarged spleen. Seed oil- Applied in rheumatism, fever, conjunctivitis, carminative, alexipharmic, etc.

Associated system(s): AFU

***Salvadora persica* L. (NT)**

Sanskrit or Ayurvedic name(s): Gudaphala, Piluh

Common name(s): Goni, Miswaak

Family: Salvadoraceae

Distribution: Habitat to arid zone and saline land of India. Cultivated.

Parts and properties used: Plant- Anti-inflammatory, hypoglycaemic, anti-bacterial. Fruit/seed- Carminative, DE obstruent, diuretic, lithotripsic. Leaf's decoction- Used in cough, and asthma. Flower- Stimulant, laxative. Applied in rheumatism. Seeds- Diuretic and purgative.

Associated system(s): AFSU

***Salvia aegyptiaca* L.**

Common name(s): Balangoo, Tukhm-Malangaa

Family: Lamiaceae/ Labiatae

Distribution: Habitat to arid areas of India.

Parts and properties used: Seed- Used in diarrhoea, and haemorrhoids.

Associated system(s): UF

***Salvia coccinea* Linn.**

Sanskrit or Ayurvedic name(s): Samudrashosha (a variety)

Common name(s): Red Sage

Family: Lamiaceae/ Labiatae

Distribution: Native of Americas. Grown in Indian garden.

Parts and properties used: Decoction- Used in renal diseases, and for lumbago.

Associated system(s): A, English

***Salvia haematodes* L. (halophyte)**

Common name(s): Behman, Behman Surkh

Family: Lamiaceae/ Labiatae

Distribution: Chinese. Grown in Indian gardens.

Parts and properties used: Roots, fruit (seed)- Aphrodisiac- Sex, cardiac, liver tonic.

Associated system(s): FU

***Salvia plebeia* R.Br.**

Sanskrit or Ayurvedic name(s): Samudrashosha

Common name(s): Kachora

Family: Lamiaceae/ Labiatae

Distribution: Habitat to stream sides & wet fields of plains and hills 1500 m.

Parts and properties used: Plant- Diuretic, astringent, anthelmintic, demulcent. Plant paste- Applied on wounds. Leaves- Used in toothache. Seeds- Used in diarrhoea, leucorrhoea, menorrhagia, haemorrhoids.

Associated system(s): AF

***Salvia sclarea* L.**

Common name(s): Behman safed

Family: Lamiaceae/ Labiatae

Distribution: From the Mediterranean to Kashmir.

Parts and properties used: Fruit (seed)- Antispasmodic and aromatic, used for digestive disorder.

Associated system(s): U

***Salvinia cucullata* Roxb. (Floating fern)**

Sanskrit, Ayurvedic and common name(s): Aakhukarni

Family: Azollaceae

Distribution: Habitat to throughout India near water.

Parts and properties used: Root- Digestive, diuretic, febrifuge anthelmintic. Used in epistasis, fever, colic, skincare.

Associated system(s): A

***Sandoricum koetjape* (Burm. f.) Merrill; syn. *S. indicum* Cav.**

Common name(s): Sentol, Sevai (Tamil)

Family: Meliaceae

Distribution: Wild in Himachal hills, Andamans. Cultivated.

Parts and properties used: Root- Astringent, carminative, antispasmodic. Bark- Anthelmintic. Treat influenza virus A and B.

Associated system(s): FS

***Sansevieria hyacinthoides* (Linn.) Druce., syn. *S. zeylanica* (L.) Willd.**

Sanskrit or Ayurvedic name(s): Naagadamani (wild relative).

Common name(s): Marul

Family: Liliaceae

Distribution: Wild in coastal regions of India.

Parts and properties used: Leaves/ rhizome- Applied in fever. Used like *S. roxburghiana* and can be a substitute.

Associated system(s): AS

***Sansevieria roxburghiana* JJ Schultes**

Sanskrit, Ayurvedic and common name(s): Naagadamani

Family: Liliaceae

Distribution: Wild on the east coast, Bengal to TN.

Parts and properties used: Plant- Given in glandular enlargement and rheumatism. Rhizomes- Used in cough. Shoot juice- Given to children with throat problem.

Associated system(s): AS

***Santalum album* L. (EN)**

Sanskrit or Ayurvedic name(s): Sveta Chandana

Common name(s): Chandan

Family: Santalaceae

Distribution: Habitat to dry peninsular India. Cultivated.

Parts and properties used: Bark/heartwood- Cooling, diuretic, diaphoretic, expectorant, antiseptic. Essential oil- Antimicrobial. Treat infections of lower urinary tract.

Associated system(s): AFHLSU

***Sapindus emarginatus* Vahl. (NT)**

Sanskrit or Ayurvedic name(s): Arishtaka

Common name(s): Reetha, Soapnut

Family: Sapindaceae

Distribution: Habitat to peninsular and south India. Cultivated.

Parts and properties used: Fruit (seed)- Astringent, emetic, detergent, anthelmintic. Pulp- Used in migraine, epilepsy, and hysteria. Pericarp- Yields Saponin.

Associated system(s): AFSU

***Sapindus laurifolius* Vahl., syn. *S. trifoliatus* L.**

Sanskrit or Ayurvedic name(s): Reethaakaranja

Common name(s): Reethaa

Family: Sapindaceae

Distribution: Habitat to south India. Cultivated in other parts.

Parts and properties used: Fruit- Astringent, emetic, detergent, anthelmintic. Pulp solution, used as nasal drops in migraine, epilepsy, hysteria. Root- Used for gout, rheumatism, and paralysis.

Associated system(s): ASU

***Sapindus mukorossi* Gaertn.**

Sanskrit or Ayurvedic name(s): Arishtaka

Common name(s): Aretha mota

Family: Sapindaceae

Distribution: Spread in Himalayas from HP to Assam. Also, cultivated.

Parts and properties used: Fruit- Emetic, expectorant; used in excessive salivation, chlorosis, epilepsy. Plant- Yield Saponin, that is antimicrobial used cosmetics. Fruit rind- Antimicrobial.

Associated system(s): AF

***Sapium sebiferum* Roxb.**

Common name(s): Tayapippali

Family: Euphorbiaceae

Distribution: Native of China. Planted allover India.

Parts and properties used: Seed oil- Vulnerary, emetic, purgative; used in skin care. Leaves latex- Vesicant. Resin- Purgative.

Associated system(s): F

***Saraca indica* L., syn. *S. asoca* (Roxb.) W.J.de Wilde (EN)**

Sanskrit, Ayurvedic and common name(s): Ashoka, Ashoka

Family: Leguminosae/ Fabaceae (Caesalpiniaceae)

Distribution: Native, grown all over India as ornamental tree.

Parts and properties used: Bark- Uterine tonic for suppressed menses, leucorrhoea, menstrual bleeding/ pain, menorrhagia, lymphadenitis, inflammations. Advised in metrorrhagia (Uterine bleeding), chronic lymphadenitis, menorrhagia, and inflammations.

Associated system(s): AFHS

***Sarcostemma viminale* (L.) R.Br., syn. *S. acidum* (Roxb.) Voigt; *S. brevistigma* W. & A.**

Sanskrit or Ayurvedic name(s): Somavalli

Common name(s): Soma, Somalatha, Pullangi tiga

Family: Asclepiadaceae

Distribution: Habitat to dry Bihar, Bengal, and peninsular India.

Parts and properties used: Plant- Insecticidal. Dried stem- Emetic.

Associated system(s): A

***Sarcostigma kleinii* W. & A.**

Sanskrit or Ayurvedic name(s): Ingudi

Common name(s): Odal

Family: Icacinaceae.

Distribution: Habitat to western Ghats. Cultivated.

Parts and properties used: Seed oil- Applied rheumatism. Bark powder- Rheumatism, neurological and skin disorders.

Associated system(s): AS

***Sauropus androgynus* Merrill.**

Common name(s): Thavasai, Murungai

Family: Euphorbiaceae

Distribution: Occur in Sikkim, Khasi Hills, and Western Ghats. Cultivated.

Parts and properties used: Plant- Source of vitamin, diuretic. Leaves- Used for weight loss. Decoction- Control bladder problem and fever.

Associated system(s): S

***Saussurea affinis* Spreng. ex DC.**

Common name(s): Ganga-muula

Family: Asteraceae/ Compositae

Distribution: Habitat to Assam, foothills of eastern Himalayas.

Parts and properties used: Root juice- Cure gynaecological diseases.

Associated system(s): F

***Saussurea costus (*Falc.) Lipsch., syn. *S. lappa* (Decne.) Sch. Bip.] (CR)**

Sanskrit or Ayurvedic name(s): Kushtha, Kusth

Common name(s): Kostum, Uplet, Rusta

Family: Asteraceae/ Compositae

Distribution: Habitat to Himalayas 2000-3300 m, Ladakh. Cultivated.

Parts and properties used: Root- Anthelmintic, carminative, antiasthma. Yields tonic. Advised in cough, bronchitis, dyspnoea, erysipelas, gout. Root essential oil- Antiseptic, disinfectant.

Associated system(s): AL(Amchi)H

***Saussurea glacialis* Herder**

Common name(s): Pansipa

Family: Asteraceae/ Compositae

Distribution: Habitat to alpine Himalaya 3800-5200 m, Ladakh.

Parts and properties used: Leaves, flowers- Used in liver, throat, heart troubles, mental disorder, intestinal complaints.

Associated system(s): Amchi

***Saussurea gossypiphora* D. Don. (EN)**

Common name(s): Kasturi kamal

Family: Asteraceae/ Compositae

Distribution: Wild in Himalayas 4200-5100 m (Garhwal-Sikkim).

Parts and properties used: Plant root decoction- Seals wound, by cease bleeding, gynaecological diseases.

Associated system(s): F

***Saussurea involucrata* Kar et Kir. (EN)**

Common name(s): Tian-Shan snow lotus

Family: Asteraceae/ Compositae

Distribution: Habitat to alpine Himalayas.

Parts and properties used: Flowers and stems- Used in rheumatoid arthritis, cough cold, stomachache, dysmenorrhea.

Associated system(s): Tibetan

***Saussurea obvallata* Wall. (VU -EN)**

Sanskrit, Ayurvedic and common name(s): Brahm Kamal, Brahmakamal

Family: Asteraceae/ Compositae

Distribution: Native to Himalayas 4200-5000 m.

Parts and properties used: Root- Antiseptic, styptic, anti-inflammatory. Applied to cuts and wounds. Plant- Hypothermic Flower- CNS active, antiviral. Used in rheumatism.

Associated system(s): F

***Scaevola frutescens* auct. non-Krause., syn. *S. koenigii* Vahl. *S. taccada* (Gaertn.) Roxb.**

Common name(s): Vella-muttangam

Family: Goodeniaceae

Distribution: Habitat to seacoast all around India.

Parts and properties used: Leaves- Digestive, carminative. Root- Used in dysentery.

Associated system(s): SF

***Schizachyrium exile* (Hochot.) Stapf**

Sanskrit, Ayurvedic and common name(s): Sprkaa, Sprkka

Family: Poaceae/ Gramineae

Distribution: Widespread grass in Bihar, Assam, Bengal to TN.

Parts and properties used: Plant- Substitute for Zalil/Yellow Delphinium, *Delphinium zalil.*

Associated system(s): A

***Schleichera oleosa* (Lour.) Oken, syn. *S. trijuga* Willd & Klein**

Sanskrit or Ayurvedic name(s): Koshaamra

Common name(s): Kusum beeja

Family: Sapindaceae

Distribution: Native to sub-Himalayan to peninsular India.

Parts and properties used: Bark- Astringent. With oil treat skin eruption. Seed oil- Applied in rheumatism, alopecia, itch, acne.

Associated system(s): AFS

***Schrebera swietenioides* Roxb. (VU)**

Sanskrit or Ayurvedic name(s): Muskakaa

Common name(s): Ghanti phal

Family: Oleaceae

Distribution: Habitat to dry forests of sub-Himalayan tracts.

Parts and properties used: Leaves- Enlarged spleen & urinary discharges. Root- Used for leprosy. Bark- Used for boils and burns.

Associated system(s): AFSL

***Schweinfurthia sphaerocarpa* (Benth.) A. Br., syn. *S. papilionacea* (L.) Boiss.**

Sanskrit or Ayurvedic name(s): Nepaal-Nimba

Common name(s): Saannipaat

Family: Scrophulariaceae

Distribution: Habitat to arid zone of Rajasthan and Gujarat.

Parts and properties used: Dried stem, leaves, fruit- Diuretic. Used in fever. Leaf- Antidiabetic.

Associated system(s): AF

***Scilla indica* Baker non-Roxb., syn. *S. hyacinthiana* (Roth) Macb.**

Sanskrit or Ayurvedic name(s): Korikanda

Common name(s): Jangli Piyaz.

Family: Liliaceae

Distribution: Habitat to central and southern India.

Parts and properties used: Bulb- Cardiotonic, stimulant, expectorant, diuretic. Used in cough, dysuria, strangury.

Associated system(s): ASU

***Scindapsus officinalis* (Roxb.) Schott.**

Sanskrit or Ayurvedic name(s): Gajapippali

Common name(s): Gaj pipal

Family: Araceae

Distribution: Habitat from China to tropical and subtropical Himalayas.

Parts and properties used: Fruit- Stimulant, carminative, diaphoretic, anthelmintic, antidiarrheal, hypoglycaemic. Given in diarrhoea, worm infestation. Decoction- Treat asthma. Dried spadix- Treat dyspnoea.

Associated system(s): AFLSU

***Scirpus articulatus* Linn.**

Sanskrit or Ayurvedic name(s): Laghu Kasheruka

Family: Cyperaceae

Distribution: Cosmopolitan. Grown in aquatic gardens.

Parts and properties used: Tubers- Prescribed in diarrhoea and vomiting.

Associated system(s): A

***Scirpus corymbosus* Roth., syn. *Schoenoplectus corymbosus* (Roth ex Roem. & Schult.)**

Sanskrit or Ayurvedic name(s): Kronchaadana

Family: Cyperaceae

Distribution: Occur in shallow waters of southwest Spain, Africa to India.

Parts and properties used: Tuber- Prescribed in diarrhoea, dysentery, and emesis.

Associated system(s): A

***Scripus grossus* L.F., syn. *S. kysoor* Roxb.**

Sanskrit or Ayurvedic name(s): Kasheruka

Common name(s): Kasuru

Family: Cyperaceae

Distribution: Habitat to all over India.

Parts and properties used: Tuber- Diuretic, antiseptic antidiarrheal, antiemetic, galactagogue, hypoglycaemic. Advised in dysuria, diabetes, genitourinary affections. Powder- Promote spermatogenesis and breast.

Associated system(s): ASF

***Scirpus tuberosus* Desf., syn. *S. maritimus* L.**

Sanskrit or Ayurvedic name(s): Raaj Kasheruka

Family: Cyperaceae

Distribution: Habitat to marshy places of India up to 3000 m.

Parts and properties used: Tuberous root- Astringent, diuretic, and laxative.

Associated system(s): A

***Scoparia dulcis* L.**

Common name(s): Mithi Patti, Ghoda Tulsi

Family: Plantaginaceae

Distribution: American. Naturalized as weed in India.

Parts and properties used: Herb- Used in diabetes, hypertension, kidney stones.

Associated system(s): S

***Scrophularia koelzii* Pennell**

Common name(s): Fogwart, Yarma

Family: Scrophulariaceae

Distribution: East Afghanistan to western alpine Himalayas, Ladakh.

Parts and properties used: Arial part- Anti-inflammatory. Given in joint pain, gout.

Associated system(s): UF

***Securinega suffruticosa* (Pall.) Rehder., *S. ramiflora* Muell., syn. *Flueggea suffruticosa* (Pall.) Baill.**

Common name(s): Dalme, Kodarsi, Pandharphali, Vellaippula (*S. virosa*).

Family: Euphorbiaceae

Distribution: Distributed widely, including East Himalayas, up to 250m.

Parts and properties used: Twig/ Leaves- Yield Securinine, stimulates central nervous system. Treat lumbago, limb numbness and indigestion.

Associated system(s): Chinese, FS

***Securinega leucopyrus* (Willd.) Mull-Arg., syn. *S. leucopyrus* (Willd.) Mull-Arg.**

Sanskrit or Ayurvedic name(s): Panduphali

Common name(s): Hartho, Humari Bhuriphali

Family: Euphorbiaceae

Distribution: Habitat to arid zone of south India.

Parts and properties used: Leaf paste- Anti-inflammatory, anti-arthritic. Applied to heal difficult wounds.

Associated system(s): AFS

***Sedum ewersii* Ledeb., syn. *Hylotelephium ewersii* (Ledeb.) H. Ohba**

Common name(s): Stone crop

Family: Crassulaceae

Distribution: Native to mountains central Asia and Himalayas (invasive to alpine, Ladakh).

Parts and properties used: Crushed herbs- Cooling, soothing. Given in dysentery.

Associated system(s): L (Amchis)

***Selaginella involvens* Spring., and *S. rupestris* Spring. (Fern)**

Sanskrit or Ayurvedic name(s): Kara-jodi-kanda

Common name(s): Hatthaa jodi

Family: Selaginellaceae.

Distribution: Occur on boulders and soil near the streams of Kerala.

Parts and properties used: Plant/decoction- Yields tonic.

Associated system(s): AF

***Selinum candollei* DC., syn. *S. tenuifolium* (NT)**

Sanskrit, Ayurvedic and common name(s): Muraa, Mura

Family: Apiaceae/ Umbelliferae

Distribution: Habitat to Himalayas.

Parts and properties used: Root- Nervine, sedative, analgesic.

Associated system(s): ASF

***Selinum vaginatum* (Edgew.) C.B. Clarke**

Sanskrit or Ayurvedic name(s): Rochanaa-Tagara

Common name(s): Butkesh

Family: Apiaceae/ Umbelliferae

Distribution: Habitat to north-west Himalayas 1800-3900 m.

Parts and properties used: Rhizome- Nervine sedative. Oil- Sedative, analgesic, hypotensive.

Associated system(s): AF

***Semecarpus anacardium* L. f.**

Sanskrit or Ayurvedic name(s): Bhallataka

Common name(s): Balave, Bhilavan, Bhilawa, Marking nuts.

Family: Anacardiaceae

Distribution: Habitat to outer Himalayas and peninsular India.

Parts and properties used: Fruit and nut- Antiatherogenic, anti-inflammatory, antimicrobial, anti-reproductive, CNS stimulant, hypoglycemic, anticarcinogenic.

Associated system(s): AFHLSU

***Senecio chrysanthemoides* DC.**

Common name(s): rGu-drus

Family: Asteraceae/ Compositae

Distribution: Habitat to alpine Himalayas, Ladakh

Parts and properties used: Crushed flower/leaf oil-Antiseptic, antirheumatic, antimicrobial.

Associated system(s): HF

***Sesamum orientale* L., syn. *S. indicum* L.**

Sanskrit, Ayurvedic and common name(s): Tila, Til

Family: Pedaliaceae

Distribution: African. Cultigen in India, widely grown in most of India.

Parts and properties used: Seeds- Laxative, diuretic emollient, lactagogue, demulcent. Powder- Used in treatment of amenorrhoea and dysmenorrhoea.

Associated system(s): ASU

***Sesbania bispinosa* W. f. Wight., syn. *S. aculeata* (Willd.) Poir.**

Sanskrit or Ayurvedic name(s): Jayanti

Common name(s): Mudchembai

Family: Leguminosae/ Fabaceae

Distribution: Habitat to western Himalaya to peninsular India.

Parts and properties used: Plant- Treat wounds. Seeds- Treat ringworm and skin diseases.

Associated system(s): AS

***Sesbania grandiflora* (L.) Poir. (VU)**

Sanskrit or Ayurvedic name(s): Agastya

Common name(s): Agatti.

Family: Leguminosae/ Fabaceae

Distribution: Distributed from Assam to eastern Ghats.

Plant- Astringent, antihistaminic, febrifuge. Used in fevers, cough catarrh, glandular enlargement.

Associated system(s): AS

***Sesbania sesban* (Linn.) Merrill., syn. *S. aegyptiaca* Pers.**

Sanskrit or Ayurvedic name(s): Jayantikaa

Common name(s): Sembai, Jainta

Family: Leguminosae/ Fabaceae

Distribution: Wild and grown all over India.

Parts and properties used: Seed and bark- Astringent, emmenagogue. Treat menorrhagia, spleen enlargement, and diarrhea. Leaves- Used in dysuria. Bark juice- Treat skin eruptions. Seed oil- Cardiac depressant, antibacterial.

Associated system(s): AFS

***Seseli diffusum* (Roxb. ex Sm.) Santapau, syn. *S. indicum* W & A**

Sanskrit or Ayurvedic name(s): Vanya-yamaani.

Common name(s): Kirmani, Ajwain

Family: Apiaceae/ Umbelliferae

Distribution: Native spread over India to Myanmar.

Parts and properties used: Seed- Stimulant, anthelmintic (Round worms), carminative, diuretic antispasmodic.

Associated system(s): A

***Seseli sibiricum* Benth. ex C. B. Clarke**

Sanskrit or Ayurvedic name(s): Bhuutakeshi

Family: Apiaceae/ Umbelliferae.

Distribution: Habitat to J & K 2300-3500 m.

Parts and properties used: Herb- Tranquilizer. Used in mental disorders. Volatile oil- Hypotensive.

Associated system(s): A

***Setaria italica* (L.) Beauv**

Sanskrit or Ayurvedic name(s): Kangunikaa

Common name(s): Kangni

Family: Poaceae/ Gramineae

Distribution: Cosmopolitan. Cultivated.

Parts and properties used: Plant- Sedative. Grain- Astringent, digestive, emollient and stomachic.

Associated system(s): AFSLU

***Shorea robusta* Gaertn.**

Sanskrit, Ayurvedic and common name(s): Shaala, Raal

Family: Dipterocarpaceae

Distribution: Distributed in north, east, central India. Cultivated.

Parts and properties used: Fruit- Diarrhoetic. Resin- Astringent, detergent; antidiarrheal, antidysentery. Resin (essential oil)- Antiseptic. Used for skin diseases.

Associated system(s): AFSLU

Sida acuta **Burm.f., syn. *S. carpinifolia* auct. non. Linn f.**

Sanskrit, Ayurvedic and Common name(s): Balaa (white flower) Bala, Chikna

Family: Malvaceae

Distribution: Habitat to all over warm India.

Parts and properties used: Root- Astringent, stomachic, febrifuge, diuretic. Used in nervous sexual debility. Leaves- Demulcent- Used in elephantiasis.

Associated system(s): AF

Sida cordifolia **L.**

Sanskrit, Ayurvedic and common name(s): Balaa, Bala

Family: Malvaceae

Distribution: Habitat to all over moist India.

Parts and properties used: Plant juice- Invigorating, spermatopoietic in spermatorrhoea. Root- Treat rheumatism, neurological disorders, polyuria, dysuria, cystitis, strangury, and uterine disorders. Leaves- Demulcent, febrifuge; used in dysentery.

Associated system(s): ALSU

Sida rhombifolia **L.**

Sanskrit or Ayurvedic name(s): Mahaabalaa

Common name(s): Bala

Family: Malvaceae

Distribution: Native to new world found all over moist India. Cultivated and traded.

Parts and properties used: Plant- Used in pulmonary tuberculosis, nervous diseases, and rheumatism. Leaves- Paste suppress swelling. Stem mucilage- Demulcent, and emollient. diuretic. Root- Advised in deficient spermatogenesis and oedema.

Associated system(s): ASU

Sida spinosa **Linn., syn. *S. alba* Linn.**

Sanskrit or Ayurvedic name(s): Naagabalaa

Common name(s): Gulasakari

Family: Malvaceae

Distribution: Habitat to all over hotter parts of India.

Parts and properties used: Root- Diaphoretic, a nervine tonic; used in debility and fevers. Decoction- Demulcent in irritable of bladder genitourinary tract. Leaves- Demulcent, refrigerant; used for scalding urine.

Associated system(s): AFS

***Sida veronicaefolia* Lam., syn. *S cordata* (Burm. f.) Borssum., *S. humilis* Cav.**

Sanskrit or Ayurvedic name(s): Raajabalaa

Common name(s): Palampasi, Farid-booti.

Family: Malvaceae

Distribution: Habitat to all over warm India.

Parts and properties used: Flower/fruit- Given in burning urination. Leaf juice- Given in diarrhoea; leaf used as poultice. Root bark- Used in leucorrhoea and genitourinary affections.

Associated system(s): AFS

***Siegesbeckia orientalis* Linn.**

Common name(s): Katampam, Latlatiaa

Family: Asteraceae/ Compositae

Distribution: Habitat to all over India up to 2000 m.

Parts and properties used: Plant- Antiscorbutic, sialagogue, cardiotonic, diaphoretic. Used in rheumatism, renal colic, and ague.

Associated system(s): FS

***Silybum marianum* (L.) Gaertn.**

Common name(s): Milk thistle

Family: Asteraceae/ Compositae

Distribution: Habitat to western Himalayas (2400 m). Grown in gardens.

Parts and properties used: Seed- Given for liver, gallbladder shielding; used in jaundice and liver diseases. Part of drug formulations.

Associated system(s): FH

***Simmondsia chinensis* (Link) C.K. Schneid.**

Common name(s): Jojoba

Family: Simmondsiaceae

Distribution: Native to USA. Introduced and cultivated in Rajasthan.

Parts and properties used: Plant, seed wax- Used in skin disorder, cancer, kidney disorders, colds, dysuria, obesity, parturition.

Associated system(s): F

***Sinapis alba* L.**

Common name(s): Safed Rai, Yellow mustard

Family: Brassicaceae/ Cruciferae

Distribution: Mediterranean native. Cultivated in many parts of India.

Parts and properties used: Seed- Antibacterial, antifungal, carminative, diaphoretic, digestive, diuretic, emetic, expectorant, rubefacient and stimulant.

Associated system(s): AHS

***Sisymbrium irio* L.**

Sanskrit, Ayurvedic and common name(s): Khaaksi Khubkalan, Khubkalan

Family: Brassicaceae/ Cruciferae

Distribution: Habitat to cultivated fields in rocky parts of India.

Parts and properties used: Seed- Expectorant, restorative, febrifuge, rubefacient, antibacterial. Used in chest congestion, rheumatism, detoxify liver/ spleen, and swelling.

Associated system(s): AU

***Smilax aspera* L. syn. = *Smilax aspera* subsp. *mauritanica* (Poir.) Arcang.**

Sanskrit or Ayurvedic name(s): Chobachini

Common name(s): Italian Sarsaparilla, subst. of Chopchini

Family: Smilacaceae

Distribution: Habitat to tropical India. Harvested from wild.

Parts and properties used: Root- Alterative, demulcent, depurative, diaphoretic, diuretic, stimulant, tonic. Substitute to *Hemidesmus indicus*.

Associated system(s): AH

***Smilax china* L.**

Sanskrit or Ayurvedic name(s): Madhusnuhi

Common name(s): Chobchini

Family: Smilacaceae

Distribution: Chinese, habitat to Assam too.

Parts and properties used: Tubers- Alterative, arthritis, rheumatoid, used as *Sarsaparilla* (Dried rhizome).

Associated system(s): AHSU

***Smilax glabra* Roxb. (CR)**

Sanskrit or Ayurvedic name(s): Dweepaantara-Vachaa

Common name(s): Chobchini, Lokhandi

Family: Smilacaceae

Distribution: Chinese, habitat to Assam, northeast hills as well.

Parts and properties used: Roots- Antiarthritic, depurative, used in syphilis, sores, venereal diseases.

Associated system(s): A

***Smilax zeylanica* L. and *S. ornata* Hook. (VU)**

Common name(s): Substitute for Chopchini, *S. aspera.*

Family: Smilacaceae

Distribution: Habitat to tropical Himalayas and northeast hills.

Parts and properties used: Root- Used in venereal diseases. Decoction- Antirheumatic.

Associated system(s): FSU

***Solanum anguivi* Lam., syn. *S. indicum* auct. non L. (?); *S. violaceum* Ortega**

Sanskrit or Ayurvedic name(s): Brihati

Common name(s): Katheli-badhi, forest bitterberry (Flowers small, white with purple veins, fruit in cluster)

Family: Solanaceae

Distribution: Native to tropical and subtropical India and cultivated.

Parts and properties used: Root- Carminative, expectorant. Used for cough, catarrh.

Associated system(s): AFSU

***Solanum dulcamara* Linn.**

Sanskrit or Ayurvedic name(s): Kaakamaachi-vishesha

Common name(s): Mako (red)

Family: Solanaceae

Distribution: Habitat to alpine Himalayas (1200-2400 m).

Parts and properties used: Twig/bark- Stimulating, hepatic expectorant, astringent, alterative, antirheumatic, antifungal. Used in bronchitis, eczema, rheumatism.

Associated system(s): AU

***Solanum ferox* Linn.**

Sanskrit or Ayurvedic name(s): Brihati, Brihatikaa

Common name(s): Raam-begun

Family: Solanaceae

Distribution: Habitat to all over warm India.

Parts and properties used: Plant, root- Digestive, carminative, expectorant, diaphoretic, stimulant, anthelmintic. Used in cough catarrh, asthma, etc.

Associated system(s): AFSU

***Solanum indicum* Linn.**

Sanskrit or Ayurvedic name(s): Brihati, Kshudrabhantaaki

Common name(s): Barahantaa, Katkari (Flower purple, fruit big; wild relative of Eggplant)

Family: Solanaceae

Distribution: Found all over tropical and subtropical India.

Parts and properties used: Root- Carminative, expectorant. Used in colic, dysuria, coughs, and catarrhal affections. Decoction is advised in parturition (Part of *Dash moola*)

Associated system(s): AFSU

***Solanum melongena* L.**

Sanskrit or Ayurvedic name(s): Bhantaki

Common name(s): Baingan, Eggplant

Family: Solanaceae

Distribution: Native and grown all over India.

Parts and properties used: Fruit- Advised for liver complaints and for amenorrhoea.

Associated system(s): AFSU

***Solanum nigrum* L.**

Sanskrit or Ayurvedic name(s): Kakamaci

Common name(s): Makoi, Kakamachi

Family: Solanaceae

Distribution: Occur throughout India up to 2100 m, including Ladakh.

Parts and properties used: Plant- Anti-inflammatory, antispasmodic, diuretic, laxative; extract used to treat inflammation, enlargement of liver and spleen and in liver cirrhosis. Berries- Antidiarrheal, antipyretic, used in cough/cold/asthma.

Associated system(s): AFHSLU

Solanum torvum **Sw.**

Sanskrit or Ayurvedic name(s): Goshtha-vaartaaku

Common name(s): Padarchunda, Ran-Baingan

Family: Solanaceae

Distribution: Habitat to tropical India, Assam to Kerala. Grown marginally.

Parts and properties used: Plant- Digestive, diuretic, sedative. Leaves- Haemostatic. Fruit- Used in liver and spleen enlargement and cough. Source of glycoalkaloid solasodine.

Associated system(s): AFS

Solanum trilobatum **L.**

Sanskrit, Ayurvedic and common name(s): Alarka, Alarka

Family: Solanaceae

Distribution: Indo-Malayan. Found from peninsular India to Western Ghats.

Parts and properties used: Berries flower decoction- Used in respiratory diseases cough, bronchial asthma.

Associated system(s): AFS

Solanum virginianum **L., syn.** ***S. surattense*** **Burm.f.;** ***S. xanthocarpum*** **Schrad. & H. Wendl.**

Sanskrit or Ayurvedic name(s): Varhat, Kantakaari

Common name(s): Kateli

Family: Solanaceae

Distribution: Found throughout India. Himalayas to Western Ghats.

Parts and properties used: Plant- Stimulant, expectorant, diuretic, laxative, febrifuge. Used in cough, bronchitis, asthma, tenacious, phlegm, and cancer. Fruit, root- Source of alkaloids (Part of *Dashmoola*).

Associated system(s): AHSU

Soymida febrifuga **(Roxb.) A. Juss.**

Sanskrit or Ayurvedic name(s): Maansrohini

Common name(s): Rohan

Family: Meliaceae

Distribution: Habitat to many parts of India.

Parts and properties used: Bark- Antipyretic, yields a tonic. Used in diarrhoea, dysentery, and for gargle.

Associated system(s): AFSLU

***Spermacoce articularis* L. f., syn. *S. hispida* Linn.; *Borreria articularis* (Linn. f.) F.N. Williams**

Sanskrit or Ayurvedic name(s): Madana Ghanti

Common name(s): Tukah

Family: Rubiaceae

Distribution: Habitat to tropical Himalayas, Assam, Gujarat, Maharashtra, TN.

Parts and properties used: Leaf's extract- Astringent, used to treat haemorrhoids gall stones. Seed- Demulcent. Given in diarrhoea and dysentery.

Associated system(s): AS

***Sphaeranthus africanus* L.**

Common name(s): Mundi

Family: Asteraceae/ Compositae

Distribution: Found in southern Western Ghats

Parts and properties used: Flower- Used in vata, pitta, epilepsy, migraine, jaundice, cough, hemorrhoids, helminthiasis, nervine tonic.

Associated system(s): A

***Sphaeranthus indicus* L.**

Sanskrit or Ayurvedic name(s): Munditika, Munditikaa

Common name(s): Gorak-mundi

Family: Asteraceae/ Compositae

Distribution: Distributed all over tropical India in rice fields.

Parts and properties used: Juice- Styptic, emollient, resolvent. Seeds/root-Anthelmintic. Decoction- Given in cough, chest diseases, and catarrhal. Dried leaf-Used to treat cervical. lymphadenitis, chronic sinusitis, migraine, epilepsy, lipid disorders, spleen problem, anaemia, dysuria.

Associated system(s): AFSLU

***Spilanthes oleracea* L., syn. *S. oleracea* Murr. var. *oleracea* C.B. Clarke**

Sanskrit or Ayurvedic name(s): Mahaaraashtri

Common name(s): Akarkara, Sarahattika, Vana-mugali

Family: Asteraceae/ Compositae

Distribution: Habitat to waste places of tropical sub-tropical India. Introductions from Brazil grown in gardens.

Parts and properties used: Plant- Ant dysenteric. Decoction- Diuretic and lithotripsic, treat scabies/ psoriasis. Root- Purgative. Flower- Treat gum, tooth, bladder, and gout. Seed- Treat xerostomia and throat infection.

Associated system(s): AF

***Spinacia oleracea* Linn., syn. *S. tetrandra* Roxb.**

Sanskrit or Ayurvedic name(s): Paalankikaa

Common name(s): Paalak

Family: Chenopodiaceae

Distribution: South-west Asian. Grown all over India.

Parts and properties used: Seeds- Cooling and laxative. Given during jaundice (?).

Associated system(s): ASU

***Spondias pinnata* (L. f.) Kurz.**

Sanskrit or Ayurvedic name(s): Aamraata

Common name(s): Amate, Amara

Family: Anacardiaceae

Distribution: Native of moist deciduous forests of India. Cultivated.

Parts and properties used: Bark, leaf, fruit- Antidysentery, antiseptic, antiscorbutic. Bark- Haemorrhagic. Bark paste- Advised in muscular rheumatism.

Associated system(s): AFSLU

***Stachytarpheta jamaicensis* Vahl., syn. *S. indica* (L.) Vahl**

Sanskrit or Ayurvedic name(s): Kariyartharani

Common name(s): Simainayuruvi, Chirchiti

Family: Verbenaceae

Distribution: Native of tropical America. Grown in Indian gardens.

Parts and properties used: Plant- Febrifuge, ant inflammatory. Applied on purulent ulcers and given in rheumatic inflammations and fever.

Associated system(s): AFS

***Stephania glabra* (Roxb.) Miers**

Sanskrit or Ayurvedic name(s): Paathaa

Common name(s): Patha, Rajapatha

Family: Menispermaceae

Distribution: Distributed in Himalayas from Assam to Northeast Hills.

Parts and properties used: Tubers- Anti-inflammatory. Used pulmonary diseases, asthma, intestinal disorders, and hyperglycaemia.

Associated system(s): AF

***Stephania japonica* var. *discolor* (Miq.) Forman, syn. S. *hernandiifolia* (Willd.) Walp.**

Sanskrit or Ayurvedic name(s): Rajapatha

Common name(s): Tubuki-lata

Family: Menispermaceae

Distribution: Habitat to Himalayas, Eastern and Western Ghats.

Parts and properties used: Root- Advised in skin diseases, pruritus, inflamed piles, internal abscesses, urinary and respiratory diseases, diarrhoea, colic.

Associated system(s): A

***Sterculia foetida* L.**

Sanskrit or Ayurvedic name(s): Vitkhadirah

Common name(s): Jangli badam, Java olive

Family: Sterculiaceae

Distribution: Pantropical, Assam to south India. Grown along roadside and in gardens.

Parts and properties used: Bark/leaf- Aperient, diuretic. Fruit- Astringent. Seed oil- Carminative, laxative. Wood + oil- Cure rheumatism.

Associated system(s): AFS

***Sterculia urens* Roxb. (EN)**

Common name(s): Karaya, Kateera

Family: Sterculiaceae

Distribution: Occur in many parts of India.

Parts and properties used: Gum- Aphrodisiacal. Substitute to *Tragacanthin* (gum of middle east gum) in throat infection.

Associated system(s): FSU

***Stereospermum chelonoides* (L.f.) DC., syn. *S. personatum* (Hassk.) Chatterjee; *S. suaveolens* (Roxb.) DC. (CR)**

Sanskrit or Ayurvedic name(s): Patalai

Common name(s): Patala, Padal

Family: Bignoniaceae

Distribution: Found throughout moist parts of India.

Parts and properties used: Root- Used in lipid disorders. Stem bark- Given in oedema, retention of urine. anti-asthmatic (Part of *Dashmoola*).

Associated system(s): AFS

***Stevia rebaudiana* Bertoni**

Common name(s): Sweet leaf or candy leaf

Family: Asteraceae/ Compositae

Distribution: South American. Introduced to semi-humid subtropical India.

Parts and properties used: Leaves- Yields natural sugar substitute. Anti-diabetes. Controls heart problems by lowering BP.

Associated system(s): South America traditional medicinal system

***Streblus asper* Lour.**

Sanskrit or Ayurvedic name(s): Shaakhotaka

Common name(s): Bajradanti

Family: Moraceae

Distribution: Habitat to drier parts of peninsular India.

Parts and properties used: Stembark- Febrifuge, antidiarrheal, used in cervical lymphadenitis, lipid disorders. Root- Treat inflammation and syphilitic, eruptions. Latex- Used in glandular swelling.

Associated system(s): AFS

***Striga asiatica* (Linn.) Kuntze.**

Sanskrit or Ayurvedic name(s): Agnivrksha, Kuranti

Common name(s): Agiyaa

Family: Scrophulariaceae

Distribution: Found all over India as field weed.

Parts and properties used: Herb- Appetizer. Treat urinary and vitiated blood disorders.

Associated system(s): AFS

***Strychnos colubrina* Linn.**

Sanskrit or Ayurvedic name(s): Kupilu-lataa

Family: Loganiaceae

Distribution: Habitat to Deccan Peninsula and Western Ghats.

Parts and properties used: Root- Purgative, febrifugal, anthelmintic. Boiled roots/ leaves- Used in rheumatic swelling. Wood- Fever and skin eruption.

Associated system(s): AF

***Strychnos nux-vomica* L.**

Sanskrit or Ayurvedic name(s): Visamusti, Kapilu

Common name(s): Kuchla, Itti beeja

Family: Loganiaceae

Distribution: Found in moist deciduous forests of tropical India.

Parts and properties used: Plant- Detoxified seeds. Bark juice- Nervine tonic, CNS stimulant. Seeds- Paralysis, facial paralysis, sciatica, and impotency. Given in dysentery, diarrhoea, and colic.

Associated system(s): AFHSLU

***Strychnos potatorum* L. f.**

Sanskrit or Ayurvedic name(s): Kataka

Common name(s): Nirmali, Nux Vomica, Cleaning nuts

Family: Loganiaceae

Distribution: Occur in foothills of peninsular India. Cultivated.

Parts and properties used: Seeds- Antidiabetic, antidysentery, emetic, Advised in dysuria, polyuria, urolithiasis, and epilepsy.

Associated system(s): AFSLU

***Strychnos rheedei* C.B. Clarke., syn. *S. cinnamomifolia* Thw., *S. wallichiana* Steud**

Common name(s): Valli Kanjiram

Family: Loganiaceae

Distribution: Found in AP and Western Ghats.

Parts and properties used: Roots- Antirheumatic, anti-inflammatory, febrifuge. Used for neurological affections, elephantiasis, and muscular pains.

Associated system(s): S

***Stylocoryna lucens* (Hook.f.) Gamble, syn. *Tarenna alpestris* (Wight) N.P. Balakr.**

Ayurvedic name(s): Paphanals

Family: Rubiaceae

Distribution: Distributed in southern Western Ghats.

Parts and properties used: Root, stem- Antibacterial, anti-inflammatory, and anthelmintic. Stem juice- Applied on bruises.

Associated system(s): A

***Styrax officinals* Linn.**

Sanskrit or Ayurvedic name(s): Silhaka, Silaarasa, Turushka

Common name(s): Silaajit, Usturak

Family: Styracaceae

Distribution: Native of Asia minor and Syria. Grown in Indian gardens.

Parts and properties used: Plant extract- Antimicrobial. Resin- Antiseptic. Balsam (Essential oil)- Used for cough and respiratory tract catarrh.

Associated system(s): AF

***Swertia alata* (Royle *ex* D. Don) C.B. Clarke**

Ayurvedic name(s): Kiratatikta

Family: Gentianaceae

Distribution: Habitat to temperate Himalayas.

Parts and properties used: Plant- Febrifuge. Used as appetite tonic.

Associated system(s): A

***Swertia angustifolia* Ham. (EN), (= *Swertia affinis* CB Clarke)**

Sanskrit or Ayurvedic name(s): Kiraatatikta, Kiraata

Common name(s): Chiraeta, Pahaari Kiretta, Shireen

Family: Gentianaceae

Distribution: Habitat to sub-tropical Himalaya, Northeast Hills & peninsular India.

Parts and properties used: Plant- Febrifuge and bitter tonic; a substitute for *S. chirayita.*

Associated system(s): AU

***Swertia chirayita* (Roxb. ex Fleming) H. Karst., syn. *S. chirata* (VU)**

Sanskrit or Ayurvedic name(s): Kiratatikta

Common name(s): Chiraiyata, Chiraitaa.

Family: Gentianaceae

Distribution: Indigenous to temperate Himalayas, Khasi Hills. Grown.

Parts and properties used: Plant- Blood purifier, and bitter tonic, used in skin diseases. Anti-inflammatory yields Oleanolic acid. Used for loss of appetite, digestive disorders, diabetes.

Associated system(s): AFH

***Swertia ciliata* (D. Don) Burtt.**

Sanskrit or Ayurvedic name(s): Shailaja

Common name(s): Kiratatikta relative

Family: Gentianaceae

Distribution: Wild in southern Western Ghats.

Parts and properties used: Plant- Contains tetra oxygenated and penta-oxygenated xanthones. Substitute for *S. chirayita.*

Associated system(s): A

***Swertia cordata* (Wallich ex G. Don) C. B. Clarke**

Common name(s): Chirayta, Tikta

Family: Gentianaceae

Distribution: Wild in temperate Himalayas, including Ladakh.

Parts and properties used: Plant extract- Anti-fever, antioxidant, antibacterial and antidiabetic. Adulterant of *S. chirayita.*

Associated system(s): AUS

***Swertia densifolia* (Griseb.) Kashyapa, syn. *S. decussata* Nimmo ex C.B.Clarke**

Sanskrit or Ayurvedic name(s): Shailaja

Common name(s): Kiratatikta relative

Family: Gentianaceae

Distribution: Wild in southern Western Ghats.

Parts and properties used: Like that of *S. chirayita* as a Substitute.

Associated system(s): A

***Swertia lawii* Burkill., (EN)**

Common name(s): Kiratatikta relative

Family: Gentianaceae

Distribution: Wild in southern Western Ghats.

Parts and properties used: Substitute for *S. chirayita.*

Associated system(s): A

***Symplocos cochinchinensis* S. Moore (NT)**

Sanskrit or Ayurvedic name(s): Lodhra

Family: Symplocaceae

Distribution: Occur in southern Western Ghats.

Parts and properties used: Plant extract- Potential anticancer. Bark- Advised in diabetes mellitus.

Associated system(s): AFS

***Symplocos paniculata* (Thunb.) Miq.**

Sanskrit or Ayurvedic name(s): Lodhra-Pattikaa

Common name(s): Lodhra, Lodhra pathani

Family: Symplocaceae

Distribution: Distributed in Himalayas, and Northeast Hills.

Parts and properties used: Leaf- Spasmolytic, antiviral, antiprotozoal, anthelmintic. Bark- Astringent, yield a tonic.

Associated system(s): AU

***Symplocos racemosa* Roxb. (VU)**

Sanskrit or Ayurvedic name(s): Lodhra

Common name(s): Pathani lodh

Family: Symplocaceae

Distribution: Distributed from northeastern India to Western Ghats.

Parts and properties used: Bark- Yield uterine tonic, used in diarrhoea, dysentery, vaginal ulcer, inflammatory affections, and liver disorders.

Associated system(s): AFSLU

***Syzygium aromaticum* (L.) Merr. & L.M. Perry, syn. *Eugenia caryophyllata* Thunb.**

Sanskrit or Ayurvedic name(s): Lavanga

Common name(s): Loung, Cloves

Family: Myrtaceae

Distribution: Indonesian. Cultivated in South India.

Parts and properties used: Plant- Carminative, anti-inflammatory, antibacterial. Dried flower bud- Antiemetic, stimulant, carminative, used in dyspepsia, gastric irritation. Oil- used for Toothache and mouth and throat inflammation.

Associated system(s): AFSLU

***Syzygium caryophyllatum* (L.) Alston, syn. *Eugenia corymbose* (EN)**

Sanskrit or Ayurvedic name(s): Kshdurajambu

Common name(s): Chota Jambul, Ran Lawang

Family: Myrtaceae

Distribution: Habitat to Western Ghats, and TN.

Parts and properties used: Leaves- Absorb of vitamin E, antibacterial, treat mouth ulcer. Used in treatment of diabetes.

Associated system(s): AFSU, Sri Lanka

***Syzygium cumini* (L.) Skeels, syn. S. *jambolanum* Lam.; *Eugenia jambolana* Lam.**

Sanskrit or Ayurvedic name(s): Jambu

Common name(s): Jamun

Family: Myrtaceae

Distribution: Native and cultivated in many parts of India.

Parts and properties used: Fruit- Stomachic, carminative, diuretic, anti-diabetic. Bark/seed- Antidiarrheal and hypoglycaemic. Leaf- Antibacterial, antidysentery.

Associated system(s): AFSHU

***Syzygium hemisphericum* (Wt.) Alston, syn. *Eugenia hemispherica* Wt.**

Common name(s): Vellai Naval

Family: Myrtaceae

Distribution: Habitat to forests of south India, and Western Ghats.

Parts and properties used: Bark- Antibilious. Used for syphilitic affections.

Associated system(s): S

***Syzygium jambos* (Linn.) Alston.**

Sanskrit or Ayurvedic name(s): Raaj-Jambu

Common name(s): Rose Apple Peru Navel

Family: Myrtaceae

Distribution: Habitat to Malabar. Cultivated in parts of Kerala.

Parts and properties used: Fruit- Liver treatment. Bark- Astringent, antidiarrheal, antidysentery. Leaves- Astringent, anti-inflammatory.

Associated system(s): AFS

***Tabernaemontana divaricata* (L.) R.Br. ex Roem. & Schult., syn. *T. coronaria* (Jacq.) Willd.; *Ervatamia coronaria* (Jacq.) Stapf, *E. divaricata* (L.) Burkill**

Sanskrit or Ayurvedic name(s): Nandivriksha

Common name(s): Chandni

Family: Apocynaceae

Distribution: Occur in the sub-Himalayan tract and Gangetic plains. Cultivated.

Parts and properties used: Plant- Topically anodyne, chewed in toothache. Used as vermicide. Wood- Refrigerant. Root- Astringent, acrid, anodyne; relieves toothache. Leaf milky juice- Anti-inflammatory, treats wounds. Flower + oil- Skin care, sore eyes.

Associated system(s): AFS

Tacca aspera **Roxb., syn.** ***T. integrifolia*** **Ker Gawl.**

Sanskrit or Ayurvedic name(s): Vaaraahikanda

Common name(s): Varahikand

Family: Taccaceae

Distribution: Distributed in Assam, Aka hills, ArP, and peninsular India.

Parts and properties used: Tuber- Nutritive, digestive; treat haemorrhagic diathesis, cachexia, leprosy; contain n-triacon- tanol, castanogenin, botulinic acid, quercetin-3-a-arabinoside, and taccalin.

Associated system(s): AF

Tacca pinnatifida **Forst. f., syn.** ***T. leontopetaloides*** **(Linn.) Kuntze.**

Sanskrit or Ayurvedic name(s): Suurana

Common name(s): Karachunai.

Family: Taccaceae

Distribution: Habitat to Bihar, Indian Peninsula.

Parts and properties used: Tuber- Astringent, carminative, anthelmintic. Treat piles, haemophilia, internal abscesses, colic, spleen enlargement, vomiting, asthma, elephantiasis, and intestinal worms.

Associated system(s): AS

Tagetes erecta **L**

Sanskrit or Ayurvedic name(s): Jhandu, Zanduga, Sandu, Sthulapushpa, Ganduga

Common name(s): African Marigold, Genda, Sandu

Family: Asteraceae/ Compositae

Distribution: Native to the Americas. Grown in tropical India.

Parts and properties used: Plant infusion- Used in cold, bronchitis, and rheumatism. Flower- Alterative. Leaves- Styptic, applied to boils, carbuncles. muscle pains. Leaf + floret- Emmenagogue, diuretic, vermifuge.

Associated system(s): ASU

Tamarindus indica **L.**

Sanskrit or Ayurvedic name(s): Cinca, Amlikaa

Common name(s): Imli

Family: Leguminosae/ Fabaceae (Caesalpiniaceae)

Distribution: African. Naturalized, and grown throughout tropical India.

Parts and properties used: Fruit pulp- Appetizer, laxative, anthelminthic, stomach disorders, jaundice, febrifuge, blood tonic, skin care, free tiredness.

Associated system(s): AFSLU

***Tamarix aphylla* (Linn.) Karst.**

Sanskrit or Ayurvedic name(s): Maacheeka

Common name(s): Laal jhaau

Family: Tamaricaceae

Distribution: Habitat to saline soils of west India.

Parts and properties used: Galls/ Bark- Astringent, have 50% and 10% tannin respectively.

Associated system(s): AFSU

***Tamarix indica* Roxb.**

Sanskrit or Ayurvedic name(s): Jhaavuka, Bahugranthikaa

Common name(s): Jhan

Family: Tamaricaceae

Distribution: Habitat to saline or water-logged soils of India.

Parts and properties used: Galls- Astringent, given in dysentery and diarrhoea. Infusion- Throat gargle. Decoction- Applied to ulcers. Alcohol extract- Antiallergic.

Associated system(s): AFSU

***Tanacetum vulgare* L.**

Common name(s): Tansy

Family: Asteraceae/ Compositae

Distribution: Escape from Europe to Himalayas.

Parts and properties used: Plant- Anthelmintic, tonic, emmenagogue. Used in migraine, neuralgia, nausea, scabies lotion.

Associated system(s): FH, English

***Taraxacum officinale* F.H. Wigg., syn. *T. sikkimensis* Hand. - Mazz**

Sanskrit or Ayurvedic name(s): Dugdh-pheni,

Common name(s): Dandalion, Khur-mang Kanphul, Dudhal, Tukiphul

Family: Asteraceae/ Compositae

Distribution: Found in alpine Himalaya and hills of India. Cultivated.

Parts and properties used: Root- Analgesic, diuretic, bile duct stimulant, circulation, choleretic, urinary antiseptic, detoxicant. Used in dyspepsia, appetite loss, and for diuresis.

Associated system(s): AFH

***Taxus baccata* L., and *T. baccata* var. *wallichiana* Zucc. (EN - CR)**

Sanskrit or Ayurvedic name(s): Sthauneya

Common name(s): Talispatra, Thuno Zarnab

Family: Taxaceae

Distribution: Native to alpine Himalaya, Northeast Hills. Now cultivated.

Parts and properties used: Herb- CNS depressant, reduces motor activity, analgesic, anticonvulsant. Leaf & fruit- Antispasmodic, emmenagogue, sedative. Used in nervousness, epilepsy, hysteria, asthma, chronic bronchitis.

Associated system(s): AF

***Tecomella undulata* (Sm.) Seem. (EN)**

Sanskrit or Ayurvedic name(s): Daadimpushpaka

Common name(s): Rohitaka

Family: Bignoniaceae

Distribution: Found in Thar desert of northwest India.

Parts and properties used: Bark- Relaxant, cardiotonic, choleretic. Used in liver, spleen diseases, leukoderma, syphilis.

Associated system(s): AL

***Tectona grandis* L. f.**

Sanskrit or Ayurvedic name(s): Shaka, Dwaaradaaru

Common name(s): Teak

Family: Verbenaceae

Distribution: Native to Bihar, Central and Peninsular India. Cultivated.

Parts and properties used: Flowers & seeds- Diuretic; flower used in bronchitis, biliousness. Wood- Anti-inflammatory, antibilious, expectorant, anthelmintic. Heartwood- Advised in lipid disorders, threatened abortion. Bark- Astringent. Treat bronchitis.

Associated system(s): AFSLU

***Tephrosia purpurea* (L.) Pers.**

Sanskrit or Ayurvedic name(s): Sharapunkhaa

Common name(s): Sarad foka, Sarpankha, Yempali.

Family: Leguminosae/ Fabaceae

Distribution: Wild all over India.

Parts and properties used: Drug- Treat inflammation of spleen and liver. Dried herbs- Diuretic, deobstruent, laxative. Used in scorpion sting, asthma, bronchitis, etc.

Associated system(s): AFSU

***Teramnus labialis* (L. f.) Spreng.**

Sanskrit or Ayurvedic name(s): Maashaparni

Common name(s): Masparni

Family: Leguminosae/ Fabaceae

Distribution: Found widespread in plains of India.

Parts and properties used: Fruit- Astringent, stomachic, febrifugal. Rejuvenating advised in fatigue, muscle waste, Vata-Pitta imbalance, and spermatorrhoea.

Associated system(s): AFSL

***Terminalia arjuna* (Roxb. Ex DC.) Wight & Arn (NT)**

Sanskrit, Ayurvedic and common name(s): Arjuna, Arjun

Family: Combretaceae

Distribution: Found greater parts of India. Cultivated.

Parts and properties used: Powdered bark- Cardioprotective, cardiotonic, and diuretic. Used externally for skin care. Advised in emaciation, chest diseases, cardiac disorders, lipid imbalances and polyuria. Produce *Arjunarin.*

Associated system(s): AFHSLU

***Terminalia bellirica* (Gaertn.) Roxb.**

Sanskrit or Ayurvedic name(s): Bibhitaka

Common name(s): Behda, Bibhitaki

Family: Combretaceae

Distribution: Habitat to deciduous forests of India.

Parts and properties used: Unripe Fruit- Purgative. Ripe fruit- Astringent, antipyretic used in diarrhoea, dyspepsia respiratory tract infections, cough, and sore throat. Fruit powder given in emesis, worm infestation, etc.

Associated system(s): AFSU

***Terminalia catappa* Linn., syn. *T. procera* Roxb.**

Common name(s): Desi Almond, Natuvadom

Family: Combretaceae

Distribution: Found all over hotter India. Cultivated.

Parts and properties used: Bark- Astringent, antidysentery. Leaf- Anti-septic, anti-inflammatory. Kernel oil- Substitute for almond oil (with oleic, linoleic, palmitic, and stearic acids).

Associated system(s): FS

***Terminalia chebula* Retz.**

Sanskrit or Ayurvedic name(s): Haritaki, Kaayasthaa

Common name(s): Harda, Himaj

Family: Combretaceae

Distribution: Native to many parts of India. Cultivated.

Parts and properties used: Mature fruit- Purgative, astringent, stomachic, ant bilious, alterative. Given in fevers, chronic fevers, anaemia, polyuria.

Associated system(s): AFHSLU

***Terminalia myriocarpa* Van Heurck & Müll. Arg.**

Sanskrit or Ayurvedic name(s): Kakubha, like Arjuna

Common name(s): Hollock

Family: Combretaceae

Distribution: Native to eastern Himalaya, Bengal, Assam.

Parts and properties used: Bark- Cardiac stimulant, diuretic. Give β-sitosterol, 18% tannins.

Associated system(s): A

***Terminalia paniculata* Roth, syn. *T. alata* Herb. Madr. Ex Wall**

Ayurvedic and common name(s): Vanmaruthu, Kinjal

Family: Combretaceae

Distribution: Native to Eastern to Western Ghats.

Parts and properties used: Bark- Diuretic, cardiotonic, have 14% tannin. Applied in parotitis.

Associated system(s): AFS

***Terminalia tomentosa* (Roxb. ex DC.) Wight & Arn., syn. *T. alata* Heyne ex Roth., *T. elliptica* Willd.**

Sanskrit or Ayurvedic name(s): Asana, Bijaka

Common name(s): Asan, Sarj

Family: Combretaceae

Distribution: Native to humid forests of Himalayas, Gangetic Plains, and Indian peninsula.

Parts and properties used: Bark- Astringent, styptic, antidiarrheal, anti-leukorrhoeal. Produce, pyrogallol & catechol, which are antiseptic, antioxidant. Used in haemorrhagic and skin diseases, erysipelas, leukoderma.

Associated system(s): AFS

***Tetracera indica* Merrill., syn. *T. assa* DC.**

Common name(s): Anaittichal

Family: Dilleniaceae.

Distribution: Habitat to Assam.

Parts and properties used: Leaf shoot infusion- Given in pulmonary hemorrhages and used as a gargle in aphthae.

Associated system(s): S

***Thalictrum foliolosum* DC. (VU)**

Sanskrit or Ayurvedic name(s): Pitarangaa

Common name(s): Mamira

Family: Ranunculaceae

Distribution: Distributed from temperate Himalaya to peninsular India.
Plant extract- Treat gout and rheumatism

Parts and properties used: Root- Diuretic, febrifuge, antiperiodic; tonic. (Unexplored)

Associated system(s): AFS

***Thespesia lampas* (Cav.) Dalz. *ex* Dalz. & Gibs.**

Sanskrit or Ayurvedic name(s): Tundikera

Common name(s): Adavi benda, jangli bhindi

Family: Malvaceae

Distribution: Found all over south Asia. Grown as ornamental.

Parts and properties used: Flower- Treat cutaneous diseases. Root, fruit- Treat gonorrhea and syphilis.

Associated system(s): AFS

Thespesia populnea **(L.) Sol. ex Correa**

Sanskrit or Ayurvedic name(s): Kapitana, Paarshvpippala

Common name(s): Phalisa-Chhal, Poovarsu

Family: Malvaceae

Distribution: Found in coastal TN, India. Grown in plantation.

Parts and properties used: Plant- Treat skin diseases. Root, fruit, leaf- Treat psoriasis, scabies, cutaneous diseases. Bark- Treat haemorrhoids chronic dysentery. Leaf- Anti-inflammatory.

Associated system(s): ASU

Thevetia peruviana **(Pers.) K. Schum.**

Sanskrit or Ayurvedic name(s): Pita-Karavira

Common name(s): Pachiyalari.

Family: Apocynaceae

Distribution: Native to Peru. Cultivated as hedge.

Parts and properties used: Bark, leaves- Bitter cathartic, emetic, poisonous. Root-Used as plaster; like *Nerium oleander*.

Associated system(s): AS

Thlaspi arvense **L.**

Common name(s): Bre-ga, Drekaa

Family: Brassicaceae/ Cruciferae

Distribution: Habitat from Eurasia to alpine Himalaya, including Ladakh.

Parts and properties used: Herb- Anti-inflammatory, febrifuge, antidote, tonic. Treat pus in the lungs, renal inflammation, appendicitis, seminal, vaginal discharges.

Associated system(s): L (Tibetan)

Thymus serpyllum **L.**

Sanskrit or Ayurvedic name(s): Ajagandhaa

Common name(s): Jangali Pudinaa, Satar Farsi

Family: Lamiaceae/ Labiatae

Distribution: Native of Mediterranean. Cultivated in Uttarakhand and Western Ghats.

Parts and properties used: Herb- Aromatic. Used in cough, cold; externally as antiseptic, antimicrobial, antispasmodic, sedative, expectorant.

Associated system(s): AFHU

***Thymus vulgaris* L.**

Common name(s): Thyme

Family: Lamiaceae/ Labiatae

Distribution: Native to Europe. Introduced to Western temperate Himalaya and Nilgiris Hills.

Parts and properties used: Herb- Used in chest congestion and induces saliva and deworming. Fresh leaves- Relieve sore throats. Known for expectorant, antispasmodic, -bronchiolitis, -tussive, anthelmintic properties.

Associated system(s): Europe

***Thysanolaena agrostis* Nees., syn. *T. maxima* (Roxb.) Kuntze.**

Sanskrit or Ayurvedic name(s): Juurnaahv

Common name(s): Junaar, Pirlu

Family: Poaceae/ Gramineae.

Distribution: Habitat to temperate northern Hemisphere. Grown in gardens.

Parts and properties used: Root decoction- Used as mouthwash during fever, after parturition.

Associated system(s): AF

***Tiliacora acuminata* (Poir.) Miers ex Hook. f. & Thomson**

Common name(s): Kappatiga, Bagamushada

Family: Menispermaceae

Distribution: Found throughout tropical India. Cultivated.

Parts and properties used: Plant- CVS, CNS active, spasmolytic, hypothermic. Used in skin diseases.

Associated system(s): FS

***Tinospora cordifolia* (Willd.) Miers ex Hook.f. & Thomson**

Sanskrit or Ayurvedic name(s): Guduuchi

Common name(s): Giloy, Galo, Amrithaballi

Family: Menispermaceae

Distribution: Habitat to tropical south India. Cultivated.

Parts and properties used: Herb, stem juice- Antipyretic, antiperiodic, anti-inflammatory, antirheumatic, spasmolytic, hypoglycaemic, hepatoprotective immunity booster. Dried stem- advised in jaundice, anaemia, polyuria, and skin diseases.

Associated system(s): AFSU

***Tinospora crispa* (L.) Hook. f. & Thomson**

Sanskrit or Ayurvedic name(s): Guduuchi

Common name(s): Giloy, Galo, Amrithaballi

Family: Menispermaceae

Distribution: Grows wild in Assam.

Parts and properties used: Plant- Diuretic, febrifuge like cinchona.

Associated system(s): AF

***Tinospora sinensis* (Lour.) Merr., syn. *T. malabarica* (Lam.) Hook.f. & Thomson (VU)**

Sanskrit or Ayurvedic name(s): Kandodbhava-guduchi

Common name(s): Amrata, Gilai

Family: Menispermaceae

Distribution: Found all over tropical India from Assam to Kerala.

Parts and properties used: Leaves, Stem- Given in chronic rheumatism. A substitute for Giloy, *T. cordifolia.*

Associated system(s): AFS

***Toddalia asiatica* (L.) Lam.**

Sanskrit or Ayurvedic name(s): Kanchana

Common name(s): Jangli kalimirch

Family: Rutaceae

Distribution: Habitat to tropical Himalayas to Western Ghats.

Parts and properties used: Plant- Febrifuge, diuretic. Leaves- Antispasmodic. Root bark- Antipyretic, diaphoretic, antiperiodic. Work as tonic in convalescence and debility.

Associated system(s): AFS

***Toona ciliata* M.Roem., syn. *Cedrela toona* Roxb.**

Sanskrit or Ayurvedic name(s): Tuni, Tuunikaa

Common name(s): Thooniyanoikam

Family: Meliaceae

Distribution: Found in sub-Himalayan tract and hills of India.

Parts and properties used: Bark- Astringent, antidysentery. Leaf- Antiperiodic, spasmolytic, hypoglycaemic, antiprotozoal.

Associated system(s): ASF

***Trachyspermum ammi* (L.) Sprague**

Sanskrit or Ayurvedic name(s): Yavani

Common name(s): Ajmo, Ajwayan Omam

Family: Apiaceae/ Umbelliferae

Distribution: Mediterranean. Extensively cultivated in most parts of India.

Parts and properties used: Fruit- Antispasmodic, carminative anticholera, antidiarrheal, stimulant. Dried fruit is given in tympanites, constipation, colic, and helminthiasis. Oil- Bronchial respiratory ailments.

Associated system(s): AFSLU

***Trachyspermum roxburghianum* (DC.) Craib**

Sanskrit or Ayurvedic name(s): Ajamodaa

Common name(s): Ajmod, Sath Ajwain, Ashamtagam

Family: Apiaceae/ Umbelliferae

Distribution: Habitat to, and widely grown all over India.

Parts and properties used: Seed- Carminative, stimulant, cardiotonic, emmenagogue. Used in dyspepsia, bronchitis, asthma.

Associated system(s): AFSU

***Tragia involucrata* L.**

Sanskrit or Ayurvedic name(s): Aagmavarta KashagnihVrscikalli

Common name(s): Barhanta, Chenthatti

Family: Euphorbiaceae

Distribution: Habitat to throughout warm India.

Parts and properties used: Root- Febrifuge, diaphoretic, alterative, blood purifier. Given in fever and applied to wounds and skin infections. Fruit paste- Used in baldness.

Associated system(s): AFS

***Tragopogon porrifolius* Linn. (trade)**

Common name(s): Oyster Plant

Family: Asteraceae/ Compositae

Distribution: Eurafrican. Grown in HP, Maharashtra for ornamental flowers and edible roots.

Parts and properties used: Root- Treat obstructions of galls in jaundice, antibilious, also given in arteriosclerosis and high BP. Latex- Chewing-gum.

Associated system(s): English

***Trapa natans* L, syn. *T. bispinosa* Roxb.**

Sanskrit or Ayurvedic name(s): Srngataka

Common name(s): Singhada

Family: Trapaceae

Distribution: Cosmopolitan. Aquatic nut-crop of Assam, Bihar.

Parts and properties used: Dried kernel- Advised in bleeding disorders, threatened abortion, dysuria, polyuria, and oedema.

Associated system(s): AFU

Trema orientalis* Blume., syn. *T. amboinensis

Common name(s): Ambaratthi

Family: Ulmaceae

Distribution: Found throughout India.

Parts and properties used: Root- Astringent, styptic. Given in diarrhoea, haematuria. Bark- Analgesic. Used as poultice.

Associated system(s): FS

***Trewia nudiflora* L.**

Sanskrit or Ayurvedic name(s): Shriparni

Common name(s): Pindara

Family: Euphorbiaceae

Distribution: Found throughout moist and hot parts of India.

Parts and properties used: Plant- Ant bilious, bechic, ant flatulent, anti-inflammatory. Root- Antirheumatic, carminative. Used as poultice.

Associated system(s): AS

***Trianthema decandra* L., syn. *Zaleya decandra* (L.) Burm.f.**

Sanskrit or Ayurvedic name(s): Varshaabhu

Common name(s): Saaranai ver

Family: Aizoaceae

Distribution: Habitat to tropical India.

Parts and properties used: Root- De-obstruent. Given in asthma, hepatitis & amenorrhoea. Plant parts source of alkaloids.

Associated system(s): AS

***Trianthema portulacastrum* L.**

Sanskrit or Ayurvedic name(s): Varshaabhu, Shwet Punarnava

Common name(s): Desert Horse, Lalsabuni, Pandhari Ghetuli

Family: Aizoaceae

Distribution: Native to tropics of the world. In India found in cultivated fields and wetlands.

Parts and properties used: Root- Antipyretic, analgesic, spasmolytic, de-obstruent, cathartic, anti-inflammatory. Dried root advised for liver, spleen, anaemia, oedema.

Associated system(s): AFS

***Tribulus alatus* Del.**

Sanskrit or Ayurvedic name(s): Gokshura (Wild)

Common name(s): Gokhru kalan

Family: Zygophyllaceae

Distribution: Wild in western Rajasthan and Gujarat.

Parts and properties used: Fruit- Diuretic, anti-inflammatory, emmenagogue. Used in kidney and sexual disorder.

Associated system(s): AFU

***Tribulus terrestris* L., syn. *T. lanuginosus* L.**

Sanskrit or Ayurvedic name(s): Gokshura

Common name(s): Gokhru, gZe-ma Gokshura

Family: Zygophyllaceae

Distribution: Wild in wastelands of India up to 5400 m, including Ladakh.

Parts and properties used: Fruit- Antirheumatic, antimicrobial diuretic, anabolic cardiotonic. Root- Stomachic. Used in sexual weakness. Cardiotonic (Part of *dashmoola*)

Associated system(s): AFHSLU

***Trichodesma indicum* R. Br.**

Sanskrit or Ayurvedic name(s): Adah-Pushpi, Gandhapushpika

Common name(s): Andhaahuli

Family: Boraginaceae

Distribution: Found in greater part of India.

Parts and properties used: Herb- Diuretic, emollient, febrifuge. Leaf- Depurative. Root- Anti-inflammatory, astringent, ant dysenteric.

Associated system(s): AFS

***Trichodesma zeylanicum* (Burm.f.) R.Br.**

Sanskrit or Ayurvedic name(s): Adah-pushpi sister species

Common name(s): Dhadhona

Family: Boraginaceae

Distribution: Wild in northeast and peninsular India.

Parts and properties used: Flower- Sudorific, pectoral. Leaves- Diuretic, emollient, demulcent. Root- Analgesic. Used in tuberculosis, stomach-ache, rheumatism.

Associated system(s): AF

***Tricholepis glaberrima* DC.**

Sanskrit or Ayurvedic name(s): Brahmadandi

Family: Asteraceae/ Compositae

Distribution: Western, central, and peninsular India.

Parts and properties used: Herb- Antiseptic, nervine tonic. Used in urinary tract disinfectant.

Associated system(s): A

***Trichosanthes bracteata* (Lam.) Voigt; *T. palmata* Roxb.**

Sanskrit or Ayurvedic name(s): Indravaaruni (red) Visala

Common name(s): Indrayan, Mahakal

Family: Cucurbitaceae

Distribution: Found as tropical, subtropical vine all over India. Variety of anguina (Snake Gourd) is cultivated.

Parts and properties used: Fruits- Cathartic, antiasthma tic, anti-inflammatory.

Associated system(s): AFS

***Trichosanthes cordata* Roxb.**

Sanskrit or Ayurvedic name(s): Bhuumikushmaanda

Common name(s): Vidari

Family: Cucurbitaceae

Distribution: Habitat to foothills of Himalayas, and northeast of India.

Parts and properties used: Tuber powder- Used in treatment of enlarged spleen, liver; applied on leprous ulcers.

Associated system(s): A

***Trichosanthes cucumerina* L. & T. *cucumerina* var. *anguina* L. (Long fruit); & *T. cucumerina* var. *lobata* Roxb. (Lobed leaves) (DD)**

Sanskrit or Ayurvedic name(s): Dadhipushpi, Amritaphala

Common name(s): Patol panchang

Family: Cucurbitaceae

Distribution: Found as tropical, subtropical vine allover India. Variety *anguina* (Snake Gourd) is cultivated.

Parts and properties used: Root/seed- Ant bilious, anthelmintic, vermifuge, antidiarrheal. Fruit- Blood purifier, improve appetite, cure alopecia, abdominal tumours, bilious, colic diarrhea.

Associated system(s): AFS

***Trichosanthes dioica* Roxb.**

Sanskrit or Ayurvedic name(s): Patola, Karkashchhada

Common name(s): Parwal

Family: Cucurbitaceae

Distribution: Found as tropical and subtropical vine in northeast India. Cultivated.

Parts and properties used: Aerial parts and Fruits juice- Hypoglycaemic, spermatorrhoea. Leaves- Febrifuge, advised in enlarged liver, spleen, alcoholism, jaundice, oedema, and alopecia.

Associated system(s): AFSH

***Trichosanthes tricuspidata* Lour. Syn. *T. palmata* Roxb, *T. bracteata* Lam., *T. puber* Blume**

Sanskrit or Ayurvedic name(s): Mahakala

Common name(s): Indrayan

Family: Cucurbitaceae

Distribution: Wild in Assam, Northeast Hills, Odisha.

Parts and properties used: Fruits- Asthma, earache, migraine, inflammation. Seed- Emetic, purgative.

Associated system(s): AFS

***Tridax procumbens* L.**

Sanskrit or Ayurvedic name(s): Jayantiveda

Common name(s): Jayanti

Family: Asteraceae/ Compositae

Distribution: Native tropical America, an invasive weed of tropical India.

Parts and properties used: Leaves- Antidiarrheal, anti-dysenteric, styptic, used for bronchial catarrh. Leaf juice- Antiseptic, insecticidal.

Associated system(s): AFS

***Trigonella corniculata* (L.) L.**

Common name(s): Kasuri Methi

Family: Leguminosae/ Fabaceae

Distribution: Occur Kashmir to Sikkim. Cultivated in parts of India.

Parts and properties used: Fruit- Astringent, bitter and styptic. Used in treatment of swellings and bruises.

Associated system(s): AUS

***Trigonella foenum-graecum* L.**

Sanskrit or Ayurvedic name(s): Methikaa

Common name(s): Methi

Family: Leguminosae/ Fabaceae

Distribution: Native of Near East. Cultivated since antiquity in most parts of India.

Parts and properties used: Seeds- Treat loss of appetite, flatulence, dyspepsia, colic, diarrhoea, liver, and spleen enlargement. Secret-lytic, hyperaemic.

Associated system(s): AFSU

***Triumfetta rhomboidea* Jacq., *T. angulata* Lam.**

Sanskrit or Ayurvedic name(s): Jhinjhireetaa

Common name(s): Ban okra, Chinese burr.

Family: Tiliaceae

Distribution: Habitat to tropical and subtropical India up to 1200 m.

Parts and properties used: Leaves/bark- Astringent, anticholera, demulcent. Used in diarrhoea. Root- Styptic, diuretic, galactogen

Associated system(s): AFS

***Tulipa stellata* Hook., syn. *T. clusiana* DC.**

Common name(s): Meethi Suranjan

Family: Liliaceae

Distribution: Habitat to west Himalayas 1500-2400 m, cultivar *Tulipa* × *gesneriana.*

Parts and properties used: Bulbs- Mitogenic. Yields a tonic. Used in rheumatism.

Associated system(s): FU

***Tussilago farfara* Linn.**

Common name(s): Fanjiyun

Family: Asteraceae/ Compositae

Distribution: Habitat to western Himalayas 1500-3500 m.

Parts and properties used: Leaves, flowers- Ant catarrhal, antitussive, expectorant, antispasmodic, demulcent, anti-inflammatory. Treat acute catarrh.

Associated system(s): U

***Tylophora fasciculata* Buch. -Ham ex Wight.**

Sanskrit or Ayurvedic name(s): Go-chandanaa

Family: Asclepiadaceae

Distribution: Habitat to sub-Himalayas, central and peninsular India.

Parts and properties used: Plant- Toxic. Plant- Toxic. A substitute for Ipecac, *Cephaelis ipecacuanha* as emetic, purgative, and febrifuge. Applied to ulcers/ wounds.

Associated system(s): A

***Tylophora indica* (Burm.f.) Merr., syn. *T. asthmatica* (L.f .) Wight & Arn.**

Sanskrit or Ayurvedic name(s): Antamuula

Common name(s): Antamul, Ipecacuahna

Family: Asclepiadaceae

Distribution: Habitat to Eastern and peninsular India.

Parts and properties used: Leaves- Used in bronchial asthma and allergic rhinitis.

Associated system(s): AFHS

***Typha elephantina* Grah., non Roxb., syn. *T. australis* K. Schum. & Thom.**

Sanskrit or Ayurvedic name(s): Potagala, Gundra

Common name(s): Anaikkorai, Sambu

Family: Typhaceae

Distribution: Found in many parts of north and central India.

Parts and properties used: Rhizome- Astringent, diuretic. Spike ash- Heals wounds. Pollen + Honey- Used in dysentery, wounds, and bleeding.

Associated system(s): AFS

Typhonium trilobatum **(L.) Schott**

Common name(s): Karunai kizhangu

Family: Araceae

Distribution: Habitat to peninsular India. Grown in south India.

Parts and properties used: Tuber- Acrid, stimulant applied as poultice on scirrhous tumours. Have carotene, folic acid, niacin, thiamine, sterols, and beta sitosterol.

Associated system(s): FS

Uncaria gambier **Thwaites**

Sanskrit or Ayurvedic name(s): Khadira

Common name(s): Chinai Katthaa

Family: Rubiaceae

Distribution: Southeast Asian, widely occur and grown in India.

Parts and properties used: Plant- Intestinal astringent, used like *Acacia catechu* and for gargling. Contains catechins up to 50%.

Associated system(s): AS

Uraria lagopodioides **(L.) DC., syn.** ***U. alopecuroides*** **(Roxb.) Sweet**

Sanskrit or Ayurvedic name(s): Prishniparni

Common name(s): Kalasi, Moovilai

Family: Leguminosae/ Fabaceae

Distribution: Found in grasslands of evergreen forests of Western Ghats.

Parts and properties used: Plant- Ant catarrhal, alterative, and relieve swelling. Extract- Used to treat intermittent fever and chest inflammation. Root- Diuretic, anti-inflammatory, expectorant. Heal bone fractures. Gives cardio and nervine tonic. (Part of Ayurvedic *Dashmoola*)

Associated system(s): AFSU

Uraria picta **(Jacq.) DC.**

Sanskrit or Ayurvedic name(s): Prishniparni,

Common name(s): Oripai

Family: Leguminosae/ Fabaceae

Distribution: Found in dry grass lands of India.

Parts and properties used: Plant decoction- Heal alcoholism, insanity, psychosis, cough, bronchitis, dyspnoea. Leaves- Antiseptics. Treat urinary discharge, genitourinary infection. Root- Treat cough and fever. (Part of Ayurvedic *dash moola*)

Associated system(s): AFSL

***Urena lobata* Linn. Mast. and *U. lobata* Linn. var. *sinuata* King.**

Sanskrit or Ayurvedic name(s): Naagabalaa

Common name(s): Ottatti, a substitute of Balaa

Family: Malvaceae

Distribution: Found throughout warm India.

Parts and properties used: Root- Diuretic, emollient, antispasmodic, antirheumatic. Flower- Pectoral and expectorant. Infusion, used as gargle in sore throat.

Associated system(s): AFS

***Urginea indica* (Roxb) Kunth. (VU), syn. *Drimia indica* (Roxb.) Jessop**

Sanskrit or Ayurvedic name(s): Vana-palaandu

Common name(s): Jangli pyaz, White squill, Indian squill.

Family: Liliaceae

Distribution: Found in western Himalayas, Bihar, peninsular India.

Parts and properties used: Bulb- Antirheumatic, diuretic, anti-asthmatic/anti-bronchitis. Substitute to sea squill/ onion, *Urginea maritima*. Given for sluggish heart and kidney.

Associated system(s): AU

***Urtica dioica* L.**

Sanskrit or Ayurvedic name(s): Vrishchhiyaa-shaaka

Common name(s): Nettle, Shisuun

Family: Urticaceae

Distribution: Habitat to northwest Himalayas 2400-3600 m.

Parts and properties used: Plant- Diuretic, astringent, urinary tract infections, stones. Supportive of rheumatic ailments.

Associated system(s): AHU

***Urtica parviflora* Roxb.**

Sanskrit or Ayurvedic name(s): Vrishchhiyaa-shaaka

Common name(s): Shisuun

Family: Urticaceae

Distribution: Wild to temperate Himalayas, and Nilgiris.

Parts and properties used: Roots- Treat fractures and dislocations. Leaves and inflorescences- Given as tonic, and as cleaning agent after parturition.

Associated system(s): AF

Urtica pilulifera **Linn.**

Ayurvedic and Common name(s): Vrishchhiyaa-shaaka, substitute Anjuraa

Family: Urticaceae

Distribution: Wild in Indian hills.

Parts and properties used: Roots- Diuretic, astringent, homeostatic. Leaves/stem- Contain alkaloids.

Associated system(s): AU

Uvaria narum **Blume.**

Common name(s): Pulichan

Family: Annonaceae

Distribution: Wild in Western Ghats. Harvested from wild.

Parts and properties used: Root/leaves- Given in treatment of intermittent fevers, biliousness, jaundice; rheumatism.

Associated system(s): S

Valeriana hardwickii **Wall. (VU)**

Sanskrit or Ayurvedic name(s): Sugandhabaalaa

Common name(s): Tagar-ganth, Nihani

Family: Valerianaceae

Distribution: Habitat to temperate Himalayas 1200-3600 m.

Parts and properties used: Rhizome/ Root- Carminative, diuretic, expectorant, nervine, stimulant, tonic, rheumatism. Used as *V. jatamansi* and *V. officinalis.*

Associated system(s): AU

Valeriana jatamansi **Jones (CR), syn.** ***V. wallichii*** **DC.**

Sanskrit or Ayurvedic name(s): Tagara

Common name(s): Jatamansi, Musakbala, Tagar

Family: Valerianaceae

Distribution: Found in temperate Himalayas 3000 m, Khasi Hills, Bengal, Nilgiris.

Parts and properties used: Rhizome, fruit (seed), essential oil-Used in treatment of eye, blood, liver ailments, hysteria, and hypochondriasis, neurological disorders of nervous unrest and emotional stress. Traded from antiquity.

Associated system(s): AF

Valeriana officinalis **Linn., syn.** ***V. dubia*** **Bunge.**

Sanskrit, Ayurvedic and Common name(s): Nata Baalaka, Sugandhabaalaa, Tagara,

Family: Valerianaceae

Distribution: Native to Eurasia, found in Kashmir at Sonamarg at 2400 - 2700 m.

Parts and properties used: Rhizome- Tranquillizer, hypnotic, relaxant. Used in neurological disorders sleeplessness, restlessness (main), migraine, intestinal cramps, spasm.

Associated system(s): AF, English

Vallaris solanacea **Kuntze (white variety).**

Sanskrit or Ayurvedic name(s): Aasphotaa, Saarivaa-utpala

Family: Apocynaceae

Distribution: Habitat to Western and Eastern Ghats. Grown in gardens.

Parts and properties used: Latex- Applied to wounds. Bark- Astringent. Seed- Yields cardiac tonic.

Associated system(s): AF

Vallisneria spiralis **Linn.**

Sanskrit or Ayurvedic name(s): Shaivaala

Common name(s): Sevaar

Family: Hydrocharitaceae

Distribution: Occur wild all over India.

Parts and properties used: Plant- Stomachic, refrigerant, demulcent.

Associated system(s): AF

Vanda spathulata **Spreng. (VU)**

Sanskrit or Ayurvedic name(s): Svarna-pushpa

Common name(s): Bandaa

Family: Orchidaceae

Distribution: Habitat to south India and southern Western Ghats.

Parts and properties used: Flower powder/juice- Used in treatment of asthma, depression.

Associated system(s): AF

Vanda tessellata **(Roxb.) Hook. ex G. Don., syn.** ***V. roxburghii*** **R. Br.**

Sanskrit or Ayurvedic name(s): Raasnaa, Baandaa-Raasnaa

Common name(s): Vanda

Family: Orchidaceae

Distribution: Found in northeast, and Gangetic plains to Western Ghats.

Parts and properties used: Root- Antipyretic, tranquilizer, anti-inflammatory, tonic to liver, laxative. Treat rheumatism, lumbago, inflammation, nervous, abdomen, and chest diseases.

Associated system(s): AFSU

Vateria indica **L. (NT)**

Sanskrit or Ayurvedic name(s): Kundura, Ajakarnah, Saraja

Common name(s): Mandadhupa, Dupa

Family: Dipterocarpaceae

Distribution: Indian Peninsula to Western Ghats. Cultivated as well.

Parts and properties used: Resin- Astringent, antibacterial. antidiarrheal, emmenagogue. Treat chronic bronchitis, skin eruptions, ulcer, amenorrhea; gonorrhoea, etc. Bark- Antidysentery. Oil resin- Antirheumatic. Exudate- Treat lipid disorders, anaemia, genitourinary diseases, and diarrhoea.

Associated system(s): AFSU

Ventilago denticulata **Willd.**

Sanskrit or Ayurvedic name(s): Raktavalli

Common name(s): Pappili

Family: Rhamnaceae

Distribution: Found throughout warm India.

Parts and properties used: Stem bark + oil- Skin diseases. Root bark- Atonic dyspepsia, fever, debility.

Associated system(s): AFS

Ventilago madraspatana **Gaertn.**

Sanskrit or Ayurvedic name(s): Taamravalli

Common name(s): Pitti, Kevati

Family: Rhamnaceae

Distribution: Occur in Maharashtra and South India.

Parts and properties used: Root bark- Carminative, stomachic, febrifuge; used in atonic dyspepsia, debility, and skin diseases.

Associated system(s): AFS

Vepris bilocularis **(Wight & Arn.) Engl.**

Sanskrit or Ayurvedic name(s): Krishna-Agaru

Common name(s): Devadaram

Family: Rutaceae

Distribution: Habitat to Western Ghats up to 1200 m.

Parts and properties used: Wood oil- Used to treat rheumatic swellings, skin diseases. Root- Treat biliousness.

Associated system(s): AS

Verbascum thapsus **L.**

Sanskrit or Ayurvedic name(s): Ban Tambaaku,

Common name(s): Dandashal, Flannel mullein

Family: Scrophulariaceae

Distribution: Native to Europe. Found in alpine Himalayas, Western Ghats, and Nilgiris Hills.

Parts and properties used: Herb- Calm asthma, frostbite, and pulmonary diseases. Treat catarrh of the respiratory tract.

Associated system(s): A

Verbena officinalis **Linn.**

Common name(s): Saal-ul-hamaam

Family: Verbenaceae

Distribution: Found in Himalayas, and Northeast Hills.

Parts and properties used: Plant- Nervine, antidepressant and anticonvulsant. Advised in liver, gall bladder, nervous and menstrual disorders.

Associated system(s): U

Vernonia anthelmintica **(L.) (LC) Willd.**

Sanskrit, Ayurvedic and common name(s): Karinjeeragum, Purple Fleabane

Family: Asteraceae/ Compositae

Distribution: Indo-Malayan. Found allover India.

Parts and properties used: Seeds- Anthelmintic, purgative. Traditionally used skin care, central nervous system, kidney, gynecology, gastrointestinal, metabolism, and general health.

Associated system(s): AHSU

Vernonia cinerea **(L.) Less., syn.** ***V. conyzoides*** **DC.**

Sanskrit or Ayurvedic name(s): Sahadevi

Common name(s): Dandotpala

Family: Asteraceae/ Compositae

Distribution: Found in waste areas of Eastern Ghats, other parts of India.

Parts and properties used: Herb- febrifuge, diaphoretic. Specific for leucorrhoea, dysuria, spasm of bladder, strangury, haematological disorders, blood purifier and styptic. Advised in fever, filariasis, pityriasis versicolor, blisters, etc.

Associated system(s): AFSLU

Vernonia javanica **DC., syn.** ***V. arborea*** **Hook. f. non-Buch. - Ham.**

Common name(s): Shutthi

Family: Asteraceae/ Compositae

Distribution: Found in northeast India and Western Ghats.

Parts and properties used: Bark- Febrifuge. Chewed as a substitute for betel leaves.

Associated system(s): S

Veronica beccabunga **Linn.**

Common name(s): Titalokiyaa, Tezhak

Family: Scrophulariaceae.

Distribution: Habitat to Western Himalayas 2700-3600 m.

Parts and properties used: Herb- Antiscorbutic, blood purifier, alterative, diuretic. Used in scurvy, scrofulous affections, swollen piles, lithiasis, skin diseases, burns, ulcers.

Associated system(s): F

Vetiveria zizanioides **(L.) Nash, syn.** ***Andropogon zizanioides*** **Linn.;** ***Chrysopogon zizanioides*** **(L.) Roberty**

Sanskrit or Ayurvedic name(s): Ushira

Common name(s): Lavancha, Khas, Vettiver

Family: Poaceae/ Gramineae

Distribution: Occur in Western Ghats, plains, dry coastal hills. Cultivated along streams and backwaters.

Parts and properties used: Root infusion- Febrifuge, diaphoretic, stimulant, stomachic, antispasmodic, emmenagogue, blood purifier. Treat fevers, colic, flatulence, vomit sperm-otorrhoea and strangury. Root and oil- Colic, dysuria.

Associated system(s): AFHSL

Viburnum foetidum **Wall., syn.** ***V. premnaceum*** **Wall.**

Common name(s): Dieng-soh-lang, So-lang-ksew

Family: Adoxaceae

Distribution: Northeast India 3,000 to 5,000 ft. Cultivated too.

Parts and properties used: Plant- Astringent. Leaf juice- Administrated for menorrhagia. Fruit- Used medicinally..

Associated system(s): A

Viburnum nervosum **Hook. f. &Thoms., syn.** ***V. grandiflorum*** **Wall. ex DC.**

Sanskrit or Ayurvedic name(s): Tilvaka

Common name(s): Telam, Tilen

Family: Caprifoliaceae

Distribution: Habitat to Himalayas 3000-4000 m.

Parts and properties used: Bark- Astringent; contain 13.1% tannin on dry basis.

Associated system(s): AF

Vicoa indica **DC., syn.** ***Pentanema indicum*** **(L.) Ling**

Sanskrit or Ayurvedic name(s): Vandhyaavari.

Common name(s): Banjhori

Family: Asteraceae/ Compositae.

Distribution: Found all over dry India up to 1800m in Himalayas and Western Ghats.

Parts and properties used: Plant- Contraceptive. Isolated Vicolides (Lactone)- Exhibited anti-inflammatory and antipyretic activity.

Associated system(s): AFS

Victoria regia **Lindl., syn.** ***V. amazonica*** **Sow.**

Sanskrit or Ayurvedic name(s): Brihat-patra Kamal

Family: Nymphaeaceae

Distribution: American. Grown in gardens. Seed- Cooling to the nervous system.

Parts and properties used: Root- Externally used as astringent, ant scrofulous. Infusion for gargle in mouth/ throat ulcers.

Associated system(s): A

Vigna mungo **(L.) Hepper, syn.** ***Phaseolus mungo*** **L.**

Common name(s): Urd

Family: Leguminosae/ Fabaceae

Distribution: Native and grown in parts of Indian Subcontinent.

Parts and properties used: Grain- Treat erectile dysfunction, premature ejaculation, low sperm count and motility etc.

Associated system(s): AFS

***Vigna radiata* (L.) R. Wilczek, syn. *Phaseolus radiatus* L.**

Common name(s): Masha

Family: Leguminosae/ Fabaceae

Distribution: Native and grown in parts of Indian Subcontinent.

Parts and properties used: Grain- Detoxicant, refresh mentality and alleviate heat stroke and swelling in the summer.

Associated system(s): AF

***Vigna sublobata* (Roxb.) Babu & Sharma, syn. *V. radiata* (L.) Wilez var. *sublobata* (Roxb.) Verdc.**

Common name(s): Masaparni

Family: Leguminosae/ Fabaceae

Distribution: Wild in South India. Marginally grown.

Parts and properties used: Grain- Substitute to *V. radiata.*

Associated system(s): AS

***Viola odorata* L.**

Common name(s): Banafasha

Family: Violaceae

Distribution: European. Cultivated in Kashmir.

Parts and properties used: Herb- Expectorant, antipyretic, anti-inflammatory, diaphoretic, diuretic. Used for catarrhal, pulmonary affections, liver, intestines diseases.

Associated system(s): AFHU

***Viola serpens* Wall, syn. *V. Pilosa* Blume**

Common name(s): Banafasha

Family: Violaceae

Distribution: Occur throughout the temperate Himalayas up to 2000 m.

Parts and properties used: Plant- Antipyretic, diaphoretic, diuretic, aperient, antipyretic, and febrifuge. Used in respiratory track congestion, asthma, sore throat, cold, coryza. Also, used in bleeding piles, constipation, fever, headache, and skin diseases.

Associated system(s): U

Viola tricolor **Linn.**

Common name(s): Banafasha, related species

Family: Violaceae

Distribution: European. Grown in gardens.

Parts and properties used: Herb- Anti-inflammatory, diuretic, antiallergic, expectorant, alterative, antirheumatic, dermatological. Used in bronchitis, rheumatism, and chronic skin disorders.

Associated system(s): U

Viscum album **Linn. (semi-parasitic)**

Sanskrit or Ayurvedic name(s): Bandaaka, Suvarnabandaaka

Common name(s): Kishmish Kaabuli

Family: Viscaceae

Distribution: Habitat to temperate Himalayas 1200-2700 m.

Parts and properties used: Plant- Vasodilator, stimulant, cardiac depressant, tranquiliser, anti-inflammatory, diuretic. Given in hypertension, degenerative inflammation of joints.

Associated system(s): AU

Vitex agnus-castus **L.**

Sanskrit or Ayurvedic name(s): Renukaa

Common name(s): Chaste tree, Sambhaalu

Family: Verbenaceae

Distribution: Native to Mediterranean. Seeds imported from Iran.

Parts and properties used: Dried fruit- Has reduce follicle-stimulating hormone (FSH) and increase luteinizing stimulating hormone (LSH). Treat premenstrual syndrome

Associated system(s): AHU

Vitex altissima **L. f.**

Common name(s): Myrole

Family: Verbenaceae

Distribution: Habitat to Assam, Meghalaya, Deccan Peninsula up to 1200 m.

Parts and properties used: Whole plant- Treat vitiated *kapha*, *vata*, inflammation, wounds, ulcers, allergy, eczema, pruritus, worm infection, urinary disorder.

Associated system(s): FS

Vitex leucoxylon **Linn. f.**

Sanskrit or Ayurvedic name(s): Paaraavata-padi

Family: Verbenaceae

Distribution: Habitat to Deccan Peninsula.

Parts and properties used: Roots- Febrifuge, astringent.

Associated system(s): A

Vitex negundo **L.**

Sanskrit or Ayurvedic name(s): Nirgundi, Renuka, Shephaalikaa

Common name(s): Neergundi

Family: Verbenaceae

Distribution: Common on riverbanks, roadsides, as hedges all over Indian Subcontinent.

Parts and properties used: Seeds- Spermatorrhoea, promoting spermiogenesis. Leaf- Anti-inflammatory, analgesic, treat vaginal discharge, edema, skin diseases, pruritus, helminthiasis, rheumatism, fever.

Associated system(s): AFSLU

Vitex peduncularis **Wall. ex Schauer.**

Sanskrit or Ayurvedic name(s): Kaakajanghaa

Common name(s): Chirai-godaa

Family: Verbenaceae

Distribution: Found in Assam, Bengal, Bihar, and South India.

Parts and properties used: Leaves- Antibacterial. Leaves/bark- Treat malarial and black water fever.

Associated system(s): AF

Vitex trifolia **Linn. (White variety) (LC)**

Sanskrit or Ayurvedic name(s): Sinduvaara, jalanirgundi

Common name(s): Karu Nochi

Family: Verbenaceae

Distribution: Widely distributed as shrub all over India.

Parts and properties used: Leaves- Febrifuge, antibacterial, anthelmintic, cytotoxic; extract inhibit TB, *Mycobacterium tuberculosis.* Root- Expectorant. Fruit- Used in amenorrhea.

Associated system(s): AFS

***Vitis vinifera* L.**

Sanskrit or Ayurvedic name(s): Draakshaa, Drdkrfi, Swaduphala, Madhurasa

Common name(s): Angoor, Draksh

Family: Vitaceae

Distribution: Mediterranean. Grown in Punjab, Kashmir, and parts of India.

Parts and properties used: Dried fruit- Used in cough and catarrh. Advised in anaemia, jaundice, dyspepsia, constipation, haemorrhagic diseases.

Associated system(s): AFSLU

***Volutarella ramosa* Roxb., syn. *V. divaricata* Beth.**

Common name(s): Baadaavard, Bhu-dandi,

Family: Asteraceae/ Compositae

Distribution: Found all over India.

Parts and properties used: Plant- De-obstruent, aperient, febrifugal, styptic. Used in liver disorders. Mucilage- Given in coughs.

Associated system(s): FU

***Wagatea spicata* Dalz.**

Sanskrit or Ayurvedic name(s): Guchh-karanja

Common name(s): Okkadi-kodi, Vaakeri

Family: Leguminosae/ Fabaceae (Caesalpiniaceae).

Distribution: Habitat to Western Ghats.

Parts and properties used: Roots- Have *Valerian* compound. Used to treat pneumonia. Bark- Skin diseases.

Associated system(s): AFS

***Wattakaka volubilis* (L.f.) Stapf [= *Marsdenia volubilis* (L.f.) T. Cooke], *Dregea volubilis* (L.f.) Benth. ex Hook.f.**

Common name(s): Akad bel, Bandi guruji, Kurinjan

Family: Asclepiadaceae

Distribution: Indo-Malayan, subtropical Himalayas to south India.

Parts and properties used: Plant- Skin diseases, diabetes, cough, jaundice, ulcer. Root juice- Given in paralysis. Source of *Murva* an, Ayurvedic drug- anti-hyperglycemic activity and neuroprotective effects.

Associated system(s): AF

Wedelia biflora **DC. (Yellow flower)**

Sanskrit or Ayurvedic name(s): Bhringaraaja (Related species)

Family: Asteraceae/ Compositae

Distribution: Habitat to seacoast of India.

Parts and properties used: Leaves- Poultice on sore, ulcers. Leaf decoction- Vulnerary and anti-scabious.

Associated system(s): A

Wedelia chinensis **(Osbeck) Merr., syn.** ***W. calendulacea*** **(L.) Less.**

Sanskrit or Ayurvedic name(s): Bhringaraaja, Pitabhrnga,

Common name(s): Bhangra

Family: Asteraceae/ Compositae

Distribution: Habitat to marshy areas of Assam, Bengal, Odisha, TN, Konkan.

Parts and properties used: Plant- De-obstruent, used in swellings. Leaf- Bechic; Juice- Used in alopecia, promote hair growth.

Associated system(s): AF

Wendlandia exserta **DC.**

Sanskrit or Ayurvedic name(s): Tilaka

Common name(s): Tiliyaa

Family: Rubiaceae.

Distribution: Found in sub-Himalayas and central India.

Parts and properties used: Bark- Used in urinary affections, and cholera.

Associated system(s): AF

Withania coagulans **Dunal. (NT)**

Common name(s): Paneer-dodi

Family: Solanaceae

Distribution: Habitat to western Rajasthan. Marginally grown.

Parts and properties used: Plant- Alterative, emetic, diuretic. Ripe fruit- Sedative,

CNS depressant, ant bilious, emetic, antiasthma tic, diuretic, anti-inflammatory. Used in liver complaints, asthma, biliousness.

Associated system(s): AU

***Withania somnifera* (L.) Dunal**

Sanskrit, Ayurvedic and common name(s): Asvagandha, Ashwagandha, Asgandha

Family: Solanaceae

Distribution: Habitat to all over the drier India. Marginally grown.

Parts and properties used: Root- Aphrodisiac, liver tonic, anti-inflammatory, asthma, ulcers, insomnia, and senile dementia. Leaf- Hepatoprotective, antibacterial. Seed- Diuretic, contains several alkaloids, including Withanine, Withananine.

Associated system(s): AFHSLU

***Woodfordia fruticosa* (L.) Kurz (LC)**

Sanskrit or Ayurvedic name(s): Dhaataki, Taamrapushpi, Bahupushpi

Common name(s): Dhaiphool, Dhavadiphool

Family: Lythraceae

Distribution: Found from Assam to Western Ghats.

Parts and properties used: Dried flower- Astringent, blood purifier; cures ulcers, erysipelas, haematemesis, dysentery, diarrhoea, menorrhagia, leucorrhoea, and yields a tonic. Bark- Uterine sedative.

Associated system(s): AFSLU

***Wrightia arborea* (Dennst.) Mabb., syn. *W. tomentosa* Roem. & Schult.**

Sanskrit or Ayurvedic name(s): Kutaja (Flower) Indra(Seed)

Common name(s): Kutajah, Palal

Family: Apocynaceae

Distribution: Found all over the warmer parts Assam to western peninsula. (Dioicous)

Parts and properties used: Seed- Bitter, advised in menstrual, renal complaints, toothache, and diarrhea. Bark- Substitute to Kurchi, *Holarrhena antidysenterica.*

Associated system(s): ASF

***Wrightia tinctoria* (Roxb.) R.Br.**

Sanskrit, Ayurvedic and common name(s): Shveta Kutaja, Indrajau

Family: Apocynaceae

Distribution: Habitat to scrub jungles and deciduous forests of Rajasthan, MP, and TN.

Parts and properties used: Bark- Antidysentery, piles and skin care. Seed- Antidysentery, astringent, febrifuge, anthelmintic. Advised in bilious and flatulence.

Associated system(s): AFHSU

***Xanthium strumarium* Linn., syn. *X. indicum* Koenig. ex Roxb.**

Sanskrit or Ayurvedic name(s): Shankheshwara, Arishta

Common name(s): Maruloomatham, Bana-okraa

Family: Asteraceae/ Compositae

Distribution: American. Naturalized to tropical India.

Parts and properties used: Root- Antitumor. Plant- Used in leukoderma, and malignant diseases. Leaves- Contains sesquiterpene lactones.

Associated system(s): AFS

***Ximenia americana* Linn., Syn. *X. spinosa* Salisb.**

Common name(s): Chiru-illantai

Family: Olacaceae

Distribution: Distributed all over tropics, including peninsular India, Andaman Islands.

Parts and properties used: Fruits/ seeds- Laxative. Root/ leaves- Decoction given in jaundice, diarrhoea, fevers. Root- Treat venereal diseases. Bark- Astringent. Applied to sores.

Associated system(s): S

***Xylia xylocarpa* (Roxb.) Taub.**

Ayurvedic or Common name(s): Trumullu

Family: Leguminosae/ Fabaceae (Mimosaceae)

Distribution: Found in peninsular India.

Parts and properties used: Bark- Anthelmintic antidiarrheal. Seed oil + bark- Antileprotic, used in ulcers and piles. Leaves- Contain β-sitosterol and t-5-hydroxypipecolic acid..

Associated system(s): AF

***Xyris indica* Linn., syn. *X. robusta* Mart.**

Sanskrit, Ayurvedic and common name(s): Daadmaari, Haabiduuba

Family: Xyridaceae

Distribution: Found in sandy salt marshes of Assam, Bengal, Eastern Ghats.

Parts and properties used: Plant- Used in deworming of ringworm, itches, leprosy.

Associated system(s): AF

Yucca filamentosa **Linn.**

Common name(s): Adam's Needle

Family: Liliaceae

Distribution: Introduced from US. Cultivated.

Parts and properties used: Rhizomes/ leaves- Treat glandular, liver, gallbladder disorders. Root tincture- Treat bilious with headache, and rheumatism.

Associated system(s): H, English

Yucca gloriosa **Linn., syn.** *Y. recurvifolia* **Salisb.**

Common name(s): Spanish Dagger-Plant

Family: Liliaceae

Distribution: Introduced from Central America. Grown in Indian gardens.

Parts and properties used: Fruit- Anti-inflammatory, blood purifier, cholagogue. Used in rheumatism, oedema, bronchitis, asthma. Source of steroidal sapogenins, precursors of sex-hormones and steroids.

Associated system(s): English

Zanonia indica **Linn.**

Sanskrit or Ayurvedic name(s): Chirpoti

Common name(s): Parpoti

Family: Cucurbitaceae

Distribution: Habitat to Northeast Hills, peninsular India, Andamans.

Parts and properties used: Fruit- Cathartic, used in cough, and asthma. Leaves- Antispasmodic- reduce inflammation. Plant- Febrifuge.

Associated system(s): AFS

Zanthoxylum acanthopodium **DC.**

Sanskrit or Ayurvedic name(s): Tumburu

Common name(s): Kabab Khandan, Tummad

Family: Rutaceae

Distribution: Habitat to sub-tropical Himalayas and Khasi Hills.

Parts and properties used: Fruit gum- Carminative, stomachic and anthelmintic and dental disorders. Used like *Zanthoxylum armatum*. Give tambulin & tambuletin.

Associated system(s): ASU

***Zanthoxylum armatum* DC., syn. *Z. alatum* Roxb.**

Sanskrit or Ayurvedic name(s): Tejovati, Tumburu

Common name(s): Tejbal, Tejyovathi

Family: Rutaceae

Distribution: Habitat to sub-tropical Himalayas, Khasi Hills, Ganjam and Vishakhapatnam

Parts and properties used: Stem bark- Used in cough, hiccup, dyspnoea, indigestion, diarrhoea, clean teeth. Stem/thorn- Hypoglycaemic. Fruit/seed- Tonic in fever, dyspepsia.

Associated system(s): AFSU

***Zanthoxylum budrunga* Wall. ex DC., syn. *Z. limonella* (Dennst.) Alston. *Z. rhetsa* DC.**

Sanskrit, Ayurvedic and common name(s): Ashvaghra, Tratechai

Family: Rutaceae.

Distribution: Habitat to Assam, Meghalaya, peninsular India.

Parts and properties used: Fruit- Used in diarrhoea, dyspepsia, asthma, bronchitis, rheumatism, mouth diseases. Pericarp- Astringent, digestive, stimulant. Essential oil- Disinfectant.

Associated system(s): AS

***Zanthoxylum rhetsa* (Roxb.) DC., syn. *Z. budrunga* DC**

Ayurvedic or Common name(s): Tejbal, Triphala

Family: Rutaceae

Distribution: Occur in Assam, Odisha, and Western Ghats mountains.

Parts and properties used: Fruit and stem- Aromatic, stimulant, astringent, stomachic, with honey in rheumatism. Used as spice too.

Associated system(s): AFS

***Zea mays* L.**

Sanskrit or Ayurvedic name(s): Mahaa-Kaaya

Common name(s): Maize

Family: Poaceae/ Gramineae

Distribution: Southern Mexico, Central America. Cultivated as cereal in most parts of India.

Parts and properties used: Corn Silk- Diuretic, urinary demulcent, antilithic. Used for cystitis, urethritis, prostatitis, urinary irritation, nephritis, uncontrollable bladder, retention of pus, bed-wetting. Yield saponins.

Associated system(s): AFHSU

***Zehneria umbellata* (Klein) Thwaites**

Common name(s): Pulivanji, Tarali, Karuvikkilanu

Family: Cucurbitaceae

Distribution: Found throughout India.

Parts and properties used: Root- Cough, cold, dysuria, and spermatorrhoea. Leaves- Treat skin inflammation

Associated system(s): FS

***Zingiber cassumunar* Roxb., syn. *Z. montanum* (Koen.) Link ex. A. Dietr. *Z. purpureum* Rosc.**

Sanskrit or Ayurvedic name(s): Vanardraka, Shringaberikaa

Common name(s): Adarakhi

Family: Zingiberaceae

Distribution: Found throughout *India.*

Parts and properties used: Rhizome- Carminative, stimulant, antispasmodic. Used in diarrhoea and colic.

Associated system(s): AF

***Zingiber officinale* Roscoe**

Sanskrit or Ayurvedic name(s): Shunthi, Aardraka

Common name(s): Soonth, Sonth

Family: Zingiberaceae

Distribution: India one of the earliest centers of cultivation and use.

Parts and properties used: Rhizome- Antiemetic, anti- flatulent, anti-inflammatory, hypocholesterolaemia, antispasmodic, expectorant, stimulant, diaphoretic, boost bioavailability of drugs.

Associated system(s): AFSU

***Zingiber zerumbet* (L.) Roscoe ex Sm.**

Sanskrit or Ayurvedic name(s): Mahaabhari-vachaa

Common name(s): Ginger, Narkachur

Family: Zingiberaceae

Distribution: Native to Northeast India. Cultivated all over India.

Parts and properties used: Rhizome- Treat cough, asthma; colic; intestinal worms, leprosy, and skin diseases. Have flavonoid glycosides and curcumin. Oil- Antiseptic.

Associated system(s): AFSU

***Ziziphus jujuba* (L.) Gaertn. Non-Mill., syn. *Z. mauritiana* Lam.**

Sanskrit or Ayurvedic name(s): Badar, Kola

Common name(s): Ber

Family: Rhamnaceae

Distribution: Native to Northeast India, found throughout India up to 1350 m. Planted too.

Parts and properties used: Fruit- Astringent, anodyne, cooling, stomachic, styptic. Ripe dried- Laxative, expectorant. Leaves- Astringent and diaphoretic have protopine and berberine. Seeds/kernel- Antidiarrheal, sedative, antispasmodic, antiemetic. Yields saponins and tubulogenic.

Associated system(s): AFSLU

***Ziziphus nummularia* (Burm. f) Wight & Arn., syn. *Z. rotundifolia* Lam.**

Sanskrit or Ayurvedic name(s): Karkandhu

Common name(s): Wild Jujube, Jharber

Family: Rhamnaceae

Distribution: Wild in Western Rajasthan. Cultivated.

Parts and properties used: Fruit- Astringent, given in bilious affections. Leaves- Used in treatment of scabies/skin diseases.

Associated system(s): AU

***Ziziphus oenoplia* Mill.**

Sanskrit or Ayurvedic name(s): Laghu-badara, Shrgaalabadari

Common name(s): Jackal jujube, Soorai, Makora

Family: Rhamnaceae

Distribution: Wild throughout warm north and peninsular India.

Parts and properties used: Fruit- Stomachic. Root- Given in hyperacidity and ascaris infection.

Associated system(s): AS

***Ziziphus rugosa* Lam.**

Sanskrit, Ayurvedic and common name(s): Charai Kattu, Ilandai, Churnaa, Torana

Family: Rhamnaceae

Distribution: Habitat to sub-Himalaya, Bihar, Assam, Gujarat.

Parts and properties used: Bark- Astringent, antidiarrheal. Flowers- Advised in menorrhagia. Stem and fruit- Hypotensive.

Associated system(s): FS

Ziziphus trinervia* Roxb., syn. *Z. glabrata

Common name(s): Karakattam

Family: Rhamnaceae

Distribution: Habitat to Gujarat and Western Ghats.

Parts and properties used: Leaf- Depurative; purify blood and alterative in chronic venereal affections.

Associated system(s): S

***Ziziphus sativa* Gaertn., syn. *Z. jujuba* Mill.; *Z. vulgaris* Lam.**

Common name(s): Chinese Tsao, Unnab

Family: Rhamnaceae

Distribution: Chinese. Found in Punjab, Himachal, Bengal up to 1950 m.

Parts and properties used: Fruit-Nutritive, emollient, antitussive, antiallergic; protects liver and stress ulcer formation. Seed- Used in cough and skin eruption. Kernel- Used in insomnia. Bark- Used in ulcers and wounds.

Associated system(s): AFU

***Ziziphus xylopyrus* (Retz.) Willd.**

Sanskrit or Ayurvedic name(s): Ghontaa

Common name(s): Ghontaphala

Family: Rhamnaceae

Distribution: Distributed wild all-over Northwest India in MP, UP and Punjab.

Parts and properties used: Fruit powder- Astringent. Advised in stomach-ache. Contains tannins.

Associated system(s): AFS

1. IUCN threat category in parenthesis after species Latin botanical name
2. Associated system(s): Ayurveda = A; Folk = F; Homoeopathy = H; Sidha = S; Sowa-Rigpa/Amchi/Tibetan = L; Unani = U

Few Sources of Information

Bhadrecha, P, Kumar Vivek and Kumar Manoj (2017) Medicinal Plant Growing under Sub-optimal Conditions in trans-Himalaya Region at High Altitude. Defense Life Science Journal 2(1): 37-45. DOI: 10.14429/dlsj.2.11107

Botanical Survey of India reports (status of threat)

Chandra Prakash Kala (2006) Medicinal plants of the high-altitude cold desert in India: Diversity, distribution and traditional uses. The International Journal of Biodiversity Science and Management 2(1): 43-56. DOI: 10.1080/17451590609618098

Khare, CP. (2007) Indian Medicinal Plants: An Illustrated Dictionary pp 812. Springer Science + Business Media, LLC. ISBN: 978-0-387-70637-5

Pant, B. (2013) Medicinal orchids and their uses: Tissue culture a potential alternative for conservation. African Journal of Plant Science 7(10): 448-467. DOI: 10.5897/AJPS2013.1031

Singh, Anurudh K. (2015) Agricultural Biodiversity Heritage Sites and Systems in India pp467. Asian Agri-History Research Foundation, Secunderabad, Telangana, India, ISBN 81-903963-4-X

Singh, Anurudh K. (2016) Exotic Ancient Plant Introductions: Part of Indian 'Ayurveda' Medicinal System. Plant Genet Resour-C. 14(4):356-369. doi:10.1017/S1479262116000368

3

Inventory of Sister and Less Used Wild Medicinal and Aromatic Plants (MAPs) Species

3.1 Introduction

India is rich in biological diversity as revealed by its floristic diversity comprising 17,926 species of angiosperm, based on which it has been considered one among the 17 mega-biodiverse countries of the world. Also, it is the centre of some of the oldest civilizations of world, starting with *Vedic* civilization. Using modern day scientific tools and technology backed up with *shashtra pramaana*, *Vedic* civilization has been calculated to be as old as 25000 years (Oak, 2018). Because of its rich ecological and biological diversity, Indian Subcontinent has inhabited a large amount of ethnic diversity, consisting of both the indigenous races and those moved from other parts of the world as settlers looking for greener pastures for better living. The major ethnic races are Caucasoid (Aryan), Dravidians, Mongoloid, Negroid, and Australoid. Anthropological Survey of India has identified 461 tribal communities (Xaxa, 1999), belonging to 227 ethnic groups (Pushpangadan *et al.*, 2018) spread over 5000 forest villages. According to the 2001 census of India, the country has 122 major languages and 1599 other languages. Figures from different sources vary, primarily due to differences in definition of the term's "language" and "dialect". The largest of the language families represented, in terms of speakers, is the Indo-Aryan language family followed by the Dravidian language family. The other language families are Austroasiatic language family, Tibeto-Burman languages, Tai-Kadai language family and Great Andamanese language family. The interaction between the ethnic diversity and the available floristic diversity has resulted in identification, use and documentation of around 8000 wild plant species by the indigenous communities with medicinal properties.

The knowledge of these communities on the use of wild plants for food, medicine and for meeting many other material requirements is now considered to be potential information for appropriate scientific and technological (S&T) intervention for developing value added commercially marketable products. The indigenous traditional knowledge (ITK) of around 4000-4500 plants used by primitive tribes and communities is predominantly oral in tradition and does not qualify for the formal intellectual property rights benefits (IPR) under the presently enforced national and international legislative systems. However, the plants and associated information documented under the classical systems of medicine, Ayurveda, Siddha, Unani, Amchi, etc., based on some principles and/or rational for use/exploitation is valid for protection. Around 2,500 plants are formally documented in various documents. For example, *Charak Samhita*, 1100 plant species (700 BC), *Sushruta Samhita*, 1270 plant species (200 BC), and *Astang Hridayam*, 1150 plant species (AD 700) (Singh *et al.*, 2016; see Fig. 1.2.1). The literature mentions/reports to date around 7000-10000 plants of ethnomedicinal significances without formal documentation of many. Therefore, a large number of plant species, particularly the ancestral and wild relatives of the commonly used MAPs, require systematic scientific investigations in search of new/alternative sources of available drugs and their documentation as per the international codes; further scientific validation and/or detailed study would provide due recognition to innovation of the indigenous people as per the present requirement of intellectual property regimes/legislations, including many even from the above recognized traditional Indian medicine systems. For this reason, an initiative was made by the Government of India with the establishment of All India Coordinated Research Project on Ethnobiology (AICRPE) 1982-1998 under the auspices of Council of Scientific and Industrial Research (CSIR) to record the number of plant species of economic importance, particularly for nutritive, medicinal, and aromatic and cosmetic properties. AICRPE recorded 10,000 plants of diverse economic potential with associated traditional knowledge (Figure 3.1.1). Out of these 8000 wild plant species are used by the tribes for medicinal purposes. About 950 are found to be new claims and worthy of scientific scrutiny (Pushpangadan *et al.*, 2018). Out of these 3500 or more wild plant species used as edible subsidiary food/vegetable by tribes, about 800 are new information and at least 250 of them are worthy of investigation. Out of these 500 plant species used as fodder, 100 are worth recommending for wider use, and out of these 325 wild species used by tribal as piscides (fish) or pesticides, and at least 175 are quite promising to be developed as safe pesticides (Pushpangadan *et al.*, 2018).

For a long time, biological/genetic resources were considered part of the heritage of mankind. This led to the development and establishment of formal Global International mechanism for conservation and utilization of plant genetic resources (IUPGR). This included a non-binding legal framework, the International Undertaking on Plant Genetic Resources (IU), and an intergovernmental forum, the Commission on Plant Genetic Resources (PGR) by FAO in 1983 with a mandate for exploration, collection, conservation, evaluation, utilization, of PGR and making them available without restriction, for plant breeding and scientific research to facilitate promotion of their availability without restriction for scientific research and commercial use, mainly to ensure food and nutritional security, and other related benefits to mankind globally.

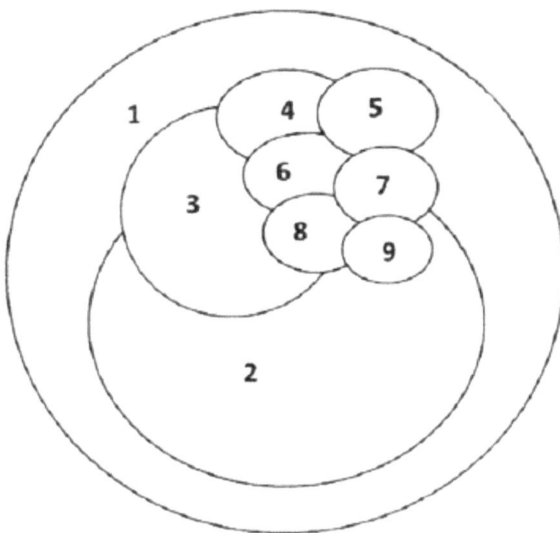

1. Total (10000)
2. Medicinal (8000)
3. Edible Use (4000)
4. Other Material and Cultural Requirements (750)
5. Fiber and Cordage (600)
6. Fodder (500)
7. Pesticides, Piscicides etc. (325)
8. Gum, Resin and Dye (300)
9. Incense and perfumes (100)

Figure 3.1.1: Wild plants used by Indian tribes for various economic use (Pushpangadan *et al.*, 2018)

This led to extensive exploitation of biological resources by the technologically rich industrial countries, earning major commercial benefits from known plant resources, for which considerable basic associated indigenous knowledge came predominantly from biodiversity rich, so-called developing countries of the old world, like India. The commercial benefits were further ensured under the auspices of intellectual property rights (IPR). However, realization of the fact, particularly in the old world, about non-recognition of the intellectual contributions of indigenous tribes/communities in identification, domestication, conservation and use over generations with significant amount of associated ITK discovered and preserved over time by them not getting fair and equitable share in the accrued commercial benefits on the plants/ biological resources of their country (origin) and of the basic information, led to discontentment. The issue of IPR on plant resources also falling within the scope of discussion in General Agreement on Trade and Tariffs (GATT) negotiations made this concern further deep, particularly for countries like India, which have been the seat of old civilizations with rich ITK-base on plants properties, which were used either under the established traditional medicinal systems, such as Ayurveda or by the local tribes and communities in their traditional medicinal practices, either informally without documentation or reported in piecemeal from time to time by different authors (Uniyal, 2006; Chaurasia *et al.*, 2008; Gupta and Gupta, 2017), or recorded in the reports of botanical survey of India, various floras, in documents like wealth of India, etc.

The basic information associated with less used resources have been the key source and the basis for further critical research and systematic commercial exploitation in the name human welfare and legally of great significance in the present day IPR regimes, as they have accrued huge commercial benefits. In fact, this encouraged biopiracy and there have been many instances where patenting attempts were made on plant-products of Indian origin. For example, patenting of the fungicidal properties of Neem [USDA and an American MNC W.R. Grace in the early 90s sought a patent No. 0426257 B from the European Patent Office (EPO) on the "method for controlling fungi on plants by the aid of hydrophobic extracted neem oil]. In last few decades numerous attempts have been made to patent plant products filing application for patent or patented, for which the basic information was derived from indigenous knowledge that is either informally associated with folk lore of local or tribal Indian communities and/or formally documented in the ancient literature, particularly Sanskrit. Therefore, there are severe concern for the protection of the rights of country of origin and the rights of indigenous tribes and communities, who were the inventors and custodians of associated knowledge regarding the properties and use of plant species and their wild relatives. Thereby demanding development of transparent mechanisms for access and ensuring fair and equitable benefit share on commercial use of the bioresources and the associated knowledge derived from the country of origin.

In last several decades starting from 80's, several international conventions and mechanisms have been developed in this regard, to facilitate access to bioresources and to protect the rights of various stakeholders, particularly of economically important plants for conservation and to promote greater use in the welfare of mankind (Singh, 2019).

3.1.1 The Convention on Biological Diversity (CBD), 1992

To address the above concerns, a task force was instituted that drafted the Convention on Biological Diversity (CBD) adopted during the Earth Summit in Rio de Janeiro in 1992, and came into force on 29 December 1993 (CBD, 1993). It aims at – i. conservation of biological diversity, ii. the sustainable use of its components, and iii. fair and equitable benefit sharing arising out of the utilization of biological/genetic resources through appropriate access and benefit sharing agreement. Article 15 of CBD on 'Access to Genetic Resources' recognizes sovereign rights of States over their natural resources and affirms that the authority to determine access to these genetic resources' rests with the national governments, as per the national legislation. Access, where granted, shall be on mutually agreed terms (MAT) and subject to prior informed consent (PIC) of the Contracting Party (CP) providing such resources.

CBD recognizes the importance of cultural diversity and value of the indigenous traditional knowledge (TK). Article 8(j) of CBD on Traditional Knowledge, Innovations and Practices calls upon the Parties to respect, preserve and maintain the knowledge, innovations and practices of indigenous and local communities embodying traditional lifestyles/culture relevant for the conservation and sustainable use of biological diversity. It also envisages promotion of their wider application with the approval and involvement of the holders of such knowledge, innovations and practices and encourages equitable sharing of the commercial benefits arising from the utilization of such knowledge, innovations, and practices.

3.1.2 Regulating Access to Bioresources (and associated ITK) in India

Despite having fair and equitable benefit sharing arising out of the utilization of genetic resources as one of the main objectives, CBD did not prescribe any mechanism to ensure this objective. In this regard, plant genetic resources (PGR) were of primary concern, because of being the building blocks in crop improvement and animal breeding efforts. Further, stimulated by unprecedented technological advances and their applications, appreciation of the monetary and non-monetary value of biological resources has grown enormously in recent years leading to increasing conflict over rights and responsibilities for these resources, including both the cultivated/domesticated forms and as well as the naturally growing wild relatives. This led to the development of a multilateral mechanism to regulate access/exchange of PGR for use in research and crop improvement and is called the International Treaty on Plant Genetic Resources for Food and Agriculture (ITPGRFA) (ITPGRFA, 2001). It recognizes the enormous contribution of farmers in generating the diversity in crops species that feed the world, and establishes a multilateral global system to provide farmers, plant breeders and scientists a facilitated access to plant genetic material and ensures that recipients share benefits they derive from the use of these genetic materials with the countries wherefrom these have originated or obtained, and the communities associated with them. It required national legislation to determine terms of access to PGR and framing policies, rules, and procedures, to regulate access to genetic/biological resources and related traditional knowledge within territorial jurisdiction.

India, being a party to CBD and several other biodiversity-related Conventions and Treaties including the ITPGRFA, which has brought 64 food and forage crops for exchange under the Multilateral System, based on the Standard Material Transfer Agreement (SMTA), has developed, and enforced necessary legislations to harness the benefits enshrined in the previsions of these international agreements.

India is also a member of international trade agreements including WTO Agreement on Trade-Related Aspects of Intellectual Property Rights (WTO-TRIPS), which cover all trading material including plants/plant products. ITPGRFA covered only 64 food and agriculture crops leaving the remaining biological diversity unregulated, reflecting that there was a need to regulate the access and use of the remaining biodiversity in a transparent manner, ensuring fair and equitable benefit sharing of commercial gains. Nagoya Protocol, which is a supplementary agreement to the

CBD, was adopted on 29 October 2010 in Nagoya, Japan to take care of access to the remaining bioresources/genetic resources. It was ratified by India in October 2012. It has placed biodiversity-rich countries like India in a better position to gain from their bioresources and to enhance their capacity to provide more incentives for identification, collection, characterization, evaluation of potential value, information documentation, conservation, and sustainable use of economically useful biodiversity offering/generating new opportunities. It provides a transparent legal framework for the effective implementation of one of the three objectives of the CBD, i.e., the fair and equitable sharing of benefits arising out of the commercial utilization of bioresources/ genetic resources. It entered into force on 12 October 2014, 90 days after the deposit of the fiftieth instrument of ratification.

The Nagoya Protocol addresses ITK associated with biological (species)/genetic resources with provisions on access, benefit-sharing, and compliance. It further goes making consent of the indigenous and local communities to be part of the established right of state to grant access to biological/genetic resources. Contracting parties (CPs) are to take measures to ensure prior informed consent of these communities and ensure fair and equitable benefit-sharing, keeping in mind the community laws and procedures as well as customary practices on use and exchange of such biological resource and the associate knowledge. More information on the Nagoya Protocol and traditional knowledge can be found on the Traditional Knowledge program of work webpage.

To fulfil the national obligations under CBD, India enacted the Biological Diversity Act, 2002 with clear specification of access regulations for domestic and foreign users of bioresources growing in India or obtained from India. This legislation, and the Biological Diversity Rules, 2004 framed under it, provide a three-tier legal framework (National Biodiversity Authority, State Biodiversity Boards and Local Biodiversity Management Committees) for documentation and regulating access to bioresources (and the associated traditional knowledge) while ensuring fair and equitable sharing of resulting benefits.

Indian citizens are free to access bioresources growing in India for research purpose, but they are required to intimate the concerned State Biodiversity Boards (SBBs) prior to obtaining them for commercial purpose. On the other hand, persons other than Indian citizens, as defined under section 3 (2), are essentially required to obtain prior approval of National Biodiversity Authority (NBA) for accessing these bioresources whether for research use or for commercial purpose. To promote benefit sharing, NBA's prior approval is also required, whenever an Indian researcher/ institution intends to transfer bioresources or results of research on them to the latter category (foreign) of users. Furthermore, no person shall apply for seeking IPR protection over any innovative process/ product, based on the use of bioresources, occurring in India or obtained from India, without prior approval of NBA and signing the agreement on benefit sharing. However, applying for protection of plant variety under the Protection of Plant Variety & Farmers' Rights Act is exempt from this provision.

Approvals are granted by NBA on a case-by-case basis, keeping in view the recommendations of an Expert Committee and imposing terms for benefit sharing in monetary or non-monetary mode. Implementing the Biological Diversity Act's provisions requires active partnership and effective coordination involving the NBA at the national level, SBBs at the state level and Biodiversity Management Committees at the local level. Approval for accessing bioresources, bio-survey & bio-utilization, transfer of research results, seeking IPR and third-party transfer of already accessed bioresource is given by NBA by signing a written agreement with the applicant as required under Rule 14(5).

3.2 Opportunities for India

Climatic diversity, floristic richness, particularly of those plants with medicinal properties and the available associated knowledge put India in a supreme position for further research and development on plant-based products, particularly herbal medicines. Global herbal medicine market was valued at $71.19 billion in 2016, It was expected to grow to $107 billion by 2017 and to increase to $ 140 billion by 2024. China is the largest supplier of herbal products to US and Europe based on 5000 plants. India's share in the global herbal medicinal market is extremely low (0.5% = $358.60 million), although it has the traditional knowledge of about ~7000 plants.

However, the market for Ayurvedic medicines in India has been expanding at the rate of 20% annually and has the potential to grow up to 90%. India needs to focus on research and development in biochemical profiling, identification of active principles in the sources, followed by molecular characterization, and developing agrotechnology and processing technology meeting the requirements of global and national IPR standards and for quality control.

3.2.1 Potential

Lately, there has been a growing concern about health and diet. It's widely proven and accepted that meat and meat products are unhealthy because of increased risk of cardiovascular diseases, obesity and cancer and added synthetic antioxidants and antimicrobials. Many crops, particularly grain legumes such soybean, pea and cereals like wheat, rice, and minor millets, besides being staple food, have significant potential as sources of protein, micro-nutrients to be ingredient(s) for fortification of basic food. Plant derivatives rich in protein and/or having antioxidant components including vitamins A, C and E, minerals, polyphenols, flavonoids and terpenoids may decrease the risk of several degenerative diseases, possibly caused by meat products. Vegetable oils and plant fibre may work as fat replacers, and spices and condiments may work as stimulators and natural antimicrobials for improved digestion and protection from microbial infection. In addition, there are several plant species that can be a source of active botanicals and herbs that can be used to maintain or improve health in general and functions, and as source of sweeteners, seasonings, colouring, and flavouring of foodstuffs. Plants can also be source of dairy alternatives, such as milk, beverages (Almond, Soy, Coconut, and Rice) and other drinks, and cultured products such as yogurt, frozen desserts, ice cream, etc.

There are ancient medicinal practices/systems like Ayurveda that have stood the test of time for centuries supporting health and wellness. In fact, many of the life-saving pharmaceuticals or drugs we rely even today are derived from plants and were discovered by indigenous communities. *Aswagandha*, *Nidrajanana* Vati, morphine, aspirin, ephedrine, etc. are a few to mention, and there are still many with untapped potential. Therefore, ITK from Indian Subcontinent can find prominent role in innovative drug discovery and production platforms.

Vegetal extracts and herbs have been used for personal care and cosmetic purposes since time immemorial, and for centuries, they were the only source to obtain colorants, fragrances, and products for soothing and protecting skin. They were replaced by synthetic materials with low prices. However, because of increasing awareness of their negative effects in the last few decades, cosmetic ingredients based on plants or plants derivatives have made a powerful comeback. Products of plant origin used in cosmetics include vegetable oils and other lipids, essential oils used as fragrances or for their antimicrobial activities, ingredients for skincare and hair care, and antioxidants, to name just a few. Plant material used to produce cosmetic ingredients comes from a variety of sources, which include not only conventional products coming from horticultural production (in field or greenhouse), but also from wild harvest in developing countries like India and biotechnological methods (e.g., tissue cultures, fermentation of genetically modified organisms, microalgae cultures, hydroponic systems, etc.) mostly in technologically rich countries. Therefore, there is a need for greater research attention to facilitate conservation and sustainable use of medicinal plant species diversity and genetic diversity within species with systematic scientific evaluation using the advance technologies, in search of new alternative sources among known and unknown species. Technological advancement must be made for safe harvest of raw materials and improved isolation and extraction techniques of principle component(s) from known species, and new species for development of innovative formulations. There are ample opportunities in this regard, for example in nutraceutical field, as per one estimate in relation to food, there is +49 percent potential for plant protein based or supplemented products, + 14 percent for meat substitute, and +20 percent dairy alternatives with value of totally around 16.3 billion.

3.2.2 Global Market

It has been revealed that in terms of value, the global plant-based meat market is expected to register a healthy Compound Annual Growth Rate (CAGR) of 5.8% during the period 2018-2026, due to various factors, regarding which Persistence Market Research (PMR) offers vital insights in detail. The global market for dairy alternative drinks only has reached around US$16.0 billion in 2018. The global market for botanical and plant-derived drugs is expected to grow from $29.4 billion in 2017 to around $39.6 billion by 2022 with an expected CAGR of almost 8%, during the 2017-2021. It will be primarily due to the low cost of herbal medicines compared to allopathic and offers new opportunities.

The domestic pharmaceutical companies and the increased public interest in herbal based products had been significantly contributing to the economic growth of plant-based medicine. For example, the global market potential of *Aloe vera,* used as medicine to treat burns, added to skin creams and cosmetics, only was estimated in the billions of dollars of which India a major producer. Devils claw used in the treatment of arthritis and other inflammatory diseases has been growing from Namibia, Africa as a major export. India is the hub of the regional trade. At the national level, 40% of the state forest-based revenues and 70% of the forest export resources come from medicinal and aromatic plants (MAPs) and non-timber forest products (NTFPs), mostly exported as unprocessed raw material. A small country like Nepal trades an estimated 20,000 tons of MAPs, worth US$ 18-20 million every year of which about 90% of are forest collection, mainly exported to India in raw forms. Therefore, in developing countries including India, there is a need for capacity building in manufacturing of final products. This can be achieved through strengthening research and development program using indigenous knowledge and skills of technologically rich countries having expertise in biotechnological and biochemical research through international cooperation, which would facilitate strengthening government's Make in India program.

For exploitation of the marginally used potential plant species, the first step could be to investigate these species for nutraceutical, medicinal, cosmetic potential, etc., based on the primary basic information available promoting their use as alternatives or adulterants to the principal species known to be the direct source or source of a product (drug), because of being ancestral or distant wild relatives. These efforts may also include other potential wild species, which have been known to be used by specific tribes or communities to overcome deficiency or an ailment, but not investigated in detail. This shall promote search for alternative sources of a medicinal component, genetic diversity (material) for genetic improvement of principal plant species and search among the sympatric plant species used for a property without detailed investigation. Such initiative can open flood gates of new opportunities for researchers, entrepreneurs, and start-ups to make the nation self-reliant (*Aatmnirbhar*) in health care.

References

CBD (Convention on Biological Diversity) (1992) Adopted during the Earth Summit in Rio de Janeiro, 1992; entered into force on 29 December 1993.

Chaurasia, OP, Khatoon, N and Singh SB. (2008) Field Guide, Floral Diversity of Ladakh pp198. Field Research Laboratory, Defence Research and Development Organization, Leh, India.

Gupta, DK, Gupta, G. (2017) Diversity of Ethno Medicinal Plant in Dist. Balod (C.G). IOSR Journal of Pharmacy and Biological Sciences 12(3):80-89. doi:10.9790/3008-1203078089 11.

International Treaty on Plant Genetic Resources for Food and Agriculture (ITPGRFA) (2001) pp 25. Food and Agriculture Organization of the United Nations, Via delle Terme di Caracalla, Rome, Italy.

International Undertaking on Plant Genetic Resources. (1983) Available at: http://www.fao.org/docrep/x5563E/X5563e0a.htm#e.%20plant%20genetic%20resources%20(follow%20up%20of%20conference%20resolution%20681 (accessed 5 January 2011).

Nagoya Protocol (2010) Nagoya Protocol on Access to Genetic Resources and the Fair and Equitable Sharing of Benefits Arising from their Utilization to the Convention on Biological Diversity pp25. A supplementary agreement to the 1992 Convention on Biological Diversity (CBD), Nagoya, Japan. ISBN 92-9225-306-9

Oak, NK. (2018) 12209 BCE Rama Ravana Yuddha, Publisher Subbu

Pushpangadan, P, George Varughese, Ijinu, TP and Chithra MA. (2018) All India coordinated research project on ethnobiology and genesis of ethnopharmacology research in India including benefit sharing. Annals of Phytomedicine 7(1): 5-12. DOI: 10.21276/ap.2018.7.1.2

Singh, Anurudh K. (2019) Nagoya protocol of CBD, mechanism to facilitate international collaborative development of plant-based products: India a case study. International Journal of Phytocosmetics and Natural Ingredients 6:1. doi:10.15171/ijpni.2019.01.

Uniyal, MR. (2006) Potential of medicinal plants from Ladakh. S & T Entrepreneurs Park, IIT Kharagpur, India.

Xaxas, Virginius (1999). Tribes as indigenous people of India. Economic and Political Weekly 34 (51): 3589-3595.

3.3 List of Less Used Sister and Wild Medicinal and Aromatic Plants (MAPs)

In this list an attempt has been made to inventories those species, which are phylogenetically close and taxonomically ancestral or sister, and the other potential species that are being marginally/locally used by the local communities belonging to the region of their geographical distribution range. Many genera have many species and therefore, the list may not cover all the taxonomical sister species, either because of lack of information or non-use and non-potential nature. Assessed medicinally potential, but new or unexplored species shall find place in the next list species based on preliminary information or experience of the local communities and are worth scientific scrutiny,

These species offer instant new opportunity to researchers and entrepreneurs, searching for the new herbal sources or drugs for overcoming ever rising problems/concerns with human health care, either because of evolution of new strains of various organisms or due to climatic or lifestyle changes.

The List with Description of MAPs Species

Acacia ferruginea **DC., syn.** ***Mimosa ferruginea*** **Roxb. (VU)**

Common name(s): Kachu, Safed Khair

Family: Leguminosae/ Fabaceae (Mimosaceae)

Distribution: Habitat to dry deciduous forests of peninsular India.

Parts and properties used: Bark- Used in skin infections, itching, leukoderma, ulcers, inflammation of the mucous lining and throat.

Source of information (system): AS

***Acampe papilliosa* (Lindl.) Lindl.**

Common name(s): Shwethuli

Family: Orchidaceae

Distribution: Wild epiphyte in Himalayas, Darjeeling & West Bengal (WB).

Parts and properties used: Root- Anti-rheumatic. Given in sciatica, neuralgia, syphilis, and uterine diseases.

Source of information (system): F

***Acantholimon lycopodioides* (Girard) Boiss.**

Common name(s): Lonze

Family: Plumbaginaceae

Distribution: Native of Afghanistan to alpine west Himalayas.

Parts and properties used: Plant ash with milk- Used to treat cardiac ailment.

Source of information (system): F

***Aconitum laciniatum* Stapf.**

Sanskrit name: Vatsanaabha (wild relative)

Common name(s): Kalo Bikhmo

Family: Ranunculaceae

Distribution: Habitat to alpine, subalpine Himalayas, 3500-4500 m Sikkim, WB. Grown in Nepal.

Parts and properties used: Root- Given in rheumatism, rheumatic fever, and headache. Mixed with *A. ferox* (Vatsanaabha) and therefore have trade potential..

Source of information (system): AF

***Aconitum spicatum* (Bruhl) Stapf**

Sanskrit name: Vatsanaabha (wild relative)

Common name(s): Bish

Family: Ranunculaceae

Distribution: Habitat to Himalayas and WB.

Parts and properties used: Root- Antipyretic, analgesic, poisonous.

Source of information (system): F

***Aconitum violaceum* Jacq. ex Stapf. (CR)**

Sanskrit name: Vatsanaabha (wild relative)

Common name(s): Mitha Telia

Family: Ranunculaceae

Distribution: Habitat to alpine Himalayas and Ladakh

Parts and properties used: Root- Antipyretic, given in fever. Produce a nervine tonic. Therefore, have trade potential.

Source of information (system): F

Acorus gramineus **Soland. ex Ait.**

Sanskrit name: Haimavati

Family: Araceae

Distribution: Native to Japan. Occasionally found in Sikkim and Khasi Hills.

Parts and properties used: Rhizome- Sedative, hypotensive, tranquilizer, analgesic, spasmolytic, anticonvulsant. Used in bronchial catarrh, chronic diarrhoea, and dysentery. Produce nervine tonic.

Source of information (system): F

Actaea spicata **Linn.**

Common name(s): Visha-phale

Family: Ranunculaceae

Distribution: Habitat to alpine Himalayas.

Parts and properties used: Root- Antirheumatic, anti-inflammatory, nerve sedative, emetic, purgative. Used in rheumatic fever.

Source of information (system): F

Actiniopteris radiata **(Sw.) Link (Fern) (R)**

Sanskrit: Mayurshikha

Common name(s): Morepankhi, Ray fern

Family: Pteridaceae

Distribution: Grasslands and dry deciduous forests of Western Ghats.

Parts and properties used: Herb- Used in acute bronchitis, to regularize periods in females, tuberculosis, diarrhoea, spermatorrhoea.

Source of information (system): F

Adiantum aethiopicum **Linn,** ***A. emarginatum*** **Bedd. (Fern)**

Sanskrit name: Hansapadi

Family: Adiantaceae.

Distribution: At higher altitude of north Canara, Nilgiris and Palani Hills.

Parts and properties used: Rhizome decoction- Abortifacient, astringent, and emetic. Emollient in cough and chest diseases and sudorific (cause sweating).

Source of information (system): A

Adiantum caudatum **Linn (Fern)**

Common name(s): Tailed Maidenhair, Walking fern, Patharchiti, Hanspadi (?)

Family: Adiantaceae

Distribution: Moist deciduous and semi-evergreen forests of Western Ghats.

Parts and properties used: Herb- Produce ointment for cuts, to cure migraine.

Source of information (system): AS

Adiantum incisum **Forsk. (Fern)**

Sanskrit or Ayurvedic name: Mayurshikhaa, Nilakantha-shikhaa

Common name(s): Maidenhair fern

Family: Adiantaceae

Distribution: Plains and hill slops of Punjab, Rajasthan, WB, Tamil Nadu (TN) & Maharashtra.

Parts and properties used: Herb- Used in hemicrania, cough, fever, and externally in skin diseases. Substitute for *A. capillus-veneris* (Maidenhair fern).

Source of information (system): A

Adiantum venustum **D. Don. (Fern)**

Common name(s): Hansraj

Family: Adiantaceae

Distribution: Habitat to Himalayas.

Parts and properties used: Fronds- Astringent, diuretic, emetic, emmenagogue, resolvent, expectorant. Produce tonic. Used in headaches.

Source of information (system): F

Adina cordifolia **(Roxb.) Brandis. (EN), syn.** ***Haldina cordifolia*** **(Roxb.) Ridsdale**

Sanskrit name: Haridru

Common name(s): Haldu

Family: Rubiaceae

Distribution: Habitat to deciduous forests of lower Himalayas plains India.

Parts and properties used: Bark extract- Antiseptic, antibacterial, antidysentery, ant bilious, febrifuge. Root- Astringent. Used in dysentery.

Source of information (system): F

***Agropyron repens* Beauv, syn *Triticum repens* L.**

Common name(s): Wheat grass

Family: Poaceae/ Gramineae

Distribution: Occur in Western Himalayas and Kashmir.

Parts and properties used: Herb- Demulcent. Used in cystitis, nephritis, aperient, a diuretic and urinary antiseptic, anticholestrolemic.

Source of information (system): F

***Ainsliaea aptera* DC.**

Common name(s): Sathjalani

Family: Asteraceae/ Compositae

Distribution: Habitat to Himalayas 2400 m to Khasi Hills.

Parts and properties used: Root powder- Diuretic. Used for relief from stomach-ache.

Source of information (system): F

***Alangium begoniaefolium* (Roxb.) Baill., syn. *A. chinense* (Lour.)**

Sanskrit name: Ankola (*A. salvifolium* relative tree).

Common name(s): Akhani.

Family: Alangiaceae

Distribution: Plains/foothills of Assam, northeast hills up to 2000 m.

Parts and properties used: Bark and roots- Sedative, anthelmintic.

Source of information (system): F

***Albizia kalkora* (Roxb.) Prain, syn. *A. coreana* Nakai, *Mimosa kalkora* Roxb.**

Common name(s): Shan huai (Siris, *A. lebbeck* wild relative)

Family: Leguminosae/ Fabaceae (Mimosaceae)

Distribution: Habitat to eastern Himalayas, Northeast Hills, and Central Highlands.

Parts and properties used: Leaves- Contain epinephrine, trans squalene, stachydrine (Constituent of *Leonurus*). Assessed medicinally potential.

Source of information (system): Locals

***Albizia odoratissima* (L. f.) Benth.**

Common name(s): Sirisa

Family: Leguminosae/ Fabaceae (Leguminosae/ Fabaceae (Mimosaceae))

Distribution: Deciduous forests and plains of India and Western Ghats.

Parts and properties used: Bark/leaf- Treat numerous inflammatory pathologies, such as leprosy, ulcers, burns and asthma.

Source of information (system): AFS

Albizia thompsoni **Brandis, (NT)**

Common name(s): Velugu Chinta

Family: Leguminosae/ Fabaceae (Mimosaceae)

Distribution: Wild in Gangetic Plains, Eastern & Western Ghats.

Parts and properties used: Stem bark- Locally used in case of skin diseases, potential.

Source of information (system): F

Alhagi maurorum **Medik.**

Common name(s): Camelthorn Oant-jhari, Javasa

Family: Leguminosae/ Fabaceae

Distribution: Arid north India, and Kashmir.

Parts and properties used: Herb- Purgative, diaphoretic, expectorant and diuretic. Locally used to treat piles, migraine, warts, and rheumatism.

Source of information (system): F

Allium carolinianum **DC**

Common name(s): Skotche, Demokh (wild relative of onion)

Family: Amaryllidaceae

Distribution: Native to central to south Asia and Himalayas (Ladakh).

Parts and properties used: Whole plant- Locally used in indigestion, reduce blood cholesterol, and produce a tonic. Aromatic (Sulphur compound).

Source of information (system): F

Althaea rosea **Cav. Diss**

Common name(s): Hollyhock

Family: Malvaceae

Distribution: Habitat to peninsular India.

Parts and properties used: Flower- Diuretic, demulcent, and emollient. Used in chest problems. Decoction treats blood circulation dysmenorrhea, and haemorrhage. Roots- Demulcent and astringent.

Source of information (system): F, Tibetan

***Amaranthus blitum* Linn. var. *oleraceus* Duthie**

Sanskrit or Ayurvedic name: Maarisha.

Common name(s): Marasaa

Family: Amaranthaceae

Distribution: Found wild throughout India.

Parts and properties used: Herb- Cooling, stomachic, emollient. Used in biliousness, and haemorrhagic diathesis.

Source of information (system): A

***Amaranthus spinosus* L.**

Common name(s): Cholai, Bhandira, Amaranths wild relative.

Family: Amaranthaceae

Distribution: Grows wild in most of India.

Parts and properties used: Plant- GA lactogenic, laxative, emollient, spasmolytic, diuretic. Used in internal bleeding, diarrhoea, excessive menstruation, boils, stomach disorders, vaginal discharges, and wounds.

Source of information (system): F

***Amorphophallus dubius* Blume (Cytotype of *A. paeoniifolius*, 2n=26)**

Common name(s): Suran (Elephant yam) substitute

Family: Araceae

Distribution: Occur wild in greater Bengal.

Parts and properties used: Tuber/Root- Stomachic and tonic. Used for boils, haemorrhoids, piles, abdominal pains, enlarged spleen, asthma, rheumatism.

Source of information (system): F

***Amorphophallus sylvaticus* (Roxb.) Kunth (Biotype of *A. paeoniifolius*)**

Common name(s): Jangali suran, wild relative of Elephant Yam, Devils Tongue

Family: Araceae

Distribution: Wild in southern Western Ghats, India.

Parts and properties used: Tuber- Antioxidant and antibacterial. Have medicinal potential.

Source of information (system): AFS

***Anaphalis busua* (Buch. -Ham.) DC.**

Common name(s): Telgang, Buki Phool

Family: Asteraceae/ Compositae

Distribution: Habitat to alpine Himalayas, Ladakh and hills of south Indian.

Parts and properties used: Herb- Locally used in cold and cough. Leaf juice- Applied on bruises, wounds, and cuts.

Source of information (system): F

***Anaphalis cuneifolia* (DC.) Hook.f.**

Common name(s): Simula

Family: Asteraceae/ Compositae

Distribution: Habitat to alpine Himalayas, and Ladakh.

Parts and properties used: Herb- Locally used in skin diseases. Assessed medicinally potential.

Source of information (system): F

***Anaphalis triplinervis* var. *monocephala* (DC.) Airy Shaw**

Common name(s): sPra-rgod

Family: Asteraceae/ Compositae

Distribution: Habitat to alpine zones of Ladakh & Uttarakhand.

Parts and properties used: Flower/ essential oil- Locally used in chronic diseases. Assessed medicinally potential.

Source of information (system): F

***Anaphalis triplinervis* var. *intermedia* (DC.) Airy Shaw, syn. *A. nepalensis* (Spreng.) Hand. -Mazz.**

Common name(s): Yaktso

Family: Asteraceae/ Compositae

Distribution: Habitat to alpine Himalayas, and Ladakh.

Parts and properties used: Leaf- Locally used in skin and genital problems. Has medicinal potential.

Source of information (system): AF

***Andrographis nallamalayana* J. L. Ellis (EN, Rare)**

Common name(s): Kalmegh (wild relative)

Family: Acanthaceae

Distribution: Occur wild in Nallamalais, Kurnool, Andhra Pradesh (AP).

Parts and properties used: Whole plant- Locally used leukoderma and mouth ulcers.

Source of information (system): F

***Andrographis serpyllifolia* (Rottler ex Vahl) Wight**

Common name(s): Kalmegh (wild relative)

Family: Acanthaceae

Distribution: Occur wild in Malabar region.

Parts and properties used: Leaf extract- Locally used in bovine mastitis.

Source of information (system): F

***Androsace rotundifolia* Hardwicke and *A. rotundifolia* var. *glandulosa* Hook. f.**

Common name(s): Round leaf Rock Jasmine wild relative

Family: Primulaceae

Distribution: Described from Himalayas, Ladakh and Kumaon.

Parts and properties used: Herb- Yields a tonic. Leaves- Stomachache, emetic. Rhizome extract- Ophthalmic.

Source of information (system): F

***Androsace villosa* L.**

Common name(s): Fujung, Roundleaf Rock Jasmine wild relative

Family: Primulaceae

Distribution: Habitat to alpine Himalayas, Ladakh. Grown in rock gardens.

Parts and properties used: Herb- Gives a tonic.

Source of information (system): F

***Anemone rupicola* Camb.**

Common name(s): Simaso

Family: Ranunculaceae

Distribution: Distributed wild from Kashmir to Sikkim in Himalayas.

Parts and properties used: Root- Locally used in stomach problems.

Source of information (system): F

***Anisochilus carnosus* (L.) Wall.**

Common name(s): Jirnya

Family: Lamiaceae/ Labiatae

Distribution: Found among grasses and sedges in Himalayas to south India.

Parts and properties used: Herb- Stimulant, expectorant, and diaphoretic. Treat gastrointestinal problems/ulcer, and skin diseases. Leaf juice- Used in urticaria and other allergic conditions. Widely used for medicinal potential.

Source of information (system): F

***Anodendron paniculatum* A. DC.**

Common name(s): Sarakkodi, Kavali

Family: Apocynaceae

Distribution: Found in Assam, Meghalaya, and Western Ghats.

Parts and properties used: Whole plant- Cytotoxic, antioxidant. Bast used to treat ulcer. Root- Used in vomiting and cough.

Source of information (system): F

***Aphananthe cuspidata* (Blume) Planch.**

Common name(s): Peenari

Family: Ulmaceae

Distribution: Found in Assam, Meghalaya, and Kerala.

Parts and properties used: Bark- Depurative. Taken as blood purifier, for itches, and skin eruptions. Medicinally promising.

Source of information (system): F

***Apium leptophyllum* (Pers.) F. Muell. ex Benth**

Sanskrit or Ayurvedic name: Ajmodaa, Dipyaka

Common name(s): Ajmod

Family: Umbelliferae

Distribution: Native to America. Cultivated in AP, Gujarat, Madhya Pradesh (MP), Karnataka.

Parts and properties used: Seeds- Anti-inflammatory, diuretic, carminative, nervine, sedative, antiemetic, antiseptic antispasmodic. Used in bronchitis, asthma, liver, and spleen diseases, emmenagogue, like *A. graveolens.* Essential oil- Antifungal.

Source of information (system): ASU

***Aquilqria khasiana* Hallier f. (CR)**

Common name(s): Agarwood, Gaharu

Family: Thymelaceae

Distribution: Endemic to Meghalaya.

Parts and properties used: Wood- Locally applied on injury, give highly aromatic resin. Source of traded agarwood.

Source of information (system): Locals

***Aquilegia fragrans* Benth. (EN)**

Common name(s): Sweet-scented columbine, Fragrant columbine

Family: Ranunculaceae

Distribution: Native to sub-alpine meadows in the western Himalayas, and Ladakh.

Parts and properties used: Underground part- used in wounds and inflammatory diseases like bovine mastitis. Seed- Antiscorbutic, diuretic and diaphoretic. Toxic. Assessed medicinally potential.

Source of information (system): F

***Arabidopsis thaliana* (L.) Hynh. (Recorded as *A. multiflorum* too)**

Common name(s): Imatso

Family: Brassicaceae/ Cruciferae

Distribution: Habitat to moist slopes, and streams in Himalayas.

Parts and properties used: Herb- Locally used in diarrhoea.

Source of information (system): F

***Arabis glandulosa* Kar. & Kir.**

Common name(s): Umakno

Family: Brassicaceae/ Cruciferae

Distribution: Found in western Himalayas, and Ladakh.

Parts and properties used: Herb- Locally used in abdominal pain.

Source of information (system): F

***Arenaria bryophylla* Fernald.**

Common name(s): Lekhum

Family: Caryophyllaceae

Distribution: Habitat to alpine Himalayan slopes and Ladakh.

Parts and properties used: Plant- Locally used to treat kidney problems.

Source of information (system): F

***Arenaria griffithii* Boiss.**

Common name(s): Sokhtam

Family: Caryophyllaceae

Distribution: Habitat to alpine high-altitude cold desert and Ladakh.

Parts and properties used: Shoot powder- Antispasmodic. Locally used in menstrual problems.

Source of information (system): F

***Aristolochia tagala* Cham. (VU)**

Common name(s): Aadutheendapalai

Family: Aristolochiaceae

Distribution: Found in Eastern India and western Ghats.

Parts and properties used: Plant- Antiproliferative, anti-inflammatory, and antioxidant.

Source of information (system): F

***Arnebia euchroma* (Royle) I. M. Johnst.**

Common name(s): Gaozaban, Ratanjot

Family: Boraginaceae

Distribution: Native and wild to alpine Himalayas, and Ladakh.

Parts and properties used: Herb- Locally used to treat Acne vulgaris (skin disease). Root- Hair tonic and blood purifier.

Source of information (system): F

***Arnebia guttata* Bunge**

Common name(s): Ratanjot, demok

Family: Boraginaceae

Distribution: Habitat wild to Himalayas, Jammu Kashmir (J&K)

Parts and properties used: Herb- Sterility, abdominal worm, tonsillitis, inflammation, blood purifier. Extract used as hair tonic like Ratanjot, *A. euchroma*. Root- Used medicinally.

Source of information (system): F

***Artemisia biennis* Willd.**

Common name(s): Mar grass, wild relative of wormwood (*A. annua*).

Family: Asteraceae/ Compositae

Distribution: Habitat wild to high-altitude Himalayas and Ladakh.

Parts and properties used: Plant extract- Locally used in obesity, stomach disorder.

Source of information (system): F

***Artemisia desertorum* Spreng.**

Common name(s): Khamlol, wild relative of *A. annua.*

Family: Asteraceae/ Compositae

Distribution: Habitat wild to alpine Himalaya, and Ladakh.

Parts and properties used: Various parts- Digestive, depurative, and locally used in wound healing.

Source of information (system): F

Artemisia dracunculus **L.**

Common name(s): Tarragon /Sanche, wild relative of *A. annua.*

Family: Asteraceae/ Compositae

Distribution: Habitat wild to alpine Himalaya, and Ladakh.

Parts and properties used: Plant extract- Diuretic. Locally used in toothache. Fresh or dried leaves have scent.

Source of information (system): F

Artemisia gmelinii **Webb ex Stechmann**

Common name(s): Russian wormwood, Khampa shridi

Family: Asteraceae/ Compositae

Distribution: Habitat wild to alpine Himalaya, and Ladakh.

Parts and properties used: Plant extract- Locally used in cold and cough. Have flavonoids, sesquiterpenes and other bio-active contents.

Source of information (system): F

Artemisia japonica **Thunb, syn.** *A. parviflora* **Roxb. ex D. Don**

Common name(s): Wild relative of *A. annua*

Family: Asteraceae/ Compositae

Distribution: Wild in western Himalayas. Grown too.

Parts and properties used: Stem, leaves, fruits, leaf extract- Aromatic. Locally used to treat malaria, eczema, and fever.

Source of information (system): F

Artemisia laciniata **Willd.**

Common name(s): Bingo, wild relative of *A. annua*

Family: Asteraceae/ Compositae

Distribution: Wild in Himalayas, and Kashmir.

Parts and properties used: Plant- Locally used in toothache. Leaves- Aromatic.

Source of information (system): F

***Artemisia parviflora* Roxb. ex D. Don**

Common name(s): Himalayan wormwood Pati

Family: Asteraceae/ Compositae

Distribution: Habitat to Himalayas, Western Ghats. And south India.

Parts and properties used: Plant- Antioxidant, digestive, febrifuge, anthelmintic, pain reliever. Locally used in treatment of abdominal pain, high blood pressure (BP), diabetes and skin diseases. Being further explored.

Source of information (system): F

***Artemisia salsoloides* Willd.**

Common name(s): Amango, wild relative of *A. annua*

Family: Asteraceae/ Compositae

Distribution: Wild in western Himalayas, and Ladakh.

Parts and properties used: Shoot- Locally used in intestinal disorder.

Source of information (system): F

***Artemisia scoparia* Waldst. & Kit.**

Common name(s): Khamptspa, paniculate wormwood, wild relative of *A. annua*

Family: Asteraceae/ Compositae

Distribution: Native to Eurasia, found in Ladakh and semi-deserted areas of Rajasthan.

Parts and properties used: Plant- Anticholesterolemic, antipyretic, antiseptic, cholagogue, diuretic, vasodilator, antibacterial. Used in jaundice, hepatitis, and gall bladder and intestine inflammation.

Source of information (system): F

***Artemisia sieversiana* Ehrh. ex Willd**

Common name(s): Khamchu, wild relative of *A. annua*

Family: Asteraceae/ Compositae

Distribution: Habitat to wild alpine Himalaya, and Ladakh.

Parts and properties used: Shoot and leaves- Anthelmintic, de-obstruent, emmenagogue, febrifuge and tonic. Externally used as antiseptic and discutient.

Source of information (system): F

***Artemisia stracheyi* Hk. f. & Th. ex Comp**

Common name(s): Rumonlo, wild relative of *A. annua*

Family: Asteraceae/ Compositae

Distribution: Found in western Himalayas, Ladakh to Tibet

Parts and properties used: Shoot- Locally used in toothache.

Source of information (system): F

Artemisia tournefortiana **Reichb.**

Common name(s): Khamchu, wild relative of *A. annua*

Family: Asteraceae/ Compositae

Distribution: Habitat to trans-Himalayas and Ladakh.

Parts and properties used: Plant extract- Antibacterial and antioxidant. Locally used in treatment of intestinal worms.

Source of information (system): F

Asparagus curillus **Buch. -Ham. ex Roxb.**

Common name(s): Jhirma, Shatawar

Family: Asparagaceae

Distribution: Temperate and tropical, central Himalayas 1000-2250 m.

Parts and properties used: Herb- Demulcent, tonic, terminate pregnancy. Used in treatment of gonorrhoea, diabetes.

Source of information (system): A

Asperugo procumbens **L.**

Common name(s): Zingsha, Madwort

Family: Boraginaceae

Distribution: Habitat to alpine Himalayas, and Ladakh.

Parts and properties used: Plant extract- Antibacterial activity, mild sedative, mood elevator. Locally used in stomach disorder.

Source of information (system): F

Asperula oppositifolia **Regel & Schmalh.**

Common name(s): Zingsha

Family: Rubiaceae

Distribution: Habitat to alpine Himalayas, and Ladakh.

Parts and properties used: Arial parts- Antioxidant, yields a tonic.

Source of information (system): F

Aster flaccidus **Bunge**

Common name(s): Wild aster

Family: Asteraceae/ Compositae

Distribution: Habitat to alpine Himalayas, and Ladakh.

Parts and properties used: Flower head- Locally used in cold, cough bronchitis, cramps, and pains.

Source of information (system): F

Astragalus multiceps **Wall. ex Benth.**

Common name(s): Wild relative of Katira, Mongolian milkvetch, *A. propinquus*

Family: Leguminosae/ Fabaceae

Distribution: Wild in north-west Himalayas and Ladakh.

Parts and properties used: Seed- Locally used in colic disorder.

Source of information (system): F

Astragalus rhizanthus **Royle ex Benth.**

Common name(s): Sarma, wild relative of *A. propinquus*

Family: Leguminosae/ Fabaceae

Distribution: Wild in alpine Himalayas, and cold arid Ladakh.

Parts and properties used: Shoot- Locally used in skin diseases.

Source of information (system): F

Astragalus strictus **Graham ex Benth.**

Common name(s): Serpang, wild relative of *A. propinquus*

Family: Leguminosae/ Fabaceae

Distribution: Found wild in Himalayas, and Ladakh.

Parts and properties used: Whole plant- Diuretic. Locally used.

Source of information (system): F

Astragalus tribulifolius **Benth.**

Common name(s): Yanglo, wild relative of *A. propinquus*

Family: Leguminosae/ Fabaceae

Distribution: Found wild in Ladakh.

Parts and properties used: Whole plant- Diuretic. Locally used.

Source of information (system): F

***Avicennia marina* (Forssk.) Vierh.**

Common name(s): Tavir

Family: Avicenniaceae

Distribution: Pleotropic to mangrove forests.

Parts and properties used: Locally used in treatment of smallpox lesions.

Source of information (system): F

***Balanophora fungosa* J.R. Forst. & J.G. Forst., syn. *B. fungosa indica* (Arn.) B. Hansen**

Common name(s): Thippali (obligate parasite on forest trees)

Family: Balanophoraceae

Distribution: Occur in tropical, subtropical rain forests, Assam to Kerala.

Parts and properties used: Dried plant powder- Used in piles, internal haemorrhage and traded. Needs further study.

Source of information (system): S

***Barleria gibsoni* Dalz.**

Common name(s): Banya, Neel koranti

Family: Acanthaceae

Distribution: Habitat to Western Ghats.

Parts and properties used: Leaf juice- Locally used in cataract, fever, gastrointestinal disorder. Leaves- Chewed in toothache.

Source of information (system): F

***Barringtonia racemosa* (L.) Spreng. (Freshwater Mangrove)**

Common name(s): Samuthira palam

Family: Baringtoniaceae

Distribution: Occur in Assam, coasts of India and Andaman.

Parts and properties used: Fruit- Has properties like Hidol, Hijjal (*B. acutangula*). Assessed medicinally potential.

Source of information (system): AFS

***Bauhinia malabarica* Roxb. (EN)**

Sanskrit or Ayurvedic name: Ashmantaka

Common name(s): Asmantaka, variety of Kachnar.

Family: Leguminosae/ Fabaceae (Caesalpiniaceae)

Distribution: Distributed from Assam, Bengal to south India.

Parts and properties used: Plant- Reported anti-dysentery. Needs more study.

Source of information (system): AS

***Begonia crenata* Dryand.**

Common name(s): Khatadi

Family: Begoniaceae

Distribution: Found in Western Ghats.

Parts and properties used: Whole plant- Used locally in gastrointestinal disorder and body heat.

Source of information (system): F

***Begonia laciniata* Roxb. var. *nepalensis* A. DC.**

Common name(s): Hooirjo, Teisu

Family: Begoniaceae

Distribution: Found in Sikkim, northeast hills, and Bengal.

Parts and properties used: Root decoction- Used in liver diseases and fever. Stalk extract- Used in venereal diseases. Fresh shoots- Used in toothache.

Source of information (system): F

***Berberis asiatica* Roxb. (Sub sp.); syn. *B. aristata* (?)**

Common name(s): Asian barberry, Daruhaldi wild relative

Family: Berberidaceae

Distribution: Found wild in Himalayas and peninsular India.

Parts and properties used: Plant, leaf buds- Antibacterial, antirheumatic, laxative and yields a tonic. Substitute to *Daruharidra*, (*B. aristata*)

Source of information (system): F

***Berberis tinctoria* Lesch.**

Common name(s): Daruhaldi/ Rasaut wild relative

Family: Berberidaceae

Distribution: Found wild in Uttarakhand.

Parts and properties used: Root- Methanol extract reported to be potential against liver disorders.

Source of information (system): F

***Berberis ulicina* Hook**

Common name(s): Khicharmaa Khizar, wild relative

Family: Berberidaceae

Distribution: Habitat to alpine Himalayas, Ladakh to Tibet.

Parts and properties used: Root extract- Haemostatic. Yield tonic with potential like *Daruharidra, (B. aristata)*

Source of information (system): F

***Biebersteinia odora* Stephan ex. Fischer.**

Common name(s): Khardung /Drakpose

Family: Geraniaceae

Distribution: Range up to alpine Himalaya, and Ladakh.

Parts and properties used: Root- Antiseptic, blood purifier. Locally used to treat urinogenital disorders, and skin sores.

Source of information (system): F

***Bistorta amplexicaulis* (D. Don) Greene**

Common name(s): Anjubar Amli, Kutrya

Family: Polygonaceae

Distribution: Habitat to Himalayan region.

Parts and properties used: Plant- Blood purifier. Cure ulcers. Leaf- Antidysentery. Cause abortion. Paste- Locally used to cure wounds, dysentery and to cause abortion.

Source of information (system): F

***Blepharis sindica* T. Anders**

Common name(s): Bhangari

Family: Acanthaceae

Distribution: Rajasthan and Gujarat.

Parts and properties used: Produce invigorating tonic, given to cattle to increase milk. Roots- Used in urinary discharge and dysmenorrhea, and as *Vajikarana* or *Vrishya chikitsa* (Ashtanga Ayurveda; Aphrodisiac) by local Vaidya's.

Source of information (system): F, local Vaidya's

***Blumea fastulosa* (Roxb.) Kurz.**

Sanskrit and Common name(s): Kukundara, Kakarondaa (wild variety)

Family: Asteraceae/ Compositae.

Distribution: Wild in tropical Himalayas and plains of India.

Parts and properties used: Pant- Locally used as diuretic. Essential oil- CNS depressant.

Source of information (system): F

***Boerhavia repanda* Willd., syn. *B. chinensis* (L.) Rottb., *Commicarpus chinensis* (L.) Heimerl**

Common name(s): Cattaranai, Punarnava, Varakanputu, *B. diffusa* wild relative

Family: Nyctaginaceae

Distribution: Recorded wild from Travancore, Kerala.

Parts and properties used: Plant, root- Edibles, contains saponins, alkaloids and flavonoids, needing investigation for medicinal use.

Source of information (system): S

***Boerhavia verticillata* Poir**

Sanskrit or Ayurvedic name: Shveta Punarnavaa

Common name(s): *B. diffusa* wild relative.

Family: Nyctaginaceae

Distribution: Found throughout plains of India.

Parts and properties used: Plant- Diuretic, anti-inflammatory, antiarthritic, spasmolytic, antibacterial. Used in nephrotic syndrome, ascites resulting from cirrhosis of liver and chronic peritonitis, dropsy with chronic Bright's disease, and for serum uric acid levels, like *B. diffusa*. Root- Anticonvulsant, analgesic, expectorant, CNS depressant, laxative, diuretic, abortifacient.

Source of information (system): A

***Boswellia ovalifoliolata* Balakr, (EN)**

Common name(s): Guggilam, Sambrani related to *B. serrata*,

Family: Burseraceae

Distribution: Wild in Tirumala, AP.

Parts and properties used: Oleo-gum-resin extract- Anti-inflammatory, used in scorpion sting. Have medicinal potential.

Source of information (system): F

***Bryonia laciniosa* Linn.**

Ayurvedic and Common name: Shivlingi

Family: Cucurbitaceae

Distribution: Occur throughout India.

Parts and properties used: Seed- Counteracting infertility, uterine tonic. *Vrishya rasayana* of Ashtanga Ayurveda.

Source of information (system): A

***Buchanania axillaris* (Desr.) Ramam.**

Sanskrit and Common name(s): Priyala variety, wild relative of *B. lanzan*.

Family: Anacardiaceae

Distribution: Habitat to dry forests of peninsular India.

Parts and properties used: Kernel- Antimicrobial. Used like *B. lanzan*.

Source of information (system): F

***Calamus tenuis* Roxb.**

Sanskrit or Ayurvedic name: Vetra variety

Common name(s): Bareilly Cane, relative of *C. rotang*

Family: Arecaceae

Distribution: Habitat to sub-Himalayan tract.

Parts and properties used: Used like Rotang, Rattan Cane, *C. rotang*.

Source of information (system): A

***Calamus travancoricus* Bedd. ex Hook. f.**

Sanskrit or Ayurvedic name: Vetra variety

Common name(s): Relative of *C. rotang*

Family: Arecaceae

Distribution: Habitat to Deccan Peninsula, and Western Ghats.

Parts and properties used: Tender leaves- Used in dyspepsia, biliousness and as an anthelmintic, and like *C. rotang*.

Source of information (system): AS

***Canella winterana* (L.) Gaertn. *C. alba* Murray**

Common name(s): Kiliyuram pattai, wild Cinnamon

Family: Canellaceae

Distribution: Caribbean native. Found in Bengaluru, Kerala.

Parts and properties used: Plant, fruit (seed)- Reported beneficial for cancer, sugar levels, pain, arthritis.

Source of information (system): F

***Capparis roxburghii* DC., syn. *C. aguba* Banks ex DC.**

Ayurvedic and Common name: Thoratti, Rudanti

Family: Capparaceae

Distribution: Habitat to southern Western Ghats.

Parts and properties used: Leaves- Heal wounds and cuts. Fruit- Used as substitute of Rudanti, *Cressa cretica.*

Source of information (system): A

***Carduus edelbergii* Rech. f., syn. *C. nutans* L.; *C. nutans* var. *indicus* DC.**

Common name(s): Musk thistle, Scotch thistle

Family: Asteraceae/ Compositae

Distribution: European, invasive species to India

Parts and properties used: Flowers- Febrifuge, purify the blood. Seeds- Rich in linoleic acid.

Source of information (system): F

***Carum strictocarpum* C.B. Clarke**

Common name(s): Substitute or alternative of Kal jeera (wild relative).

Family: Apiaceae/ Umbelliferae

Distribution: Wild in MP, Chhattisgarh, Odisha.

Parts and properties used: Fruit- Substitute of *C. carvi* Caraway (Kal jeera). Also, of *saunf* /fennel.

Source of information (system): F

***Cassia montana* Heyne ex Roth; syn. *Senna montana* (Roth) V. Singh**

Common name(s): Malaikkondrai

Family: Leguminosae/ Fabaceae (Caesalpiniaceae)

Distribution: Found wild in peninsular India.

Parts and properties used: Plant/stem extract- Analgesic and anti-inflammatory. Have medicinal potential.

Source of information (system): F

***Cassia obovata* (L.) Collad**

Sanskrit or Ayurvedic name: Maarkandikaa, Svarnapatri

Common name(s): Sonaamukhi, Cassia/Senna wild relative

Family: Leguminosae/ Fabaceae (Caesalpiniaceae)

Distribution: Found in most of tropical India.

Parts and properties used: Leaves and seeds- Purgative and anthelmintic (an adulterant of the true senna).

Source of information (system): F

Ceropegia hirsuta **Wt. and Arn.**

Common name(s): Khutti

Family: Asclepiadaceae

Distribution: Habitat to Western Ghats.

Parts and properties used: Tubers- Used gastrointestinal disorder, also, as vegetable.

Source of information (system): F

Ceropegia spiralis **Wight. (VU)**

Common name(s): Nimmati gadda

Family: Asclepiadaceae

Distribution: Endemic to peninsular India.

Parts and properties used: Tuber- Locally used to relieve indigestion. Grown as ornamental.

Source of information (system): F

Chaerophyllum reflaxum **Lindl.**

Common name(s): Lcha-wa, *C. bulbosum* relative.

Family: Apiaceae/ Umbelliferae

Distribution: Habitat to alpine Himalayas, and Ladakh.

Parts and properties used: Whole Plant- Locally used in stomach disorder. Essential oil- Have medicinal potential.

Source of information (system): F

Chaerophyllum villosum **Wall. ex DC.**

Common name(s): Chyaum, *C. bulbosum* wild relative

Family: Apiaceae/ Umbelliferae

Distribution: Habitat to Himalayas 2100-3500 m.

Parts and properties used: Leaf, seed- Locally eaten. Herb- Source of essential oil. Have medicinal potential.

Source of information (system): F

***Chenopodium botrys* L.**

Common name(s): Jangaddi Sagni, *C. album* wild relative.

Family: Chenopodiaceae

Distribution: Occur wild, from Europe to alpine Himalayas.

Parts and properties used: Arial parts- Anthelmintic, laxative, stimulant, diuretic, carminative, antispasmodic, emmenagogue, pectoral (Sugandh-*vaastuuka*). Seed-Toxic.

Source of information (system): F

***Chrysanthemum cinerariifolium* (Trevis) Vis., syn. *Tanacetum cinerariifolium* Sch. Bip.**

Common name(s): Genus *Pyrethrum* sister spp. Some put it in *Tanacetum.*

Family: Asteraceae/ Compositae

Distribution: European. Cultivated.

Parts and properties used: Flower- Source of pyrethrin's.

Source of information (system): West, F

***Chrysanthemum coronopifolium* (Vill.) Vill., syn. *Leucanthemum coronopifolium* Vill.**

Common name(s): Chrysanthemum, Guldaudi

Family: Asteraceae/ Compositae

Distribution: Found in Khasi and Janita Hills, Meghalaya.

Parts and properties used: Flower- To be explored for angina, high BP, diabetes, fever, cold, headache, dizziness, swelling, etc.

Source of information (system): F

***Chrysanthemum griffithii* C.B. Clarke**

Common name(s): Serpan, wild Guldaudi

Family: Asteraceae/ Compositae

Distribution: Habitat to west Himalayas and Ladakh.

Parts and properties used: Herb- Locally used in menses regulation.

Source of information (system): F

***Chrysanthemum pyrethroides* (Kar. & Kir.) B. Fedsch.**

Common name(s): Serpan, wild Guldaudi

Family: Asteraceae/ Compositae

Distribution: Habitat wild to alpine Himalayas and Ladakh.

Parts and properties used: Herb- Locally used in antirheumatism and fever.

Source of information (system): F

Chrysanthemum tibeticum **Hook. f. & Thomson**

Common name(s): Phemantso, wild Guldaudi

Family: Asteraceae/ Compositae

Distribution: Habitat wild to west Himalayas and Ladakh.

Parts and properties used: Herb- Locally used as antiseptic.

Source of information (system): F

Cinchona ledgeriana **Moens ex Trimen.,** *C. pubescens* **Vahl. and** *C. calisaya* **Wedd.**

Common name(s): Cinchona

Family: Rubiaceae

Distribution: South American. Marginally grown in India.

Parts and properties used: Bark- Source of Quinine to treat malaria. Has trade potential.

Source of information (system): F

Cicer microphyllum **Benth.**

Common name(s): Seri, wild relative of Chickpea.

Family: Leguminosae/ Fabaceae

Distribution: Habitat wild to alpine Himalayas and Ladakh.

Parts and properties used: Herb- Locally used as apoptogenic.

Source of information (system): F

Cinnamomum glaucenscens **(Nees) Hand. -Mezz, syn.** *C. cecidodaphne* **Meisn.**

Common name(s): Sugandhakokila

Family: Lauraceae

Distribution: Habitat to centre to east Himalayas and Northeast Hills.

Parts and properties used: Fruit pericarp- Give essential oil, used in perfumery and as analgesic, astringent, and carminative.

Source of information (system): F

Cinnamomum malabatrum **(Lam.) J. Presl.**

Common name(s): Dalchini, Tejpatta. wild relative of *C. tamala*.

Family: Lauraceae

Distribution: Habitat wild to Southwestern Ghats.

Parts and properties used: Bark, Leaf- Aromatic. Used in culinary as adulterant of *C. tamala*. Has medicinal potential.

Source of information (system): F

Cinnamomum sulphuratum **Nees (VU)**

Common name(s): Dalchini, Tejpatta. wild relative of *C. tamala.*

Family: Lauraceae

Distribution: Found wild in Assam, Kerala, Nilgiris Hills.

Parts and properties used: Leaf and bark paste- Locally used to treat cough and headache. Have medicinal potential.

Source of information (system): F

Cipadessa baccifera **(Roth) Miq.**

Common name(s): Adusoge soppu

Family: Meliaceae

Distribution: Endemic to Western Ghats.

Parts and properties used: Plant- Used to treat diabetes, dysentery, malaria, rheumatism, piles, headache, psoriasis. Has medicinal potential.

Source of information (system): FS

Cirsium wallichii **D.C., Like** ***C. verutum*** **(D. Don) Spreng. (Having yellow to white heads)**

Common name(s): Bungsee, Kakar

Family: Asteraceae/ Compositae

Distribution: Habitat to alpine Himalayas and Ladakh.

Parts and properties used: Plant- Antibacterial, antifungal, antioxidant. Has medicinal potential.

Source of information (system): F

Clausena excavata **Burm. f.**

Common name(s): Bengjari

Family: Rutaceae

Distribution: Occur in subtropical Himalayas and north India.

Parts and properties used: Plant- Astringent, antioxidant. Decoction of root, leaf-Used in abdominal disorder, cold, malaria.

Source of information (system): F

Clematis tibetana **Kuntze**

Common name(s): Clematis, Zakgic

Family: Ranunculaceae

Distribution: Habitat wild to alpine Himalayas and Ladakh.

Parts and properties used: Stem and flowers- Locally used to treat indigestion. Has medicinal potential.

Source of information (system): F

Cleome vahliana **Fresen., syn.** ***C. brachycarpa*** **Vahl** ***ex*** **DC.**

Common name(s): Kasturi, Noli, Nodi

Family: Cleomaceae

Distribution: Northwest India. Punjab to Gujarat.

Parts and properties used: Leaves infusion- Used in removing worms and insects from camel ears.

Source of information (system): F

Codonopsis clematidea **(Schrenk) C.B. Clarke**

Common name(s): Burrkutang

Family: Campanulaceae

Distribution: Habitat to Tibet, and Ladakh.

Parts and properties used: Root, leaves- Anti-rheumatic with medicinal potential.

Source of information (system): F

Codonopsis ovata **Benth.**

Common name(s): Ludut, Kashmir Bonnet

Family: Campanulaceae

Distribution: Habitat to Himalayas (Kashmir-Ladakh).

Parts and properties used: Root, leaves- Anti-rheumatic. Locally used for bruises, ulcers, and wounds with medicinal potential.

Source of information (system): F

Coffea travancorensis **Wight & Arn., syn.** ***Psilanthus travancorensis*** **(Wight & Arn.) J. F. Leroy**

Common name(s): Katu-mulla, Tsjeru-mulla

Family: Rubiaceae

Distribution: Habitat wild to Western Ghats.

Parts and properties used: Root- Locally sold as adulterant of Pushkarmool (*Inula racemosa* Hook. f.) in Kerala. Needs further investigations.

Source of information (system): F

***Coleus vettiveroides* K.C. Jacob (DD)**

Sanskrit or Ayurvedic name: Hrivera

Common name(s): Iruveli, Vettiver, Kuruver

Family: Lamiaceae/ Labiatae

Distribution: Endemic to TN, widely occur in other parts, India.

Parts and properties used: Plant- Bitter cooling, diuretic. Leaves- Cooling and carminative. Roots- Yields essential oil.

Source of information (system): AL

***Combretum decandrum* Roxb. non-Jacq.**

Common name(s): Korakukundi, Rurel

Family: Combretaceae

Distribution: Habitat to deciduous forests of India.

Parts and properties used: Leaves- Used in diarrhoea, gastric troubles. Seed oil- Used to treat eczema. Has medicinal potential

Source of information (system): F

***Combretum indicum* (L.) De Filipps, syn. *Quisqualis indica* L.**

Common name(s): Tameshvar

Family: Combretaceae

Distribution: Found wild in peninsular India.

Parts and properties used: Root/fruit decoction- Anthelmintic, expel parasitic worms. Also, used locally for gargling.

Source of information (system): F

***Commiphora caudata* (Wight & Arn.) Engler**

Common name(s): Kiluvai, Kondamavu

Family: Burseraceae

Distribution: Found wild in Eastern Ghats.

Parts and properties used: Stem paste- Externally applied to foot cracks. Has medicinal potential.

Source of information (system): FS

Corispermum tibeticum **Iljin, looks similar** ***C. hyssopifolium*** **L.**

Common name(s): Seimso

Family: Chenopodiaceae

Distribution: Habitat wild to alpine Himalayas and Ladakh.

Parts and properties used: Plant- Locally used in kidney disorder. Promising medicinal plant.

Source of information (system): F

Corydalis cashmeriana **Royle.**

Common name(s): Ralchatnakpo

Family: Papaveraceae

Distribution: Habitat wild to alpine Himalayas 2800-5500 m and Ladakh.

Parts and properties used: Root, tuber- Locally known for antiperiodic, diuretic properties.

Source of information (system): F

Corydalis crassissima **Jacquem. ex Camb**

Common name(s): sKyes-pa, dong-btags

Family: Papaveraceae

Distribution: Wild to subalpine Ladakh and north-western India.

Parts and properties used: Root, tuber- Locally used in lung disorder. Help in sleep.

Source of information (system): F

Corydalis meifolia **Wall.**

Common name(s): Tonzil

Family: Papaveraceae

Distribution: Habitat to damp moist slopes of alpine Himalayas 3900-5200 m.

Parts and properties used: Tuber or stem- Locally known for diuretic properties.

Source of information (system): F

Corydalis thyrsiflora **Prain**

Common name(s): Nyetapu

Family: Papaveraceae

Distribution: Wild to alpine northwest Himalayas 3500-5100 m and Ladakh.

Parts and properties used: Whole plant- Locally used in skin diseases. Protopine has been isolated.

Source of information (system): F

***Corylus jacquemontii* Decne., syn. *C. colurna* Linn.**

Common name(s): Findak, Curri/ Langura, Bhotia Badaam

Family: Betulaceae

Distribution: Habitat to alpine western Himalayas.

Parts and properties used: Leaf's flavonoids- Potent antiperoxidative and oxygen radical scavenger. Substitute of Hazelnuts *Phindak* (dry fruit without shell).

Source of information (system): F

***Costus igneus* Nak.**

Common name(s): Kostum, Insulin plant

Family: Costaceae

Distribution: Introduced from South and Central Americas. Grown in Himalayas.

Parts and properties used: Leaves- Anti-diabatic. Lower blood glucose levels.

Source of information (system): F

***Cousinia falconeri* J. D. Hooker**

Common name(s): Kreatising

Family: Asteraceae/ Compositae

Distribution: Habitat wild to trans-Himalayas, and Ladakh.

Parts and properties used: Herb- Diuretic. Have medicinal potential.

Source of information (system): F

***Cousinia thomsonii* C.B. Clarke**

Common name(s): Megtham

Family: Asteraceae/ Compositae

Distribution: Habitat to wild trans-Himalayas and Ladakh.

Parts and properties used: Leaves and shoots- Locally used in body pain, and swelling, diuretic. Have medicinal potential.

Source of information (system): F

***Cremanthodium ellisii* (Hook.f.) Kitam.**

Common name(s): Ming-chen-serpo/Lukumentok

Family: Asteraceae/ Compositae

Distribution: Native to Himalayas and Ladakh.

Parts and properties used: Leaves (herbal tea)- Locally used to treat contagious disease and fever.

Source of information (system): F (Tibetan)

Cremanthodium reniforme **(DC.) Benth.**

Common name(s): Ming-chen-serpo/Lukumentok

Family: Asteraceae/ Compositae

Distribution: Native to Himalayas and Ladakh.

Parts and properties used: Leaves- Locally used to treat contagious disease and fever.

Source of information (system): F (Tibetan)

Crepis flexuosa **(DC.) Benth., syn.** ***Youngia glauca*** **Edgew.**

Common name(s): Hawksbeard, Remang

Family: Asteraceae/ Compositae

Distribution: Habitat to alpine Himalayas, Ladakh and Nepal.

Parts and properties used: Fresh plant juice- Locally used in jaundice and muscle pain.

Source of information (system): F

Crotalaria burhia **Buch. -Ham.**

Common name(s): Saniya

Family: Leguminosae/ Fabaceae

Distribution: Arid regions of India (Thar desert).

Parts and properties used: Plant- Cooling. Juice used to treat gout, eczema, hydrophobia, pain and swellings, wounds and cuts, infection, kidney pain, abdominal problems, rheumatism.

Source of information (system): F

Croton joufra **Roxb.**

Common name(s): Makhudi, Dieng-lamosuh

Family: Euphorbiaceae

Distribution: Assam, and northeast region.

Parts and properties used: Plant- Contains alkaloids and tannins. Bark and leaves- Used for fermenting liquor by local tribe. Seeds- Used as arrow poison.

Source of information (system): F (Khasi tribe)

Croton scabiosus **Bedd. (VU)**

Common name(s): Silver-Leaf Croton

Family: Euphorbiaceae

Distribution: Habitat wild to Eastern Ghats.

Parts and properties used: Bark decoction- Anthelmintic, cytotoxic, with anticancer potential. Locally used in leucorrhoea. Have medicinal potential.

Source of information (system): F

Cryptocoryne spiralis **(Retz.) Fisch. ex Wydler**

Common name(s): Natti-ati-Vasa

Family: Araceae

Distribution: Found wild in Kerala and Bengal.

Parts and properties used: Root- A ofsubstitute of Ativisha, Atish (*Aconitum heterophyllum*)

Source of information (system): F

Cryptostegia grandiflora **Roxb. ex R.Br.**

Common name(s): Palay rubber vine

Family: Apocynaceae

Distribution: Found in Assam, Bihar, Gujarat, Maharashtra. Also, grown.

Parts and properties used: Leaves- Anti-inflammatory, antimicrobial. Latex- Source of rubber. Have medicinal potential.

Source of information (system): F

Curcuma pseudomontana **Grahm. (VU)**

Common name(s): Adavi Pasupu (wild relative of turmeric)

Family: Zingiberaceae

Distribution: Found wild in Eastern Ghats.

Parts and properties used: Tuber- Locally used to treat wound swellings.

Source of information (system): F

Cuscuta approximate **Bab. =** ***C. europaea*** **L. (Parasitic)**

Common name(s): Large dodder, Amarlata, related to *C. reflexa*.

Family: Convolvulaceae

Distribution: Habitat to alpine Himalayas and Ladakh.

Parts and properties used: Stem- Locally used to treat skin diseases.

Source of information (system): F

***Cuscuta capitata* Roxb., (Parasitic)**

Common name(s): Large dodder, Amarlata, related to *C. reflexa*.

Family: Convolvulaceae

Distribution: Habitat to alpine Himalayas and Ladakh

Parts and properties used: Plant- Locally used in kidney disorder.

Source of information (system): F

***Cuscuta epithymum* Linn**

Sanskrit or Ayurvedic name: Aakaashvalli

Common name(s): Aphatemun

Family: Convolvulaceae

Distribution: Imported from Iran.

Parts and properties used: Plant- Hepatic, laxative, carminative, psychometric. Used in urinary, spleen and liver disorders.

Source of information (system): AFU

***Cuscuta europaea* L. (Parasitic)**

Common name(s): Large dodder, Amarlata, related to *C. reflexa*

Family: Convolvulaceae

Distribution: Habitat to alpine Himalayas and Ladakh.

Parts and properties used: Stem- Laxative, aphrodisiac, kidney, and liver stimulant.

Source of information (system): F

***Cuscuta gigantea* Griffith (Parasitic)**

Common name(s): Hande-thapa, related to *C. reflexa*

Family: Convolvulaceae

Distribution: Habitat to central Asia to Himalayas (Ladakh).

Parts and properties used: Stem- Antiseptic. Plant- Related to *C. epithymum*.

Source of information (system): F

***Cycas beddomei* Dyer, (EN)**

Common name(s): Paireetha

Family: Cycadaceae

Distribution: Confined to Seshachalam Hills, AP

Parts and properties used: Pith- Locally used in debility (Physical weakness).

Source of information (system): F

***Cymbopogon khasianus* (Hack.) Stapf ex Bor.**

Common name(s): Wild relative of *C. flexuosus*

Family: Poaceae/ Gramineae

Distribution: Found in Meghalaya.

Parts and properties used: Essential oil- Aromatic, source of methyl eugenol. Have medicinal potential.

Source of information (system): F

***Cymbopogon pendulus* (Nees ex Steud.) W. Watson**

Common name(s): Lemon grass

Family: Poaceae/ Gramineae

Distribution: Wild in Meghalaya. Grown in Jammu.

Parts and properties used: Essential oil- Rich in elemicin, which is anti-bacterial. Have medicinal potential.

Source of information (system): F

***Daemonorops jenkinsianus* Mart., syn. *Calamus jenkinsianus* Griff.**

Ayurvedic and Common name(s): Vetra, relative of *Calamus tenuis* Roxb.

Family: Palmae, Aracaceae

Distribution: Found in Sikkim, Assam, and Khasi Hills.

Parts and properties used: Plant- Used in intrinsic haemorrhage.

Source of information (system): A

***Dendrobium nobile* Lindl. (EN)**

Common name(s): Orchid

Family: Orchidaceae

Distribution: Habitat to tropical and sub-tropical forests and Assam.

Parts and properties used: Plant- Astringent, analgesic, antipyretic, and anti-inflammatory. Have substances to nourish stomach, enhance body fluids. Source of tonic.

Source of information (system): F

***Derris scandens* (Roxb.) Benth.**

Common name(s): Gonj

Family: Leguminosae/ Fabaceae

Distribution: Grows in deciduous forests all over India.

Parts and properties used: Stem- Used as diuretic, antitussive, expectorant, antidysentery, and for muscle pain treatment.

Source of information (system): FS Thai system

Desmodium pulchellum **(L.) Benth.**

Common name(s): Lodhra, Kheri

Family: Leguminosae/ Fabaceae

Distribution: Weed to moist forest of Assam and Western Ghats.

Parts and properties used: Plant- Antimicrobial with medicinal potential.

Source of information (system): F

Dianthus anatolicus **Boiss.**

Common name(s): Related to carnation, *Dianthus caryophyllus.*

Family: Caryophyllaceae

Distribution: Habitat to wild alpine Himalayas (Ladakh).

Parts and properties used: Whole plant- Antipyretic. Locally used in stomach disorder. Have medicinal potential.

Source of information (system): F

Dianthus jacquemontii **Edgew.**

Common name(s): Khenchpa, related to carnation.

Family: Caryophyllaceae

Distribution: Habitat wild to alpine Himalayas and Ladakh.

Parts and properties used: Whole plant- Locally used to treat stomach disorder. Have medicinal potential.

Source of information (system): F

Digitalis lanata **Ehrh.**

Sanskrit or Ayurvedic name(s): Hritpatri (yellow variety),

Common name: yellow variety of *Digitalis purpurea*

Family: Scrophulariaceae

Distribution: European. Occurs wild as well as grown in Kashmir (*Yarikhah*).

Parts and properties used: Plant- Source of digoxin. Digitalis foxgloves glycosides increases contraction, forcing heart without increasing the oxygen consumption and slow the heart rate when auricular fibrillation is present (given under supervision).

Source of information (system): A

***Dioscorea anguina* Roxb.**

Ayurvedic name(s): Kaasaalu

Common name(s): Koshakanda, wild relative of *D. alata*.

Family: Dioscoreaceae

Distribution: Habitat to wet Himalayas, Assam & Bengal.

Parts and properties used: Used like *D. alata*. Alternative with similar potential.

Source of information (system): A

***Dioscorea hispada* Dennst., syn. *D. daemona* Roxb., syn. *D. hirsuta* Dennst.**

Ayurvedic name(s): Hastyaaluka

Common name(s): Karukandu, Kolo (Bihar).

Family: Dioscoreaceae

Distribution: Occur in Himalayas, Sikkim to Khasi Hills 1500 m.

Parts and properties used: Plant parts- Used in whitlow, sores, boils. Tubers- Used for ulcer, and to kill worms in wounds.

Source of information (system): AFS

***Dioscorea prazeri* Prain & Burkill, syn. *D. clarkei* Prain & Burkill, *D. sikkimensis*, Prain & Burkill, *D. deltoidea* Wall. var. *sikkimensis* Prain**

Ayurvedic name(s): Neelaalu

Family: Dioscoreaceae

Distribution: Occur in Himalayas from Nepal to Bhutan to Naga Hills 1500m.

Parts and properties used: Rhizome tuber- Antipatharia, used for hair wash for killing lice.

Source of information (system): A

***Diospyros buxifolia* (Blume) Hiern**

Ayurvedic and Common name(s): Elichevian, Ebony

Family: Ebenaceae

Distribution: Habitat to evergreen forests of Western Ghats.

Parts and properties used: Reported with properties like antidiabetic, antifatigue, needs validation. Have medicinal potential.

Source of information (system): AF

***Diospyros candolleana* Wight (VU)**

Ayurvedic and Common name(s): Kare Mara, Karimaram, Kari

Family: Ebenaceae

Distribution: Endemic to peninsular India and Western Ghats.

Parts and properties used: Root-bark decoction- Used in rheumatism and swellings.

Source of information (system): AFS

***Diospyros paniculata* Dalz. (VU)**

Ayurvedic and Common name(s): Ilakatta, Kari, Karunthuvarei

Family: Ebenaceae

Distribution: Common in peninsular India and Western Ghats.

Parts and properties used: Plant- Used in burns, gonorrhoea, biliousness, blood poisoning, rheumatism, and ulcer.

Source of information (system): AFS

***Dodonaea angustifolia* (L. f.) Benth., syn. *D. viscosa* Jacq. var. *angustifolia* (L. f.) Benth.**

Sanskrit, Ayurvedic and Common name(s): Sanatta, Rasna, Aliar

Common name(s): Aliar, jangli-anar, Sinatha, Vilayati mehndi

Family: Sapindaceae

Distribution: Pantropical. Occur in shola, evergreen, and dry deciduous forests of India.

Parts and properties used: Leaf- Used in Ethiopian traditional medicine system to treat malaria.

Source of information (system): AFS, ENVIS

***Dracocephalum stamineum* Kar. & Kir.**

Common name(s): Zinkzer

Family: Lamiaceae/ Labiatae

Distribution: Habitat to alpine Himalayas.

Parts and properties used: Whole plant- Locally used in cold and cough.

Source of information (system): F

***Elaeocarpus tuberculatus* Roxb.**

Sanskrit, Ayurvedic and Common name(s): Rudraaksha variety, Rudirak

Family: Elaeocarpaceae

Distribution: Southern Western Ghats, Kanara coast southwards.

Parts and properties used: Bark- Stomachic, antibailout. Used in hematemesis. Nut - Antiepileptic, anti-rheumatic.

Source of information (system): AFS

***Eleutherococcus senticosus* (Rupr. & Maxim.) Maxim, syn. *Acanthopanax senticosus* (Rupr. et Maxim.) Maxim**

Common name(s): Siberian Ginseng

Family: Araliaceae

Distribution: Native to Siberia, China, Japan. Extended to northeast India.

Parts and properties used: Root extract- It's a nutraceutical. Have trade potential.

Source of information (system): F

***Enhydra fluctuans* Lour.**

Common name(s): Helchi, Helonchi

Family: Asteraceae/ Compositae

Distribution: Hills of Bihar, West Bengal (WB) and Assam.

Parts and properties used: Herb- Locally used to treat diseases of the nervous system. Leaves- laxative, antibilious, etc.

Source of information (system): F

***Ephedra equisetina* Bunge.**

Common name(s): Somalatha, relative of *E. gerardiana*

Family: Ephedraceae

Distribution: Recorded in Himalayas above 4000 m.

Parts and properties used: Stem, leaf, flower, fruit- Used as substitute of Ma Huang, *E. gerardiana.*

Source of information (system): F

***Ephedra intermedia* Schr. & Mey.**

Common name(s): Asmania, Chhapat. Relative of Ma Huang *E. gerardiana*

Family: Ephedraceae

Distribution: Habitat to alpine Western Himalayas (Ladakh).

Parts and properties used: Stem, leaf, flower, fruit- Diuretic, vasoconstrictor, and vasodilator. Locally used in treatment of asthma, hay fever and allergy.

Source of information (system): F

Ephedra sinica **Stapf. (EN)**

Common name(s): Relative of *E. gerardiana* (EN - CR)

Family: Ephedraceae

Distribution: Chinese. Cultivated in India.

Parts and properties used: Stem, leaf, fruit (seed)- Anti-inflammatory and anti-arthritic.

Source of information (system): F

Epipactis helleborine **(L.) Crantz**

Common name(s): Pengilo

Family: Orchidaceae

Distribution: Habitat to alpine Himalayas and Ladakh.

Parts and properties used: Leaves- Blood purifier. Used in gout, head, and stomachache. Have medicinal potential.

Source of information (system): F

Eranthennum roseum **(Vahl) R Br.**

Common name(s): Thandikarav

Family: Acanthaceae

Distribution: Found in Western Ghats.

Parts and properties used: Roots- Antimicrobial. Have medicinal potential.

Source of information (system): F

Erigeron alpinus **L.**

Common name(s): Repan, Bashakar

Family: Asteraceae/ Compositae

Distribution: European, found in alpine western Himalayas and Ladakh.

Parts and properties used: Plant- Astringent, diuretic, yields a tonic. Locally used in cold and cough, diabetes, dropsy, kidney diseases, diarrhoea dysentery. Have medicinal potential.

Source of information (system): F

Erythraea roxburghii **G. Don., syn.** ***Centaurium roxburghii*** **(G. Don) Druce.**

Sanskrit or Ayurvedic name(s): Kiraat-tikta

Common name(s): Khet chirayata

Family: Gentianaceae

Distribution: Temperate and subtropical regions of India.

Parts and properties used: Plant- Produce bitter tonic. A substitute for *Swertia chirayita.*

Source of information (system): AF

Eupatorium triplinerve **Vahl., syn.** ***E. ayapana*** **Vent.,** ***Ayapana triplinervis*** **(Vahl) R.M. King & H. Rob.**

Sanskrit name: Vishalyakarni

Common name(s): Ayapana

Family: Asteraceae/ Compositae

Distribution: Native to America. Found in wet areas of Assam, Gangetic Plains, Western Ghats.

Parts and properties used: Herb- Analgesic, anticoagulant, antianorexic, antiparasitic, anthelmintic, sedative. Locally used as tonic, cardiac stimulant.

Source of information (system): F

Euphorbia hispida **Boiss.**

Common name(s): Therno

Family: Euphorbiaceae

Distribution: Native to Himalayas & northwest India.

Parts and properties used: Latex- Locally used to treat skin eruption.

Source of information (system): F

Euphorbia lathyris **L.**

Common name(s): Caper spurge

Family: Euphorbiaceae

Distribution: Native of Europe/China, naturalized in north India.

Parts and properties used: Plant- Poisonous, laxative. Latex- Treat cancer, corns, and skin problems, sore throats, diarrhoea, and gangrene.

Source of information (system): FU

Euphorbia royleana **Boiss., syn.** ***E. pentagona*** **Royle**

Sanskrit or Ayurvedic name(s): Snuhi, Snuk, Gudaa (Substitute)

Common name(s): Thuuhar, Danda Thor; siyuri or siudi in Nepal.

Family: Euphorbiaceae

Distribution: Western Himalayas, Kumaon – Nepal to Eastern Ghats.

Parts and properties used: Latex- Cathartic, anthelmintic. May be source of *Shilajit*, due to similarity. A substitute, equated with *Euphorbia acaulis* Roxb.

Source of information (system): AFU

***Euphorbia stracheyi* Boiss.**

Common name(s): Khron-bu

Family: Euphorbiaceae

Distribution: Habitat to alpine western Himalayas and Ladakh

Parts and properties used: Whole Plant- Locally used to treats boils. Have medicinal potential.

Source of information (system): F

***Euphrasia laxa* Pennell.**

Common name(s): Eyebright, Kangchuk. *E. officinalis* relative.

Family: Scrophulariaceae

Distribution: Habitat to alpine western Himalayas and Ladakh.

Parts and properties used: Plant- Produce eye drops for eye ailments.

Source of information (system): FH

***Euphrasia vulgaris* Gueldenst. ex Ledeb.**

Common name(s): Bepamatso. *E. officinalis* relative.

Family: Scrophulariaceae

Distribution: Habitat to alpine Himalayas and Ladakh.

Parts and properties used: Plant- Produce eye drops for eye ailments. Assessed medicinally potential.

Source of information (system): FH

***Fagopyrum dibotrys* (D. Don) Hara**

Common name(s): Kathu

Family: Polygonaceae

Distribution: Perennial grows in forests and cultivated parts of northeast Himalayas.

Parts and properties used: Plant- Anodyne, anthelmintic, antiphlogistic, carminative, depurative and febrifuge. Decoction is used in the treatment of traumatic injuries, lumbago, menstrual irregularities, purulent infections.

Source of information (system): F

Fagopyrum tataricum **Gaertn.**

Sanskrit or Ayurvedic name: Ukhal

Common name(s): Tatary buckwheat, Kutu variety.

Family: Polygonaceae

Distribution: Habitat to alpine Ladakh, Zaskar and west Tibet. Grown.

Parts and properties used: Herb/grain- Used to treat fragile capillaries, chilblains and for strengthening varicose veins, high blood pressure. Leaves, flowers- Source of Rutin (flavonoid).

Source of information (system): AF

Ficus cordifolia **Roxb., syn.** ***F. rumphii*** **Blume.**

Sanskrit or Ayurvedic name(s): Ashmantaka (variety)

Common name(s): Gajanaa, Golden Rumph's

Family: Moraceae

Distribution: Throughout Indian hills, up to 1700 m.

Parts and properties used: Fruit juice and latex- Antiasthmatic and vermifuge.

Source of information (system): AF

Ficus gibbosa **Blume, syn.** ***F tinctoria*** **var.** ***gibbosa*** **(Blume) Corner.**

Common name(s): Ithitholi

Family: Moraceae

Distribution: Evergreen in the moist deciduous forests of India.

Parts and properties used: Plant- Antioxidant. Traditionally used in skin diseases, ulcers, diabetes, inflammation.

Source of information (system): FS

Ficus microcarpa **Linn. f., syn. -** ***F. retusa*** **auct. non. Linn.**

Sanskrit or Ayurvedic name(s): Plaksha

Common name(s): Kal Ichi, Itti. Wild relative of *Ficus lacor/ F. virens*.

Family: Moraceae

Distribution: Found in Bengal, Bihar, peninsular India, and Andaman. Grown as avenue tree.

Parts and properties used: Bark- Antibailout. Leaf- Antispasmodic. Root, bark, leaf- Used for preparation of oils and ointments for ulcers, skin care, oedema, and inflammations.

Source of information (system): AFS

Ficus mollis **Vahl.**

Common name(s): Kaaduatthi

Family: Moraceae

Distribution: Found wild in Western Ghats.

Parts and properties used: Leaves, stem, root- Locally used in dysentery, snakebite, skin diseases and constipation. Assessed medicinally potential.

Source of information (system): F

Ficus palmata **Forsk,** ***F. caricoides*** **Roxb.,** ***F. virgata*** **Wall. ex Roxb.**

Sanskrit or Ayurvedic name(s): Phalgu, Anjiri

Common name(s): Manjimedi

Family: Moraceae

Distribution: Occur in northwest India, Kashmir to Nepal, up to 1000m.

Parts and properties used: Fruit - Demulcent and laxative. Latex- Applied on pimples. Ripe fruits- Hypotensive.

Source of information (system): AS

Flacourtia sepiaria **Roxb.**

Sanskrit or Ayurvedic name: Vikankata,

Common name(s): Kondai. *F. sapida* / *F. indica* Merr., wild relative.

Family: Flacourtiaceae

Distribution: Dry Forest of Kumaon, WB, Bihar, Odisha, and south India.

Parts and properties used: Bark- Triturated in sesame oil and used as liniment in gout and rheumatism.

Source of information (system): AF

Flemingia macrophylla **(Willd.) Kuntze** ***ex*** **Merr.**

Ayurvedic and Common name(s): Varrus / Varus

Family: Leguminosae/ Fabaceae

Distribution: Habitat to sub-humid to humid tropics, Assam to Kerala.

Parts and properties used: Herbal extract- Locally used to treat rheumatism.

Source of information (system): AF

Flemingia semialata **Roxb. ex W.T. Aiton**

Common name(s): Substitute of Salpan

Family: Leguminosae/ Fabaceae

Distribution: Native to Himalayas to northeast and Western Ghats, Andamans. Grown as host for lac insect.

Parts and properties used: Root tuber- Anthelmintic, vermifuge and vermicide, anthelmintic, vermifuge and vermicide. Assessed medicinally potential.

Source of information (system): F

***Flickingeria fugax* (Rchb. f.) Seidenf., syn. *Dendrobium fugax* Rchb. f., epiphytic orchid (EN)**

Common name(s): Jivanti

Family: Orchidaceae

Distribution: Habitat to central and northeast Himalayas.

Parts and properties used: Root powder- Stimulant, aphrodisiacs, tonic. Locally used in burn and bones dislocation.

Source of information (system): F

***Fumaria vaillantii* Loisel**

Sanskrit or Ayurvedic name(s): Parpata

Common name(s): Shaahtaraa

Family: Fumariaceae

Distribution: Occur all over India on the hills.

Parts and properties used: Plant- Detoxifying, laxative, diuretic, diaphoretic. Substitute of *F. parviflora.*

Source of information (system): AFU

***Galium pauciflorum* Bung.**

Common name(s): Phomongo

Family: Rubiaceae

Distribution: Habitat to alpine Himalayas.

Parts and properties used: Bark- Locally used to treats intestinal disorder.

Source of information (system): F

***Garcinia cowa* Roxb., syn. *G. kydia* Roxb.**

Sanskrit or Ayurvedic name(s): Paaraavata

Common name(s): Kujithekera

Family: Guttiferae/ Clusiaceae

Distribution: Tropical forests of Assam, WB, Odisha and Andamans.

Parts and properties used: Sundried fruit slices- Locally used in dysentery. Latex- Used as febrifuge. Gum-resin- A drastic cathartic.

Source of information (system): AF

Garcinia hanburyi **Hook. f.**

Sanskrit or Ayurvedic name(s): Kankushtha

Common name(s): Usaar-e-revand

Family: Guttiferae/ Clusiaceae

Distribution: Found in evergreen forests of Assam and Khasi Hills.

Parts and properties used: Gum-resin- Drastic hydragog, cathartic, toxic. Locally used in dropsical conditions. Gum-resin consists of 70-75% of α- and β-garcinol acids with gambogic acids and 20–25% gum.

Source of information (system): AU

Gentiana algida **Pall.**

Common name(s): Titkas, Arctic Gentian

Family: Gentianaceae

Distribution: Habitat wild to Himalayas and Ladakh.

Parts and properties used: Root beverages - Analgesic, helps in bronchitis. Assessed medicinally potential.

Source of information (system): F

Gentiana aquatica **L.**

Common name(s): Titkas

Family: Gentianaceae

Distribution: Habitat wild to alpine Himalaya and Ladakh.

Parts and properties used: Whole plant, root beverages- Yields tonic. Assessed medicinally potential.

Source of information (system): F

Gentiana carinata **(D. Don) Griseb.**

Common name(s): Pang-yankhar-bo

Family: Gentianaceae

Distribution: Habitat wild to alpine Himalaya, and Ladakh.

Parts and properties used: Root beverage- Locally used as tonic. Assessed medicinally potential.

Source of information (system): F

***Gentiana humilis* Steven**

Common name(s): Pang-yankhar-bo

Family: Gentianaceae

Distribution: Habitat wild to alpine Himalaya, and Ladakh.

Parts and properties used: Root beverage- Locally given in fever. Assessed medicinally potential.

Source of information (system): F

***Gentiana squarrosa* (Ledeb.) V. Zuev.**

Common name(s): Ziang

Family: Gentianaceae

Distribution: Found wild in Ladakh.

Parts and properties used: Root- Diuretic. Assessed medicinally potential.

Source of information (system): F

***Gentiana stracheyi* (Clarke) Kitam**

Common name(s): Ziang

Family: Gentianaceae

Distribution: Habitat wild to alpine Himalaya, and Ladakh.

Parts and properties used: Root beverage- Locally used as tonic. Assessed medicinally potential.

Source of information (system): F

***Gentianella moorcroftiana* (Wall. ex Griseb.)**

Common name(s): Span-gain-karpo

Family: Gentianaceae

Distribution: Habitat to alpine Himalayas, and Ladakh

Parts and properties used: Plant extract- Febrifuge. Assessed medicinally potential.

Source of information (system): F

***Gentianella paludosa* (Hk.) H. Smith.**

Common name(s): Kags-ti-nagpo

Family: Gentianaceae

Distribution: Habitat to alpine Himalayas and Ladakh.

Parts and properties used: Plant- Febrifuge. Locally used as tonic.

Source of information (system): L (Tibetan)

Geophila reniformis **D.Don, syn. *G. repens* (L.) I.M. Johnst.**

Common name(s): Kakamaci

Family: Rubiaceae

Distribution: Found in Assam, Western Ghats, and Andaman.

Parts and properties used: Root- Antiprotozoal, expectorant, emetic.

Source of information (system): F

Geranium collinum **Steph. ex Willd.**

Common name(s): Legkatin, wild geranium

Family: Geraniaceae

Distribution: Habitat to west Himalayas, and Ladakh.

Parts and properties used: Plant- Locally used in rheumatism, gout, dysentery, bleeding, and skin disorder.

Source of information (system): F

Geranium pratense **L.**

Common name(s): Meadow Crane's bill, Gad-ur

Family: Geraniaceae

Distribution: Habitat to western Himalayas, Ladakh. Grown.

Parts and properties used: Plant- Cooling, analgesic, febrifuge, and anti-inflammatory.

Source of information (system): L

Geranium sibiricum **L.**

Common name(s): Eyamlomentok, wild geranium

Family: Geraniaceae

Distribution: Habitat to alpine Himalayas, and Ladakh.

Parts and properties used: Plant- Antioxidant. Used locally in dysentery.

Source of information (system): F

Geranium tuberaria **Camb.**

Common name(s): Yusiang, wild Geranium

Family: Geraniaceae

Distribution: Habitat to alpine Himalayas, and Ladakh.

Parts and properties used: Herb- Locally used in kidney problems.

Source of information (system): F

***Geranium wallichianum* D. Don ex Sweet.**

Common name(s): Laal Jadi

Family: Geraniaceae

Distribution: Habitat to Western Himalayas 2500-3700 m. Cultivated.

Parts and properties used: Plant- Astringent. A substitute of Ratanjot.

Source of information (system): F

***Geum urbanum* L.**

Common name(s): Wood avens, Herb Bennet, Colewort

Family: Rosaceae

Distribution: European. Found in alpine western Himalayas.

Parts and properties used: Herb and root- Astringent, styptic, stomachic, febrifuge. Rootstock- Used in fever, etc.

Source of information (system): F, English

***Gisekia pharnaceoides* Linn.**

Sanskrit or Ayurvedic name(s): Variety of Elavaaluka (*Prunus cerasus*) an aromatic herb

Common name(s): Baalu-ka-saag, Morang

Family: Aizoaceae

Distribution: Found in drier northwest India and Deccan Peninsula.

Parts and properties used: Plant- Anthelmintic. Fresh herb locally used is used for taenia.

Source of information (system): AF

***Gnaphalium stewartii* (Holub) C.B. Clarke ex Hook. f.**

Common name(s): Peo

Family: Rosaceae

Distribution: Habitat wild to alpine Himalayas, and Ladakh.

Parts and properties used: Arial parts- Locally used to treat skin disorder. Assessed medicinally potential.

Source of information (system): F

***Gossypium barbadense* Linn.**

Sanskrit or Ayurvedic name(s): Karpasa

Common name(s): Kapaas (sister sp.)

Family: Malvaceae

Distribution: South American. Cultivated in India.

Parts and properties used: Root- Emmenagogue, oxytocic, abortifacient, parturient, lactogogue. Leaf- Hypotensive, antirheumatic. Flower- Used in hypochondriasis and bronchial inflammations. Seed and leaf- Antidysentery. Seed- Galactagogue, pectoral, febrifuge. Seed oil- Used in skin care.

Source of information (system): F

***Gouania microcarpa* DC.; *G. leptostachya* DC**

Common name(s): Sinhanbali

Family: Rhamnaceae

Distribution: Occur from Uttar Pradesh (UP) to Assam, northeast and Andaman.

Parts and properties used: Leaf- Locally used as poultice for sores. Bark- Used in washing hair, and to check vermin.

Source of information (system): F

***Grewia elastica* Royle = *G. eriocarpa* Juss.**

Common name(s): Phalsa chhaal, relative of phalsa (*G. asiatica*)

Family: Tiliaceae

Distribution: Grows wild in Assam. Edible.

Parts and properties used: Bark- Locally used with spirit to treat skin. Assessed medicinally potential.

Source of information (system): F

***Grewia microcos* L., syn. *G. nervosa* (Lour.)**

Common name(s): Asolin, Panigrahi.

Family: Tiliaceae

Distribution: Habitat to Deccan Plateau.

Parts and properties used: Locally used for treating indigestion, eczema and itch, smallpox. Needs investigation.

Source of information (system): F

***Gymnocladus assamicus* Kanjilal ex P.C. Kanjilal (CR)**

Common name(s): Mewangmanba-shi or Minkling

Family: Leguminosae/ Fabaceae

Distribution: Habitat to northeast India.

Parts and properties used: Locally used to cure dermatological disorders and to get rid of leaches.

Source of information (system): F

Gynandropsis gynandra **(Linn.) Briq., syn.** ***Cleome gynandra*** **Linn.**

Sanskrit or Ayurvedic name(s): Tilaparni, Ajagandha

Common name(s): Thaivelai, related to mustard.

Family: Capparidaceae

Distribution: Habitat to waste lands all over warm India.

Parts and properties used: Leaves/seed- Locally used as mustard adulterant. Assessed medicinally potential.

Source of information (system): AS

Gypsophila cerastioides **D. Don**

Common name(s): Chicktkpa

Family: Caryophyllaceae

Distribution: Habitat to alpine Himalayas, 3000-4700 m. Grown in gardens.

Parts and properties used: Plant- used in cold and cough. Assessed medicinally potential.

Source of information (system): F

Habenaria roxburghii **(Pers.) R. Br.**

Common name(s): Leena Gadda, Chekku dumpa

Family: Orchidaceae

Distribution: Habitat to forests shady places in Eastern Ghats and south India.

Parts and properties used: Tuber paste- Applied to treat syphilis and wounds. Assessed medicinally potential.

Source of information (system): F

Halenia elliptica **D. Don**

Common name(s): lCags-tig-ramgo-ma

Family: Gentianaceae

Distribution: Native range extends to Central Asia to China and Himalaya

Parts and properties used: Plant- Antioxidant. Locally used in jaundice, virus hepatitis, stomach disorder.

Source of information (system): F

Heliotropium strigosum **Willd.**

Common name(s): Chitiphul, related garden heliotropes

Family: Boraginaceae

Distribution: Habitat to moist places all over north India.

Parts and properties used: Plant (juice)- Fragrant laxative and diuretic.

Source of information (system): F

Hemionitis arifolia **(Burm.) Moore (Tongue Fern)**

Common name(s): Akhukarni, Pattsjivi-Maravara, Mule Fern

Family: Hemionitidaceae

Distribution: Occur wild all-over south India in rain forests.

Parts and properties used: Fonds- Antibacterial. Locally used in burns, febrifuge. Assessed medicinally potential.

Source of information (system): F

Heracleum grande **(Dalzell & Gibson) Mukh**

Common name(s): Bafali

Family: Apiaceae/ Umbelliferae

Distribution: Endemic to hill slopes of Western Ghats.

Parts and properties used: Root leaves decoction- Locally used to treat stomach pain. Assessed medicinally potential.

Source of information (system): F

Heracleum candolleanum **(Wight & Arn.) Gamble (VU)**

Common name(s): Kattumalli substitute

Family: Apiaceae/ Umbelliferae

Distribution: Wild in forests of southern Western Ghats.

Parts and properties used: Fruit (seed)- Anti-inflammatory, aphrodisiac. Locally used in nervous disorders.

Source of information (system): F

Heracleum pinnatum **C.B.Clarke**

Common name(s): sPru-ma

Family: Apiaceae/ Umbelliferae

Distribution: Wild in alpine Himalayas, Drass and Ladakh.

Parts and properties used: Plant- Yield *Xanthotoxin*. Locally used to treat leukoderma, menses disorder.

Source of information (system): F

Herminium monorchis **(Linn.) R. Br.**

Common name(s): Musk orchid

Family: Orchidaceae

Distribution: Habitat to alpine Himalayas, Ladakh and Bhutan.

Parts and properties used: Roots- Locally tonic is extracted. Assessed medicinally potential.

Source of information (system): F

Hibiscus cannabinus **L.**

Common name(s): Kenaf, Deccan hemp

Family: Malvaceae

Distribution: Southern Asia.

Parts and properties used: Leaves- Used in dysentery and bilious, blood and throat disorders. Seeds- Applied to aches and bruises.

Source of information (system): AF

Hibiscus mutabilis **Linn.**

Sanskrit or Ayurvedic name(s): Sthala-Padam

Common name(s): Sembarattai

Family: Malvaceae

Distribution: Native to China. Planted as hedges in Indian gardens.

Parts and properties used: Leaf and flower- Expectorant, bechic, anodyne. Used in menorrhagia, dysuria, swellings, fistulae, wounds, and burns. Flower- Used in pectoral and pulmonary affections.

Source of information (system): AS

Hippophae salicifolia **D. Don, syn.** ***H. rhamnoides*** **subsp.** ***salicifolia*** **D. Don.**

Common name(s): Substitute/ wild relative of Sea buckthorn.

Family: Elaeagnaceae

Distribution: Himalayan region.

Parts and properties used: Fruit juice- Nutritive, antioxidant, antimicrobial, anti-cancer, anti-inflammatory, immunomodulatory, radio-protective, adaptogen, anti-cancer, anti-atherosclerosis, and anti-sterility.

Source of information (system): F

***Hippophae tibetana* Schlecht,**

Common name(s): Seabuckthorn.; Wild Sea buckthorn.

Family: Elaeagnaceae

Distribution: Habitat to alpine northwest Himalayas and Ladakh.

Parts and properties used: Fruit- Antitussive, blood purifier and expectorant. Locally used in disorder of lungs, etc.

Source of information (system): F

***Hippuris vulgaris* L.**

Common name(s): Dam-bu-karamachong

Family: Hippuridaceae

Distribution: Habitat to alpine northwest Himalayas and Ladakh.

Parts and properties used: Plant juice- Astringent, antiseptic, febrifuge. Assessed medicinally potential.

Source of information (system): F

***Hodgsonia macrocarpa* (Bl.) Cogn., syn. *Trichosanthes macrocarpa* Bl.**

Common name(s): Topou-guti

Family: Cucurbitaceae

Distribution: East Himalaya, Assam, & Northeast Hills.

Parts and properties used: Boiled or ashes of leaves- Locally used in nose complaints and to reduce fevers.

Source of information (system): F

***Homalomena aromatica* (Roxb.). Schott (VU)**

Common name(s): Gandh Kochu

Family: Araceae

Distribution: Habitat to marshy humus rich moist forest of Assam.

Parts and properties used: Rhizome- Anti-inflammatory, analgesic, antidepressant, antiseptic, sedative, antispasmodic. Used to treat joint pain, and skin infections. Assessed medicinally potential.

Source of information (system): A (locals)

***Hydrocotyle javanica* Thunb.**

Sanskrit or Ayurvedic name(s): Manduukaparni

Common name(s): Brahma manduki

Family: Apiaceae/ Umbelliferae

Distribution: Occur from Himalayas to northeast Hills to Western Ghats.

Plant- Blood purifier. Locally used in indigestion, dysentery, and nervousness. Substitute to Brahmi, *Centella asiatica.*

Source of information (system): F

***Hygroryza aristata* (Retz.) Nees ex Wight & Arn.**

Common name(s): Janli dal, Asian watergrass

Family: Poaceae/ Gramineae

Distribution: Assam, Punjab, Maharashtra, & Karnataka.

Parts and properties used: Root- Antioxidant, anti-inflammatory. Assessed medicinally potential.

Source of information (system): F

***Hyoscyamus pusillus* L.**

Common name(s): Ajwain, Khursani Gaylangthang, Henbane

Family: Solanaceae

Distribution: Habitat wild to alpine Himalayas and Ladakh.

Parts and properties used: Whole plant- Have Hyoscyamine. Locally used to treat asthma, stomach disorder, manic psychosis, etc. Assessed medicinally potential.

Source of information (system): F

***Inula rhizocephala* var. *rhizophaloides* (C.B. Clarke) Kitam.**

Common name(s): Riamko

Family: Asteraceae/ Compositae

Distribution: Habitat to moist meadows of Himalayas 2500-4500m, (Ladakh).

Parts and properties used: Plant- Locally used in chest pain and cold. Root- Expectorant, cough, chest pains. Have medicinal potential.

Source of information (system): F

***Inula royleana* DC.**

Ayurvedic and Common name(s): Pushkarmool, Rupmak

Family: Asteraceae/ Compositae

Distribution: Habitat to alpine Himalaya, Drass-Kargil; Chamba, Lahul Spiti.

Parts and properties used: Root- Dermatitis. Adulterant of "Kuth" *Saussurea lappa.* Have medicinal potential.

Source of information (system): A

***Ipomoea bona-nox* Linn, syn. *I.* alba Linn. *Calonyction bona-nox* Bojer. *C. aculeatum* (Linn.) House**

Sanskrit or Ayurvedic name(s): Chandrakaanti

Common name(s): Chaandani, Naganamukkori

Family: Convolvulaceae

Distribution: Occur throughout India.

Parts and properties used: Root bark- Purgative. Leaves- Used in filariasis.

Source of information (system): AFS

***Ipomoea eriocarpa* R. Br., syn. *I. hispida* Roem. & Schult. *I. sessiliflora* Roth.**

Sanskrit or Ayurvedic name(s): Nikhari

Common name(s): Buta, wild relative of Akhukarni, *I. reniformis*, syn. *Merremia emarginata*

Family: Convolvulaceae

Distribution: Occur throughout India.

Parts and properties used: Plant- Antirheumatic, antiphallic, antiepileptic and antileprotic.

Source of information (system): AF

***Ipomoea pes-tigridis* L.**

Ayurvedic or common name(s): Pulichuvadan

Family: Convolvulaceae

Distribution: Pantropic. Habitat to deciduous forests of Kerala.

Parts and properties used: Plant- Antidote dog bite, boils/ carbuncles. Root- Purgative. Assessed medicinally potential.

Source of information (system): AFS

***Ipomoea turpethum* (L.) R. Br. (= *Operculina turpethum* (*L.*) Silva Manso (VU)**

Sanskrit or Ayurvedic name(s): Triputa, Nishotra, Trivrutha

Common name(s): Indian Jalap, Nishodhi

Family: Convolvulaceae

Distribution: Found throughout India in moist places.

Parts and properties used: Herb- Useful in colic constipation, dropsy, vitiated *vata*, paralysis, myalgia, arthralgia, bronchitis, obesity, helminthiasis, gastropathy, ascites, inflammations, intermittent fever, tumours, jaundice, ophthalmia.

Source of information (system): AFLU

***Jatropha multifida* Linn. = *Baliospermum montanum* Muell.Arg.**

Sanskrit or Ayurvedic name(s): Brihat-Danti

Common name(s): Danti (variety)

Family: Euphorbiaceae

Distribution: Native to South American, naturalized in India.

Parts and properties used: Leaves- Used for scabies. Latex- Applied to wounds/ ulcers. Seeds- Purgative, emetic. Fruits- Poisonous.

Source of information (system): AF

***Juniperus recurva* Buch. -Ham. Ex D.Don,**

Common name(s): Himalayan juniper, Sukpa, *J. communis* relative.

Family: Asteraceae/ Compositae

Distribution: Native to Himalayas. Grown as ornamental.

Parts and properties used: Berries- Used like *J. communis*. Purgative, incense. Assessed medicinally potential.

Source of information (system): F

***Jurinea ceratocarpa* (Decne.) Benth. ex Clarke.**

Common name(s): Chholmong Ispangchot Turjit

Family: Asteraceae/ Compositae

Distribution: Habitat wild to alpine Himalayas, and Ladakh.

Parts and properties used: Plant, root- Locally used in bronchitis, colic, and puerperal fever. Leaf extract- Cure wounds, joint pains. Assessed medicinally potential.

Source of information (system): F

***Kochia indica* Wt.**

Common name(s): Bui-chholi

Family: Chenopodiaceae

Distribution: Occur in north-west & peninsular India.

Parts and properties used: Herb- Works as cardiac stimulant. Assessed medicinally potential.

Source of information (system): F

***Kydia calycina* Roxb.**

Common name(s): Pola, Puli, Boranga

Family: Malvaceae

Distribution: Habitat to dry deciduous forests of sub-Himalayas, and Ghats.

Parts and properties used: Plant- Anti-inflammatory, febrifuge. Leaf paste- Poultice for body pain. Leaf/root- Antirheumatic. Assessed medicinally potential.

Source of information (system): F

Lactuca lessertiana **(DC.) C.B.C.I.**

Common name(s): Wild salad

Family: Asteraceae/ Compositae

Distribution: Habitat to alpine Himalayas and Ladakh.

Parts and properties used: Herb- Sedative, hypnotic, expectorant, cough, purgative, demulcent, diuretic, antiseptic.

Source of information (system): F

Lactuca intybacea **Jacq., syn.** ***L. remotiflora*** **DC,** ***L. runcinata*** **DC.**

Common name(s): Wild salad

Family: Asteraceae/ Compositae

Distribution: Occur in dry, evergreen to deciduous forests of Gangetic plain, TN, Maharashtra.

Parts and properties used: Herb- Antimicrobial. Locally used to treat stomach problems, pain relief, and inflammation.

Source of information (system): F

Lagotis glauca **Gaertn.**

Common name(s): Adulterant of Kutki, *Picrorhiza kurroa.*

Family: Scrophulariaceae

Distribution: Habitat wild to Himalayas 3300-4500 m.

Parts and properties used: Root- Used to treat wounds and gastric pains. Plant- Stomachic, carminative, stimulant, and sudorific. Plant extract- An antiviral drug.

Source of information (system): FL (Tibetan) U

Lancea tibetica **Hook. f. & Thomson**

Common name(s): Spa-yang-rtsa-ba

Family: Scrophulariaceae

Distribution: Habitat to alpine Himalayas and Ladakh.

Parts and properties used: Fruit, leaves- Locally used in pulmonary and heart disorders, healing wound. Assessed medicinally potential.

Source of information (system): F

Larix griffithiana **Carr.**

Common name(s): Boargasella

Family: Pinaceae

Distribution: Native to eastern Himalayas 2400-3600 m.

Parts and properties used: Balsam- Antiseptic, hyperaemic. Needs further exploration.

Source of information (system): F

Lemanea australis **Alkins (CR)**

Common name(s): Nungsham

Family: Leimaneaceae

Distribution: Found in eastern India.

Parts and properties used: Whole plant extract- Locally used as abortifacient and aphrodisiac.

Source of information (system): Local tribes

Leptadenia pyrotechnica **(Forssk.) Decne.**

Common name(s): Kheep

Family: Asclepiadaceae

Distribution: Widespread in Thar desert.

Parts and properties used: Plant- Antifungal, antibacterial, anticancer, antioxidant, anthelmintic, antiatherosclerosis, hypolipidemic, antidiabetic, wound healing, and hepatoprotective.

Source of information (system): Used by desert people.

Leontopodium leontopodium **Hand-Mazz.**

Common name(s): Palloo

Family: Asteraceae/ Compositae

Distribution: Habitat to alpine Himalayas and Ladakh.

Parts and properties used: Plant- Antiseptic. Have medicinal potential.

Source of information (system): F

Leontopodium nanum **(Hook.f. & Thomson ex CB Clarke) Hand -Mazz.**

Common name(s): Palloo

Family: Asteraceae/ Compositae

Distribution: Habitat to alpine Himalayas and Ladakh.

Parts and properties used: Plant- Antiseptic. Have medicinal potential.

Source of information (system): F

Leptadenia reticulata **(Retz.) Wight & Arn.**

Common name(s): Jivanti

Family: Apocynaceae

Distribution: Found all over India.

Parts and properties used: Plant- Used in tuberculosis, haematopoiesis, emaciation, cough, dyspnoea, fever, control burning sensation, cancer, and dysentery.

Source of information (system): AFSU

Leucas lavandulaefolia **Rees., syn.** ***L. linifolia*** **Spreng**

Sanskrit or Ayurvedic name: Dronpushpi related species

Common name(s): Tumbaa, Guumaa.

Family: Lamiaceae/ Labiatae

Distribution: Found as weed all over India ascending to 1800 m in the Himalayas.

Parts and properties used: Leaf's decoction- Used as a sedative in nervous disorders. Also, as a stomachic and vermifuge. Crushed leaves- Applied as poultice on sores/ wounds as dermatosis.

Source of information (system): AF

Linum perenne **L.**

Common name(s): Wild relative of flax.

Family: Linaceae

Distribution: Habitat wild to alpine Himalayas and Ladakh.

Parts and properties used: Seed oil- Locally used as emollient, expectorant.

Source of information (system): F

Litsea cubeba **(Lour.) Pers.**

Ayurvedic and Common name(s): May chang, Mountain Pepper

Family: Lauraceae

Distribution: South China native. Found in East Himalayas and Northeast

Parts and properties used: Plant- Used in gastro-intestinal ailments, diabetes, oedema, cold, arthritis, asthma, and traumatic injury. Essential oils- Body aches.

Source of information (system): A, Chinese system

Lloydia serotina **(L.) Rchb., syn.** ***Gagea serotina*** **(L.) Ker Gawl.**

Common name(s): rTsa-awa

Family: Liliaceae

Distribution: Habitat to alpine Himalayas, and Ladakh.

Parts and properties used: Bulb, leaf- Blood purifier, diuretic. Have medicinal potential.

Source of information (system): F

Lobelia leschenaultina (**Persl) Skottsb., syn.** ***L. excelsa*** **Lesch**

Sanskrit or Ayurvedic name(s): Nala

Common name(s): Devanala's variety

Family: Campanulaceae, Lobeliaceae

Distribution: Found in hills of south India at above 1800 m.

Parts and properties used: Leaves filtered solution- Used to control aphids, tingids and mites on vegetable and other crops. Cured and smoked leaves are used as tobacco. Plant- Poisonous (man/livestock).

Source of information (system): AF

Lobelia nicotianaefolia **Roth** ***ex*** **Roem. & Schult.**

Sanskrit or Ayurvedic name(s): Devanala, Mahanala

Common name(s): Kattuppugaiyila (related to *L. inflata*)

Family: Lobeliaceae

Distribution: Wild in hills of South India.

Parts and properties used: Leaves- Substitute to Indian tobacco, (*L. inflata*). Assessed medicinally potential.

Source of information (system): AFS

Lomatogonium rotatum **(L.) Fr. ex Fernald**

Common name(s): Tsemrang

Family: Gentianaceae

Distribution: Found in Ladakh.

Parts and properties used: Herb- Locally used in cold, cough, and fever. Assessed medicinally potential.

Source of information (system): F

Lonicera spinosa **(Decne.) Jacquem. ex Walp.**

Common name(s): Spiny honeysuckle

Family: Caprifoliaceae

Distribution: Native of Afghanistan to Himalayas, Tibet. Cultivated.

Parts and properties used: Plant- Expectorant and laxative. Fruit syrup- Used in asthma. Leaf's decoction- Used in liver and spleen diseases.

Source of information (system): F

Loranthus pentandrus **Linn., syn.** ***Dendrophthoe pentandra*** **(Linn.) Miq.**

Sanskrit or Ayurvedic name(s): Bandaaka variety

Common name(s): Baandaa

Family: Loranthaceae

Distribution: A parasite found on trees in east India.

Parts and properties used: Leaves- Used as poultice for sores and ulcers.

Source of information (system): AF

Lychnis himalayensis **(Rohrb.) Edgew. & Hook., syn.** ***Silene himalayensis*** **(Rohrb.) Majumdar**

Common name(s): Gayangarlo

Family: Caryophyllaceae

Distribution: Endemic to west Himalayas, and Ladakh.

Parts and properties used: Plant- Locally used in cold and cough. Assessed medicinally potential.

Source of information (system): F

Lychnis nutans **Benth., syn.** ***L. ciliata*** **Wall.;** ***L. indica*** **(Roxb. ex Otth) Benth.,** ***Silene nutans*** **L.**

Common name(s): Lappu

Family: Caryophyllaceae

Distribution: Occur in Ladakh, Himachal Pradesh (HP).

Parts and properties used: Plant- Applied on swollen skin to release pus. Leaf juice- Locally used in dysentery. Assessed medicinally potential.

Source of information (system): F

Lycopsis arvensis **L., syn.** ***Anchusa arvensis*** **(L.) M. Bieb.**

Common name(s): Bugloss

Family: Boraginaceae

Distribution: Habitat to alpine Himalayas and Ladakh.

Parts and properties used: Plant- Antidiabetic. Yields a tonic. Have medicinal potential.

Source of information (system): F

***Madhuca butyracea* Macr, related to *M. indica* J.F. Gmel.**

Sanskrit or Ayurvedic name(s): Madhuuka

Common name(s): Mahuaa

Family: Sapotaceae

Distribution: Habitat to sub-Himalayas, Kumaon to Bhutan.

Parts and properties used: Fat- Locally used as ointment in rheumatism and skin moisturizer during winter.

Source of information (system): F

***Mappia foetida* (Wight) Miers, syn. *Nothapodytes nimmoniana* (J. Graham) Mabb. (VU)**

Common name(s): Ghanera

Family: Icacinaceae

Distribution: Habitat to Western Ghats.

Parts and properties used: Plant/root- Antibacterial, source of camptothecin (CPT). Locally used in the treatment of cancer and HIV-I. Needs further study.

Source of information (system): Locals

***Maytenus emarginata* (Willd.) Ding Hou**

Common name(s): Kattangi

Family: Celastraceae

Distribution: Common in coast of Eastern Ghats and Karnataka.

Parts and properties used: Tender shoots, leaf- Cytotoxic, promising anti-tumour, and -mouth ulcers. Medicinally Promising.

Source of information (system): F

***Medicago lupulina* L.**

Common name(s): Bu-su-hang-pho, related to *M. sativa*.

Family: Leguminosae/ Fabaceae

Distribution: Native to alpine Himalayas and Ladakh.

Parts and properties used: Extract- Antibacterial. Locally used in lung disorder and cough.

Source of information (system): F

***Melastoma malabathricum* L. (rhododendron)**

Common name(s): Palore, Indian rhododendron, Nakkukappan

Family: Melastomataceae

Distribution: Found in many parts of India.

Parts and properties used: Leaf- Antidiarrheal, antiseptic. Locally used to treat dysentery, diarrhoea, haemorrhoids, cuts wounds, tooth, and stomach-ache. Flower- Astringent, and ant leucorrhoea.

Source of information (system): F

***Melia composita* Willd., syn. *M. azedarach* L.**

Sanskrit or Ayurvedic name(s): Arangaka

Common name(s): Hill Neem

Family: Meliaceae

Distribution: Spread wild from Assam to Deccan to Western Ghats.

Parts and properties used: Fruit- Anthelmintic. Used in skin diseases.

Source of information (system): A

***Merremia umbellata* (L.) Hallier f.**

Common name(s): Kolia lota, Motia. Jalpa substitute.

Family: Convolvulaceae

Distribution: Found in wild Assam to Kerala.

Parts and properties used: Root- Antibacterial, cytotoxic, antioxidant, anti-inflammatory. Assessed medicinally potential.

Source of information (system): F

***Micromeria biflora* (Buch. -Ham. Ex D. Don) Benth.**

Common name(s): Indian Wild Thyme

Family: Lamiaceae/ Labiatae

Distribution: Habitat wild to Himalayas, 1000-3000 m.

Parts and properties used: Root- Locally used in toothache. Plant- Used in nose bleeding, and as poultice. Have medicinal potential.

Source of information (system): F

Monochoria hastata **(L.) Solms**

Common name(s): Karinkuvalu, Launkia

Family: Pontederiaceae

Distribution: Tropical aquatic herb of South America and India.

Parts and properties used: Herb- Cooling. Locally used as alterative, tonic.

Source of information (system): F

Morina longifolia **Wall ex Candolle**

Common name(s): Kim

Family: Dipsacaceae, Caprifoliaceae

Distribution: Habitat wild to Himalayas and Ladakh.

Parts and properties used: Herb- Digestive, emetic, and stomachic. Used in stomach disorders and worm infestation.

Source of information (system): FL (Tibetan)

Mucuna monosperma **Wight**

Sanskrit or Ayurvedic name(s): Khatavangi

Common name(s): Kaagadolia, Periyattalargai

Family: Leguminosae/ Fabaceae

Distribution: Found in evergreen forest of northeast and Andamans.

Parts and properties used: Seeds- Sedative, restorative, expectorant, have been used in coughs and asthma.

Source of information (system): AFS

Mukia maderaspatana **(L.) M. Roem. [=** ***Melothria maderaspatana*** **(L.) Cogn.]**

Common name(s): Mucukkai

Family: Cucurbitaceae

Distribution: Paleotropical on hedges in most parts of India.

Parts and properties used: Plant extract- Antidiabetic, antibacterial, antioxidant, larvicidal, antiulcerogenic, hypolipidemic, antihypertensive, immunomodulatory.

Source of information (system): AFS

Myricaria squamosa **Desv.**

Common name(s): Umbo

Family: Tamaricaceae

Distribution: Habitat to west Himalayas and Ladakh.

Parts and properties used: Plant- Antitussive and febrifuge, localize poison, ripens pimples, dries up serous fluids. Locally used in inflammation. Assessed medicinally potential.

Source of information (system): F

Myxopyrum serratulum **A.W. Hill**

Common name(s): Chathuravalli

Family: Oleaceae

Distribution: Native to Western Ghats.

Parts and properties used: Leaves- Locally used in cough, asthma, chest diseases, nervous complaints, and rheumatism. Oil- For massage in fever and ache.

Source of information (system): F

Naregamia alata **Wight & Arn.**

Common name(s): Nilanaaringa

Family: Meliaceae

Distribution: Habitat to Western Ghats up to 1000 m.

Parts and properties used: Root- Emetic, cholagogue, expectorant, antidysentery. Plant- Antirheumatic. Leaf and stem- Antibilious.

Source of information (system): F

Naringi crenulata **(Roxb.) Nicolson**

Ayurvedic and Common name(s): Kadunimba Okarikavela, Narinarakam

Family: Rutaceae

Distribution: Indomalaya, found in Western Ghats.

Parts and properties used: Root extract- Locally used for vomiting, dysentery, and colic disorders.

Source of information (system): AF

Nepeta coerulescens **Maxim.**

Common name(s): Wild Kharu,

Family: Lamiaceae/ Labiatae

Distribution: Habitat to trans-/alpine Himalaya and Ladakh.

Parts and properties used: Plant- Locally used in dysentery.

Source of information (system): F

***Nepeta discolor* Royle ex Benth.**

Common name(s): Shamlolo, grown as ornamental.

Family: Lamiaceae/ Labiatae

Distribution: Habitat to trans-/alpine Himalaya and Ladakh.

Parts and properties used: Plant- Locally used in cold and cough.

Source of information (system): F

***Nepeta erecta* (Royle ex Benth.) Benth.**

Common name(s): Eripantso wild

Family: Liliaceae

Distribution: Habitat to trans-/alpine Himalaya and Ladakh.

Parts and properties used: Plant- Locally used in dysentery.

Source of information (system): F

***Nepeta eriostachya* Benth.**

Common name(s): Zimathikle wild

Family: Lamiaceae/ Labiatae

Distribution: Habitat to trans-/alpine Himalaya and Ladakh.

Parts and properties used: Plant- Locally used in eye complaints.

Source of information (system): F

***Nepeta floccosa* Benth.**

Common name(s): Shangukaram, wild

Family: Lamiaceae/ Labiatae

Distribution: Habitat to trans-/alpine Himalaya and Ladakh.

Parts and properties used: Plant- Locally used in cold and cough.

Source of information (system): F

***Nepeta glutinosa* Benth.**

Common name(s): Jatukpa

Family: Lamiaceae/ Labiatae

Distribution: Habitat to trans-/alpine Himalaya and Ladakh.

Parts and properties used: Plant- Locally used in diarrhoea.

Source of information (system): F

***Nepeta laevigata* (D. Don) Hand. -Mazz.**

Common name(s): Jatukpa

Family: Lamiaceae/ Labiatae

Distribution: Habitat to trans-/alpine Himalaya and Ladakh.

Parts and properties used: Plant- Locally used in pneumonia.

Source of information (system): F

***Nepeta longibracteata* Benth.**

Common name(s): Prainku /Chhagnamgo, wild

Family: Lamiaceae/ Labiatae

Distribution: Habitat to trans-/alpine Himalaya and Ladakh.

Parts and properties used: Plant- Locally used in stomach-ache.

Source of information (system): F

***Nilgirianthus heyneanus* (Nees) Bremek, syn. *Strobilanthes heyneanus* Nees**

Common name(s): Substitute of Kurinji, *N. ciliates*

Family: Acanthaceae

Distribution: Habitat wild to evergreen forests of Western Ghats.

Parts and properties used: Herb- Anti-inflammatory, antiarthritic, antimicrobial. Widely used as substitute to Saha Chara, *N. ciliates*, equated to *Barleria prionitis*. Ayurvedic *Vatahara* drug (eliminating wind).

Source of information (system): F

***Ochna obtusata* DC, (wild relative of *O. jabotapita* Linn., *O squarrosa* Linn.)**

Common name(s): Chilanti, Kanaka Champaa

Family: Ochnaceae

Distribution: Found in wild Assam, Odisha.

Parts and properties used: Bark- Yields digestive tonic. Root- Anti-asthmatic/ anti-bronchitis; decoction used in asthma, tuberculosis, and menstrual disorders.

Source of information (system): FS

***Olea dioica* Roxb., (wild relative of *O. europaea*)**

Common name(s): Koli, Itala, Edala, Rose Sandal Wood

Family: Oleaceae

Distribution: Found wild in Kerala.

Parts and properties used: Leaves, bark, root, fruits- Used to cure skin disease, rheumatism, fever, and cancer.

Source of information (system): FS

***Ophioglossum vulgatum* Linn., (fern)**

Common name(s): Adder's-tongue

Family: Ophioglossaceae

Distribution: Grasslands and semi-evergreen forests of north Himalayas & hills of India.

Parts and properties used: Fern- Antiseptic, styptic, vulnerary, detergent, emetic. Yields an ointment. Have medicinal potential.

Source of information (system): F

***Orchis mascula* (L.) L. (orchid)**

Sanskrit or Ayurvedic name(s): Salabmisri

Common name(s): Salam misri, Salabmisri

Family: Orchidaceae

Distribution: Native to UK, imported from UK and cultivated in India.

Parts and properties used: Plant- Antioxidant, astringent, demulcent, expectorant, nutritive. Used as aphrodisiac and a nervine tonic.

Source of information (system): AFSU

***Orobanche hansii* A.Kern.**

Common name(s): sGro-shangprtse

Family: Orobanchaceae

Distribution: Parasitic herb of alpine Himalayas 3300-3700 m.

Parts and properties used: Herbal extract- Locally used to treat boils. Have medicinal potential.

Source of information (system): F

***Oxyria digyna* (L.) Hill**

Common name(s): Mountain sorrel

Family: Polygonaceae

Distribution: Habitat to Himalayas 3000-6000 m, and Ladakh.

Parts and properties used: Plant- Refrigerant, antiscorbutic digestive, purgative.

Source of information (system): F

***Oxytropis chiliophylla* Royle ex Benth.**

Common name(s): King of Herbs, "Er-Da-Xia" (Chinese Pharmacopoeia)

Family: Leguminosae/ Fabaceae.

Distribution: Occur in Ladakh, and northwest India.

Parts and properties used: Plant- Cytotoxic, dimeric chalcones. Reported to have medicinal potential.

Source of information (system): L (Amchi), Tibetan medicine system

***Oxytropis lapponica* (Wahl.), Gay**

Common name(s): Rushkum

Family: Leguminosae/ Fabaceae

Distribution: Habitat wild to Himalayas.

Parts and properties used: Plant- Antiseptic. Have medicinal potential.

Source of information (system): F

***Oxytropis microphylla* (Pall.) DC.**

Common name(s): sTag-sha nagpo

Family: Leguminosae/ Fabaceae

Distribution: Habitat wild to alpine Himalayas, and Ladakh

Parts and properties used: Plant- Antiseptic. Strengthen skin and blood clotting. Have medicinal potential.

Source of information (system): F

***Pachygone ovata* (Poir.) Miers ex Hook. f. & Thomson**

Common name(s): Kadukkodi, Pedda Dusar

Family: Menispermaceae

Distribution: Native to south India.

Parts and properties used: Leaves- Locally found pain and inflammation reliever due to high alkaloid. Have medicinal potential.

Source of information (system): F

***Papaver nudicaule* L.**

Common name(s): Orange Iceland Poppy

Family: Papaveraceae

Distribution: Distributed in Himalayas, 3000-5000 m & northeast hills. Grown.

Parts and properties used: Leaves- Source of vitamin C. Capsule- Antioxidant analgesic, with toxic alkaloids. Medicinally equivalent to other *Papaver* spp.

Source of information (system): F

Parmelia kamstchadalis **Ach.**

Common name(s): Charelaa, Dagad Phool substitute

Family: Parmeliaceae

Distribution: Himalayas and Kashmir Hills.

Parts and properties used: Whole plant- Locally found useful in biliousness.

Source of information (system): F

Parmelia perforata **Ach., American lichen, syn.** ***Parmotrema perforatum*** **(Jacq.) A. Massal.**

Sanskrit or Ayurvedic name(s): Shilapushp, Shailaya in Samhita

Common name(s): Stone Flowers Charelaa, Dagad phool substitute.

Family: Parmeliaceae

Distribution: Common in the eastern USA. Grown in Eastern Himalayas.

Parts and properties used: Stone flowers are an important ingredient in herbal compositions for seminal weakness and male sexual debility. Yields drug- *Chharila.*

Source of information (system): AU

Parthenocissus himalayana **(Royle) Planch.**

Common name(s): Laderi, Philunaa

Family: Vitaceae

Distribution: Endemic to wild of Himalayas.

Parts and properties used: Bark and twigs- Astringent and expectorant. Assessed medicinally potential.

Source of information (system): F

Passiflora incarnata **L.**

Common name(s): Passionflower

Family: Passifloraceae

Distribution: American. Introduced, now grown in Western Ghats.

Parts and properties used: Stem, flower, fruit- Sedative. Assessed medicinally potential, it suppresses anxiety and insomnia, induce sleep.

Source of information (system): H

***Pedicularis brevifolia* D. Don., (root parasite)**

Common name(s): Luramenthag

Family: Scrophulariaceae

Distribution: Found wild in western to central Himalayas.

Parts and properties used: Plant- Locally used as sedative. Have medicinal potential.

Source of information (system): F

***Pedicularis cheilanthifolia* Schrenk**

Common name(s): Lug-ru-karpo

Family: Scrophulariaceae

Distribution: Habitat wild to west Himalayas, and Ladakh.

Parts and properties used: Leaves- Locally used in stomach disorder. Have medicinal potential.

Source of information (system): F

***Pedicularis longiflora* ssp. *tubiformis* (Klotzsch) Pennell.**

Common name(s): Lug-ru-karpo

Family: Scrophulariaceae

Distribution: Habitat wild to west Himalayas and Ladakh.

Parts and properties used: Tuber- Diuretic, febrifuge. Locally used to treat inflammation of liver, gallbladder, and seminal/vaginal discharges.

Source of information (system): F

***Pedicularis mollis* Wall ex Benth.**

Common name(s): Lug-ru-karpo

Family: Scrophulariaceae

Distribution: Habitat wild to west Himalayas, and Ladakh.

Parts and properties used: Herb- Antiseptic, febrifuge. Locally used to treat fevers, leucorrhoea, rheumatism, and help maintain vitality. Assessed medicinally potential.

Source of information (system): F

***Pentanema indicum* (L.) Ling**

Common name(s): Aggikoora chettu

Family: Asteraceae/ Compositae

Distribution: Found in Eastern Ghats, Odisha, and Kerala.

Parts and properties used: Leaf juice- Locally used in abdominal pain. Root decoction- Treat insect stings, and fever.

Source of information (system): F

Pentatropis spiralis **Decne, syn.** ***P. cynanchoides*** **R. Br.**

Common name(s): Ambarvel (related to Kaakanaasa, *Martynia annua*)

Family: Asclepiadaceae

Distribution: A climber habitat to northwest India.

Parts and properties used: Root- Astringent, antigonorrhoeic, alterative. Plant- Emetic, purgative.

Source of information (system): F

Pericampylus glaucus **(Lam.) Merr.**

Common name(s): Barakkanta

Family: Menispermaceae

Distribution: Found in central India and Assam.

Parts and properties used: Leaf infusion- Locally used in asthma, fever. Sap- Produce an eye drop.

Source of information (system): F

Periploca calophylla **Falc.**

Common name(s): Krishna Saraivaequal

Family: Apocynaceae

Distribution: Distributed in Himalayas, Meghalaya, and Assam.

Parts and properties used: Plant- Used in rheumatism, urinary and skin diseases, in place of *Cryptolepis buchanani* and *Hemidesmus indicus*. Needs further investigations.

Source of information (system): A

Perovskia abrotanoides **Kar., syn.** ***Salvia abrotanoides*** **(Kar.) Sytsma**

Common name(s): Russian Sage, Skiling, wild lavender , Brazemble (Persian).

Family: Lamiaceae/ Labiatae

Distribution: Habitat wild to alpine Himalayas and Ladakh.

Parts and properties used: Plant- Locally used as fortifier, antiseptic, anti-inflammatory, anti-rheumatic pains, leishmaniasis, anthelmintic and laxative.

Source of information (system): F

Peucedanum grande **(Dalzell & A. Gibson) C.B. Clarke**

Common name(s): Baphli, Duku, wild carrot.

Family: Apiaceae/ Umbelliferae

Distribution: Found in Western Ghats, peninsular Indian hills.

Parts and properties used: Fruit- Carminative, diuretic, stimulant. Infusion is locally used in gastric intestinal trouble.

Source of information (system): F

Phoenix pusilla **Gaertn. =** ***P. farinifera*** **Roxb. (LC)**

Sanskrit or Ayurvedic name(s): Parushaka

Common name(s): Chitteenth, Kalangu

Family: Palmae/ Arecaceae

Distribution: Habitat to South Indian coastal regions.

Parts and properties used: Plant extract/fruit- Cooling, used in fever, cardiac debility, peptic ulcer, weakness, and Pitta, Vata dosh. Gum- Treat diarrhoea, genitourinary. Sap- Laxative.

Source of information (system): AFS

Pholidota articulata **Lindl.**

Common name(s): Substitute of Jivanti, *Leptadenia reticulata*

Family: Orchidaceae.

Distribution: Found in Himalayas from Uttarakhand to Assam.

Parts and properties used: Herb- Stimulant, produce age-sustaining restorative tonic. Used in healing fractures.

Source of information (system): A

Phyllanthus debilis **Klein ex Willd.**

Common name(s): Bhumiamla

Family: Phyllanthaceae

Distribution: Found in plains & moist places, Maharashtra, TN.

Parts and properties used: Plant decoction- Locally used in liver disorder, diabetes mellitus and skin diseases.

Source of information (system): F

Phyllanthus fraternus **G.L. Webster.**

Sanskrit or Ayurvedic name(s): Bhumyamalaki

Common name(s): Bhumiamla

Family: Phyllanthaceae

Distribution: Occur all over India.

Parts and properties used: Plant extract- Diuretic, laxative given in dysentery and to treat gonorrhoea and malaria.

Source of information (system): AF

***Phyllostachys bambusoides* Sieb. & Zucc (bamboo)**

Common name(s): Tulatipati, Tankari

Family: Solanaceae

Distribution: Chinese. Occur in Sikkim.

Parts and properties used: Exudate- Locally used to treat infection. Assessed medicinally potential.

Source of information (system): F

***Physalis divaricata* L.**

Common name(s): Chirpoti

Family: Solanaceae

Distribution: Weedy annual to western & central Himalayas.

Parts and properties used: Plant- Diuretic. Have medicinal potential.

Source of information (system): F

***Physochlaina praealta* (Decne.) Miers, syn. *P. dubia* (VU)**

Common name(s): Bharangi, Kadavi, Langthang

Family: Solanaceae

Distribution: Habitat to slopes of Himalayas 3300-4600 m and Ladakh.

Parts and properties used: Plant- Anthelmintic, narcotic, and source of atropine. Needs further investigations.

Source of information (system): F

***Picrorhiza scrophulariiflora* Pennell.**

Common name(s): Kutki, wild relative of *P. kurroa.*

Family: Scrophulariaceae

Distribution: Found in southwestern China to east Himalayas & UP.

Parts and properties used: Root- Used as substitute for *P. kurroa*. Needs more investigations.

Source of information (system): A

***Pimpinella tenera* (DC.) Benth. ex C.B. Clarke**

Common name(s): Aromatic herb

Family: Apiaceae/ Umbelliferae

Distribution: Habitat to alpine Himalayas and Ladakh.

Parts and properties used: Root, plant- Carminative, diuretic. Have medicinal potential.

Source of information (system): F

***Pimpinella saxifraga* Linn. var. *dissectifolia* C.B. Clarke, non-Boiss. (ecotype)**

Common name(s): Burnet-saxifrage

Family: Apiaceae/ Umbelliferae

Distribution: Native to Europe to wild habitat of Kashmir at 3900 m.

Parts and properties used: Root- Carminative, stimulant, expectorant, cholagogue, diuretic, emmenagogue. Used in catarrhs of the upper respiratory tract and diarrhoea.

Source of information (system): English

***Pimpinella tirupatiensis* Balakr. and Subram. (VU- EN)**

Common name(s): Aromatic herb Konda Kottimeera

Family: Apiaceae/ Umbelliferae

Distribution: Habitat to Seshachalam Hills of Eastern Ghats.

Parts and properties used: Plant, root paste- Aphrodisiac, stomach-ache. Locally used to treat sexual debility, and ulcers.

Source of information (system): F

***Pinus khasya* Royle., syn. *P. insularis* Endl.**

Sanskrit or Ayurvedic name(s): Sarala variety

Common name(s): Digsaa

Family: Pinaceae

Distribution: Native to Assam introduced to Bengal hills.

Parts and properties used: Plant- Spasmolytic, antimicrobial. Oleo resin- Considered superior to that of *P. roxburghii* for turpentine.

Source of information (system): AF

***Piper mullesua* Buch. -Ham. ex D. Don. (VU), syn. *P. brachystachyum* C. DC.**

Sanskrit or Ayurvedic name(s): Kattukurumulagu

Family: Piperaceae

Distribution: Evergreen and shola forests of peninsular and northeast India.

Parts and properties used: Boiled fruit juice- Used to treat coughs and colds.

Source of information (system): A

***Piper peepuloides* Roxb.**

Common name(s): Pippali. Wild relative and a substitute of long pepper

Family: Piperaceae

Distribution: China, Assam, northeast India, and Myanmar.

Parts and properties used: Seed powder- Used to treat cough. Extract- Anti-inflammatory, antioxidant, and anticancer.

Source of information (system): AF

***Piper sylvaticum* Roxb.**

Sanskrit or Ayurvedic name(s): Vana-Pippali

Common name(s): Pahaari Pipali. Long pepper wild relative.

Family: Piperaceae

Distribution: Found in wet places of China and northeast India.

Parts and properties used: Fruit- Carminative. Arial part- Diuretic.

Source of information (system): AF

***Piper wallichii* (Miq.) Han. -Mazz.**

Common name(s): Renukbeej, Nagodbeej. Pepper wild relative

Family: Piperaceae

Distribution: From China to northeast Hills of India.

Parts and properties used: Fruit- Uterine stimulant. Used in arthritis, inflammatory diseases, cerebral infarction, and angina.

Source of information (system): AF

***Plantago depressa* Willd.**

Common name(s): Tha-ram

Family: Plantaginaceae

Distribution: Habitat wild to alpine Himalayas and Ladakh.

Parts and properties used: Seed husk- Anti-diarrheal, anthelmintic, stimulant. Assessed medicinally potential.

Source of information (system): FU

Plectranthus barbatus **Andrews [=** ***Coleus forskohlii*** **(Poir.) Briq.], syn.** ***C. barbatus*** **(EN)**

Sanskrit or Ayurvedic name(s): Gandhamulika

Common name(s): Gandhira

Family: Lamiaceae/ Labiatae

Distribution: Grasslands and dry deciduous forests of Western Ghats.

Parts and properties used: Tuberous roots- Source of Forskolin (coleonol), drug used in hypertension, congestive heart failure, eczema, colic, respiratory disorders, painful urination, insomnia, and convulsions.

Source of information (system): AF

Plectranthus mollis **Spreng., syn.** ***P. incanus*** **Link**

Common name(s): Laal-Aghaadaa Related to ornamental spp.

Family: Lamiaceae/ Labiatae

Distribution: Habitat wild to alpine Himalayas to Western Ghats.

Parts and properties used: Leaves- Styptic, febrifuge. Assessed medicinally potential.

Source of information (system): F

Plectranthus vettiveroides **(Jacob) N.P. Singh & B.D. Sharma**

Sanskrit or Ayurvedic name(s): Hriberam, Hribera

Common name(s): Vettiver

Family: Lamiaceae/ Labiatae

Distribution: Endemic to south Indian plains. Grown too.

Parts and properties used: Root- Aromatic. Essential oil- Used in 75 different Ayurvedic formulations.

Source of information (system): AS

Pogostemon cablin **(Blanco) Benth.**

Sanskrit or Ayurvedic name(s): Patchouli

Family: Lamiaceae/ Labiatae

Distribution: Widely grown in India.

Parts and properties used: Leaves- Source of patchouli oil. Used in soaps, perfumes and as medicine.

Source of information (system): A

Polygonum hydropiper **L.**

Common name(s): Paakur-muula, ornamental *Polygonum* spp. relative.

Family: Polygonaceae

Distribution: Found throughout wet India.

Parts and properties used: Herb, leaves- Haemostatic, anti-inflammatory, astringent, carminative, diaphoretic, diuretic, emmenagogue, stomachic, styptic.

Source of information (system): F

***Polygonum punctatum* Buch. -Ham. ex D. Don, also var. *alatum*?, syn. *Persicaria nepalensis* (Meisn) H. Gross**

Common name(s): Kangani-machan-pillu, ornamental spp. relative

Family: Polygonaceae

Distribution: Weed of Indian Himalayas to Assam.

Parts and properties used: Plant- Used to treat skin infections, dysentery, haemorrhoids, insomnia, and heart diseases. Leaves- Cure swelling.

Source of information (system): FH

***Portulaca tuberosa* Roxb.**

Sanskrit or Ayurvedic name(s): Bichhuu-buuti

Common name(s): Sanjivani, Jangali Gaajar, related to purslane.

Family: Portulacaceae

Distribution: Habitat wild to peninsular and coastal India. Cultivated.

Parts and properties used: Herb- Diuretic, analgesic, cellulolytic, antipyretic. Leaf's infusion- Given in dysuria, erysipelas.

Source of information (system): AF

***Potentilla atrosanguinea* G. Lodd. ex D. Don**

Common name(s): Chisheing, Himalayan Cinquefoil

Family: Rosaceae

Distribution: Habitat wild to alpine Himalaya and Ladakh.

Parts and properties used: Flower and leaf- Locally given in fever. Assessed medicinally potential.

Source of information (system): F

***Potentilla fruticosa* L.**

Common name(s): Cinquefoil Khiangar

Family: Rosaceae

Distribution: Habitat to alpine Himalayas and Ladakh. Cultivated.

Parts and properties used: Leaves- Astringent. Root juice- Indigestion. Have medicinal potential.

Source of information (system): F

Potentilla multifida **L.**

Common name(s): Thakto

Family: Rosaceae

Distribution: Habitat wild to alpine Himalayas and Ladakh.

Parts and properties used: Flower and leaf- Locally used to cure sleeplessness. Assessed medicinally potential.

Source of information (system): F

Potentilla nepalensis **Hook.**

Common name(s): Substitute of Ratanjot, Dori ghaas

Family: Rosaceae

Distribution: Habitat to west and central Himalayas.

Parts and properties used: Rootstock- Depurative. Locally applied on burns. Assessed medicinally potential.

Source of information (system): F

Premna flavescens **Ham.**

Common name(s): Arnimool, substitute of Gambhari.

Family: Lamiaceae/ Labiatae

Distribution: Sikkim.

Parts and properties used: Plant- Locally used to cure skin diseases.

Source of information (system): F

Premna latifolia **Roxb.**

Sanskrit or Ayurvedic name(s): Agnimantha, Ganikarnika

Common name(s): Bakar

Family: Lamiaceae/ Labiatae

Distribution: Found in Bihar, WB & peninsular India.

Parts and properties used: Leaves- Diuretic, spasmolytic. Bark- Hypoglycaemic.

Source of information (system): AFS

***Premna tomentosa* Willd.**

Sanskrit or Ayurvedic name(s): Agnimantha

Common name(s): Gineri, Sonachal, Chambara, Moria, Bastard Teak

Family: Lamiaceae/ Labiatae

Distribution: Found in Bihar, Odisha & peninsular India.

Parts and properties used: Bark/root essential oil- Used in stomach disorder. Leaves-Diuretic, vulnerary.

Source of information (system): AFS

***Primula microphylla* (Hook. f.) Kuntze.**

Common name(s): Relative of ornamental Primrose sp.

Family: Primulaceae

Distribution: Habitat wild to Himalayas, and Ladakh.

Parts and properties used: Herb- Locally used in cough.

Source of information (system): F

***Primula rosea* Royle.**

Common name(s): Relative of ornamental Primrose sp.

Family: Primulaceae

Distribution: Habitat wild to Himalayas 2700-4000 m and Ladakh.

Parts and properties used: Plant- Locally used in muscular pains. Assessed medicinally potential.

Source of information (system): F

***Prosopis juliflora* DC., syn. *P chilensis* Stuntz.**

Common name(s): Vilaayati Kikar

Family: Leguminosae/ Fabaceae (Mimosaceae)

Distribution: American. An invasive species adapted to India.

Parts and properties used: Plant- Antimicrobial. Produces inferior gum, Juliflorinine and other alkaloids. Assessed medicinally potential.

Source of information (system): F

***Prosopis stephaniana* Kunth.**

Sanskrit or Ayurvedic name(s): Samudra-shami,

Common name(s): Khejaraa. Shami variety

Family: Leguminosae/ Fabaceae (Mimosaceae)

Distribution: Occur in parts of Punjab and Gujarat.

Parts and properties used: Pods and roots- Astringent, styptic, antidysentery.

Source of information (system): AF

***Prunus amygdalus* Btsch var. *amara*, syn. *Prunus dulcis* var. *amara* (DC.), Buchheim**

Common name(s): Bitter almond, Badam,

Family: Rosaceae

Distribution: Cultivated in Kashmir, HP, Uttarakhand.

Parts and properties used: Unripe fruit- Astringent, treat gum. Kernels- Nutritious, demulcent & stimulant. Gives nervine tonic. Oil- Nutritive, demulcent, laxative.

Source of information (system): AFH

***Prunus cerasoides* D. Don**

Sanskrit or Ayurvedic name(s): Padmaka

Common name(s): Padamkasht, Wild cherry

Family: Rosaceae

Distribution: Habitat to alpine Himalayas. Also, to Ooty Hills.

Parts and properties used: Stem- Refrigerant, antipyretic. Kernel- Antilithic.

Source of information (system): AL

***Prunus cerasus* Linn.**

Sanskrit or Ayurvedic name(s): Elavaaluka

Common name(s): Sour Cherry

Family: Rosaceae

Distribution: Origin Eurasian. Cultivated in Kashmir, HP, Kumaon.

Parts and properties used: Fruit- Diuretic, ant inflammatory. Used in genitourinary, cystitis, inflammations, urine retention Bark- Febrifuge, antidiarrheal. Heartwood- Used in skin problems.

Source of information (system): AF

***Quercus incana* Roxb.**

Common name(s): Shilaa Supaari, Phanat

Family: Fagaceae

Distribution: Habitat to cold humus places in Himalayas 1000-2400.

Parts and properties used: Fruit (acorn/nut)- Diuretic, astringent. Used in indigestion and diarrhoea.

Source of information (system): UF

***Ranunculus brotherusii* Freyn, Bull.**

Common name(s): Maoking, related to ornamental Buttercup sp.

Family: Ranunculaceae

Distribution: Habitat wild to alpine Himalayas, Ladakh and Lahaul Spiti.

Parts and properties used: Plant decoction- Digestive. Locally used to treat ulcer. Assessed medicinally potential.

Source of information (system): F

***Ranunculus laetus* Wall. ex Hook. f. & Thoms.**

Common name(s): Related to ornamental Buttercup sp.

Family: Ranunculaceae

Distribution: Habitat to wild alpine Himalayas and Ladakh.

Parts and properties used: Plant- Stimulant, anti-inflammatory, analgesic. Assessed medicinally potential.

Source of information (system): F

***Ranunculus lobatus* Jacquem. ex Camb.**

Common name(s): Related to ornamental Buttercup sp.

Family: Ranunculaceae

Distribution: Habitat wild to alpine Himalayas and Ladakh.

Parts and properties used: Plant- Locally used in gum inflammation and toothache. Assessed medicinally potential.

Source of information (system): F

***Ranunculus trichophyllus* Chaix**

Sanskrit or Ayurvedic name(s): Kaandira

Common name(s): Rengo

Family: Ranunculaceae

Distribution: Habitat wild to alpine Himalayas and Ladakh.

Parts and properties used: Herb- Locally used in fevers, rheumatism, asthma, diarrhoea. Assessed medicinally potential.

Source of information (system): F

***Ranunculus tricuspis* Maxim.**

Common name(s): Chu-rug-sbal-lag

Family: Ranunculaceae

Distribution: Occur in J & K, Ladakh and Bhutan.

Parts and properties used: Herb- Locally used to cure heat, tendons ligament, eye inflammation. Assessed medicinally potential.

Source of information (system): F

***Rauvolfia densiflora* (Wall.) Benth. ex Hook.f.**

Common name(s): Sarpagandha, Pagal Buti

Family: Apocynaceae

Distribution: Habitat wild in Western Ghats.

Parts and properties used: Root- Locally used to treat malaria and typhus.

Source of information (system): F

***Rauvolfia micrantha* Hook. f (population degraded)**

Common name(s): Sarpagandha, Pagal Buti, *R. serpentina* substitute

Family: Apocynaceae

Distribution: Distributed wild in Southwestern Ghats, TN.

Parts and properties used: Root- Locally used to treat nervous disorders.

Source of information (system): F

***Rauvolfia tetraphylla* L.**

Common name(s): Barachandrika

Family: Apocynaceae

Distribution: West Indian. Naturalized to Assam, WB, Bihar, and western peninsular India.

Parts and properties used: Root- Sedative, hypotensive, expectorant cathartic. Locally used to treat dropsy and eyes. Plant juice + castor oil- Applied to skin problem.

Source of information (system): F

***Rhamnus virgatus* Roxb.**

Common name(s): Tadru, Chadolaa

Family: Rhamnaceae

Distribution: Found throughout Himalayas & hills of India.

Parts and properties used: Ripe fruit- Purgative, emetic.

Source of information (system): F

***Rhizophora mucronata* Lam.**

Common name(s): Bhara Ganjan

Family: Rhizophoraceae

Distribution: Found in mangrove forests of Bengal to Kerala, India.

Parts and properties used: Plant- Used as an astringent and to treat angina, haemorrhaging. Seedling extracts- Used in diarrhoea, diabetes, hepatitis, inflammation, wounds, and ulcers.

Source of information (system): FS

***Rhodiola crenulata* (Hook. f. & Thomson) H. Ohba.**

Common name(s): Golden arctic root, Shrolo; *R. rosea* relative.

Family: Crassulaceae

Distribution: Wild in sub-alpine Himalayas Ladakh, HP Lahul Spiti.

Parts and properties used: Root- Adaptogens, anti-stress, memory restorer. Produce a tonic.

Source of information (system): F

***Rhodiola heterodonta* (Hook. f., & Thomson) Boriss.**

Common name(s): Golden arctic root, Shrolo; *R. rosea* relative.

Family: Crassulaceae

Distribution: Wild in alpine Himalayas 3300-5000 m and Ladakh

Parts and properties used: Root- Adaptogens, anti-stress, memory restorer. Produce a tonic.

Source of information (system): F

***Rhodiola imbricata* Edgew.**

Common name(s): Golden arctic root, Shrolo; *R. rosea* relative

Family: Crassulaceae

Distribution: Habitat to Trans- Himalayas, and Ladakh.

Parts and properties used: Plant- Locally known for nutritional and medicinal value. Root- Adaptogens, anti-stress, memory restorer. Promising radioprotector.

Source of information (system): F

***Rhodiola tibetica* (Hook. f. & Thomson) S.H. Fu.**

Common name(s): Golden arctic root, Shrolo; *R. rosea* relative.

Family: Crassulaceae

Distribution: Wild to sub-alpine Himalayas and Ladakh, HP.

Parts and properties used: Rhizome- Adaptogens, anti-stress, memory restorer. Produce a tonic.

Source of information (system): F

***Rhododendron campanulatum* D. Don.**

Common name(s): Cherailu

Family: Ericaceae

Distribution: Habitat to Himalayas 2400-5200 m.

Parts and properties used: Leaf- Locally used in rheumatism, sciatica, as snuff in cold, and hemicrania. Assessed medicinally potential.

Source of information (system): F

***Rhododendron cinnabarinum* Hook. f.**

Common name(s): Balu, Sanu

Family: Ericaceae

Distribution: Habitat wild to eastern Himalaya.

Parts and properties used: Plant- Vasodepressor. Assessed medicinally potential.

Source of information (system): F

***Rhododendron lepidotum* Wall.**

Common name(s): Talisapatra (*Abis webbiana*) substitute.

Family: Ericaceae

Distribution: Habitat wild to eastern Himalaya.

Parts and properties used: Leaf juice- Blood purifier. Assessed medicinally potential.

Source of information (system): F

***Rhus chinensis* Mill., syn. *R. javanica* Linn. *R. semialata* Murr. (VU)**

Common name(s): Tatri, Arkhar

Family: Anacardiaceae

Distribution: Habitat to alpine Himalayas 1300-2400 m.

Parts and properties used: Galls- Astringent, expectorant. Used in ointments of haemorrhoids. Fruits- Spasmolytic. Locally used in colic, diarrhoea, and dysentery.

Source of information (system): F

Rhynchosia minima **(L.) DC.**

Common name(s): Raan-ghevaraa

Family: Leguminosae/ Fabaceae

Distribution: Occur wild all over plains.

Parts and properties used: Leaves- Abortifacient. Seeds- Bitter and toxic. Assessed medicinally potential

Source of information (system): F

Rhynchosia rufescens **(Willd.) DC**

Common name(s): Walmoyda

Family: Leguminosae/ Fabaceae

Distribution: Found wild in Assam, TN, Western Ghats.

Parts and properties used: Herb- Anti-inflammatory. Assessed medicinally potential.

Source of information (system): F

Rhynchosia suaveolens **(L. f.) DC.**

Common name(s): Karu Kandi

Family: Leguminosae/ Fabaceae

Distribution: Habitat wild to Eastern, Western Ghats.

Parts and properties used: Roots- Analgesic, heal wounds, locally used as tonic. Assessed medicinally potential.

Source of information (system): F

Rivea loatica **Ostster**

Common name(s): Phang, phangnta

Family: Convolvulaceae

Distribution: Habitat to Western Ghats.

Parts and properties used: Root extract + olive oil- Locally used in dysentery.

Source of information (system): Warli tribe

Rosa **x** *damascene* **Mill. (hybrid)**

Common name(s): Damask rose

Family: Rosaceae

Distribution: Cultivated.

Parts and properties used: Flowers- Fragrant, source of rose oil.

Source of information (system): Locals

***Rosa indica* L.**

Common name(s): Wild Gulab.

Family: Rosaceae

Distribution: Chinese, found in Assam to Western Ghats.

Parts and properties used: Flower- Locally used in blood purification, treatment of the intestinal ulcer, diarrhoea, and for anti-inflammatory action. Cosmetic.

Source of information (system): FS

***Roscoea alpina* Royle**

Common name(s): Kshirkakoli

Family: Zingiberaceae

Distribution: Native to Himalayas. Grown as ornamental.

Parts and properties used: Root powder- Locally used in cure impotency, diabetes, diarrhoea. Assessed medicinally potential.

Source of information (system): F

***Rubia manjith* Roxb. ex Fleming (VU)**

Common name(s): Indian Madder

Family: Rubiaceae

Distribution: Distributed in Indochina and Himalayan region.

Parts and properties used: Root- Alterative, anodyne, antiphlogistic, antitussive, astringent, diuretic, emmenagogue, expectorant, styptic, tonic, and vulnerary. Used to lower BP.

Source of information (system): F

***Rumex dentatus* Linn.**

Common name(s): Jangali Paalak

Family: Polygonaceae

Distribution: Habitat wild to Himalayas up to 300 m.

Parts and properties used: Plant- Astringent. Locally used in cutaneous disorders. Potentially Medicinal.

Source of information (system): F

***Rumex hastatus* D. Don., *R. dissectus* Leveille**

Common name(s): Arrow leaf Dock, Churki, Khatti Buti

Family: Polygonaceae

Distribution: American. Naturalized in north Indian hills 700-2500 m.

Parts and properties used: Plant juice- Astringent, locally used to treat dysentery, anti-diabetic. Tuber- Chewed to relieve aches and throat.

Source of information (system): F

***Rumex patientia* ssp. *tibeticus* (Rech. f.) Rech. f.**

Common name(s): Shoma

Family: Polygonaceae

Distribution: Occur wild in west Himalayas and Ladakh.

Parts and properties used: Root- Anthelmintic, purgative. Infusion- Poultice for skin disorder. Assessed medicinally potential.

Source of information (system): F

***Rumex nepalensis* Spreng.**

Common name(s): Kulli

Family: Polygonaceae

Distribution: Habitat to alpine Himalayas, Western Ghats (Nilgiris).

Parts and properties used: Root- Purgative, astringent, swollen gums. Substitute of Da Huang (*R. palmatum*). Have medicinal potential.

Source of information (system): F

***Salacia reticulata* Wight (EN)**

Common name(s): Anukudu-chettu, Pitila, Koranti

Family: Celastraceae

Distribution: Occur in Odisha, AP, Western Ghats.

Parts and properties used: Plant- Antiseptic. Root bark- Locally used in diabetes, rheumatism, gonorrhoea, and skin diseases.

Source of information (system): F

***Salix acmophylla* Boiss.**

Sanskrit name: Jala-vetasa

Common name(s): Wild relative of *S. alba*.

Family: Salicaceae

Distribution: Native to sub-Himalayas.

Parts and properties used: Bark- Locally used as febrifuge.

Source of information (system): F

Salvia moorcroftiana **Wall. *ex* Benth.**

Common name(s): Tuk marian, Tuth. *S. officinalis* wild relative

Family: Lamiaceae/ Labiatae

Distribution: Native to Kashmir to Kumaon 2000-3000 m.

Parts and properties used: Root- Bechic. Leaf- Antitussive; used in skincare. Seed- Antispasmodic, emetic. Locally used for colic, dysentery.

Source of information (system): F

Sapium indicum **Willd.**

Common name(s): Hurnaa, Pencolum

Family: Euphorbiaceae

Distribution: Habitat to moist parts of India, sea cost and back waters.

Parts and properties used: Root bark- Emetic, acrid, and locally used in purgative.

Source of information (system): FS

Saussurea affinis **Spreng. ex DC.**

Common name(s): Ganga-muula

Family: Asteraceae/ Compositae

Distribution: Found in Assam, foothills of East Himalayas.

Parts and properties used: Root juice- Locally used to cure gynaecological diseases.

Source of information (system): F

Saussurea bracteata **Decne. (R, Rare)**

Sanskrit or Ayurvedic name(s): Prerak Mul

Common name(s): Pseudo Brahmakamal, spanrtsa-do-bo.

Family: Asteraceae/ Compositae

Distribution: Habitat wild to Himalayas, and Ladakh.

Parts and properties used: Plant- Locally used to treat boils. Medicinally promising.

Source of information (system): F

Saussurea fastuosa **(Decne.) Sch. Bip.; syn. *S. forrestii* Diels**

Common name(s): Singamindro

Family: Asteraceae/ Compositae

Distribution: Distributed wild in Himalayas, and Sikkim.

Parts and properties used: Aerial part- Locally used in cut and bleeding.

Source of information (system): F

Saussurea gnaphalodes **Royle.**

Common name(s): Yuliang

Family: Asteraceae/ Compositae

Distribution: Wild to alpine west Himalayas and Ladakh.

Parts and properties used: Plant- Locally used in cough/ cold. Root- Used in kidney, arthritic problems.

Source of information (system): F

Saussurea hypoleuca **Spreng. ex DC.**

Sanskrit name: Kushtha

Common name(s): Substitute of Kuth (*S. costus*).

Family: Asteraceae/ Compositae

Distribution: Occur wild in Himalayas, Ladakh to Sikkim.

Parts and properties used: Plant, root- CNS depressant and hypothermic. Leaves- Locally used in treatment of syphilis.

Source of information (system): F

Saussurea jacea **C.B. Clarke**

Common name(s): Pashuk

Family: Asteraceae/ Compositae

Distribution: Wild in alpine Himalayas, Nubra Valley, Saichen and Ladakh.

Parts and properties used: Plant- Locally used as diuretic.

Source of information (system): F

Saussurea sacra Edgew.**, syn. *S. simpsoniana* (Field & Gard.) Lipsch. (EN)**

Common name(s): Phen Kamal, Jogipaadshaah Ghuggi

Family: Asteraceae/ Compositae

Distribution: Habitat wild to Himalayas 4000 m and above.

Parts and properties used: Plant- Locally used in nervous debility. Root- Used in gynaecological disorders.

Source of information (system): F

Saxifraga cernua **Linn.**

Common name(s): Bri-rta sa-dzin dman-pa

Family: Saxifragaceae

Distribution: Habitat to alpine Himalayas, Ladakh, HP, Uttarakhand.

Parts and properties used: Plant- Locally yields tonic. Used in kidney stones. Medicinally promising.

Source of information (system): F

Saxifraga jacquemontiana **Decne.**

Common name(s): Sasomantso

Family: Saxifragaceae

Distribution: Habitat to alpine Himalayas, Ladakh & HP.

Parts and properties used: Plant- Locally used as liver tonic, and in kidney stones. Assessed medicinally potential.

Source of information (system): F

Saxifraga pulvinaria **Harry Sm.**

Common name(s): Sum-cu-tig

Family: Saxifragaceae

Distribution: Wild to alpine Himalayas, Ladakh to Sikkim.

Parts and properties used: Plant- Locally used as liver tonic, and in kidney stones. Assessed medicinally potential.

Source of information (system): F

Saxifraga stenophylla **Royle**

Common name(s): Man-pa

Family: Saxifragaceae

Distribution: Wild to alpine Himalayas, Ladakh to Sikkim.

Parts and properties used: Plant- Locally used as blood purifier, and for removal of kidney stones. Assessed medicinally potential.

Source of information (system): F

Schima wallichii **(DC.) Korth., Choicy.**

Common name(s): Makria

Distribution: Theaceae

Distribution: Habitat to eastern Himalayas & hills of India.

Parts and properties used: Stem bark- Anthelmintic (worms), rubefacient. Aerial parts- Antifungal. Assessed medicinally potential.

Source of information (system): F

Scorzonera virgata **DC.**

Common name(s): Rtsa-mkhris, Thumbu

Family: Asteraceae/ Compositae

Distribution: Habitat to stony slops of Himalayas.

Parts and properties used: Herbal grass- Locally used in constipation. Assessed medicinally potential.

Source of information (system): Locals

Scrophularia dentata **Royle ex Benth.**

Common name(s): Hamchi

Family: Scrophulariaceae

Distribution: Habitat to alpine Himalayas, and Ladakh.

Parts and properties used: Plant- An appetizer, anti-inflammatory. Assessed medicinally potential.

Source of information (system): F

Scutellaria galericulata **Linn., syn. S** ***lateriflora*** **Linn.**

Common name(s): Skullcap, Scurvy grass.

Family: Lamiaceae/ Labiatae

Distribution: Habitat to Kashmir 1500-2400 m.

Parts and properties used: Herb- Sedative, diuretic, sleep promoter. β-Elemene present in the herb and had anti-cancer potential.

Source of information (system): F

Scutellaria prostrata **Jacq. ex Benth.**

Common name(s): Haunching

Family: Lamiaceae/ Labiatae

Distribution: Habitat to alpine Himalayas and Ladakh.

Parts and properties used: Root- Antimicrobial. Locally advised in jaundice.

Source of information (system): F

Senecio tibeticus **Hook. f.**

Common name(s): Niyangar

Family: Asteraceae/ Compositae

Distribution: Native to Tibet and Ladakh.

Parts and properties used: Herb- Diuretic. Assessed medicinally potential.

Source of information (system): F

***Sesbania cannabina* (Retz.) Poir.**

Common name(s): Yellow Pea Bush

Family: Leguminosae/ Fabaceae

Distribution: Occur all over India.

Parts and properties used: Leaves- Has aperient, diuretic, emetic, emmenagogue, febrifuge, laxative, properties. Yields tonic.

Source of information (system): F

***Sesbania speciosa* Taub., syn. *S. pubescens* DC. var. *grandiflora* (VU)**

Common name(s): Seemai agathi

Family: Leguminosae/ Fabaceae

Distribution: Grown in coast of south India.

Parts and properties used: Whole plant- Locally used in catarrh, headache, and epilepsy.

Source of information (system): F

***Shorea tumbuggaia* Roxb. (CR)**

Common name(s): Thamba Jalari

Family: Dipterocarpaceae

Distribution: Endemic to Eastern Ghats.

Parts and properties used: Leaf juice- Locally advised in duodenal ulcers, amoebic, dysentery, earache. Assessed medicinally potential.

Source of information (system): F

***Sibbaldia parviflora* Willd.**

Sanskrit name(s): Bajradanti.

Family: Rosaceae

Distribution: Native to Eurasia, naturalized to Garhwal area.

Parts and properties used: Herb- Locally used as a tooth powder.

Source of information (system): A

***Silene vulgaris* (Moench) Garcke.**

Common name(s): Bladder companion

Family: Caryophyllaceae

Distribution: Habitat to alpine Himalayas and Ladakh.

Parts and properties used: Plant juice- Locally used in skin disorder, ophthalmia. Assessed medicinally potential.

Source of information (system): F

Skimmia laureola **(DC.) Zucc. ex Walp.**

Common name(s): Ner, Patrang, Barru

Family: Rutaceae

Distribution: Common in west Himalayas.

Parts and properties used: Leaves- Treat stomachic disorders, smallpox. Assessed medicinally potential.

Source of information (system): F

Smilax chinensis **L.**

Common name(s): Chinaroot, Jin Gang Ten

Family: Liliaceae, Smilacaceae

Distribution: Native to East Asia, including India, climber in temperate biome.

Parts and properties used: Herb- Analgesic, treat various diseases, also anti-inflammatory and reduce body weight.

Source of information (system): Locals

Smilax ovalifolia **Roxb., syn.** ***S. macrophylla*** **Roxb.**

Common name(s): Maitri (Ushbaa, *Smilax medica* wild relative)

Family: Liliaceae

Distribution: Tropical forests of India.

Parts and properties used: Leaf/root infusion- Used in diarrhoea. Roots- Substitute of Anantha Moola (*Hemidesmus indicus*). Treat venereal diseases, urinary infections, rheumatism, dysentery.

Source of information (system): AFU

Solanum aculeatissimum **Jacq.**

Ayurvedic or Common name(s): Brihati (*S. indicum*) related sp.

Family: Solanaceae

Distribution: Habitat wild to damp areas of Assam, Kerala

Parts and properties used: Fruit/leaves- Contains glycoalkaloid solanine. Used like *S indicum*

Source of information (system): A

Solanum albicaule **Kotschy ex Dunal.**

Common name(s): Brihati (*S. indicum* related sp.), Narkanta

Family: Solanaceae

Distribution: Found wild in Saurashtra, Rajasthan.

Parts and properties used: Plant decoction- Locally advised in ulcer. Used like *S indicum*.

Source of information (system): AF

Solanum aviculare **Forst. f.**

Ayurvedic or Common name(s): Kantakaari related sp.

Family: Solanaceae

Distribution: Australian. Introduced to Kashmir.

Parts and properties used: Source of solasodine. Used like *S. xanthocarpum*.

Source of information (system): A

Solanum viarum **Dunal, syn.** ***S. khasianum*** **C.B. Clarke var.** ***Chatterjeeanum*** **Sengupta**

Common name(s): Khasi Kateri, Nightshade, Tropical soda apple, relative of *Solanum* spp.

Family: Solanaceae

Distribution: Native South America. Occur in Khasi Hills, and Western Ghats.

Parts and properties used: Berries- Source of Solasodine. The alkaloid used to manufacture of corticosteroid hormones. Assessed medicinally potential.

Source of information (system): F

Sophora moorcroftiana **(Benth.) Benth. ex Baker.**

Common name(s): Singtik

Family: Leguminosae/ Fabaceae

Distribution: Habitat to alpine Himalayas, Ladakh, Tibet.

Parts and properties used: Seeds- Diuretic and stomachic. Seed paste- Used in gastric trouble. Assessed medicinally potential.

Source of information (system): F

Sonchus arvensis **Linn.**

Ayurvedic and Common name(s): Sahadevi (Vigorous variety)

Family: Asteraceae/ Compositae

Distribution: Occur as weed all over India in waste places.

Parts and properties used: Plant- Sedative, hypnotic, anodyne, diuretic expectorant. Seed- Treat nervous debility, bronchitis, asthma.

Source of information (system): A

***Sonchus asper* Hill**

Sanskrit or Ayurvedic name(s): Sahadevi

Common name(s): Didhi

Family: Asteraceae/ Compositae

Distribution: Occur as weed all over India in waste places.

Parts and properties used: Herb- Emollient. Pounded and applied to wounds and boils.

Source of information (system): AF

***Sonchus oleraceus* Linn.**

Common name(s): Duudhi, Jangli buti

Family: Asteraceae/ Compositae

Distribution: Occur as weed all over India in waste places.

Parts and properties used: Herb- Galactagogue, febrifuge, sedative, vermifuge. Used to treat indigestion, and liver ailments.

Source of information (system): F

***Sonneratia caseolaris* (L.) Engl.**

Common name(s): Apple Mangrove

Family: Lythraceae

Distribution: Western Ghats.

Parts and properties used: Fruit juice- Help arrest haemorrhage. Old fruit wall- Vermifuge.

Source of information (system): Locals

***Spatholobus parviflorus* (DC.) Kuntze, syn. *S. roxburghii* Benth.**

Common name(s): Bando, Mula, Malini

Family: Leguminosae/ Fabaceae

Distribution: Indo-Malayan. Habitat to Western Ghats forests.

Parts and properties used: Bark decoction- Treat dropsy, worms, bowel movement.

Source of information (system): FS

Sphaeranthus amaranthoides **Burm.**

Common name(s): Cevayam, Sivakaranthai

Family: Asteraceae/ Compositae

Distribution: Endemic to south India.

Parts and properties used: Herb- Rejuvenator, antibiotic, anti-inflammatory, analgesic, antibacterial. Assessed medicinally potential.

Source of information (system): FS

Spilanthes paniculata **Wall. ex DC, syn.** ***S. calva*** **DC.;** ***S. pseudoacmella*** **auct. non (L.) Murr.**

Sanskrit or Ayurvedic name(s): Marahattikaa

Common name(s): Huhuni Sak, Bhringara,

Family: Asteraceae/ Compositae

Distribution: Found degraded forest of Karnataka and Kerala.

Parts and properties used: Flower heads- Locally used as local anaesthetic and in toothache, etc.

Source of information (system): F

Strophanthus wightianus **Wight.**

Common name(s): Nerivalli, Naithal kizhangu

Family: Apocynaceae

Distribution: Endemic to southern Western Ghats.

Parts and properties used: Root- Used to treat arteriosclerosis, heart, and high BP problems. Plant- Yield's glycosides. Have medicinal potential.

Source of information (system): F

Strychnos minor **Dennst. (DD)**

Sanskrit, Ayurvedic and Common name: Cherukanjiravally

Family: Loganiaceae

Distribution: Tropical Asia to Santa Cruz Islands in evergreen forests and sacred groves.

Parts and properties used: Wood, bark, and roots decoction-Emmenagogue, treat throat. Stem- Carminative, antipyretic and cure stomachic activity.

Source of information (system): AF

Swertia paniculata **Wall., syn.** ***Ophelia paniculata*** **(Wall.) D. Don**

Common name(s): Wild relative *S. chirayita*

Family: Gentianaceae

Distribution: Habitat to Himalayas 1500-2400 m.

Parts and properties used: Root- Gives xanthones, flavone- C-glycosides-swertisin and homoorientin. Substitute of (Chirayata) *S. chirayita.*

Source of information (system):? -

***Swertia petiolata* D. Don**

Common name(s): Wild Chirayta, Tikta

Family: Gentianaceae

Distribution: Habitat to alpine Himalayas and Ladakh.

Parts and properties used: Shoot, bark, leaves- Given in fever, as tonic- febrifuge, vermifuge, digestive, appetizer, laxative, anthelmintic. Substitute of (Chirayata) *S. chirayita.*

Source of information (system): AUS

***Swertia thomsonii* C.B. Clarke**

Common name(s): Wild Chirayta, Tikta

Family: Gentianaceae

Distribution: Habitat to alpine Himalayas and Ladakh.

Parts and properties used: Plant- Anthelminthic (Xanthone Dixylopyranoside have been isolated).

Source of information (system): F

***Syzygium alternifolium* (Wight) Walp., (EN)**

Common name(s): Thamba Jalari, Mogi. *S. aromaticum* relative

Family: Myrtaceae

Distribution: Cosmopolitan. Wild all over Eastern Ghats, AP.

Parts and properties used: Shoot leaf juice, fruit- Locally used to treat dysentery, cough. Seeds- Used to treat diabetes. Stem bark- To treat gastric ulcer.

Source of information (system): F

***Tabernaemontana dichotoma* Roxb., syn. *Ervatamia dichotoma* Blatter.**

Common name(s): Kandalaippalai

Family: Apocynaceae

Distribution: Native, wild in Western Ghats.

Parts and properties used: Seed, leaves, bark- Purgative. Latex- Cathartic.

Source of information (system): FS

Tagetes minuta **L.**

Common name(s): Stinking roger

Family: Asteraceae/ Compositae

Distribution: South American. Cultivated in Indian plains.

Parts and properties used: Flower- Anthelmintic, antispasmodic, aromatic. Used as diaphoretic, diuretic, purgative.

Source of information (system): F

Tamarix dioica **Roxb.**

Ayurvedic and Common name: Maacheeka, Nirumari, wild relative

Family: Tamaricaceae

Distribution: Habitat to coast and riverbed of TN.

Parts and properties used: Twigs and galls- Astringent. Tannin- leaves have 8%, twig-bark 10%, galls 50%.

Source of information (system): AFS

Tamarix ericoides **Rottl.**

Ayurvedic and Common name: Maacheeka, Jhaau, wild relative

Family: Tamaricaceae

Distribution: Habitat to south Indian riverbed.

Parts and properties used: Galls- Astringent. Leaf's decoction- Treat enlarged spleen and cough.

Source of information (system): AF

Tanacetum dolichophyllum **(Kitam.) Kitam.**

Common name(s): Khampa serpo

Family: Asteraceae/ Compositae

Distribution: Habitat wild to alpine Himalayas and Ladakh.

Parts and properties used: Plant essential oil- Anthelmintic. Locally used in stomach pain, and indigestion.

Source of information (system): F

Tanacetum fruticulosum **Ledeb.**

Common name(s): Khamchu. Tansy (*T. vulgare*) relative.

Family: Asteraceae/ Compositae

Distribution: Habitat wild to alpine Himalayas and Ladakh.

Parts and properties used: Plant essential oil- Locally used in stomach pain, new guaianolide isolated, anti-tumour.

Source of information (system): F

***Tanacetum gracile* Hook. f. & Thomson**

Common name(s): Khamchu, Tansy relative

Family: Asteraceae/ Compositae

Distribution: Habitat wild to alpine Himalayas and Ladakh.

Parts and properties used: Plant essential oil- Locally used as anthelmintic, aromatic.

Source of information (system): F

***Tanacetum nanum* C.B. Cl., syn. *Chrysanthemum nanum* (C.B. Clarke) B. Fedtsch.**

Common name(s): Khamchu, Tansy relative

Family: Asteraceae/ Compositae

Distribution: Habitat wild to alpine Himalayas and Ladakh.

Parts and properties used: Herb- Locally used as anthelmintic.

Source of information (system): F

***Tanacetum tibeticum* Hook.f. & Thomson ex C.B. Clarke., syn. *Ajania tibetica*, *Chrysanthemum tibeticum* C.B. Clarke**

Common name(s): Khamchu, Tansy relative

Family: Asteraceae/ Compositae

Distribution: Habitat wild to alpine Himalayas and Ladakh.

Parts and properties used: Herb- Locally used as anthelmintic.

Source of information (system): F

***Terminalia citrina* Roxb. ex Flem.**

Common name(s): Haritaki, wild relative of *T. chebula*

Family: Combretaceae

Distribution: Habitat wild to foothills of Himalayas.

Parts and properties used: Bark- Diuretic, cardiotonic. Fruits- Used like *T. chebula*.

Source of information (system): F

***Terminalia crenulata* Roth**

Sanskrit or Ayurvedic name: Kari Maruthu

Common name: Asan, wild relative of *T. chebula*

Family: Combretaceae

Distribution: Habitat to wild deciduous forests of Karnataka.

Parts and properties used: Bark- Bitter, astringent, cooling, antifungal.

Source of information (system): AFS

***Terminalia pallida* Brandis (EN)**

Common name(s): Tella Karakkaya

Family: Combretaceae

Distribution: Endemic to Eastern Ghats, AP.

Parts and properties used: Bark/Fruit powder- Locally used in dysentery.

Source of information (system): SF

***Thalictrum alpinum* Linn.**

Common name(s): Alpine rue, Chakhiya

Family: Ranunculaceae

Distribution: Habitat wild to alpine Himalayas and Ladakh.

Parts and properties used: Herb- Locally used to treat fever.

Source of information (system): F

***Thalictrum foetidum* L.**

Common name(s): Lesser meadow-rue, sNgo-sprin

Family: Ranunculaceae

Distribution: Habitat wild to alpine Himalayas and Ladakh.

Parts and properties used: Parts and properties used: Herb- Antiemetic, diuretic. Locally given in fever. Medicinally promising.

Source of information (system): F

***Thermopsis barbata* Benth.**

Common name(s): Black pea

Family: Leguminosae/ Fabaceae

Distribution: Range from China to Himalayas.

Parts and properties used: Herb- Antibacterial. Used as pain killer.

Source of information (system): L (Tibetan)

***Thlaspi alpestre* L., syn. *T, caerulescens* J. Presl & C. Presl**

Common name(s): Bre-ga

Family: Brassicaceae/ Cruciferae

Distribution: Habitat to alpine Himalayas and Ladakh.

Parts and properties used: Herbal decoction- Locally used improve digestion, fever, phytoremediation.

Source of information (system): Chaurasia *et al.*, 2008

***Tragopogon gracilis* D. Don**

Common name(s): Goat's beard/ Tharno

Family: Asteraceae/ Compositae

Distribution: Habitat wild to west Himalayas, and Ladakh.

Parts and properties used: Plant- Anticough, astringent, vulnerary, and locally used in skin wound healing.

Source of information (system): F

***Tragopogon pratensis* Linn.**

Common name(s): Goat's beard/ Tharno

Family: Asteraceae/ Compositae

Distribution: Habitat wild to west Himalayas, Ladakh along roadside.

Parts and properties used: young shoots and roots, juice- Locally used in diabetic, wound healing.

Source of information (system): F

***Tribulus rajasthanensis* M.M. Bhandari & V.S. Sharma (CR)**

Common name(s): Substitute to Gokhru

Family: Zygophyllaceae

Distribution: Endemic to arid, semi-arid of Rajasthan.

Parts and properties used: Fruit- Locally used like *T. alatus*.

Source of information (system): AU

***Tribulus subramanyamii* P. Singh, G.S. Giri & V. Singh**

Common name(s): Gokhru substitute

Family: Zygophyllaceae

Distribution: Endemic to south India.

Parts and properties used: Fruit- Antimicrobial, used like *T. alatus*.

Source of information (system): AU

***Tricholepis angustifolia* DC.**

Sanskrit, Ayurvedic and Common name: Brahmadandi, Uuntakataaraa (*Echinops*) wild relative.

Family: Asteraceae/ Compositae

Distribution: Habitat to coastal southwestern Ghats

Parts and properties used: Herb- Diuretic, and bechic.

Source of information (system): AF

***Trichopus zeylanicus* Gaertn. (CR)**

Sanskrit or Ayurvedic and Common name: Arogyapacha, chathan kalanji

Family: Dioscoreaceae

Distribution: Native to Agastya Hills, southwest of mountains in Western Ghats, India.

Parts and properties used: Root- Boost immune system, stamina, and help lose weight. Fruit- Energetic, used by Kaani tribe.

Source of information (system): F (Kaani tribe)

***Trichosanthes nervifolia* Linn., syn. *T. caudata* Willd.**

Common name(s): Patoli (Wild relative of *T. dioica,* Patola)

Family: Cucurbitaceae

Distribution: Recorded from Andaman and Nicobar Islands

Parts and properties used: Herb- Local tonic, febrifuge. Root- Purgative. Fruit- Used as dentifrice.

Source of information (system): FS

***Trigonella incisa* Benth.**

Common name(s): Sainji (related to *Medicago*/ *Trigonella* spp.)

Family: Leguminosae/ Fabaceae

Distribution: Afghanistan to West Himalayas. Grown in Punjab.

Parts and properties used: Seeds- Locally used as antidiarrhoeic.

Source of information (system): F

***Triumfetta malabarica* Koen ex Rottb.**

Common name(s): Pivla lepta

Family: Tiliaceae

Distribution: Wild on hill slopes, Kolhapur Maharashtra.

Parts and properties used: Herb- Locally known for blood clotting or as coagulum.

Source of information (system): F

Turnera ulmifolia **Linn.**

Common name(s): Bhinjir

Family: Turneraceae

Distribution: South American. Naturalized to Bengal, Odisha and peninsular India.

Parts and properties used: Herb- Locally advised in indigestion, biliousness, chest ailments and rheumatism.

Source of information (system): F

Typha elephantina **Roxb.**

Sanskrit or Ayurvedic name: Gundra (Gundraka wild relative)

Common name(s): Elephant Grass, Gondapateraa

Family: Typhaceae

Distribution: Widespread in brackish water from Kashmir to Assam.

Parts and properties used: Rhizomes- Astringent, diuretic. Used in dysentery. Pollen- Used in bleeding, including of uterine and nose.

Source of information (system): AFS

Typha laxmanni **Lepech.**

Sanskrit or Ayurvedic name: Airakaa

Common name(s): Pizh

Family: Typhaceae

Distribution: Habitat to Gilgit and J & K 2700 m.

Parts and properties used: Stamens- Astringent and styptic. Locally used externally.

Source of information (system): AF

Ulmus wallichiana **Planch. =** ***U. fulva***

Common name(s): Himalayan Elm, Umbok.

Family: Ulmaceae

Distribution: American. Found in northwest Himalayas and Ladakh.

Parts and properties used: Bark- Astringent, demulcent, emollient, expectorant, diuretic. Used to treat wounds, fractured bones hair-tonic.

Source of information (system): H, Chaurasia *et al.*, 2008

Uraria alopecuroides **Wight.**

Sanskrit or Ayurvedic name: Prishniparni

Common name(s): Wild relative of Prisni prani, *U. picta.*

Family: Leguminosae/ Fabaceae

Distribution: Habitat to forest of Western Ghats and most India.

Parts and properties used: Pods/roots- Used against ringworm by Ayurveda practitioners.

Source of information (system): AF

Uraria crinita **Desv.**

Sanskrit or Ayurvedic name: Prishniparni

Common name(s): Wild relative of Prisni prani, *U. picta.*

Family: Leguminosae/ Fabaceae

Distribution: Habitat to Himalayas and northeast hills.

Parts and properties used: Herb- Advised in dysentery, diarrhoea and in case of enlarged spleen and liver by Ayurveda practitioners.

Source of information (system): A

Uraria rufescens **(DC.) Schindl., syn.** ***U. hamosa*** **Sweet**

Common name(s): Salaparni, substitute of *U. lagopodies.*

Family: Leguminosae/ Fabaceae

Distribution: Found wild in greater part of Karnataka, India. Grown too.

Parts and properties used: Root/plant- Febrifuge, expectorant, diuretic, used in oedema, diarrhoea, cardiac disease, pain, emaciation, dysphonia, polyuria, piles, vomiting, rejuvenator. Locally used medicinally.

Source of information (system): F

Urtica hyperborean **Jacquin ex Weddell**

Common name(s): Stinging hairs, Zazot

Family: Urticaceae

Distribution: Habitat wild to Himalayas, 3000-6000 m and Ladakh.

Parts and properties used: Plant- Locally used as diuretic and antirheumatic.

Source of information (system): F

Usnea longissima **Ach. (Lichen) (Rare)**

Common name(s): Ushnaa

Family: Usneaceae

Distribution: Habitat to alpine Himalayas.

Parts and properties used: Herb- Local use it as expectorant and to treat ulcers. Assessed medicinally potential.

Source of information (system): Locals

***Uvaria gandiflora* Roxb., syn. *U. purpurea* Blume**

Common name(s): Cowherb

Family: Annonaceae

Distribution: Southeast Asian forests and India. Grown too.

Parts and properties used: Root decoction- Locally against flatulence and stomach-aches. Assessed medicinally potential.

Source of information (system): F

***Vaccaria pyramidata* (L.) Medik, *Saponaria vaccaria* L.**

Common name(s): Musna, Saabuni

Family: Caryophyllaceae

Distribution: Occur as weed all over India.

Parts and properties used: Roots- Locally used for cough, asthma, respiratory disorders, jaundice, liver, and spleen diseases. Mucilaginous sap- Used in scabies.

Source of information (system): F

***Valeriana pyrolifolia* Decne. = *V. pyrolaefolia* Decne.**

Sanskrit, Ayurvedic and common name: Dhyaamaka, Sugandhabala, *V. jatamansi* substitute

Family: Valerianaceae

Distribution: Habitat to alpine Himalayas, from Kashmir to Bhutan.

Parts and properties used: Roots and rhizomes extract- Has exhibited anti-anxiety and anti-depressant properties. Included into aromatic drugs of *Jatamansi* group and used similarly.

Source of information (system): AF

***Vandellia pedunculata* Benth., syn. *V. erecta* Benth.; *Lindernia cordifolia* (Colsmann) Merrill.**

Common name(s): Gadaga-vel, Vakapushpi

Family: Scrophulariaceae

Distribution: Habitat to alpine Himalayas 1200-1700 m.

Parts and properties used: Herb- Locally used against sexually transmitted diseases and urethral discharges. Dried flower powder- Given for asthma, consumption, mood-swings and psychosomatic.

Source of information (system): AF

***Vernonia roxburghii* Less.**

Sanskrit, Ayurvedic and common name: Sahadevi, Doraa-baahaa, *V. cinerea* wild relative.

Family: Asteraceae/ Compositae

Distribution: Found in HP, UP, Bihar, Bengal states of India.

Parts and properties used: Roots- Locally used in joint rheumatism.

Source of information (system): AF

***Viburnum coriaceum* Blume.**

Sanskrit or Ayurvedic name: Tilvaka

Common name(s): Telam, Tilen, *V. nervosum* wild relative.

Family: Caprifoliaceae

Distribution: Habitat to Himalayas, Punjab, Nilgiris 1200-2500 m. Grown too.

Parts and properties used: Root, stem bark- Antispasmodic, antioxidant, anti-inflammatory, uterine sedative (*in vitro* rat uterus). Promising for drug.

Source of information (system): AF

***Viburnum prunifolium* Linn.**

Sanskrit or Ayurvedic name: Tilvaka,

Common name(s): Black Haw, *V. nervosum* wild relative

Family: Caprifoliaceae

Distribution: American. Grown in Nilgiris Hills.

Parts and properties used: Herb- Uterine sedative, spasmolytic. Used in treatment of miscarriage.

Source of information (system): A

***Vigna vexillata* (L.) A. Rich.**

Ayurvedic name: Kattupayar, wild relative of cultigen *Vigna* spp.

Family: Leguminosae/ Fabaceae

Distribution: Habitat wild to Himalayas and hills & foothills of Northeast.

Parts and properties used: Herb- Induce tissue formation and sustainment. Seeds- Contain L-Dopa used for Parkinson's.

Source of information (system): A

Vincetoxicum hirundinaria **Medik.**

Common name(s): sNgo-dug-monyung

Family: Asclepiadaceae

Distribution: Eurasian, extend to Himalayas and Ladakh.

Parts and properties used: Herb- Cooling. Seeds- Locally used in dysentery, lung diseases. Assessed medicinally potential.

Source of information (system): F

Viola biflora **Linn.**

Common name(s): Banafasha wild relative.

Family: Violaceae

Distribution: Habitat to alpine Himalayas 1800-3000 m, and Ladakh.

Parts and properties used: Herb- A substitute of *V. odorata*. Leaves- Laxative, emollient. Flowers- Antiseptic, pectoral, diaphoretic. Root- Emetic. Used in skin care.

Source of information (system): UF

Viola canescens **Wall.**

Common name(s): Himalayan white violet Banafasha, wild relative.

Family: Violaceae

Distribution: Habitat to west Himalayan region.

Parts and properties used: Herb- Locally used in cough, cold, flu, fever, malaria, and give anti-cancerous drug. Assessed medicinally potential.

Source of information (system): F

Viola cinerea **Boiss., syn.** ***V. stocksii*** **Boiss.**

Common name(s): Banafasha wild relative

Family: Violaceae

Distribution: Occur in northwest India.

Parts and properties used: Root- Emetic. Used as a substitute to *V. odorata*. Assessed medicinally potential.

Source of information (system): UF

Viola patrinii* var. *laotiana* H Boissieu, syn. *V. betonicifolia

Common name(s): Banafasha wild relative

Family: Violaceae

Distribution: Habitat to grassland of Western Ghats.

Parts and properties used: Herb- Applied to bruised/ ulcers. Dried flowers- Treat coughs/ colds.

Source of information (system): U

***Viola pilosa* Blume.**

Common name(s): Banafasha wild relative

Family: Violaceae

Distribution: Indomalaya. In India found from Himalaya- Kerala.

Parts and properties used: Herb- Febrifuge. Roots, Flower- Emetic. Used in lung and infant disorder. Substitute to *V. odorata.*

Source of information (system): FU

***Viola sylvestris* Lam.**

Common name(s): Banafasha wild relative

Family: Violaceae

Distribution: Habitat to Kashmir 1200-2400 m.

Parts and properties used: Plant- Pectoral, bechic. Used in chest troubles. Stem/leaf/ flower- Applied to foul sores/ wounds.

Source of information (system): U

***Viscum articulatum* Burm.**

Sanskrit or Ayurvedic name: Kaamavriksha

Common name(s): (*Dendrophthoe falacta*) Bandaaka, (related sp.), Jivantikaa

Family: Viscaceae

Distribution: Found in most of India.

Parts and properties used: Plant- Febrifuge, aphrodisiac. Paste- Applied to bone fractures.

Source of information (system): AF

***Viscum monoicum* Roxb. ex DC. (A parasite herb over *nux-vomica*)**

Sanskrit name: Pashumohanikaa.

Common name(s): Kuchleikaa-malang

Family: Viscaceae

Distribution: Found wild in Sikkim.

Parts and properties used: Properties like *Strychnos nux-vomica*. Substitute to *Strychnos nux-vomica*/ strychnine.

Source of information (system): FS

***Viscum orientale* Willd. (Wild parasite over *nux-vomica*).**

Common name(s): Baandaa

Family: Viscaceae

Distribution: Indo-Malayan. Found in Western Ghats.

Parts and properties used: Substitute *Strychnos nux-vomica* Leaf's poultice- Used in neuralgia; ash is applied to skin diseases. Assessed medicinally potential.

Source of information (system): F

***Waldheimia glabra* (Decne.) Regel.**

Common name(s): Palu, ghaan-poe

Family: Asteraceae/ Compositae

Distribution: Habitat to alpine Himalayas, and Ladakh.

Parts and properties used: Herb/ Essential oil- Antibacterial (Have wound healing properties, and for anti-influenza drug).

Source of information (system): F (Tibetan)

***Waldheimia stoliczkai* (CB Clarke) Ostenf., syn. *Allardia stoliczkai* CB Clarke**

Common name(s): Palu

Family: Asteraceae/ Compositae

Distribution: Habitat to alpine Himalayas, and Ladakh.

Parts and properties used: Plant- Locally used to heal septic wounds. Assessed medicinally potential.

Source of information (system): F

***Waldheimia tomentosa* (Decne.) Regel.**

Common name(s): Palu

Family: Asteraceae/ Compositae

Distribution: Habitat to high altitude Himalayas and Ladakh.

Parts and properties used: Plant- Locally used wound healing, antibacterial. Unexplored, appears promising needing further study.

Source of information (system): Locals

Walsura trifoliata **(A. Juss.) Harms**

Common name(s): Cheddavokko, Valsura

Family: Meliaceae

Distribution: Habitat to Karnataka, and Western Ghats.

Parts and properties used: Bark- Stimulant, expectorant, emmenagogue, emetic. Kills vermin.

Source of information (system): FS

Waltheria indica **Linn., syn.** ***W. americana*** **Linn.**

Common name(s): Shembudu, Khar-Duudhi

Family: Sterculiaceae

Distribution: Found all over tropical India.

Parts and properties used: Plant- Emollient, bechic, febrifuge, purgative, abortifacient. Root- Locally used to treat internal haemorrhages.

Source of information (system): FS

Wendlandia tinctoria **DC.**

Sanskrit or Ayurvedic name: Tilaka (*W. exserta* wild relative)

Family: Rubiaceae

Distribution: Sub Himalayan tract from central to Khasi Hills.

Parts and properties used: Bark- Used for cramps in cholera patients.

Source of information (system): A

Wikstroemia indica **Mey, syn.** ***W. viridiflora*** **Meissn**

Common name(s): Salago

Family: Thymelaeaceae

Distribution: Native to Assam, TN.

Parts and properties used: Root bark- Diuretic, vesicant. Used locally as purgative and piscicide.

Source of information (system): F

Xylocarpus granatum **Koen., syn.** ***Carapa granatum*** **(Koen.) Alston.**

Common name(s): Somanthiri, Pussur

Family: Meliaceae

Distribution: Habitat to coastal India.

Parts and properties used: Bark- Astringent, antidysentery, Febrifuge.

Source of information (system): FS

***Xylopia parviflora* Hook.f. Thoms**

Common name(s): Saanthu, Kalpottan

Family: Annonaceae

Distribution: Habitat to evergreen forests of Kerala.

Parts and properties used: Root bark- Antiseptic, treat ulcers. Root bark, flowers, fruits- Used locally for oral hygiene.

Source of information (system): F

***Xylosma longifolium* Clos.**

Common name(s): Sallu

Family: Flacourtiaceae

Distribution: Habitat to western to central Himalayas.

Parts and properties used: Herb- Antispasmodic, narcotic and sedative. Locally advised in dysentery, restlessness, insomnia.

Source of information (system): F

***Xyris commplanata* R. Br.**

Common name(s): Kochelachi-pullu

Family: Xyridaceae

Distribution: Habitat to marshy areas of peninsular India and Western Ghats.

Parts and properties used: Herb- Antiseptic. Leaf juice/powder- Locally used to treat itches, leprosy, skin diseases, ringworm.

Source of information (system): F

***Yucca aloifolia* Linn.**

Common name(s): Yerum Lei

Family: Liliaceae

Distribution: North American. Naturalized to northeast, Assam, and central India.

Parts and properties used: Flower- Yields Aloifoline- effective against lung-tumour. Fruit- Locally used as a purgative.

Source of information (system): F

***Zanthoxylum nitidum* (Roxb.) DC.**

Common name(s): Kumkumada, Tez-mui

Family: Rutaceae

Distribution: Forests of Assam, Northeast Hills, WB, Andaman.

Parts and properties used: Stem bark- Locally used in toothache, gingivitis, fever, colic vomiting, diarrhoea, and cholera. Assessed medicinally potential.

Source of information (system): F

***Zanthoxylum oxyphyllum* Edgew., syn. *Xanthoxylon violaceum* Wall**

Common name(s): Mezenga

Family: Rutaceae

Distribution: Habitat to Himalayas to Khasi Hills.

Parts and properties used: Bark- Stimulant, stomachic, sudorific; used in colic and fevers. Fruits- Locally advised in dyspepsia. Assessed medicinally potential.

Source of information (system): F

***Zornia diphylla* (L.) Pers.**

Common name(s): Samraapani

Family: Leguminosae/ Fabaceae

Distribution: Habitat to grassy slopes of Western Ghats.

Parts and properties used: Herb- Locally used in treatment of dysentery. Root- Given to induce sleep. Assessed medicinally potential.

Source of information (system): FS

***Zygophyllum simplex* Linn., syn. *Tetraena simplex* L. (L.) Beier & Thulin**

Common name(s): Alethi, Lunio

Family: Zygophyllaceae

Distribution: Arid, semi-arid Rajasthan, and Gujarat.

Parts and properties used: Leaves, seeds infusion- Locally applied in ophthalmia and glaucoma. For gout, asthma, and inflammation in Arabia. Seeds- Anthelmintic. Assessed medicinally potential.

Source of information (system): F

1. IUCN threat category in parenthes is after the species Letin botanical name
2. Source of information (system) : Ayurveda = A; Folk = F; Homoeopathy = H; Sowa-Rigpa/Amchi/Tibetan = L; Sidha = S; Unani = U

Some other Sources of Information

Bhadrecha, P, Kumar, V and Kumar M. (2017) Medicinal Plant Growing under Sub-optimal Conditions in trans-Himalaya Region at High Altitude. Defence Life Science Journal 2(1):37-45. DOI: 10.14429/dlsj.2.11107

Botanical Survey of India reports (status of threat)

Chaurasia, OP, Khatoon, N and Singh SB. (2008) Field Guide: Floral Diversity of Ladakh pp 198. Defense Institute of High-Altitude Research (DIHAR), Defence Research and Development Organization (DRDO), Leh, Ladakh, India.

Jain, BJ, Kumane, SC and Bhattacharya S. (2006) Medicinal flora of Madhya Pradesh and Chhattisgarh- A review. Indian Journal of Traditional Knowledge 5(2): 237-242.

Pant, B. (2013) Medicinal orchids and their uses: Tissue culture a potential alternative for conservation. African J Plant Sci 7(10):448-467. DOI:10.5897/AJPS2013.1031

Singh, Anurudh K. (2017) Wild Relatives of Cultivated Plants in India: A Reservoir of Alternative Genetic Resources and More pp 310. Springer Nature Singapore Pte Ltd. ISBN 978-981-10-5115-9

4

Inventory of Medicinal and Aromatic Plants (MAPs) Species of Trade Value

4.1 Introduction

The adverse effect of non-vegetarian diet, particularly red meat, on human health and on environment due to destruction/clearing of forest for raising more pasture and to increase the industrial production of grains for animal feed have led to greater greenhouse gas emissions, and environmental pollution due to extensive use of inorganic fertilizers, fungicides, pesticides. Clearing of forest has also caused extinction of plant species. For these reasons, greater interest has been generated in vegetarian diet based on the direct consumption of whole grains, legumes, nuts, fruits, and vegetables that are easily assimilated by the human body, nutritionally more efficient, and have least impact on the environment. In fact, the healthfulness of vegetarian diets has been found to be strongly correlated with sustainability of agriculture because of easy affordability of plant-based foods. Vegetarian people have been often referred to as vegan and pescatarian (including fish and other seafood). Most pescatarians include dairy and eggs as well. Similarly, allopathic medicines, which are synthetic inorganic concentrated drugs, have adverse effect on non-targeted organs, besides being expensive. These reasons have generated further interest in the scientific evaluation of traditional herbal medicines, reviving their use with a stronger science base. Though, they have slower courses of action but are sustainable, cheaper, and affordable for economically poor masses of rural and interior areas. In this regard, the time-tested herbal medicines belonging to Indian traditional medicinal systems, such as Ayurveda have been attracting global attention and are even being exploited/used in other parts of the world, particularly in west. This scenario has led to greater interest and cultivation of plant species for food and other economic exploitation, particularly for production of nutraceuticals, herbal medicines, cosmetics, and other plant-based products.

Recent investigations on the early history of various cultures and civilizations have revealed that food science and herbal medicinal systems predominantly originated and voluminously developed in India and then travelled to the east in China and to the west in the middle east/Arabian world and southern Europe. Food science and herbal medicinal systems spread beyond Indian Subcontinent through cultural exchange of visitors and students seeking knowledge from China and through trade with the Arabs and Greeks. From the Arabs, it spread to Europe via southern Europe (Greece). But the Europeans rarely formally acknowledged it with due formal acknowledgement or citation in references the source of information and material out of greed to claim credit for the innovations and avoid sharing intellectual credit with others. Nevertheless, there are some writings, where Greek and Latin writers have referred to the Lower Gangetic Bengal region as *Gangaridai* (*Ganga Rashtra* in Sanskrit, meaning Nation on the River Ganga), the present-day Bengal existing around 300 BCE [Ptolemy: Describing Ancient India (John W. McCrindle. p.172); Periplus: The Erythraean Sea (Wilfred H. Schoff. pp.47-48)]. The main articles received from this region were the finest of silk and muslin, Gangetic spikenard [*jatamansi* (*Nardostachys jatamansi*)], a medicinal plant, and pearls (Arthasastra 4th century BCE, Periplus/1st century CE) (Singh, 2012, 2015).

According to *Upanishads* and Ayurveda, the food steers/guides our physical, temperamental, and mental states. Therefore, a stable and healthy diet is essential for human wellness. Accordingly, foods were classified as '*Sativk'* food consisting of juicy fruit, vegetables, milk, and honey, which induces wiseness, calm, and goodness; '*Rajsik*' food consisting of fats like butter, vegetable oil, and may be fresh meat with appropriate condiments and spices in appropriate quantity, providing greater energy, passion, and desire; and '*Tamasik*' food consisting of meat, garlic, and overly cooked spiced and stale food, inducing intense desire and/or provoking passion, and laziness. These facts are being revealed in recent times with the survey of literature and are being published with hard data/evidence, generating greater interest in the Indian food, nutraceuticals, and traditional medicinal systems, particularly Ayurveda, Yoga and Naturopathy, with strong recommendations for their adoption in daily lifestyle, thereby, generating new local and international markets for Indian plant-based products, such as fortified food/supplements providing wellness and health care, medicinal drugs, cosmetics, etc. This scenario demands a comprehensive scientific revisit to the components of Indian diet for a balance source of various nutritional requirements for wellness and healthy growth and development. Further, it also requires validation and detailed scientific characterization and evaluation of the commonly used traditional herbal medicines used in cure/treatment of various ailments to generate information required as per the requirement of national and international patent laws and marketing of products. This will provide great opportunities to Indian researchers, entrepreneurs (start-ups), and traders for economic exploitation and gains.

4.2 Commercial Potential of MAPs and Plant-based Products

There has been a growing global concern about nutritional diet and health. In general, people believe that meat and meat products are unhealthy because of increased risk of cardiovascular diseases, obesity and cancer and added synthetic antioxidants and antimicrobials. On the other hand, fruits, vegetables, pulses, and grains provide a docile and nutritionally balanced diet as discussed above. Nevertheless, many crops, particularly grain legumes, such soybean, pea and cereals like wheat, rice, and minor millets, besides being staple food, have significant potential as sources of protein, micro-nutrients to be ingredient(s) for fortification of food. Plant derivatives, which are rich in protein and/or have antioxidant components including vitamins A, C and E, minerals, polyphenols, flavonoids and terpenoids, may decrease the risk of several degenerative diseases caused by meat products. Vegetable oil and plant fibre may work as fat replacers and spices and condiments may work as natural antimicrobials agents. In addition, there are several plant species that can be sources of active botanicals and herbs that can be used to maintain or improve health in general, and function as source of sweeteners, seasonings, colouring, and flavouring of foodstuffs. Plants can also be source of dairy alternatives, such as milk, beverages (Almond, Soy, Coconut, and Rice) and other drinks, and cultured products such as yogurt, frozen desserts, ice cream, etc. All these economically rich plant-based bioresources are of trade significance and can play a significant role in the country's economy.

Similarly, there are ancient plant-based medicinal practices/systems, like Ayurveda that have stood the test of time in health care and overall wellness for millenniums. In fact, many of the life-saving pharmaceuticals or drugs we rely on even today are derived from plants and were discovered by indigenous Indian communities. *Aswagandha*, *Nidrajanana* Vati, morphine, aspirin, ephedrine, etc. are a few to mention. Therefore, traditional knowledge about the medicinal properties of plants from Indian Subcontinent can find prominent role in innovative drug discovery and production platforms. Further, based on uncoded preliminary information there are still many under-utilized or primitively used, and unexplored plant species with untapped potential. They deserve critical research attention for scientific and economic evaluation, promoting commercial exploitation and use.

Vegetal extracts and herbs have been used for personal care and cosmetic purposes since time immemorial. For centuries, they were the only source to obtain colorants, fragrances, and products for soothing and protecting skin. Subsequently, they were replaced by synthetic materials, which were low in price. However, because of negative effects of synthetics, in the last few decades plant-based or plants derivatives-based cosmetic ingredients have made a powerful comeback. Products of plant origin used in cosmetics include vegetable oils and other lipids, essential oils used as fragrances or for their antimicrobial activities, ingredients for skincare and hair care, and antioxidants, are just a few to name. Plant material used to produce cosmetic ingredients comes from a variety of sources, which include not only conventional products coming from horticultural production (in field or greenhouse), but also from wild harvest in developing countries like India and biotechnological methods/process (e.g., tissue cultures, fermentation of genetically modified organisms, microalgae cultures, hydroponic systems, etc.) in developed. However, there is a need for greater research attention, and conservation to promote sustainable use of such biodiversity/ bioresources including genetic diversity in search of alternative sources with desired molecules and for efficient use of known source particularly for improved isolation with precise extraction techniques. Also, research is required for the post-harvest safety and evaluation of raw materials, and finally for the development of innovative formulations. As per one estimate in relation to food, there is +49 percent potential for plant protein based or supplemented products, + 14 percent for meat substitute, +20 percent dairy alternatives of totally worth of around $16.3 billion.

4.3 Global Market Perspective and Opportunities

It has been revealed that the market for global plant-based meat (in texture and taste) is expected to register a healthy Compound Annual Growth Rate (CAGR) of 5.8% during the period 2018-2026. This is due to various factors, such as similar potential nutritional value, health safety, and lower cost. Persistence Market Research (PMR) offers vital insights in detail regarding these. Similarly, the global market for plant-based dairy alternative drinks has reached around US$16.0 billion in 2018.

The global market for botanical and plant-derived drugs is growing; from $29.4 billion in 2017 it is expected to grow around to $39.6 billion by 2022 with an expected CAGR of almost 8%, during the 2017-2021 period. It will be primarily due to the low cost of herbal medicines compared to allopathic and offers of new opportunities.

Global market for herbal medicine was valued at $71.19 billion in 2016. It was expected to grow to $107 billion by 2017 and to increase to $ 140 billion by 2024. The market for medicinal plants in India stood at Rs. 4.2 billion (US$ 56.6 million) in 2019 and is expected to increase at a CAGR of 38.5% to Rs. 14 billion (US$ 188.6 million) by 2026. The export of herbs and value-added extracts of medicinal herbs has been gradually increasing over years (Press, Dec 17, 2020). The total world herbal trade is currently assessed at US$ 120 billion. India's share in the global export of herbs and herbal products is low due to unsophisticated agricultural and low-quality control procedures, lack of quality processing, research and development, standardization of products and regulatory framework in trade of medicinal plants. Nevertheless, the export of herbs and value-added extracts of medicinal herbs from India has been gradually increasing over years. In 2017-2018, India exported US$ 330.18 million worth of herbs at a growth rate of 14.22% over the previous year. Also, exports of value-added extracts of medicinal herbs and herbal products in 2017-2018 stood at US$ 456.12 million, recording a growth rate of 12.23% over the previous year. The demand for herbal/value-added extracts of medicinal herbs is gradually increasing in foreign countries, especially in European and other developed countries.

There is a huge gap between the supply and demand of medicinal plants for manufacturing Ayurvedic medicines in India. According to the 'All India Trade Survey of Prioritized Medicinal Plants, 2019', demand for high-value medicinal plants increased by 50%, while the availability declined by 26%. This has led to increase in pressure on habitat degradation and levels of over-exploitation of medicinal plants from nature by pharmaceutical industries, threatening the populations and causing species genetic erosion and demanding development of cultivation practices to ensure quality raw material.

However, China acted appropriately on these concerns and has taken advantage of the present world scenario. It is the largest supplier of herbal products (to US and Europe) based on 5000 plants. Contrary to this, India has been lagging far behind with a share of only 0.5% ($358.60 million) of the global herbal medicinal market, as government informed the Lok Sabha in November 2016. This is despite having traditional knowledge about the diverse properties of more than 9000 plant species, including medicinal properties of more than 7000 plant species.

The market for Ayurvedic medicines has been expanding at the rate of 20% annually in India. The floristic richness of the plant species found in different biogeographical regions (10+1) of India (Singh, 2017), particularly of those with medicinal properties and available associated knowledge, puts India in a supreme advantageous position to concentrate on further research and development of plant-based products to take the advantage of prevailing scenario of market demand and harness the maximum economic benefits. This will also help to address some of the recent concerns and apprehension expressed about the competition with China and to take research and trade advantage in case medicinal plant species found in Indochina region, particularly in the areas bordering in Ladakh region in case of plants species, such as, *Berberis aristata* DC. (Daruhaldi, Rasnajan, Zarisuk), *Hippophae rhamnoides* L. (Sea Buckthorn), *Reinwardtia indica* Dumort. (Basanti, Yellow flax), *Rhodiola rosea* (Sanjeevani, Solo), *Salacia oblonga* (Kadalainjil, Saptrangi), which are in demand.

4.4 Protection of Commercially Important MAPs Species

4.4.1 Regulatory Mechanisms

There have been threats and concerns about the commercially significant medicinal and aromatic plants harvested from nature/wild. Many of these are threatened and are at the verse of extinction because of over- or -unskilled harvesting. Therefore, they need immediate attention for protection to promote conservation and their sustainable use. As per one estimate, around in 65 species (i.e., 10% of the total species) fall into the critically endangered, endangered, vulnerable, and nearly threatened categories, such as *Nardostachys jatamansi*, *Aconitum* spp., etc. Many are reported at the verge of extinction from specific areas, because of over collection and utilization, for example *Rauwolfia serpentina* (sarpgandha) from Bastar, *Blepherispermum subsessile* (Rasna) from Chhattisgarh, *Hedychium coronaicum* (Gulbakavali) from Amarkantak and *Curcuma caesia* (Kali haldi) from Sal Forest (Kotwal and Banerjee,1998). Therefore, they need to be protected in nature using *in situ* or *ex situ* conservation strategies, restricting their erosion or loss due to over exploitation. Government of India through implementation of the Wildlife Protection Act (1972) provides guidelines to protect these plants under the Minor Forest Product category. Under minor forest produce category, 29 plant species have been banned from the trade to outside India (Table 4.4.1).

Table 4.4.1: Banned 29 MAPs species acquired from wild and traded for economic use

S.No.	Botanical name	Trade name
1	*Aconitum* species	Atis
2	*Aquilaria malaccensis*	Agarwood
3	*Ceropegia* species	Ornamental, hedulo, khadula
4	*Coptis teeta*	Mamira
5	*Coscinium fenestrum*	Calumba wood, Daruharidra
6	*Cyatheaceae* species	Tree ferns
7	*Cycadacea* species	Cycad
8	*Cycas beddomei*	Beddomes cycade

S.No.	Botanical name	Trade name
9	*Dactylorhiza hatogirea*	Salampanja, Orchid
10	*Dioscorea deltoidea*	Elephant's foot, Shingli mingli
11	*Euphorbia* species	Euphorbias
12	*Frerea indica*	Shindal mankundi
13	*Gentiana kurroo*	Kuru, Kutki
14	*Gnetum* species	Joint fir, edible
15	*Kampheria galenga*	Kachora, Narkachur, Kapoor Kachri
16	*Nardostachys grandiflora*	Balchad, Jatamansi
17	*Nepenthes khasiana*	Pitcher plant
18	*Orchidaceae* species	Orchids
19	*Panax pseudoginseng*	Ginseng (subst.)
20	*Paphiopedilium* species	Ladies' slipper orchid
21	*Picrorhiza kurrooa*	Kutaki
22	*Podophyllum hexandurm*	Emodi, Indian podophyllum, Bankakri, Vanatrapusi
23	*Pterocarpus santalinus*	Red sanders, Raktachandana, Lal chandan
24	*Rauwolfia serpentina*	Sarpagandha, Pagal Buti
25	*Renanthera imschootiana*	Red vanda
26	*Saussurea costus*	Kuth, Uplet
27	*Swertia chirata*	Chirayatah
28	*Taxus wallichiana*	Common Yew or Birmi leaves, Talispatra
29	*Vanda coerulea*	Blue vanda

However, the export of plant-based finished products is allowed. In 2000, above list of 29 banned plants were further expanded creating a list a total of 113 medicinal plants, regulating their harvest from nature, and requiring legal procurement certificate from the concerned forest officer to accompany the export consignment and for cultivation of plant material as per MoEF circular dt. 4.10.2000 (Table 4.4.2).

Table 4.4.2: Threatened MAPs species requiring 'Certificate of Cultivation' or 'Legal Procurement Certificate' from the designated authorities.

S.No.	Botanical/ Trade name
1	*Aconitum* spp.
2	*Acorus* spp.
3	*Adhatoda beddomei*
4	*Ampelocissus indica*
5	*Angelica glauca*
6	*Angiopteris* species
7	*Aquilaria malaccensis*
8	*Aristolochia* spp.
9	*Arnebia benthamii*
10	*Artemisia* spp.
11	*Arundinaria jaunsarensis*

S.No.	Botanical/ Trade name
12	*Atropa* spp.
13	*Balanophora* spp.
14	*Berberis aristata*
15	*Berberis kashmirana*
16	*Berberis lycium*
17	*Berberis petiolaris*
18	*Bunium persicum*
19	*Cayratia pedata*
20	*Ceropegia* spp.
21	*Colchicum luteum*
22	*Commiphora whightii*
23	*Coptis* spp.
24	*Coptis teeta*
25	*Coscinium eenestratum*
26	*Coscinium fenestratum*
27	*Costus speciosus*
28	*Craterostigma plantagineum*
29	*Curcuma caesia*
30	*Cyatheaceae* spp.
31	*Cycadacea* spp.
32	*Cycas beddomei* (Beddomes cycad)
33	*Cyclea fissicalyx*
34	*Decalepis hameltonii*
35	*Didymocarpus pedicellata*
36	*Dioscorea deltoidea*
37	*Diptreocarpus indicus*
38	*Dolomiaea pedicellata*
39	*Drosera* spp.
40	*Dyzoxylum malabaricum*
41	*Epedra* spp.
42	*Eulophia cullenii*
43	*Eulphia ramentacea*
44	*Euphorbia* spp.
45	*Frerea indica*
46	*Fritillaria roylei*
47	*Garcinia travancorica*
48	*Gentiana kurroo*
49	*Gloriosa superba*
50	*Gnetum* spp.
51	*Gymnema khandalense*
52	*Gymnema montanum*
53	*Gynocardia odorata*

S.No.	Botanical/ Trade name
54	*Hedychium coronarium*
55	*Hedychium spicatum*
56	*Hellotrophium keralense*
57	*Humboldtia vahliana*
58	*Hydnocarpus alpina*
59	*Hydnocarpus* spp.
60	*Hyoscyamus niger*
61	*Ilex khasiana*
62	*Inula racemosa*
63	*Iphignia indica*
64	*Janakia arayalpathra*
65	*Kampferia galanga*
66	*Kingiodendron pinnatum*
67	*Lamprachaenium microcephalum*
68	*Luvunga scandens*
69	*Madhuca diplostemon*
70	*Madhuca longifolia*
71	*Meconopsis aculeate*
72	*Meconopsis betonicifolia*
73	*Nardostachys* species
74	*Nepenthes khasiana* (Pitcher plant)
75	*Nervilia aragoana*
76	*Niligirianthus ciliatus*
77	*Orchidaceae* spp.
78	*Osmunda* spp.
79	*Panax pseudoginseng*
80	*Paphiopedilium* spp. (Ladies slipper orchid)
81	*Physochlaina praealta*
82	*Picrorhiza kurroa*
83	*Piper barberi gamble*
84	*Podophyllum hexandrum*
85	*Praltia serpumlia*
86	*Prezwalskia tangutica*
87	*Pterocarpus santalinus*
88	*Rauvolfia serpentina*
89	*Renanthera imschootiana* (Red vanda)
90	*Rheum emodi*
91	*Rheum nobile*
92	*Rhododendron* species
93	*Salacia oblonga*
94	*Salacia reticulata*
95	*Saussurea abvallata*

S.No.	Botanical/ Trade name
96	*Saussurea gossyphora*
97	*Saussurea lappa* (Kuth)
98	*Saussurea simpsoniana*
99	*Shorea tumbugaia*
100	*Strychhnos aenea*
101	*Strychnos potatorum*
102	*Swertia chirata*
103	*Swertia lawii*
104	*Syzygium travancoricum*
105	*Taxus wallichiana*
106	*Trichopus zeylanicus*
107	*Trichosanthes anamalaiensis*
108	*Urginea* spp.
109	*Utleria salicifolia*
110	*Valeriaia leschenaultii*
111	*Valeriana iatamansi*
112	*Vanda coerulea* (Blue vanda)
113	*Vateria macrocarpa*

In addition, Forest Act 1927, Import and Export Act (Control) 1947, Wildlife (Protection) Act (Amended) 1991, and Spices Board Act 1986 are some other mechanisms evolved to help conservation of biodiversity/medicinal plants. There are many other laws in India for protection of wild flora, out of which Convention on International Trade in Endangered Species of Wild Flora and Fauna (CITES), Foreign Trade (Development and Regulation) Act 1992, Export Import Policy, Plant Fruit and seeds (regulation of import into India) order 1989, and Convention on Biological Diversity (CBD) are the main, protecting plant species (jharenvis.nic. in › WriteReadData › CMS PDF Endangered Plant Protection Laws). Further, the National Biodiversity Act 2004, controlling access to national biodiversity, exempts only Vaidyas and Sidhas from collection of plant diversity, and bans others, thereby, promoting their *in-situ* conservation. A major part of these plant species belongs to extreme ecologies, such as high Himalayan range and western arid and semi-arid zone.

4.4.2 Domestication and Cultivation of Commercial/Traded MAPs

The other mechanism to protect the MAPs of commercial and traded value to facilitate conservation and sustainable use is to promote domestication and start cultivating to ensure raw material. The Ayurvedic medicines can be grown and harvested, based on their life cycle, to harness the plants' raw materials. For example, herbs can be grown and harvested in a period of one year, while medicinal shrubs and trees may take more than 2 to 10 years, respectively, to get ready for harvesting. Therefore, engaging in developing cultivation practices as per their breeding behaviour and life cycle of plant species is important. It must be, followed by characterization, potential evaluation, and research and development to harvest principal/ commercial components of medicinal plants effectively and efficiently. This would avoid over extraction of these plants from their natural habitat, facilitating dynamic *in situ* conservation and ensuring the availability of raw material in a sustainable manner with uniformity in quality.

As per one estimate, only around 14 medicinal species are cultivated on regular basis and another 14 on marginal basis. This number is exceptionally low compared to the number of plant species reported with medicinal value or commonly used in traditional medicines. Therefore, one of the approaches to facilitate conservation and promote use can be to develop cultivation practices for common medicinal plant species, particularly those of trade value. The National Medicinal Plant Board (NMPB) under the Ministry of Health and Family Welfare has established state medicinal plant board for different states and had initially identified 28 medicinal plant species of high demand in market and industry for development of cultivation practices and good quality planting material for distribution to keen farmers. Therefore, efforts are being made for promoting cultivation of medicinal plants so that the industrial sector is served with true potent quality raw material for manufacturing of quality product. However, various approaches must be complemented to help in conservation of threatened species by providing protection to populations growing in diverse natural habitats responding to changing climate and generating/evolving diversity that may help meet the future requirements of genetic enhancement and for development of cultivation of practices for the species of trade value to ensure availability of raw material facilitating sustainable use.

Cultivation of medicinal plants in a commercial mode can also be one of the most profitable agri-businesses for Indian farmers to increase their income. If anyone (group) has sufficient land and knowledge of herb marketing, then they can earn a high income with moderate investments either independently or in partnership. Cultivation of medicinal herbs, such as *shankhapushpi, atis, kuth, kutki, kapikachhu* and *karanja* are changing the Indian agrarian ayurvedic scenes and providing extraordinary opportunities for farmers to increase their incomes. According to the traditional treatment health centre, there are about 25 significant medicinal plants that are always in full demand. These plants include the Indian Barberry, Liquorice, *Bael, Isabgol, Atis, Guggal, Kerth, Aonla, Chandan, Senna, Baiberang*, Long Pepper, *Brahmi, Jatamansi,* and *Madhunashini* (combination of herbs), *Kalmegh, Satavari, Ashwagandha, Chirata, Katki, Shankhpushpi, Ashoka, Giloe, Kokum,* and *Safed Musli*. Table 4.4.3 lists some of the medicinal aromatic plants under commercial cultivation. Many more can be added to this list as differentiated in the main table with species of commercial trade significance.

Table 4.4.3: List of MAPs under commercial cultivation

S.No.	Botanical name	Trade name
1	*Abelmoschus moschatus*	Muskdana
2	*Acacia catechu*	Katha
3	*Acacia farnesiana*	Vilaayati Kikar
4	*Acacia nilotica*	Babul, Keekar
5	*Achillea millefolium* L.	Brinjasif
6	*Acorus calamus*	Bach
7	*Adhatoda zeylanica*	Adusa
8	*Aegle marmelos*	Bael
9	*Albizia lebbeck*	Shirish
10	*Allium* spp.	Garlic/lahsan, Onion, Shellot
11	*Aloe barbedensis*	Kumari
12	*Alpinia calcarata*	Chittartha, Aratha
13	*Alpinia galangal*	Kulinjan, Rasnamool
14	*Amaranthus* spp.	Ramdana, Lal sag
15	*Amomum* spp.	Elachi
16	*Atropa* spp.	Belladonna
17	*Azadirachta indica*	Neem
18	*Caesalpinia sappan*	Pathimugam
19	*Cassia angustifolia*	Sonamukhi
20	*Catharanthus roseus*	Periwinkle, Sadabahar
21	*Cichorium intybus*	Kasani
22	*Crataeva religiosa*, syn. *C. nurvula*	Varun chhal
23	*Croton tiglium*	Jamalghota
24	*Curcuma angustifolia*	Tikhur
25	*Curcuma zerumbet,* syn. *C. zedoaria*	Kachur
26	*Cymbopogon flexuosus*	Lemongrass
27	*Cymbopogon martini*	Ginger grass
28	*Cymbopogon winterianus*	Chitronella
29	*Digitalis purpurea*	Foxglove, Tilpushpi
30	*Dioscorea* spp.	Yam, Kand
31	*Emblica officinalis*	Amla
32	*Ficus benghalensis*	Vada Chhal
33	*Ficus religiosa*	Arali chakki
34	*Gloriosa superba*	Kalihari
35	*Indigofera tinctoria*	Nili
36	*Inula racemosa*	Pushkarmool
37	*Jatropha curcas*	Nepalam seed
38	*Kaempferia galanga*	Kacholum
39	*Lawsonia inermis*	Henna
40	*Lepidium sativum*	Halim
41	*Mentha* spp.	Mentha, Poudina

S.No.	Botanical name	Trade name
42	*Ocimum basilicum*	Sweet basil
43	*Ocimum tenuiflorum*	Tulasi
44	*Papaver somniferum* L.	Opium poppy, Postdana
45	*Piper longum*	Pippali
46	*Plantago ovata*	Isabgol
47	*Plectranthus barbatus*	Gandhira
48	*Pongamia pinnata*	Karanj
49	*Premna corymbosa*, syn. *P. integrifolia, P. mucronata*; *P. serratifolia*	Arni, Munnai
50	*Prunus armeniaca*	Chuli
51	*Rauvolfia serpentina*	Sarpagandha
52	*Saussurea costus*	Kuth
53	*Silybum marianum*	Milk Thistle
54	*Simmondsia chinensis*	Jojoba
55	*Solanum* spp.	For Solasodine
56	*Stevia rebaudiana* Bertoni	Stevia, recent introduction/cultivation
57	*Trachyspermum ammi*	Ajwain
58	*Vetiveria zizanioides* (L.) Nash, syn. *Andropogon zizanioides* Linn.	Vetivers
59	*Vitex negundo*	Neergundi
60	*Withania somnifera*	Ashvagandha
61	*Ziziphus sativa*, syn. *Z. jujuba, Z. vulgaris*	Ber, Chinese Tsao, Unnab

On the other hand, to lay strong scientific foundation and widen the scope of traditional systems, India needs to intensify identification of alternative resources among the documented plant species with economic and medicinal potential and their sister/ancestral relatives, with focus on research and development in biochemical profiling, identification of active principles (in the available/reported plant resources), followed by molecular profiling and characterization of principal components. This needs to be followed with domestication and development of cultivation practices for effective harnessing of economic component for exploitation/use. In case plant species of trading value, priority must be on development of best agrotechnology for their cultivation, restricting their harvesting/exploitation from nature- meeting the market demand for raw material and large scale facilitated use. This action would help overcoming inevitable threat to the natural habitat of important potential medicinal plant species promoting their use in the health sector. This must be supported with improved processing technology meeting the requirements of global and national IPR standards and the quality control.

4.5 Present and Emerging Markets for Commercial/ Traded MAPs

The export of herbs and value-added extracts of medicinal herbs has been gradually increasing over years. In 2017-2018, India exported US$ 330.18 million worth of herbs at a growth rate of 14.22% over the previous year. Also, exports of value-added extracts of medicinal herbs and herbal products in 2017-2018 stood at US$ 456.12 million, recording a growth rate of 12.23% over the previous year. The demand for herbal/value-added extracts of medicinal herbs is gradually increasing in foreign countries, especially in European and other developed countries.

For these reasons, manufacturing of herbal-based medicines by multinational and domestic pharmaceutical companies and the increased public interest to use of herbal-based products had been significantly increasing, contributing to the economic growth of the plant-based medicine in recent times. For example, the global market potential of *Aloe vera*, used as medicine to treat burns and added to skin creams and cosmetics, was estimated in the billions of dollars for which India a major producer. The export of many other medicinal plants and plant products from India has substantially increased in the last few years. India is the second largest producer of castor seeds in the world, producing about 1,25,000 tons per annum. Other major pharmaceuticals exported from India in the recent years are Isabgol (Psyllium husk), opium alkaloids, Senna derivatives, Vinca (Periwinkle) extract, Cinchona alkaloids, Ipecac root alkaloids, solasodine (Solanaceous plants), Diosgenine/16DPA, Menthol (Mint), gudmar (Madhunashini) herb, mehndi (Henna) leaves, papain (papaya), Rauvolfia, guar gum, jasmine oil, Agarwood (*Aquillaria*) oil, sandal wood oil, etc., (Kokate *et al.*, 2005). The turnover of herbal medicines in India as over-the-counter products, ethical and classical formulations, and home remedies of traditional systems of medicine is about 1 billion US$ and export of herbal crude extract is about $ 80 million (Kamboj, 2000). Also, India is the hub of the regional trade. At the national level, 40% of the state forest-based revenues and 70% of the forest export resources come from MAPs and non-timber forest products (NTFPs), mostly exported as unprocessed raw material (lost opportunities). In addition, India receives about 90% of the forest collection as (common) raw material from Nepal, which includes an estimated 20,000 tons of MAPs, worth US$ 18-20 million every year. This scenario offers India a lot of opportunities for capacity building in manufacturing of final products rather than exporting raw material. This can be achieved either through strengthening research and development programs using indigenous knowledge and modern technological skills in areas like biotechnological and biochemical profiling research through skill development or international support and cooperation. This would facilitate strengthening government's Make in India program and making India manufacturing hub of herbal-based products and drugs.

4.5.1 Indian Perspective of MAPs in National and International Trade

Considering enormous resources of medicinal and herbal plants in India with associated knowledge, Indian perspective from trade point of view is quite bright. However, non-exploration and/or non-effective exploitation of the pre-historic knowledge of Ayurveda and its applications to cure illnesses to its fullest potential with the solid science-based research reflects that there is lot of scope and opportunities for future research and development. The MAPs of the Indian traditional medicinal systems need revisit with the help of modern technologies, to compete in the national and international market, facilitating conservation and promoting global use. If this happens successfully, India could gain an incredibly significant competitive edge in the global market, especially in the wellness pharma, healthcare, and beauty care segments. There is a lot of scope for India to achieve global leadership through export of quality produce and products of MAPs and other related plant-based products.

Government of India has expressed its support and encouragement for the Traditional Indian Medicine (TIM) with establishment of a separate department for Indian Systems of Medicine and Homeopathy now known as AYUSH (Ayurveda, Yoga, Unani, Siddha, Homoeopathy) in March 1995 to strengthen and promote the indigenous systems. Priorities laid in the program of AYUSH include education, standardization of drugs, enhancement of availability of raw materials, research and development, information, communication, and larger involvement in the national system for delivering health care. The Central Council of Indian Medicine oversees teaching and training institutes, while Central Council for Research in Ayurveda and Siddha deals with interdisciplinary research.

To facilitate and promote cultivation, and export of MAPs, the Government of India has taken several measures. The National Medicinal Plants Board (NMPB) offers up to 75% subsidy to farmers; formulates schemes and guidelines for financial assistance in various zones of MAPs divisions secured under promotional and commercial plans, which are relevant for government and non-government associations.

The Department of Commerce has set up export promotion councils (EPCs) to promote exports of various product groups and has assigned Shellac and Forest Products Export Promotion Council (SHEFEXIL) with a mandate of exports of herbs and medicinal plants. The export promotion of several herbal products has been assigned to Pharmaceuticals Export Promotion Council (PHARMEXCIL). The EPCs facilitate the exporting community and undertake various promotional measures to promote exports of their products.

Under the Market Access Initiative (MAI) Scheme of the Department of Commerce, the EPCs/trade bodies are provided with financial assistance to participate and organize trade fairs, buyer–seller meets (BSMs), reverse buyer–seller meets (RBSMs), research and product development, market studies, etc.

The Merchandise Exports from India Scheme (MEIS) provides incentives to the exporting community for specified goods in order to overcome infrastructural inefficiencies and the associated costs of exporting products manufactured in India. The Scheme gives special emphasis to those products which are of India's export interest and have the capability to generate employment and enhance the country's competitiveness in the global market.

The International Cooperation Scheme of the Ministry of AYUSH provides financial assistance to exporters to help them participate in trade fairs, organize international business meets and conferences, and avail product registration reimbursements.

In November 2017, the National Medicinal Plants Board (NMPB), Ministry of AYUSH, Government of India launched a 'Voluntary Certification Scheme for Medicinal Plants Produce (VCSMPP)' to encourage good field collection practices (GFCPs) and good agricultural practices (GAPs) for cultivation and in trading of MAPS. The VCSMPP enhances the availability of certified quality medicinal plants and raw materials in the country. It also boosts exports and India's share in the global export of herbal material.

The Ministry of AYUSH through its quality certification program, such as 'AYUSH' and 'Premium' marks has been assisting the industry in setting up quality standards. In addition, it has also entered into a Memorandum of Understanding (MoU) with a few countries to promote trade and exchange of traditional medicines.

It is estimated that there are about 25,000 licensed pharmacies of Indian system of medicine in India. Presently, about 1000 single drugs and about 3000 compound formulations are registered. Herbal industry in India uses about 8000 MAPs. However, very few pharma companies have standardized herbal medicines using active compounds as markers linked with confirmation of bioactivity of herbal drugs (active principle) in experimental animal models (Kamboj, 2000). From about 8000 drug manufacturers in India, there are, however, not more than 25 that can be classified as large-scale manufacturers. The current annual turnover of Indian herbal industry is estimated to be around US $ 300 million in Ayurvedic and Unani medicine. The industry started with a turnover of about US $ 27.7 million. In 1998-1999, it went up to US $ 31.7 million and in 1999-2000, the total turnovers were US $ 48.9 million of Ayurvedic and herbal products. Export of herbal drugs from India is around $ 80 million (Anonymous, 1996). Some of the highly consumed and demanded in trade MAPs are highlighted with superscript in following main table of the section (Kalia, 2005). However, over-harvesting of many important medicinal plant species from natural/ wild habitat, particularly by private sector, has led to increased habitat degradation of many medicinal plant species putting them into threat for long term survival, which must be discouraged with alternative regeneration technologies as discussed earlier.

An analysis of most frequently used plant-based therapies in the Ayurvedic system revealed that 43% of them have been tested on humans, while 62% have been the subject of one or more animal studies. Among these, drugs having sufficient clinical data are Guggul, Brahmi, Ashwagandha, Amlaki, Guduchi, Kutki, Shatavari, and Shunthi (Aneesh *et al.*, 2009). Pharmacopoeia of India (1996) covers few botanical monographs, like on Clove, Guggul, Opium, Mentha, Senna and Ashwagandha. The Ayurvedic Pharmacopoeia of India gives monographs for 258 different Ayurvedic drugs (Fabricant and Farnsworth, 2001). Presently, Indian systems of medicine use more than 1100 medicinal plants, majority of which are collected from the wild. More than 60 species, which are in great demand, are highlighted in the main table. The tribal belts of India are rich in these plants and the trade of raw material of these plants is the main source of income for the tribes to support their livelihood. Because of lack of education and skills in identification actual plant species and unethical motives due to poverty, there are ample opportunities/practices for adulteration and contamination of the final product in the process. Thus, the adequate availability of quality raw materials free from adulterants at reasonable prices has become a big problem for the industry, while the demand is increasing every year. Very few efforts have been made either by government or by industry to seriously study the supply and demand chain and to strengthen it with capacity building through state-of-the-art infrastructure development, meeting international standards. Therefore, skill development, supporting and sustainable price regimes with assured market, to make collection and cultivation more remunerative, enhancing the potential value of herbal plant material, and plant biodiversity at large are very much the need of the hour.

India needs to follow good agricultural practices (GAPs) to ensure the use of correct raw materials and cover the entire cycle including harvesting, processing, transportation, and storage. The selection of the correct germplasm using modern DNA fingerprinting and chemo-profiling techniques is also a priority (De Smet, 2002). The domestic herbal market has now crossed the Rs.5000 crore mark and is set to reach further heights in coming years. Herbal exports from India are worth Rs.450 crore, whereas even a decade back the amount was barely Rs.100 crore. This may not be a huge sum, when compared to the Chinese herbal exports market, which is worth Rs.2000 crore, but the sheer potential of the Indian herbal companies should make people to sit up and take notice. The biggest stumbling block facing the Indian herbal care segment is still the lack of trust it generates in the developed markets like Europe and USA. There is a wide impression that Indian herbal extracts do not undergo rigorous toxicity tests. Indian firms would have to be better in promoting themselves internationally, especially in regulated markets. Indian herbal extracts will have to undergo stricter toxicity tests to win the confidence of the developed world (Anonymous, 2001).

Numerous drugs have entered the international pharmacopoeia via the study of ethnopharmacology and traditional medicines. For traditional medicines, newer guidelines of material standardization, manufacture, and quality control with rigorous scientific research (scrutiny) meeting internationally accepted scientific standards will be required to promote use of traditional treatments. Traditional medical practices can offer a more holistic approach to drug design and myriad of possible targets for scientific analysis. Powerful new technologies such as automated separation techniques, high-throughput screening, and combinatorial chemistry are revolutionizing drug discovery, and this will help to rediscover the drug discovery process.

Numerous drugs have entered the international market through exploration of ethnopharmacology and traditional medicines. Progress in genomics and proteomics has opened new gateways in therapeutics and drug discovery and development. Better understanding of the human genome has helped in understanding the scientific basis of individual variation. Traditional Indian Medicine (TIM) and Traditional Chinese Medicine (TCM) carry many generations' observations that have been organized and documented. However, a review on competitive ability of the two in global market reflects that the TIM (Ayurveda) and TCM remain the most ancient and dynamic yet living traditions. Because of the increased global interest in traditional medicine, efforts have been made or are underway to monitor and regulate herbal drugs and traditional medicine. However, China has been successful in promoting its therapies with more research and science-based approach, while TIM (Ayurveda) still needs more extensive scientific research and evidence base (Patwardhan *et al.*, 2005; Aneesh *et al.*, 2009). Consequently, China has successfully promoted its own therapies and drugs like ginseng (*Panax notoginseng*), mahuang (*Ephedra vulgaris*) and gingko (*Ginkgo biloba*) with scientific evidence acceptable to the global community. The approach of integrative medicine by selective incorporation of elements of TCM alongside the modern methods of diagnosis has achieved a great success in China. It may be right to say that an 'Herbal Revolution' by India is just waiting to happen (Aneesh *et al.*, 2009). India could truly become a global leader in the herbal medicine category by inventing and patenting medicines for several ailments by using a combination or mixture of herbal formulations (Phillipsion, 1989; Bhattacharya *et al.*, 1999). Although herbal medicines have been used for thousands of years, however, the basic research programs are still needed to be focused on identification of principal component(s), molecular characterization, and mode of action, to provide strong science base, scientific rational and the quality assurance to practitioners and users.

To overcome contamination from pesticide residues and heavy metals, there should be control measures to implement necessary standard operating procedure (SOP) at source. Good laboratory practices (GLP) and good management practices (GMPs) are also needed to produce good quality medicinal products. Without all these measures, it is impossible to realize the dream of having a major share of global herbal drug industry despite having gold mine of well-documented and well-practiced knowledge of traditional herbal medicines. The TCM examples would help India at various levels including policies, quality standards, integration practices, research models, and the complementary integration, where public health is kept at the central position. Both TIM and TCM are great traditions with strong time-tested faith, imperative experience and philosophical basis and could play an important role in new therapies, drug discovery and development processes.

4.5.2 Strengthening of Herbal Industry for Trade

Considering great demand for herbal medicine in the developed as well as developing countries, because of their wide biological activities, higher safety margin than the synthetic drugs, and lesser costs (Gadre *et al.*, 2001), new vistas have opened for the herbal industry. Nutraceutical (medicinally or nutritionally functional foods) and MAPs play a great role in food supplements for wellness as well as in personal care of the mankind alongside the therapeutically active substances. Thus, plant-based industry is a promising sector with enormous economic growth potential. Nutraceuticals are in great demand in the developed world particularly USA and Japan. The nutraceutical market in USA alone is about $ 80-250 billion with a similar market size in Europe. In Japan, it is worth $ 1.5 billion (Kokate *et al.*, 2005). After passing of the Dietary Supplement Health Education Act in USA in 1994, a huge market has arisen, which permits unprecedented claims to be made about food or the dietary supplement's ability about health benefits including prevention and treatment of certain deficiencies and diseases. This act has motivated pharma industry not to include only drugs (compounds) isolated from fauna and flora, but herbal medicines as nutraceuticals. The Indian herbal pharma companies must see this as a good opportunity for producing and marketing of such products (Satakopan, 1994). Further, the importance of MAPs in the national economy and their potential for the rapid growth of herbal products, such as perfumery and allied industry in India has been emphasized from time to time (Kokate, *et al.*, 2005). New trends are emerging in the standardization of herbal raw materials whereby it is carried out to reflect the total content of phytoconstituents like polyphenols, which can be correlated with biological activity (Sapna and Ravi, 2007). The major players in the traditional medicine pharma industry are, namely, Himalaya, Zandu, Dabur, Hamdard, Maharishi, Patanjali, etc. They have been standardizing their herbal formulations by chromatography techniques like thin-layer chromatography (TLC)/ High-performance thin-layer chromatography (HPTLC), DNA finger printing, etc. (Borris, 1996).

Despite the long-standing knowledge and understanding of using medicinal plants in its codified medical systems as well as through their use in highly diverse folk traditions; having experience and the knowledge about essential features of plant for identification, methods of collection and processing information about biological activities and uses (that were transmitted both in written form, under the traditional medical cultures and/or orally with substantial domestic and international trade from antiquity), India has not been able to capitalize on this position to promote use in developed country markets to its full potential. Though, the Indian herbal medicine market is growing at a steady pace of between 15% and 20% every year. To achieve improvement in this area, India must identify products, which may be relevant to diseases found in the developed world and for which no medicine or only palliative therapy is currently available. This would enable rapid access of these herbal medicines into developed country markets. However, there are major challenges in tapping the substantial potential for utilizing medicinal, aromatic, and natural dyes plants (MADPs) nationally available in India as well as in export markets. At the forefront of the problems is the issue of ensuring consistency and acceptability of quality. Traditionally, in the past, the usage of medicinal plants was a part of a local community's culture and health practices, and quality was more manageable. However, this tradition was being rapidly eroded to meet the growing demand, and to further usage more widely. The control of medicinal plant research and development and the onus for maintaining quality and consistency has rightly moved to the industry and needs greater attention.

4.5.3 Meeting the International Standards

Every herbal formulation must be standardized as per WHO guidelines (Anonymous, 1991, WHO, 1998). The objective of WHO guidelines is to define basic criteria for the evaluation of quality, safety, and efficacy of herbal medicine drugs (Phillipsion, 1989). One report says, India is one of the world's twelve leading biodiversity centres, of which about 15,000-20,000 plants have good medicinal properties of which only about 7,000-7,500 are being used by traditional practitioners. Under the coded Indian traditional medicinal systems, the Siddha system uses around 600, Ayurveda 700, Unani 700, and modern medicine about 30 plants species. Chaudhri (1996) made projection that after information technology, herbal technology will be India's biggest revenue earner. Therefore, India has a great role to play, as supplier of herbal products not only to meet the domestic needs, but also to take advantage of the tremendous export potential. However, to be a global supplier of herbal medicines conforming to international specification, the following aspects still demand attention:

- Development of taxonomic keys to ensure botanical identification of targeted medicinal plants associated with Indian System of Medicine, particularly the one of commercial significance.
- Development of descriptors for characterization of commercially important plant species and genetic diversity, which may be of significance.
- All herbal ingredients in preparation are to be specified by their botanical names besides their popular/common names. This would avoid adulteration of material with sister species or any other source.
- Develop good agricultural cultivation practices (GAPs) for most commercially important medicinal plant species for sustainable production of raw material, restricting harvesting from wild/nature. This would lead to *in-situ* conservation of natural population facilitating their continued dynamic evolution. In these cultivation practices, integration of precautions to avoid microbial contamination must be an important component.
- In most commercially significant medicinal plant species attempts should be made for their genetic improvement, particularly to improve productivity and quality of principal component(s) of commercial value..
- Development of systematic and scientific post-harvesting processing protocols of plants in a clean, economic, and safe way, ensuring quality, consistency of products/drugs at par with modern synthetic drugs.
- Biochemical profiling, isolation, molecular characterization of principal component(s) along with estimation of possible toxic substances and inorganic constituents, providing full spectrum.
- There is an urgent need for studies to quantify the frequency and potential risk of heavy metal poisoning from Ayurvedic medicines. Also needed is 'culturally appropriate education' that can inform the public of the potential for toxicity associated with many different products associated with this practice.
- Standardized manufacturing process, the pattern and concentration of constituents should be kept as constant as possible, as this is a prerequisite for reproducible therapeutic results. Product standards must match international regimes and must be adjusted to regulatory requirements and consumer demands.
- Pharmacological and clinical studies to ascertain their efficacy and safety.
- Standardization to ensure uniformity- The use of medicinal plants in combination to be limited to facilitate analysis and to apply quality control and standardization parameters to herbal drug preparations.
- Documentation of research, as such scientific aspects will project herbal medicine in a proper perspective and help in sustained global market (Handa and Kapoor, 1995).

An estimated number of about 80% of world population depends on natural products for their health care, because of absence of side effects and high cost of modern medicine (WHO, Satakopan, 1994). These natural drugs are safe and can be made available at low cost, therefore, the World Health Organization currently recommends and encourages traditional herbal remedies in natural health care programs. People's faith in such remedies is reflected by a whooping turnover in about 450 Crores Indian Rupees annually in herbal market besides a healthy 11% annual growth rate and the increasing export potential. This has attracted several large and medium scale pharmaceutical industries and even multinationals to jump on to the band wagon (Handa and Kapoor, 1995).

4.6 Future of MAPs Industry in India

In India, the cultivation and production of medicinal plants is mostly unorganized. Considering small land holdings and poor economic resources, formation of farmer associations and a well-equipped supply chain management will improve the production and ensure sales of medicinal plants raw material in the country and abroad. The sector has observed establishment of industrial houses and recent entries of start-ups bringing in technology upgradation, which is an encouraging sign for trade of MAPs. These start-ups are using precise farming techniques by integrating artificial intelligence (AI) and data analytics for crop profiling, seed analysis for better germination, crop production, protection, and post-harvest management to enable harnessing of potential economic yield to a larger extent. The information compiled in this section must provide impetus to these initiatives.

References

Aneesh TP, Mohamed Hisham, Sonal Sekhar M, Manjusree Madhu, Deepa, TV. (2009) International market scenario of traditional Indian herbal drugs – India declining. International Journal of Green Pharmacy 184-190. DOI:10.4103/0973-8258.56271

Anonymous (1991) Guidelines for the Assessment of Herbal Medicines, Document No. WHO/TRM/91, 4, World Health Organization, Geneva.

Anonymous (1996) Sectoral Study an Indian Medicinal Plants-status, perspective and strategy for growth. Biotech Consortium India Ltd., New Delhi.

Anonymous (2001) Legal status of traditional medicine, complementary/alternative medicine: A worldwide review. World Health Organization, Geneva.

Bhattacharya, AA, Chatterjee, S Ghosal and Bhattacharya, SK. (1999) Antioxidant Activity of Active Tannoid Principles of Embilica officinalis. Indian Journal of Experimental Biology 37: 676-680.

Borris, J. (1996) Natural Products Research Perspectives from a Major Pharmaceutical Company. Merck Research Laboratories. J. Ethnopharmcol 51: 29.

Chaudhri, RD. (1996) Herbal Drugs Industry. Eastern Publisher, New Delhi, 1st Edn.

De Smet PA. (2002) Herbal remedies. New England journal of medicine 347:2046–56.

Fabricant, DS and Farnsworth NR. (2001) The value of plants used in traditional medicine for drug discovery. Environ Health Perspect. 109:69–75.

Gadre, AY, Uchi, DA, Rege, NN and Daha SA. (2001) Nuclear Variations in HPTLC Fingerprint Patterns of Marketed Oil Formulations of Celastrus Paniculates. Ind. J. of paniculatue.. 33: 124-45.

Handa, SS and Kapoor, VK. (1995) Textbook of Pharmacognosy. Vallabh Prakshan, Delhi, 2nd Edn.

Kalia, AN. (2005) Textbook of Industrial Pharmacognosy. Oscar publication.

Kamboj, VP. (2000) Herbal Medicine. Current Science 78 (1): 35-39.

Kokate, CK. Purohit, AP and Gokhale SB. (2005) Pharmacognosy. Nirali Prakashan, 30th Edn.

Kotwal, PC and Banerjee, S. (1998) Biodiversity conservation in forests and protected areas: Practical problems and prospects. In: Biodiversity Conservation in managed forests and protected areas (Eds. Kotwal PC and Banerjee S) pp 1-10. Agro Botanica, Bikaner.

Patwardhan, B, Warude Dnyaneshwar, Pushpangadan, P and Bhatt Narendra (2005) Ayurveda and Traditional Chinese Medicine: A Comparative Overview. Advance Access Publication eCAM 2(4)465–473. doi:10.1093/ecam/neh140

Phillipsion, JD. (1989) Ethnopharmacology and Western Medicine. J. Ethnopharmacol 25: 61-72.

Sapna, S and Ravi, TK. (2007) Approaches Towards Development and Promotion of Herbal Drugs. Pharmacognosy Review 1 (1): 180-184.

Satakopan, S. (1994) Pharmacopeial Standards for Ayurvedic, Siddha and Unani Drugs pp 43. In: Proceedings of WHO Seminar on Medicinal Plants and Quality Control of Drugs Used in ISM. Ghaziabad.

Sharma, Alok, Shanker, C, Tyagi, LK, Singh, M and Rao, ChV. (2008) Herbal Medicine for Market Potential in India: An Overview. Academic Journal of Plant Sciences 1 (2): 26-36.

Singh, Anurudh K. (2012) Probable Agricultural Biodiversity Heritage Sites in India: XIII. Lower Gangetic Plain or Delta Region. Asian Agri-History 16 (3):237-260.

Singh, Anurudh K. (2015) Agricultural Biodiversity Heritage Sites and Systems in India. Asian Agri-History Research Foundation, Secunderabad, Telangana, India pp 467. ISBN 81-903963-4-X

Singh, Anurudh K. (2017) Revisiting the Status of Cultivated Plant Species Agrobiodiversity in India: An Overview. Proc Indian Natn Sci Acad 83 (1):151-174 (DOI: 10.16943/ptinsa/2016/v82/48406)

WHO (World Health Organization) (1998) Quality Control Guidelines for Medicinal Plant Materials. Geneva World Health Organization Available from: http://www.who.int/medicines/library/ trm/medicinalplants/qualitycontrolmeth.pdf

WHO (World Health Organization) (1998) Quality control methods for medicinal plant materials. World Health Organization. https://apps.who.int/iris/handle/10665/.

WHO (World Health Organization) (2007) WHO Library Cataloguing-in-Publication Data WHO monographs on selected medicinal plants Vol. 3. 1. Plants, Medicinal. 2. Angiosperms. 3. Medicine, Traditional. I. WHO Consultation on Selected Medicinal Plants (3rd: 2001: Ottawa, Ont.) II. World Health Organization. ISBN 978 92 4 154702 4

4.2 List of Medicinal and Aromatic Plants (MAPs) species of trade value

As discussed previously, there are many medicinal and aromatic plants, which are being used directly or in formulations of various herbal drugs (collected from natural habitats or through their cultivation), have significant commercial market. There has been a gradual increase in recent times in market demand of these species. Therefore, there is an urgent need to take notice of these and produce an inventory to attract entrepreneurs and investment to facilitate establishment large production units and organized trade. This has become even more essential in the present time of intellectual property rights (IPR) and related issues enhancing the importance of inventorying such resources along with associated indigenous knowledge.

At present only 250-300 species are grown or collected commercially and traded at the global level (Pareek *et al.*, 2005). However, the global demand for herbal products has increased manifold and there has been a significant increase in industrial turnover of plant-based products in the past few decades. A notice of these facts has been taken by National Medicinal Plant Board (NMPB), Ministry of Ayush, Government of India and several steps has been taken to promote research institutes, growers, industries in India in the efforts of commercial exploitation of these important resources to improve India's position in the international trade. This includes development of a checklist of medicinal and aromatic plant for trade (www.nmpb.nic.in › content › ongoing-and-completed-...). Here an attempt has been made to produce an updated list of these plants with basic information on trade name, habit, source of raw material (wild/cultivated) and associated systems (knowledge) to strengthen trading efforts. The proceeding pages present this list.

4.7 The list of MAPs species of Trade Value

Botanical Name[1 & 2]	**Trade Name**	**Habit**	**Material Source[3]**	**Associate Systems[4]**
Abelmoschus esculentus (L.) Moench [= *Hibiscus esculentus* L.][2]	Bhendi	Shrub	C	AFU
Abelmoschus ficulneus (L.) Wight & Arn. [= *Hibiscus ficulneus* L.]	Kattu vendai	Shrub	W	AF
Abelmoschus moschatus Medik. [= *Hibiscus abelmoschus* L.][1]	Muskdana, Latakasturi	Shrub	C/W	AFU
Abies pindrow (Royle ex D. Don) Royle	Granthiparna, Talisa	Tree	W	A
Abies spectabilis (D. Don) Spach [= *Abies webbiana* Wall. ex D. Don][2]	Talispatra	Tree	W	A
Abrus precatorius L.[2]	Chirmati, Chinnoti, Gundumani	Climber	C/W	AFHSTU
Abutilon indicum (L.) Sweet subsp. *guineense* (Schumach.) Borss.	Tutti	Shrub	W	AFSTU
Acacia caesia (L.) Willd.	Incha, Indu, Singapore pattai	Shrub	W	AF
Acacia catechu (L.f.) Willd.[1]	Katha	Tree	W	AFSTU
Acacia chundra (Roxb. ex Rottler) Willd.	Khadir	Tree	W	A
Acacia farnesiana (L.) Willd.	Irmed	Tree	C	AFS

Botanical Name[1 & 2]	**Trade Name**	**Habit**	**Material Source[3]**	**Associate Systems[4]**
Acacia nilotica (L.) Willd. ex Delile subsp. *indica* (Benth.) Brenan [=*A. arabica* (Lam.) Willd. var. *indica* Benth.][2]	Babul, Keekar	Tree	W/C	AS
Acacia sinuata (Lour.) Merr. [= *A. concinna* (Willd.) DC.][2]	Chikakai (Shikakai)	Climber	W	AF
Acalypha fruticosa Forssk.	Cinni	Shrub	W	FS
Acalypha indica L.	Khokali, Arishtamanjari	Herb	W	AFHS
Achillea millefolium L.	Brinjasif	Herb	C/W	HSUW
Achyranthes aspera L[2].	Apamarga	Herb	W	AFHSTU
Achyranthes bidentata Blume	Apamarga	Herb	W	AF
Aconitum balfourii Stapf	Vatsnabha	Herb	W	A
Aconitum bisma (Ham.) Rapaics	Bikhma	Herb	W	AF
Aconitum chasmanthum Stapf	Vatsnabha	Herb	W	AH
Aconitum deinorrhizum Stapf	Vatsnabha	Herb	W	A
Aconitum falconeri Stapf	Vatsnabha	Herb	W	A
Aconitum ferox Wall. ex Ser[1]	Vachhnag, Atis	Herb	W/C	AFHSU
Aconitum heterophyllum Wall. ex Royle[1]	Atis, Aconite, Ativish	Herb	W/C	AFSTU
Aconitum laciniatum Stapf., syn. *A. ferox* var. *laciniata* Brühl	Kalo Bikhmo	Herb	W	F
Aconitum spicatum (Bruhl) Stapf., *A. ferox* var. *spicata* Bruhl	Bish	Herb	W	F
Aconitum violaceum Jacq. ex Stapf	Mitha Telia	Herb	W	F
Acorus calamus L.[1]	Vach, Vaj, Calamus roots, Vekhand	Herb	C/W	AFSTUW
Actiniopteris dichotoma Mett.	Mayurpankhi	Herb	W	AS
Adansonia digitata L.	Gorakimli	Tree	C	AFSU
Adenanthera pavonina L.	Kamboji, Tamaraka	Tree	C/W	AFS
Adenia hondala (Gaertn.) W.J. de Wilde	Vidari	Climber	W	AF
Adhatoda beddomei C.B. Clarke [= *Justicia beddomei* (C.B. Clarke) Bennet]	Vasa	Shrub	W	A
Adhatoda zeylanica Medik. [= *A. vasica* Nees; *Justicia adhatoda* L.][1]	Adusa, Vasaka	Shrub	W/C	AFHS
Adiantum capillus-veneris L.	Hanraj, Parshoshan	Herb	W	FU
Adiantum philippense L.f. [= *A. lunulatum* Burm.]	Hansraj	Herb	W	AF
Adiantum venustum D. Don.	Hansraj	Herb	W	F

Botanical Name[1 & 2]	Trade Name	Habit	Material Source[3]	Associate Systems[4]
Aegle marmelos (L.) Correa[2]	Bael, Belgiri, Bilva, Vilvam	Tree	C/W	AFHSTU
Aerva lanata (L.) Juss[2]	Cheroola	Herb	W	AFS
Aesculus indica (Wall. ex Cambess) Hook.	Indian Horse Chesnut, Bankhor pangar	Tree	C/W	F
Ageratum conyzoides L.	Visamustih, Goat weed, Visadodi	Herb	W	AFS
Aglaia elaeagnoidea (A. Juss.) Benth. syn. *A. roxburghiana* Miq. Hiern	Priyangu	Tree	W	AF
Aglaia odorata Lour.	Pishthparni, Pithavan	Tree	C	F
Ailanthus excelsa Roxb.	Aralu	Tree	C/W	AFS
Ailanthus triphysa (Dennst.) Alston	Guggula Dhup	Tree	C/W	AF
Ainsliaea aptera DC., syn. *Liatris cordata* Royle ex C.B. Clarke	Sathjalani	Herb	W	F
Ajuga bracteosa Benth.	Nilkanti	Herb	W	F
Alangium salvifolium (L. f.) Wang. syn. *A lamarckii* Thw.	Ankodah, Dirghakilaka	Tree	W	AFST
Albizia amara (Roxb.) Boivin[2]	Krishnasirish	Tree	W	AFS
Albizia lebbeck (L.) Benth.	Vaakaveru, Shirish	Tree	C/W	AFSU
Alhagi pseudalhagi (M. Bieb.) Desv.	Durlabha, Yavasaka	Shrub	W	ASTU
Allium ampeloprasum L.	Leek, Levant garlic	Herb	C	FW
Allium cepa L.	Onion	Herb	C	AFHSTUW
Allium sativum L.[1]	Lasan, Velathulli	Herb	C	AFHSTUW
Allium stracheyi Baker	Jambu	Herb	W/C	A
Aloe barbadensis Mill. [= *A. vera* (L.) Burm.f.][1]	Ghikanvar, Kumari	Herb	C/W	AFSTUW
Alpinia calcarata (Haw.) Roscoe[2]	Chittaratha, Aratha	Herb	C	A
Alpinia galanga (L.) Willd.	Kulinjan, Rasnamool, Koshtakulinjan	Herb	C/W	AFSU
Alpinia officinarum Hance	Khulinjan	Herb	I	ASU
Alstonia scholaris (L.) R.Br.[2]	Saptaparnachal (Satveen)	Tree	W	AFHSTU
Alstonia venenata R.Br.	Anadana, Raja - adana	Shrub	W	AF
Alternanthera sessilis (L.) R.Br. ex DC.	Ponnanganni	Herb	W	AFST

Botanical Name[1 & 2]	**Trade Name**	**Habit**	**Material Source[3]**	**Associate Systems[4]**
Althaea officinalis L.	Khatmi, Resha-Khatami, Shilaras	Herb	C/W	AFSU
Altingia excelsa Noronha	Silaras	Tree	C/W	AFSU
Amaranthus paniculatus L.	Chaulai, Ramdana	Herb	C	AS
Amaranthus spinosus L.	Cholai, Bhandira	Herb	W	AFST
Amaranthus tricolor L. [= *A. gangeticus* L.; *A. mangostanus* L.]	Lal sag, Alpamarisa	Herb	C	AS
Ambroma augusta (L.) L.f., syn. *Abroma augusta* Jacq.	Ulatkambal	Shrub	C/W	AFH
Ammannia baccifera L.	Kurand ghas	Herb	W	AS
Ammi majus L.	Ammi	Herb	C	FW
Amomum aromaticum Roxb.	Bengal cardamom, Brhdela	Herb	W	A
Amomum kepulaga Sprague, syn. *A. compactum* Sol. ex Maton	Elachi	Herb	I/C	F
Amomum subulatum Roxb.	Elachi Badi	Herb	C/W	AFSTU
Amorphophallus paeoniifolius (Dennst.) Nicolson var. *campanulatus* (Blume ex Decne.) Sivad. [= *A. campanulatus* Blume ex Decne.]	Arsghna, Surankand, Zaminkand	Herb	C/W	AFSU
Amorphophallus sylvaticus (Roxb.) Kunth	Tubers	Herb	W	AFS
Anacardium occidentale L.[2]	Kaju	Tree	C	AFHS
Anacyclus pyrethrum (L.) Cass.[2]	Akarkara	Herb	I/W	ASU
Anamirta cocculus (L.) Wight & Arn.	Rakthala	Climber	W	AFHSUW
Anchusa strigosa Labill.	Gaozaban (Subs.)	Herb	I	A
Andrographis paniculata (Burm.f.) Wall. ex Nees[1]	Kalmegh	Herb	W/C	AFHSUW
Anethum graveolens L. [= *A. sowa* Roxb. ex Flem.]	Sowa	Herb	C	AF
Angelica archangelica L.	Conda, Angelica	Herb	W	AW
Angelica glauca Edgew.	Chuara, Choru	Herb	W	AT
Anisomeles malabarica (L.) R.Br. ex Sims	Karimthumpa	Shrub	W	AFS
Anodendron paniculatum A. DC.	Sarakkodi, Kavali	Climber	W	F
Anogeissus latifolia (Roxb. ex DC.) Wall. ex Guill. & Perr. [2]	Dhawada	Tree	W	AFSTU
Anthemis nobilis L.	Babuna, Gulbabuna	Herb	C/W	HSU
Anthocephalus cadamba (Roxb.) Miq.	Kadam	Tree	C/W	AS

Botanical Name[1 & 2]	**Trade Name**	**Habit**	**Material Source[3]**	**Associate Systems[4]**
Aphanamixis polystachya (Wall.) Parker [= *Amoora rohituka* (Roxb.) Wight & Arn.]	Rakta Rohida, Rohitak	Tree	C/W	AHS
Aphananthe cuspidata (Blume) Planch.	Peenari	Tree	W	F
Apium graveolens L.	Ajmoda, Celery	Herb	C/W	AHSTU
Aquilaria agallocha Roxb.[1], syn. *A. malacaccensis* Lam.[1]	Agar kala, Akil	Tree	I/W	AFSTU
Arctium lappa L.	Phaggarmul	Herb	W	FH
Areca catechu L.	Supari	Tree	C	AFHSTUW
Argemone mexicana L.	Brhami Dandi, Kusme Beeja	Herb	W	AFHSU
Argyreia speciosa (L. f.) Sweet [= *A. nervosa* (Burm.f.) Bojer]	Samudraphal, Samundra Sokh	Shrub	W	ASTU
Aristolochia bracteolata Lam.	Kidamari	Herb	W	AFS
Aristolochia indica L.	Ishwar Mool	Climber	W	AFSTU
Aristolochia rotunda L	Zarawand	Climber	W	FU
Aristolochia tagala Cham.	Aadutheendapalai	Climber	W	F
Arnebia benthamii (Wall. ex G. Don) I.M. Johnst.	Gauzaban	Herb	W	A
Arnebia euchroma (Royle) I.M. Johnst.	Gaozaban	Herb	W	F
Arnebia nobilis Rech. f.	Ratan jyothi / Ratan jot	Herb	I	FU
Artabotrys hexapetalus (L. f.) Bhandari	Harichampa, Madanmast	Shrub	C/W	AF
Artemisia absinthium L.	Afsanteen	Shrub	W	AHSUW
Artemisia maritima L. ex Hook.f.	Ajavayana, Chauhara, Cina	Herb	W	AHW
Artemisia nilagirica (C.B. Clarke) Pamp. syn. *A vulgaris* Linn. var. *nilagirica* Clarke	Thavanam	Herb	W	AF
Artemisia pallens Wall. ex DC.	Davna	Herb	C	FS
Artemisia parviflora Roxb. ex D. Don	Pati	Shrub	W	F
Artocarpus heterophyllus Lam. syn. *A. integrifolia*	Kathal, Panasa	Tree	C/W	AFS
Arundo donax L.	Dhamana, Baranal, Nala	Shrub	W	AHS
Asparagus adscendens Roxb.[2]	Musali Safed, Satawar	Climber	W	ASU
Asparagus racemosus Willd.[1]	Shatavari, Satawar	Climber	W/C	AFSTU
Asparagus sarmentosus L.	Satawari, Shakakul	Climber	W	ASU

Botanical Name[1 & 2]	**Trade Name**	**Habit**	**Material Source[3]**	**Associate Systems[4]**
Atalantia monophylla (L.) DC.	Kurandubeeja	Tree	W	AFS
Atropa acuminata Royle ex Lindl. syn. *A. belladonna* auct. non L.	Jharka	Herb	C/W	FU
Atropa belladonna L.	Belladona	Herb	I	AHSUW
Atylosia goensis (Dalzell) Dalzell	Katttupaayar	Climber	I/C	AF
Atylosia scarabeoides (L.) Benth.	Ban kulatha	Climber	W	F
Azadirachta indica A. Juss.[1]	Neem	Tree	C/W	AFHSTU
Azima tetracantha Lam.	Kanta-gur-kamai, Kantangur, Kundali	Shrub	W	AFS
Bacopa monnieri (L.) Wettst.[1]	Brahmi, Jal Brahmi, Nirbrahmi	Herb	W/C	AFHS
Balanites aegyptiaca (L.) Delile	Hingan, Hingol, Ingua	Tree	W	AFTU
Balanophora fungosa J.R. Forst. & J.G. Forst. syn. *B. fungosa indica* (Arn.) B. Hansen	Thippali	Herb	W	S
Baliospermum montanum (Willd.) Mull. Arg. [= *B. solanifolium* (Geiseler) Suresh][2]	Dantimool, Nagadanti	Shrub	W	AFSTU
Bambusa bambos (L.) Voss. *B. arundinaceae* (Retz.) Willd.[2]	Bansalochan, Tabashir	Tree	W	AFSU
Bandeiraea simplicifolia (M. Vahl ex DC.) Benth. [= *Griffonia simplicifolia* (M. Vahl ex DC.) Baill.]	African Griffoni [5-hydroxytryptophan (5-HTP) source]	Climber Shrub	I/C/W	Africa
Barleria cristata L.	Raktapushpa, Sweta Saireyaka	Shrub	W	AST
Barleria prionitis L.	Vajradanti	Shrub	C/W	AFS
Barleria strigosa Willd.	Bala, Dasi, Amli	Shrub	W	AST
Barringtonia acutangula (L.) Gaertn.	Samudraphal	Tree	W	AFSTU
Barringtonia racemosa (L.) Spreng.	Samuthira Palam	Tree	W	AFS
Basella alba L.	Pasalai, Vasalacheera pacha	Climber	C	AF
Bauhinia malabarica Roxb.	Amlosa, Asmantaka	Tree	W	AS
Bauhinia purpurea L.	Sonachal	Tree	C/W	AFST
Bauhinia racemosa Lam.	Kachnar	Tree	W	ASU
Bauhinia tomentosa L.	Kachnar, Kanchana	Shrub	W	AFS
Bauhinia vahlii Wight & Arn.	Adda	Climber	W	AF

Botanical Name[1 & 2]	**Trade Name**	**Habit**	**Material Source[3]**	**Associate Systems[4]**
Bauhinia variegata L.	Kachnar	Tree	C/W	AFS
Benincasa hispida (Thunb.) Cogn.[2]	Kumpalanga pacha	Climber	C	AFSTU
Berberis aristata DC.[1]	Daruhaldi, Rasnajan, Zarisuk	Shrub	W/C	AFSTU
Berberis asiatica Roxb.	Daru haldi, Rasout	Shrub	W	AFU
Berberis chitria Lindl., syn. *B. aristata* auct. Hook. f. &Thoms.	Daruhaldi/Rasaut	Shrub	W	AF
Berberis lycium Royle	Dar-hald, Chitra	Shrub	W	AU
Berberis tinctoria Lesch.	Daruhaldi/Rasaut	Shrub	W	F
Berberis umbellata Wall. ex G. Don	Daruhaldi/Rasaut	Shrub	W	F
Bergenia ciliata (Haw.) Sternb., syn. *B. ligulata* (Wall.) Engl.[2]	Pashnabheda	Herb	W	A
Bergenia stracheyi (Hook. f. & Thoms.) Engl.	Pashnabheda	Herb	W	A
Betula utilis D. Don	Bhojpatra	Tree	W	AST
Biophytum sensitivum (L.) DC.	Mukkutti	Herb	W	AFS
Bixa orellana L.	Sindhuri, Latkan dana, Annato	Shrub	C	AFHS
Blepharis edulis auct. non (Forssk.) Pers. [= *B. persica* (Burm. f.) O. Kuntze.]	Uttangan	Herb	I	AU
Boerhavia diffusa L.[1]	Punarnava raktha, Punarnava	Herb	W	AFHSU
Boerhavia repanda Willd.	Cattaranai, Varakanputu	Herb	W	S
Bombax ceiba L. [= *B. malabaricum* DC.]; *Salmalia malabarica* (DC) Schott & Endl.[2]	Mochras	Tree	W	AFS
Borassus flabellifer L.	Palmyra palm, Tad	Tree	C/W	AFSTU
Borreria hispida (L.) Schum. [= *Spermacoce hispida* L.], syn. *B. articularis* (Linn. f.) F. N. Williams.	Thaarthaaval	Herb	W	AF
Boswellia carteri Birdw.	Sali guggul	Tree	I	F
Boswellia frereana Birdw.	African elemi	Tree	I	F
Boswellia serrata Roxb.[2]	Kandur, Guggul, Dupa	Tree	W	ASU
Brassica alba (L.) Rabenh	Safed rai, Kadugu	Herb	C	AHSU
Brassica juncea (L.) Czern.	Kaduku, Sasuve Bili	Herb	C	AFS
Briedelia montana (Roxb.) Willd.	Gondni, Geia	Tree	W	AF

Botanical Name[1 & 2]	**Trade Name**	**Habit**	**Material Source[3]**	**Associate Systems[4]**
Buchanania lanzan Spreng., syn. *B. cochiinchinensis* (Lour.)[1]	Chironji	Tree	W	AFSTU
Bunium persicum (Boiss.) B. Fedts. ch	Kala-zirah, Shah-Zira	Herb	C/W	AF
Butea monosperma (Lam.) Taub. [= *B. frondosa* Willd.][2]	Palash, Tesu Phool, Lac	Tree	W	AFSTU
Butea superba Willd.	Phul kesu, Tesu Phool	Climber	W	AFS
Caccinia crassifolia (Vent.) C.Koch	Gaozaban	Climber	I	F
Cadaba fruticosa (L.) Druce [= *C. indica* Lam.]	Kodham, Pulika	Shrub	W	FS
Caesalpinia bonduc (L.) Roxb.	Sagargota, Gatran, Nataphal	Shrub	W	AFHS
Caesalpinia digyna Rottler.	Teripods, Teri Beej	Climber	W	A
Caesalpinia pulcherrima (L.) Sw.	Gultora, Krishnachura, Peacock flower	Shrub	C	AFS
Caesalpinia sappan L.[2]	Pathimugam	Tree	C	AFSU
Calamus rotang L.	Pirapan Kizhangu, Bet	Shrub	W	AFSU
Callicarpa macrophylla Vahl	Priyangu	Tree	W	AT
Calophyllum apetalum Willd.	Cherupunna	Tree	W	FS
Calophyllum inophyllum L.	Punnappoovu	Tree	C/W	AFS
Calotropis gigantea (L.) R.Br.	Erukkin veru	Shrub	W	AFHSTU
Calotropis procera (Aiton) W.T. Aiton	Akada Phool	Shrub	W	AFSTU
Calycopteris floribunda (Roxb.) Poir.	Pullaaniyila	Climber	W	AF
Canarium strictum Roxb.	Dhoop, Raldhoop	Tree	W	AFS
Cannabis sativa L.[1]	Maya, Bhang	Shrub	C/W	AFHSTUW
Canscora decussata Schult. & Schult.f.	Shankhuli, Nakuli, Shankahpuspi	Herb	W	ASU
Capparis divaricata Lam.	Turatti	Tree	W	S
Capparis moonii Wight	Rudanti	Climber	W	A
Capparis roxburghii DC.	Rudanti (Substitute)	Climber	W	A
Capparis sepiaria L.	Karungurai	Shrub	W	AFS
Capparis spinosa L.	Kanther	Shrub	W	AU
Capparis zeylanica L. [= *C. horrida* L.]	Sivappu Boomi, Sakkarai Kizhangu	Climber	W	AFS
Cardiospermum halicacabum L.[2]	Mudakkathan	Climber	W	AFHSU

Botanical Name[1 & 2]	**Trade Name**	**Habit**	**Material Source[3]**	**Associate Systems[4]**
Careya arborea Roxb. [2]	Vaaikumbha, Kumbhiphoo	Tree	W	AFS
Carica papaya L. [2]	Papaya, Papita	Tree	C	AFHSUW
Carissa carandas L.	Christ's thorn, Karanda	Shrub	W/C	AFS
Carthamus tinctorius L.	Kusum Phool	Shrub	C	AFSTU
Carum carvi L.	Shahjeera, Kalazira	Herb	C/W	AHSTU
Casearia esculenta Roxb.	Saptarangi	Shrub	W	A
Cassia absus L. [2]	Chaksoo	Herb	W	AFSTU
Cassia alata L. [= *Senna alata* (L.) Roxb.]	Dadmurdan, Dat-ka-pat	Shrub	C/W	AFS
Cassia angustifolia Vahl [= *C. senna* L.; *Senna alexandrina* Gars. ex Mill.] [1]	Sona Patta, Sonamukhi, Senna	Herb	C	AHSUW
Cassia auriculata L. [= *Senna auriculata* (L.) Roxb.]	Avarai	Shrub	W	AFS
Cassia fistula L.[1]	Amalthasphali, Sonari chal	Tree	W/C	AFSTU
Cassia obtusifolia L. [= *Senna obtusifolia* (L.) H.S. Irwin & Barneby]	Panevar, Taga	Shrub	C/W	A
Cassia occidentalis L. [= *Senna occidentalis* (L.) Link] [2]	Kasondi, Kasmardah	Herb	W	AFST
Cassia siamea Lam. [= *Senna siamea* (Lam.) H.S, Irwin & Barneby]	Kassod	Tree	C	FS
Cassia sophera L. [= *Senna sophera* (L.) Roxb.]	Kasodi, Ponthakaram	Shrub	W	AFHS
Cassia tora L. [= *Senna tora* (L.) Roxb.] [2]	Chakoda Beeja, Pawad, Chankuda	Herb	W	AFSTU
Cassytha filiformis L.	Astimu, Akash Bel	Climber	W	AFS
Catharanthus roseus (L.) G. Don [= *Vinca rosea* L.] [1]	Sadabahar, Vinca rosea	Herb	C	AFW
Catunaregum spinosa (Thunb.) Tirveng. [= *Xeromphis spinosa* (Thunb.) Keay]	Mainphal, Maggare	Shrub	W	A
Cayratia carnosa (Wall.) Gagnep. *ex* Wight	Amal Ved, Amal bel, Gutt	Climber	W	AF
Cayratia pedata (Lam.) A. Juss. ex Gagnep.	Suvaha, Gummatige	Climber	W	AFS
Cedrus deodara (Roxb.) G. Don[2]	Deodar	Tree	W	AFSTU

Botanical Name[1 & 2]	**Trade Name**	**Habit**	**Material Source[3]**	**Associate Systems[4]**
Ceiba pentandra (L.) Gaertn. [= *Eriodendron pentandrum* (L.) Kurz]	Safed semal	Tree	C	AFSU
Celastrus paniculatus Willd.[2]	Malkangani, Jyothismathi, Bavanthi beeja	Climber	W	AFSTU
Celtis philippensis Blanco	Kakamushti (Subs.)	Tree	W	FS
Centaurea behen L.	Behmen Safed	Herb	I	FU
Centella asiatica (L.) Urb.[1]	Brahmi, Brahmi booty	Herb	C/W	AFHSTW
Centratherum anthelminticum (L.) Kuntze [= *Vernonia anthelmintica* L.][2]	Kali zeera	Herb	W	AHSU
Cephaelis ipecacuanha (Brot.) A. Rich. [= *Psychotria ipecacuanha* (Brot.) Standl.]	Ipecacuanha	Shrub	C	HSUW
Chlorophytum arundinaceum Baker, syn. *Anthericum tuberosum* Hook.f., nom. illeg.[1]	Safed musali	Herb	C/W	AFTU
Chlorophytum borivilianum Santapau & R.R. Fern.[1]	Safed musali	Herb	C/W	F
Chlorophytum tuberosum Baker[2]	Safed musali	Herb	W/C	AS
Chonemorpha fragrans (Moon) Alston	Murva	Climber	W	AF
Chrozophora prostrata Dalzell & A. Gibson, syn. *C. plicata* Hook f.	Neelkanthi	Herb	W	AU
Chrysanthemum cinerariifolium (Trevis) Vis.	Pyrethrum	Herb	C	FW
Cichorium endivia L.	Kasini, Endive	Herb	C	A
Cichorium intybus L.[1]	Kasani	Herb	C/W	AFHSUW
Cinchona ledgeriana Bern. Moens ex Trimen	Cinchona	Tree	I/C	FW
Cinchona officinalis L. *C. ledgeriana* Moens ex Trimen, *C. pubescens* Vahl. and *C. calisaya* Wedd. (Grown in India)[2]	Cinchona bark	Tree	I/C	AHU
Cinnamomum camphora (L.) Presl	Kapur	Tree	C	AFHSTUW
Cinnamomum cassia Nees ex Blume	Dalchini	Tree	I	ASTU
Cinnamomum macrocarpum Hook. f.	Lavang	Tree	W	AFS

Botanical Name[1 & 2]	**Trade Name**	**Habit**	**Material Source[3]**	**Associate Systems[4]**
Cinnamomum malabathrum (Lam.) J. Presl	Dalchini, Tejpatta	Tree	W	F
Cinnamomum sulphuratum Nees[1]	Dalchini, Tejpatta	Tree	W	F
Cinnamomum tamala (Buch. -Ham.) T. Nees & Eberm.[2]	Tamal Patra, Tejpatra	Tree	W	AFSTU
Cinnamomum verum J. Presl [= *C. zeylanicum* Blume] [2]	Dalchini	Tree	C	AFH
Cinnamomum wightii Meisn.	Dalchini, Tejpatta	Tree	W	AS
Cipadessa baccifera (Roth) Miq.	Adusoge soppu	Shrub	W	FS
Cissampelos pareira L. var. *hirsuta* (Ham. ex DC.) Forman	Pahad Mool	Climber	W	AFHSTUW
Cissus quadrangularis L.	Pirandai	Climber	W	AFS
Citrullus colocynthis (L.) Schrad.[2]	Indrayan, Indravaruni	Climber	W	AFHSTU
Citrullus lanatus (Thunb.) Matsumara & Nakai	Terbuz, watermelon	Climber	C	AH
Citrus aurantiifolia (Christm.) Swingle (*C. hystrix* x *C. medica*), syn. *C. acida* Pers.	Limbu, Nemu-tenga, Key lime	Tree	C	AFSU
Citrus aurantium L.	Narangi	Tree	C	AFSU
Citrus bergamia, Risso., syn. *Citrus* x *bergamia* Risso.	Limbu chaal, Jambeeram, Bergamot	Tree	C	AU
Citrus limon (L.) Burm. f.	Lemon	Tree	C	AFHTUW
Citrus medica L.	Matunga, Mahnimbu	Tree	W	ASU
Citrus reticulata Blanco	Santra	Tree	C	ATU
Clausena dentata (Willd.) Roem.	Mahasindur	Shrub	W	FS
Cleistanthus collinus (Roxb.) Benth.	Kutaja, Garbar	Tree	W	AS
Clematis gouriana Roxb. ex DC.	Morvel	Climber	W	A
Clematis triloba Heyne	Morvel	Climber	W	AFU
Cleome gynandra L. [= *Gynandropsis pentaphylla* (L.) DC.]	Cat whiskers, Ridhi	Herb	W	AFST
Cleome viscosa L., syn *C. icosandra* L.	Nai kadugu	Herb	W	AFSU
Clerodendrum indicum (L.) Kuntze	Bharangi	Shrub	W/C	AF
Clerodendrum inerme (L.) Gaertn.	Nir-Notsjil, Sangan-Kuppi, Agnimanth	Shrub	C/W	AFS
Clerodendrum phlomidis L.f.[2]	Arnimul	Shrub	W	AFSU

Botanical Name[1 & 2]	**Trade Name**	**Habit**	**Material Source[3]**	**Associate Systems[4]**
Clerodendrum serratum (L.) Moon	Bharangi	Shrub	W	AFSTU
Clitoria ternatea L.	Kajli	Climber	C/W	AFSTU
Coccinia grandis (L.) Voigt [= *C. indica* W. & A.]	Bhrngaraj, Bimb	Climber	C/W	AHS
Cocculus hirsutus (L.) Theob.	Vasanvel	Climber	W	AFSU
Cochlospermum religiosum (L.) Alston, syn. *C. gossypium* DC.	Katira	Tree	W/C	AFSU
Cocos nucifera L.[2]	Nariyal	Tree	C	AFSTU
Coffea arabica L.	Coffee (Stimulant, nervine)	Shrub	C	AFHSUW
Coix lacryma-jobi L.	Sankhlu, Dabhir	Herb	W	AFSU
Colchicum luteum Baker	Suranjan	Herb	W	ASU
Coleus zeylanicus (Benth.) Cramer, and *C. vettiveroides* K. C. Jacob., syn. *Plectranthus vettiveroides* (K. C. Jacob) Singh & Sharma	Valakah	Herb	W	A
Combretum decandrum Roxb. non. Jacq.	Korakukundi	Climber	W	F
Commiphora caudata (Wight & Arn.) Engler (Potential)	Kiluvai, Kondamavu	Tree	W	FS
Commiphora myrrha (T.Nees) Engl.	Hirabol	Tree	I	ATU
*Commiphora wightii (*Arn.) Bhandari[1] [= *C. mukul* (Hook. ex Stocks) Engl.]; *Balsamodendron mukul* Hook. ex Stocks[1]	Guggul, Guggulu	Shrub	I/W	A
Convolvulus arvensis L.	Chandvel (Prasarni)	Climber	C/W	AHSU
Convolvulus microphyllus Sieb. ex Spreng. [= *C. prostratus* Forssk.; *C. pluricaulis* Chois.] [1]	Shankapushpi	Herb	W	A
Convolvulus scammonia L.	Saqmunia	Climber	I	HSU
Coptis teeta Wall.	Mamira	Herb	C/W	ASU
Corallocarpus epigaeus (Rottl. & Willd.) C.B. Clarke	Akasgaddah, Patalagaruda	Climber	W	AST
Corchorus depressus Stocks	Bahuphali	Herb	W	AU
Corchorus trilocularis L.	Arenukam	Shrub	W	A
Cordia dichotoma G. Forst.	Lasora	Tree	W/C	AFSU
Cordia monoica Roxb.	Narivari	Tree	W	FS
Cordia wallichii G. Don [= *C. obliqua* Willd. var. *wallichii* (G. Don) C.B. Clarke]	Sapistan	Tree	W	F

Botanical Name[1 & 2]	Trade Name	Habit	Material Source[3]	Associate Systems[4]
Cordyceps sinensis (Berk.) Saac (mushroom)	Yarsa Gomba	Herb	W	F
Coriandrum sativum L.[2]	Dhaniya, Dhana	Herb	C	AFSTU
Coscinium fenestratum (Gaertn.) Colebr.[1]	Maramanjal, Daruharidra	Climber	W	AFST
Costus speciosus (J. Konig) Sm.	Kustha, Koshtum, Kuth	Shrub	W	AFSU
Crataegus oxyacantha L.	Hawthorn	Tree	C/W	FH
Crataeva religiosa G. Forst. [= *C. nurvula* Buch. - Ham.]	Varun chhal, Varunah	Tree	W/C	AFS
Crocus sativus L.[1]	Kesar	Herb	C	AFHSTU
Crotalaria juncea L.	Dhanahari, Patashan, Sann	Herb	C/W	AFSU
Crotalaria verrucosa L.	Banshana, San, Sanpushpi	Herb	W	AFS
Croton oblongifolius Roxb.	Bhutankusha	Tree	W	AFST
Croton tiglium L.[1]	Jamlghota, Japala	Tree	C	AFHSTU
Cryptocoryne spiralis (Retz.) Fisch. ex Wydler	Natti-ati-Vasa	Herb	W	AFS
Cryptolepis buchananii R.Br. ex Roem. & Schult.	Medaksinghi, Sariva	Climber	W	AFS
Cucumis trigonus Roxb.	Jangli-Indrayan	Climber	C	AFS
Cucurbita maxima Duch.	Pumpkin	Climber	C	AFSU
Cucurbita moschata (Lam.) Duchesne *ex* Poir.	Kaddu, Kumrah	Climber	C	FSU
Cuminum cyminum L.[1]	Jeera, Shah jeera	Herb	C	AFSTU
Curculigo orchioides Gaertn.[2]	Kali musali, Musli shiya	Herb	W	AFSTU
Curcuma amada Roxb.	Amba haldi	Herb	C/W	AFSU
Curcuma angustifolia Roxb.[2]	Tikhur	Herb	C/W	AFSU
Curcuma aromatica Salisb.	Kasturi manjal, Kasturi arishna, Kapu-kachri	Herb	C/W	AFSU
Curcuma caesia Roxb.	Nar-kachura, Kala-haldi	Herb	W	F
Curcuma longa L., syn. *C. domestica* Valeton[1]	Arishna, Haldi, Kari manjal	Herb	C	AFHSTU
Curcuma zerumbet Roxb. [= *C. zedoaria* (Christm.) Roscoe][2]	Kachur, Kachari	Herb	C/W	T
Cuscuta reflexa Roxb.[2]	Aftimoon, Tukhme-Kasus	Climber	W	AFSU
Cyathula prostrata (L.) Blume	Kadalaavanakkin veru	Herb	W	AF

Botanical Name[1 & 2]	**Trade Name**	**Habit**	**Material Source[3]**	**Associate Systems[4]**
Cycas circinalis L.	Madhana Kama	Tree	W	AFS
Cyclea peltata (Lam.) Hook.f. & Thomson, syn. *C. arnotii* Miers[2]	Paada kizhangu	Climber	W	AFS
Cydonia oblonga Mill., syn. *Pyrus cydonia* L.	Bee dana, Quince	Tree	C	F
Cymbopogon citratus (DC.) Stapf	Serai, Rohisha	Herb	C/W	AFSU
Cymbopogon flexuosus (Nees ex Steud.) Wats.[1]	Lemon grass, Cochin oil	Herb	W	F
Cymbopogon martinii (Roxb.) Wats	Ginger grass	Herb	W	AST
Cynodon dactylon (L.) Pers.[2]	Durva, Karuka	Herb	C	AFHSTU
Cyperus esculentus L.[2]	Musta	Herb	W	AU
Cyperus rotundus L.[1]	Nagarmotha, Mustha	Herb	W	ASTU
Cyperus scariosus R.Br.	Nagarmotha	Herb	W	AFSU
Dactylorhiza hatagirea (D. Don.) Soo (orchid)	Salampanja	Herb	W	AFSTU
Dalbergia latifolia Roxb.	Kala Sheeshan	Tree	W	AFS
Dalbergia sissoo DC.	Shisham	Tree	C/W	ASTU
Datura arborea L.	Datura, Umathi	Shrub	I	H
Datura innoxia Mill.	Datura	Herb	W	A
Datura metel L.[1]	Duttura	Herb	W	ASW
Datura stramonium L.	Datura, Umathi	Herb	W	AFHSU
Daucus carota L. var. *sativa* DC.	Gaajar Beej	Herb	C	AFSU
Decalepis hamiltonii Wight & Arn.[2]	Magali	Climber	W	AFS
Delonix elata (L.) Gamble	Sanchal	Tree	C	AFS
Delphinium denudatum Wall. *ex* Hook. f. & Thomson	Jadwar, Jadavar kath, Nirbishi	Herb	W	AU
Dendrophthoe falcata (L.F) Blume	Bandaka	Shrub	W	AF
Derris scandens (Roxb.) Benth.	Gonj	Climber	W	FS
Desmodium gangeticum (L.) DC.[1]	Salparni, Prasniparni	Herb	W	AFHSTU
Desmodium pulchellum (L.) Benth.	Lodhra, Kheri	Shrub	W	A
Desmodium triflorum (L.) DC.	Hamsapapdi	Herb	W	AFS
Desmostachya bipinnata (L.) Stapf	Dharbha	Herb	W	ASTU
Dichrostachys cinerea (L.) Wight & Arn.	Virtuli	Shrub	W	AFST

Botanical Name[1 & 2]	**Trade Name**	**Habit**	**Material Source[3]**	**Associate Systems[4]**
Didymocarpus pedicellata R.Br.	Shilapushpi, Pathar phori, Pasanphodi	Herb	W	A
Dillenia indica Linn., *D. pentagyna* Roxb.	Nagkesaram	Tree	W	AFS
Dioscorea alata L.	Sewalli kodi	Climber	C/W	AFS
Dioscorea bulbifera L.	Varahi kand	Climber	C/W	AFSTU
Dioscorea deltoidea Wall. ex Griseb.	Shingli mingli	Climber	C/W	FW
Dioscorea oppositifolia L.	Sarpahya	Climber	W	AFS
Dioscorea pentaphylla L.	Kanta Alu	Climber	W	AFS
Diospyros buxifolia (Blume) Hiern	Elichevian	Tree	W	F
Diospyros ebenum Koenig	Abnus	Tree	W	FSU
Diospyros lotus L.	Amlok	Tree	W	F
Diospyros melanoxylon Roxb.	Tendu	Tree	W	AFSU
Diplocyclos palmatus (L.) C. Jeffrey	Shivling beej	Climber	W	AFS
Dolichos biflorus L. [= *Macrotyloma uniflorum* (Lam.) Verdc.; *Vigna unguiculata* (L.) Walp.]	Kulthi, Muthira	Herb	C	AFSTU
Dorema ammoniacum D. Don	Ushaq	Herb	I	AHU
Drosera peltata Smith	Drosera	Herb	W	FS
Dryobalanops aromatica Gaertn.	Baraas (Bhimseni Kapoor)	Tree	I	A
Dryopteris filix-mas (L.) Schott	Shield fern, Hirvi	Herb	W	HS
Drypetes roxburghii (Wall.) Hurus.[= *Putranjiva roxburghii* Wall.]	Putrjivak	Tree	C/W	AFS
Dysoxylum malabaricum Bedd. ex Hiern	Vellakil	Tree	W	AFS
Ecbolium viride (Forssk.) Alston [= *E. linneanum* Kurz]	Udajati, Sahacarah	Herb	W	AF
Echinops echinatus Roxb.	Utkanta	Herb	W	AFSU
Eclipta prostrata (L.) L. [= *E. alba* Hassk.][1]	Bhringaraj	Herb	W	AFS
Elaeocarpus sphaericus (Gaertn.) K. Schum. [=*E. ganitrus* Roxb.]	Rudraksh	Tree	C/W	AF
Elephantopus scaber L.	Gjihiva	Herb	W	AFS
Elettaria cardamomum Maton [2]	Elachi Chhoti, Ilaychi	Herb	C/W	A

Botanical Name[1 & 2]	Trade Name	Habit	Material Source[3]	Associate Systems[4]
Eleutherococcus senticosus (Rupr. & Maxim.) Maxim, syn. *Acanthopanax senticosus* (Rupr. *et* Maxim.)	Ginseng	Shrub	I	W
Embelia ribes Burm.f.[1]	Vaividang, Vidanga	Climber	W	AFHSTU
Embelia tsjeriam-cottam (Roem. &Schult.) DC. [= *E. basaal sensu* (Roem. & Schult.) DC., non Mez]; *E. robusta* CB Clarke, non-Roxb.[2]	Vaividang	Shrub	W	AF
Emblica officinalis Gaertn. [= *Phyllanthus emblica* L.][1]	Amla, Anwala (part of Triphala)	Tree	W/C	AFSTU
Emilia sonchifolia (L.) DC.	Muyalcheviyan	Herb	W	AFS
Enicostemma axillare (Lam.) Raynal [= *E. littorale* non-Blume; *E. hyssopifolium* (Willd.) Verd.]	Mamejava	Herb	W	AFS
Entada pursaetha DC. [=*E. rheedii* Spreng.; *E. scandens* auct. non Benth.]	Yaanai Kazharchi Kaai	Climber	W	F
Ephedra gerardiana Wall. ex J.A. Mey[1]	Somalatha	Shrub	W	AH
Ephedra intermedia Schr. & Mey	Somalatha	Shrub	W	F
Ephedra sinica Stapf	Ephedra	Shrub	W	F
Equisetum arvense L.	Bottle brush, Horse tail	Herb	W	FW
Erythrina suberosa Roxb.	Pangra, Vellaimurukku	Tree	W	FS
Erythrina variegata L. [= *E. indica* Lam.]	Murikkila	Tree	C/W	AFST
Eucalyptus citriodora Hook.	Eucalyptus	Tree	C	F
Eucalyptus globulus Labill[2]	Eucalyptus	Tree	C	AFHSW
Eulaliopsis binata (Retz.) C.E. Hubb.	Bal vaja	Herb	W	A
Eulophia campestris Wall. ex Lindl.	Salam mishri	Herb	W	AU
Euphorbia antiquorum L.	Kallippalaveru	Tree	W	AFSU
Euphorbia hirta L.	Dudhi, Dudhika	Herb	W	AFSU
Euphorbia neriifolia L.	Pattankisend, Nanda	Shrub	C/W	AFSTU
Euphorbia nivulia Buch. -Ham.	Sehunda	Tree	W	AFS
Euphorbia thymifolia L.	Choti - Dudhi	Herb	W	AFU
Euphorbia tirucalli L. or *E. Pilosa* L.	Modu Kalli	Shrub	C	AFSU

Botanical Name[1 & 2]	Trade Name	Habit	Material Source[3]	Associate Systems[4]
Euphorbia tortilis Rottler ex Ainslie	Tirugakalli, Vajratunda	Shrub	W	AS
Euryale ferox Salisb. ex K.D. Koenig	Phool makhana	Herb	C/W	ASU
Evolvulus alsinoides (L.) L.	Shankhavali, Shankhpush	Herb	W	AFSTU
Excoecaria agallocha L.	Agaru, Tejbala	Tree	W	AFS
Fagonia cretica L.	Dhamasa	Herb	W	ATU
Ferula foetida (Bunge) Regel., syn. *F. assa-foetida* L. (Potential)[2]	Hingu, Hing hira (Baandani)	Shrub	I/W	AFHTSU
Ferula jaeschkeana Vatke	Hingupatri, Hingu	Shrub	W	A
Ferula narthex Boiss.	Sahasravedhi	Herb	W	AFHU
Ficus arnottiana (Miq.) Miq. [2]	Kallaltholi	Tree	W	AFS
Ficus benghalensis L.[2]	Vadachhal, Vatathwak	Tree	W/C	AFHSTU
Ficus carica L.	Anjeer, Anjoora	Tree	C	AFSU
Ficus gibbosa Blume	Ithitholi	Tree	W	FS
Ficus mollis Vahl.[2]	Kaaduatthi	Tree	W	AF
Ficus racemosa L.	Umbar chal, Lakh pipal	Tree	C/W	AFSTU
Ficus religiosa L.[1]	Lakh pipal, Arayaalin tholi	Tree	C/W	AFHSTU
Ficus tsiela Roxb. ex Buch. -Ham.	Plaksha	Tree	W	AFS
Ficus tsjakela Burm.f.	Pilkhan, Kapitanah	Tree	W	A
Flacourtia indica (Burm. f.) Merr.	Kattar	Shrub	W	AFST
Flacourtia jangomas (Lour.) Raeusch., syn. *F. cataphracta* Roxb.	Talish Pattar, Brahmi Talispatar	Tree	W	AS
Flemingia chappar Buch. -Ham. *ex* Benth.	Salpan (Substi.)	Shrub	W	F
Flemingia grahamiana Wight & Arn.	Warrus/ Varus	Herb	W	F
Flemingia macrophylla (Willd.) Kuntze ex Merr.	Varrus / Varus	Shrub	W	AF
Flickingeria macraei (Lindl.) Seidenf. [= *Ephemerantha macraei* (Lindl.) Hunt & Summerh]	Jivanti	Herb	W	AF

Botanical Name[1 & 2]	**Trade Name**	**Habit**	**Material Source[3]**	**Associate Systems[4]**
Flickingeria nodosa (Dalz.) Seidenf. [= *Dendrobium macraei* auct. non Lindl.]	Jivanti	Herb	W	AF
Flueggea virosa (Roxb. ex Willd.) Royle, syn. *F. microcarpa* Blume	Dalme, Patali	Shrub	W	A
Foeniculum vulgare Gaertn.	Badiyan Khatal, Saunf, (Variyali)	Herb	C	AFSTUW
Fritillaria roylei Hook.	Kakoli	Herb	W	A
Fumaria indica (Hauskn.) Pugsley [= *F. parviflora* Lam.][2]	Shahtara	Herb	W	A
Fumaria officinalis L.	Pittapapdo ghas, Pitpapra	Herb	I	AFU
Garcinia gummi-gutta (L.) N. Robson [= *G. cambogia* (Gaertn.) Desr.][1]	Kokam, Kodampuli, Camboge	Tree	C/W	A
Garcinia indica (Dup.) Choisy[1]	Kokam, Cambogie	Tree	W/C	AT
Garcinia morella (Gaertn.) Desr.	Kadukaai puli	Tree	W	AFHSTU
Garcinia pedunculata Roxb. ex Buch. -Ham.	Amlavetasa	Tree	W	AT
Gardenia gummifera L.f.	Dekamalli, Dikamali	Tree	W	AFSTU
Gardenia resinifera Roth [2]	Dikamali	Tree	W	AS
Garuga pinnata Roxb.	Ghoghar, Kharpat	Tree	W	AFST
Gentiana kurroo Royle[2]	Trahimaan, Kadu, Katuki	Herb	W	ATU
Geophila reniformis D. Don [= *G. repens* (L.) I.M. Johnst]	Kakamaci	Herb	W	F
Girardinia heterophylla Decne. [= *G. diversifolia* (Link) Friis]	Gaddanelli, Anachoriyan	Herb	W	F
Givotia rottleriformis Griff. ex Wight	Polki	Tree	W	FS
Glinus lotoides L.	Gandhi-Buti	Herb	W	AS
Glinus oppositifolius (L.) A. DC.	Ushnasundara	Herb	W	AFS
Gloriosa superba L.[1]	Kalihari	Climber	C/W	AFSTU
Glossocardia bosvallea (L.f.) DC.	Parpataka	Herb	W	A
Glycosmis arborea (Roxb.) DC.	Potali, Kupilah	Shrub	W	A
Glycyrrhiza glabra L.[1]	Mulathi, Jeshtimadhu Gulegafis	Shrub	W/C/I	AFSTUW
Glycyrrhiza uralensis DC.	Substitute- Mulathi, Jeshtimadhu Gulegafis	Shrub	I	F

Botanical Name[1 & 2]	Trade Name	Habit	Material Source[3]	Associate Systems[4]
Gmelina arborea Roxb. [2]	Ghambar chal, Shivan (Ghambari) (Chal)	Tree	W/C	AFST
Gmelina asiatica L.	Kapas Beej	Shrub	W	AFS
Gossypium arboreum L.	Kapas	Shrub	C	AFSU
Gossypium herbaceum L.	Kapas	Shrub	C	AFHSTU
Gossypium hirsutum L.	Kapas	Shrub	C	ASW
Gouania microcarpa DC., *G. leptostachya* DC	Sinhanbali	Climber	W	F
Grewia asiatica auct. non L., syn *G. subinaequalis* DC.	Phaalsaa	Shrub	W/C	AFSTU
Grewia elastica Royle [= *G. asiatica* L. var. *vestita* Mast.], syn. *G. eriocarpa* Juss.	Phalsa chhaal	Tree	C/W	A
Gymnema sylvestre R.Br. ex Schult.[1]	Gudmar, Sirukurinjan	Climber	W	AFHSTU
Habenaria edgeworthii Hook. f. ex Collett	Riddhi	Herb	W	A
Hedychium coronarium J. Konig	Garland flower, Common Ginger Lily	Herb	W	S
Hedychium spicatum Sm.[2], syn. *H. acuminatum* Roscoe	Kapoor Kachri	Herb	W	AFTU
Hedyotis corymbosa (L.) Lam. [= *Oldenlandia corymbosa* L.][1]	Pitpapra	Herb	W	AF
Hedyotis herbacea L., syn. *Oldenlandia herbacea* (Willd.) Roxb.	Pippapada	Herb	W	AFH
Hedyotis puberula (G. Don) Arn. [= *Oldenlandia umbellata* L.]	Impural	Herb	W	AF
Helianthus annuus L.	Sunflower	Shrub	C	AHSU
Helicteres isora L.[2]	Murudshing, Marodphali	Shrub	W	AFSU
Heliotropium indicum L.	Thekkada	Herb	W	AFSU
Helleborus niger L.	Katuka Rohini	Herb	I	AFHSU
Hemidesmus indicus (L.) R. Br. [1]	Anatmool, Sariwa, Sarasaparilla roots	Climber	W	AFSTU
Hemionitis arifolia (Burm.) Moore	Akhukarni, Pattsjivi-Maravara	Herb	W	AF
Heracleum candolleanum (Wight & Arn.) Gamble, syn. *Tetrataenium rigens* (Wall. ex DC.) I.P. Mandenova	Substitute of Kattumalli	Herb	W	F

Botanical Name[1 & 2]	**Trade Name**	**Habit**	**Material Source[3]**	**Associate Systems[4]**
Heracleum lanatum Michx., syn. *H. candicans* Wall ex DC.; *H. nepalensis*	Krandel	Herb	C/W	F
Heracleum rigens DC.	Cittrelam	Herb	W	FS
Hibiscus rosa-sinensis L.	Jashwanti, Jaswand, Japa	Shrub	C	AFSU
Hibiscus surattensis L.	Ranbhindi	Shrub	W	FS
Hibiscus tiliaceus L.	Parutti	Herb	W	AFS
Hippophae rhamnoides L.	Sea Buckthorn	Shrub	W/C	F
Hippophae salicifolia D. Don, syn. and *H. rhamnoides* subsp *salicifolia* (D. Don) Servettaz.	Substitute of Sea Buckthorn	Shrub	C/W	F
Holarrhena pubescens (Buch.-Ham.) Wall. ex G. Don [= *H. antidysenterica* (Wall. ex DC.) Wall.] [1]	Inderjao Kadwa, Kudachal, Kutja	Shrub	W	AFH, Chinese
Holoptelea integrifolia (Roxb.) Planch. [2]	Aavitholi	Tree	W	AFS
Holostemma ada-kodien Schult. [= *H. annulare* (Roxb.) K. Schum.; *H. rheedii* Wall.] [2]	Jeevanti	Climber	W	AF
Homonoia riparia Lour.	Serna, Pashanbhedaka	Shrub	W	AS
Hugonia mystax L.	Kamsamrah	Shrub	W	AFS
Hybanthus enneaspermus (L. f.) F. r. Muell.	Ratanpurus, Padmacarini	Herb	W	AFS
Hydnocarpus kurzii (King) Warb.	Nirati, Chaulmoogra	Tree	W	AFS
Hydnocarpus pentandrus (Buch. -Ham.) Oken., syn. *H. laurifolia* (Dennst.) Sleum	Marotti	Tree	W	AFS
Hygrophila schulli (Buch. -Ham.) M.R. & S.M. Almeida [= *H. auriculata* (Schumach.) Heine; *H. spinosa* T. Anderson; *Asteracantha longifolia* Nees] [2]	Tal makhana, source of the Ayurvedic drug 'Kokilaaksha' and the Unani Talimakhana	Herb	W	AF
Hygroryza aristata (Retz.) Nees ex Wight & Arn.	Janli dal, Nivara	Herb	W	AF
Hymenodictyon excelsum (Roxb.) Wall.	Kuthan	Tree	W	AF
Hyoscyamus muticus L.	Parasikaya	Herb	I	A
Hyoscyamus niger L.	Khursani Ajwain	Herb	C/W	AHSUW
Hypericum perforatum L.	St. John's wart	Herb	W	HW
Hyssopus officinalis L.	Zoofa, Jupha	Shrub	C/W	FU

Botanical Name[1 & 2]	Trade Name	Habit	Material Source[3]	Associate Systems[4]
Ichnocarpus frutescens (L.) W.T. Aiton	Dudhi, Saraiva	Climber	W	AFST
Illicium griffithii Hook.f. & Thomson	Star Anise (Substi.)	Tree	W	AS
Illicium verum Hook.f.	Star Anise	Tree	I	AS
Imperata cylindrica (L.) Raeusch.	Thatch grass	Herb	W	AFS
Indigofera tinctoria L.[2]	Akika, Avuri	Shrub	C	AFHSTU
Inula racemosa Hook.f.[2]	Pushkarmool, Pokharmool	Herb	C	AFT
Inula royleana DC.	Pushkarmool	Herb	W	A
Ipomoea mauritiana Jacq. [= *I. digitata* L.][2]	Palmudhukkan kizhangu	Climber	W	AF
Ipomoea nil (L.) Roth.[2]	Kaladana	Climber	W	AFS
Ipomoea obscura Ker-Gawl.	Laksmana	Climber	W	AFS
Ipomoea pes-caprae (L.) R. Br. [= *I. biloba* Forssk.]	Dopatilata	Climber	W	AFS
Ipomoea pes-tigridis L.	Pulichuvadan	Climber	W	AFS
Ipomoea petaloidea Choisy	Vrddhadaru	Herb	W	F
Ipomoea reptans Poir.	Kalashaka	Herb	C/W	A
Ipomoea sepiaria Roxb.	Thiruthaali	Climber	W	AFS
Iris ensata Thunb. = *I. germanica* L.	Puskaramulam	Herb	W	AHTU
Ixora coccinea L.[2]	Thechippoovu	Shrub	W	AFS
Jasminum angustifolium (L.) Thunb.	Ban-Mallika	Climber	C/W	AFS
Jasminum grandiflorum L.	Chamel	Climber	C/W	AFSTU
Jasminum officinale L.	Ban Chameli	Climber	C/W	AFHU
Jasminum sambac (L.) Aiton	Mallika, Mogra	Climber	C/W	AFSTU
Jatropha curcas L.[2]	Napalm seed	Shrub	C	AFHSU
Juglans regia L.	Akhrot	Tree	C/W	AFHSTU
Juniperus communis L.[2]	Hauber	Tree	W	AHTUW
Juniperus macropoda Boiss.[2]	Hapusa	Tree	W	A
Jurinea macrocephala DC. [= *J. dolomiaea* Boiss.]	Dhoop	Herb	W	AF
Justicia gendarussa Burm.f. [= *Gendarussa vulgaris* Nees]	Nilani gundi	Shrub	C/W	AFSU
Kaempferia galanga L.[1]	Kachora, Narkachur, Kapoor Kachri No 1	Herb	C	AFS
Kaempferia rotunda L.	Bhuichampa	Herb	C	AFS
Kalanchoe laciniata (L.) DC.	Hemsagara	Herb	W	AS

Botanical Name[1 & 2]	**Trade Name**	**Habit**	**Material Source[3]**	**Associate Systems[4]**
Kedrostis rostrata (Rottler) Cogn., syn. *K. foetidissima* (Jacq.) Cogn.	Arunkovai	Climber	W	FS
Lagenaria siceraria (Molina) Standley	Bottle gourd	Climber	C	AFSTU
Lagerstroemia speciosa (L.) Pers., syn. *L. flos-reginae* Retz.	Jarul	Tree	W	AF
Lagotis glauca Gaertn.	Adulterant of Kutki, *Picrorhiza kurrooa*	Herb	W	F
Lallemantia royleana (Wall. ex Benth) Benth.	Tukhme-Balanga	Herb	C/W	FU
Lannea coromandelica (Houtt.) Merr.[2]	Jingini	Tree	C/W	AF
Lantana camara L.	Lantana	Shrub	W	A
Laurus nobilis L.	Hub-ul-ghar	Tree	I	FU
Lavandula stoechas L.	Ustakuthus	Herb	W	U
Lawsonia inermis L.[1]	Henna, Mehndi	Shrub	C	AFSTU
Lepidium sativum L.[2]	Kurassaani, Halam	Herb	W	AFSU
Leptadenia reticulata (Retz.) Wight & Arn.	Jivanti	Climber	W	AFST
Leucas aspera (Willd.) Link	Dronpushpi, Dharanpushpi	Herb	W	AFHS
Lilium polyphyllum D. Don	Kakoli	Herb	W	A
Limonia acidissima L. [= *Feronia elephantum* Correa; *F. limonia* (L.) Swingle]	Wood apple	Tree	C/W	AFSU
Linum usitatissimum L.	Alsi	Herb	C	AFHSTUW
Litsea glutinosa (Lour.) C.B. Rob. [= *L. chinensis* Lam.][2]	Menda Lakadi, Naramamidi, Maidachal	Tree	W	AFS
Lobelia nicotianaefolia Roth ex Roem. & Schult.[2]	Lobelia leaves (wild tobacco)	Shrub	W	AFS
Lodoicea maldivica (J.F. Gmel.) Pers. ex H. Wendl.	Dariai narial	Tree	I	AFU
Luffa echinata Roxb.	Loofah	Climber	W	AFHST
Luvunga scandens (Roxb.) Buch. -Ham. ex Wight & Arn., and *L. eleutherandra* Dalz (unresolved)	Sugandh-Kokila	Shrub	W	A
Lycopodium clavatum L.	Foxtail	Herb	W	FH
Lygodium flexuosum (L.) Sw.	Tsjeru-Valli-Panna	Climber	W	F
Madhuca indica J.F. Gmel. [= *Bassia latifolia Roxb.*][1]	Madhuka, Mahua	Tree	W	AFST

Botanical Name[1 & 2]	**Trade Name**	**Habit**	**Material Source[3]**	**Associate Systems[4]**
Madhuca longifolia (J. Konig ex L.) J.F. Macbr., syn. *Bassia latifolia* Roxb.	Mahua Phool	Tree	C/W	AFS
Mahonia leschenaultia (Wall. *ex* Wight & Arn.) Takeda ex Gamble	Daruharidra	Shrub	W	F
Malaxis acuminata D. Don., syn. *Microstylis wallichii* L.	Jivak	Herb	W	A
Malaxis muscifera (Lindl.) Kuntze, syn. *Microstylis musifera* Ridley	Rsabhakah	Herb	W	A
Mallotus philippensis (Lam.) Mull. -Arg.	Kameela	Tree	W	AFHSU
Malva sylvestris L.	Khawaji, Khubaji	Herb	C/W	FU
Mammea suriga (Buch. -Ham. ex Roxb.) Kosterm.	Substitute of Nagakesr	Tree	W	AF
Mappia foetida (Wight) Miers [= *Nothapodytes nimmoniana* (J. Graham) Mabb.]	Ghanera	Tree	W	W
Maranta arundinacea L.	Citalapattiri	Herb	C	AFS
Marsdenia roylei Wight	Murva	Climber	W	A
Marsdenia tenacissima (Roxb.) Moon	Nishod sufed	Climber	W	A
*Marsdenia volubilis (*L.f.) T. Cooke [= *Wattakaka volubilis* (L.f.) Stapf*; Dregea volubilis* L.f.]	Murva, Jukti	Climber	W	A
Marsilea quadrifolia L.	Cupatiya	Herb	W	AS
Martynia annua L.	Kaknasa	Herb	W	AFS
Maytenus emarginata (Willd.) Ding Hou	Kattangi (Popular)	Shrub	W	F
Meconopsis aculeata Royle	Kunda	Herb	W	F
Melastoma malabathricum L.	Palore, Nakkukaruppan	Shrub	W	AFS
Melia azedarach L.	Bakayan phal	Tree	W/C	AFSTU
Melocanna bambusoides Trin.	Bansalochan	Tree	W	F
Mentha aquatica L.	Water mint (Substi.)	Herb	I	F
Mentha arvensis L.	Pudina, Podina pati	Herb	C	AFSUW
Mentha piperita L.	Menthol, Peppermint	Herb	C	HUW
Mentha viridis (L.) L., syn. *M. spicata* Linn. emend. Nathh.	Pudino	Herb	C	F
Merremia emarginata (Burm. f.) Hallier f.	Underkarni, Mooshkarni	Herb	W	AS

Botanical Name[1 & 2]	**Trade Name**	**Habit**	**Material Source[3]**	**Associate Systems[4]**
Merremia hederacea (Burm. f.) Hallier f.	Kudici-Valli	Climber	W	FS
Merremia tridentata (L.) Hallier f.[2]	Prasarani	Herb	W	AFS
Merremia umbellata (L.) Hallier f.	Hogvine, Jalap, *Ipomoea purga*, (English) (Substi)	Climber	C	F
Mesua ferrea L. [= *M. nagassarium* (Burm. f.) Kosterm.][2]	Naga Kesari, Nag Keshar	Tree	W	AFHSTU
Michelia champaca L.	Champaka	Tree	C/W	AFST
Micromeria biflora (Buch. -Ham. ex D. Don) Benth.	Indian wild scented Thyme	Herb	W	F
Mimosa pudica L.	Lajwanti, Lajjalu	Herb	W	AFST
Mimusops elengi L.[2]	Bakul	Tree	W/C	AFSTU
Mirabilis jalapa L.	Gulabash	Herb	C/W	AFSU
Mitragyna parvifolia (Roxb.) Korth.	Kadamb	Tree	C/W	AFS
Mollugo cerviana Ser.	Parpata	Herb	W	AFS
Mollugo pentaphylla L.	Turapoondu	Herb	W	FS
Momordica charantia L.[1]	Karela	Climber	C	AFSTU
Momordica dioica Roxb. ex Willd.	Jangli-Kareli	Climber	W/C	AFS
Monochoria vaginalis (Burm. f.) C. Presl.	Karinkuvalum	Herb	W	AFS
Morchella esculenta Pers. (mushroom)	Guchhi	Herb	W	F
Morinda citrifolia L. [2]	Canary wood	Tree	W	AFS
Morinda pubescens J.E.Sm. [= *M. coreia* Buch. -Ham.; *M. tinctoria* Roxb., var. *tomentosa* (Heyne ex Roth) Hook. f.][2]	Manjanatthi	Tree	W	AFS
Moringa concanensis Nimmo ex Dalzell & A. Gibson	Kadvo saragvo	Tree	W	AST
Moringa oleifera Lam.	Sahen Jana	Tree	C/W	AFSTU
Morus alba L.	Toot	Tree	C	AFSU
Mucuna pruriens (L.) DC. [= *M. prurita* Hook.] *M. pruriens* var. *utilis* W.[1]	Kaunch beej (source of L-dopa)	Climber	W/C	AFHSTU
Mukia maderaspatana (L.) M. Roem. [= *Melothria maderaspatana* (L.) Cogn.]	Mucukkai	Climber	W	AFS
Murraya koenigii (L.) Spreng.	Kariveppila	Tree	C/W	AFS

Botanical Name[1 & 2]	Trade Name	Habit	Material Source[3]	Associate Systems[4]
Musa sapientum L.[2]	Kadali	Herb	W	ASU
Mussaenda frondosa L.	Sribati	Shrub	W	AFS
Myrica esculenta Buch. -Ham. ex D. Don [= *M. nagi* auct. non Thunb.]	Katphala, Kaiphal	Tree	W	AFSTU
Myristica dactyloides auct. non Gaertn., syn. *M. malabarica* Lam.	Jaiphal, Javatri, Nutmeg	Tree	W	AS
Myristica fragrans Houtt.[2]	Jatipatre, Jaiphal = Nutmeg and Mace	Tree	C	AFHSTUW
Myristica malabarica Lam.	Rampatri	Tree	W	AFS
Myrsine africana L.	Chapra, Vidanga	Tree	W	AU
Myxopyrum serratulum A.W. Hill	Chathuravalli	Shrub	W	AF
Nardostachys jatamansi (D. Don) DC., syn. *N. grandiflora* DC.[1]	Balchad, Jatamansi	Herb	W	A
Naringi crenulata (Roxb.) Nicolson	Kadunimba	Tree	W	AF
Nelumbo nucifera Gaertn.	Kamal phul	Herb	W/C	AFSTU
Nepeta cataria L.	Badranjboya	Herb	C/W	FUW
Nepeta hindostana (B. Heyne ex Roth) Haines	Badranjboya	Herb	W	FU
Nerium oleander L. [= *N. indicum* Mill.]; *N. odoratum* Soland	Oleander, Kaner	Shrub	W/C	AFH
Nervilia aragoana Gaudich.	Sthalapadma	Herb	W	A
Nervilia plicata (Andrews) Schltr.	Padmacarini	Herb	W	AF
Nicotiana tabacum L.	Tambaku	Herb	C	AFHSU
Nigella sativa L.	Jeera (Black), Kalongi	Herb	C	AFSU
*Nilgirianthus ciliatus (*Nees) Bremek. [= *Strobilanthes ciliatus* Nees][2]	Kurinji, Karvi	Shrub	W	A
Nilgirianthus heyneanus (Nees) Bremek.	Substitute of Kurinji	Shrub	W	F
Nyctanthes arbor-tristis L.	Paruatak pan	Tree	C/W	AFHS
Nymphaea alba L.	Neel Kamal, Neelofar	Herb	W	ASTU
Nymphaea nouchali Burm. f., syn. *N. rubra* Roxb. ex Salisb	Niloth phal	Herb	W/C	AFS
Nymphaea stellata Willd.	Nilofar	Herb	W	AFST

Botanical Name[1 & 2]	**Trade Name**	**Habit**	**Material Source[3]**	**Associate Systems[4]**
Ocimum americanum L.[2]	Kali Tulsi, Bantulsi	Herb	W	AFHS
Ocimum basilicum L.[2]	Sweet basil, Kali tulsi, Bantulsi	Herb	C/W	A
Ocimum gratissimum L.	Ram Tulasi	Shrub	C/W	AFHSU
Ocimum kilimandscharicum Gurke	Karpoora Tulasi	Shrub	C	A
Ocimum tenuiflorum L. [= *O. sanctum* L.][1]	Tulasi, Holy basil	Herb	C	AFHS
Olea dioica Roxb.	Koli, Itala, Edala	Tree	W	FS
Onosma bracteatum Wall.	Gule-Gazbaan	Herb	I/W	ASU
Onosma echioides CB Clarke non-L., syn. *O. hispidum* Wall. ex G. Don[2]	Ratanjot	Herb	W	AU
Operculina turpethum (L.) J. Silva Manso [= *Merremia turpethum* (L.) Shah & Bhat][2]	Nishoth	Climber	W	AFHSTU
Orchis mascula (L.) L.	Salam misri	Herb	I	FSU
Origanum majorana L., syn. *Majorana hortensis* Moench	Gandhira, Sweet Marjoram	Herb	C	AHSU
Oroxylum indicum (L.) Benth. ex Kurz[2]	Tetuchaal, Arlu	Tree	W	AFSTU
Orthosiphon glabratus Benth. syn. *O. tomentosus* Benth. var. *glabratus* Hook. f.	Pratanika	Herb	W	F
Oxalis corniculata L., and *O. acetosella* Linn.	Puliyaarila	Herb	W	AFSU
Pachygone ovata (Poir.) Miers ex Hook. f. & Thomson	Kadukkodi, Pedda Dusar tree	Climber	W	F
Paederia foetida L., syn. *P. scandens* (Lour.) Merr.	Prasaarani, Skunk Vine	Climber	W	AF
Paeonia emodi Wall. ex Royle	Ud saleev	Shrub	W	AU
Panax assamicus Banerjee	Ginseng (Subst.)	Herb	W	F
Panax ginseng C.A. Mey.	Ginseng	Herb	I	W
Panax pseudoginseng Wall.	Substitute- Ginseng	Herb	W	FHW
Pandanus fascicularis Lam.	Kewada, Ketaki	Shrub	W	AF
Papaver somniferum L.	Postdana, Khaskhas	Herb	C	AFHSUW
Paris polyphylla Sm.	Daiswa paris, Svetavaca	Herb	W	A
Parmelia perforata (Jacq.) Ach., sister spp. *P. kamstchadalis* Ach.	Substitute- Chadela Dagad Phool	Herb	W	AU
Parmelia perlata (Huds.) Ach. (Lichen)[2]	Charelaa, Dagad Phool, Stone flower	Herb	W/C	ASTU
Passiflora foetida L.	Foul passiflora	Climber	W	AFS

Botanical Name[1 & 2]	Trade Name	Habit	Material Source[3]	Associate Systems[4]
Passiflora incarnata L.	Passionflower	Climber	C/W	HW
Pavonia odorata Willd.	Moramasi, Sugandh Bala	Herb	W	AFST
Pavonia zeylanica (L.) Cav.	Sitranmuttiver	Herb	C/W	AFS
Pedalium murex L.	Gokhru bada	Herb	W	AFSU
Peganum harmala L.[1]	Harmal	Herb	W	AFSU
Pentapetes phoenicea L.	Dopohoria, Pushparakta	Herb	C/W	AFS
Pergularia daemia (Forssk.) Choiv., syn. *P. extensa*	Atrilal	Climber	W	AFS
Pericampylus glaucus (Lam.) Merr.	Barakkanta	Climber	W	F
Peristrophe bicalyculata (Retz.) Nees	Atrilal	Herb	W	F
Peucedanum grande (Dalzell & A. Gibson) C.B. Clarke	Baphli, Duqu	Herb	W	AU
Phoenix dactylifera L.	Khajur	Tree	C	AFSU
Phoenix loureirii Kunth [= *P. humilis* Royle var. *pedunculata Beccari*]	Sitreechu	Tree	W	S
Phoenix pusilla Gaertn. [= *P. farinifera* Roxb.]	Chitteenth	Shrub	W	AF
Phoenix sylvestris (L.) Roxb.	Khajur	Tree	C/W	AFST
Phyla nodiflora (L.) Greene, syn, *Lippia nodiflora* Rich.	Chota-okra, Jalpippali	Herb	W	AFS
Phyllanthus amarus Schumach. & Thonn.[2] [= *P. fraternus* Webst.], syn. *P. niruri* var. *amarus* (Schumach. & Thonn.) Leandri	Bhumiamla	Herb	W/C	AFS
Phyllanthus maderaspatensis L.	Kanocha	Herb	W	AFSU
Phyllanthus reticulatus Poir. [= *Kirganelia reticulata* (Poir.) Baill.]	Buinowla	Shrub	W	AFS
Phyllanthus urinaria L.	Lal-Bhuin-Anvalah	Herb	W	AS
Phyllanthus virgatus G. Forst., syn. *P. simplex* Retz.	Niruri	Herb	W	F
Phyllostachys bambusoides Sieb. & Zucc.	Banslochan (Substi.)	Herb	W	F
Physalis minima L.	Tulatipati, Tankari	Herb	W	AFST
Physochlaina praealta (Decne.) Miers	Lal tang, Bajar Bang	Herb	W	F

Botanical Name[1 & 2]	**Trade Name**	**Habit**	**Material Source[3]**	**Associate Systems[4]**
Picrasma quassioides (D. Don) Benn.	Bharangi, Kadavi	Tree	W	AU
Picrorhiza kurroa Royle ex Benth.[1]	Kutaki	Herb	W	AFSTU
Picrorhiza scrophulariiflora Pennell	Substitute of Kutaki	Herb	W	A
Pimpinella anisum L.	Ajamoda, Badiyan Khathayi	Herb	C/W	AHSUW
Pinus roxburghii Sarg., and *P. excelsa* Wall. ex D. Don.; *P. khasya* Royle, *P. wallichiana* A. B. Jackson	Gandabiroja	Tree	W	AFSTU
Piper betle L.[2]	Betel	Climber	C	AFSTU
Piper chaba Hunter [= *P. retrofractum* Vahl.] [2]	Kabab Chini, Chavya, Gajpippal	Climber	I	ATU
Piper cubeba L. f. [2]	Cubeb, Tailed pepper	Climber	I	AFHSTUW
Piper longum L.[1]	Pipal, Thippili, Pippali	Shrub	W/C	AFSTU
Piper nigrum L. [2]	Pipal Gol, Kalimirch	Climber	C/W	AFHSTU
Piper wallichii (Miq.) Han. -Mazz.	Renukbeej (Nagod Beej)	Climber	W	AF
Pistacia integerrima Stew. ex Brand. [= *P. chinensis* Bunge var. *integerrima* Brandis] [2]	Kakarsinghi	Tree	W	A
Pistacia lentiscus L.	Rumimastagi, Mastangi	Shrub	I	ASU
Pistacia vera L.	Magaj Pista Irani	Shrub	I	AU
Plantago lanceolata L.	Baltanga, English plantain	Herb	W	AU
Plantago major L.[2]	Lahuriya	Herb	W	AHSU
Plantago ovata Forssk.[1], syn. *Plantago orbignyana* Steinh. ex Decne[1].	Isabgol	Herb	C	AFSUW
Plectranthus amboinicus (Lour.) Spreng. [2]	Patharcur	Shrub	C	A
Plectranthus barbatus Andrews [= *Coleus forskohlii* (Poir.) Briq.] [2]	Gandhira	Herb	C/W	AF
Plectranthus vettiveroides (Jacob) N.P. Singh & B.D. Sharma	Vetiver	Herb	C	AS
Pleurospermum angelicoides (DC.) C.B. Clarke	Native to Himalaya, Uttarakhand & China	Herb	W	F

Botanical Name[1 & 2]	**Trade Name**	**Habit**	**Material Source[3]**	**Associate Systems[4]**
Pluchea lanceolata (DC.) Oliv. & Hiern[1]	Rasna	Herb	W	AFT
Plumbago indica L.	Chitrak	Herb	C/W	AFS
Plumbago zeylanica L.[1]	Chitrak, Doctorbush	Herb	W	AFSTU
Podophyllum hexandrum Royle [= *P. emodi* Wall. ex Honig.]	Bankakri, Vanatrapusi	Herb	W	A
Pogostemon cablin (Blanco) Benth.	Patchouli	Herb	C	A
Pogostemon heyneanus Benth.	Patchouli	Herb	W/C	AFS
Polyalthia longifolia (Sonn.) Thwaites	Ashok (Substi.)	Tree	C	AFS
Polygonatum cirrhifolium (Wall.) Royle	Meda-m-meda	Herb	W	A
Polygonatum multiflorum (L.) All.	Solomnons seal	Herb	W	F
Polygonatum verticillatum (L.) All.	Meda-m-meda	Herb	W	AT
Polygonum alatum Buch. -Ham. ex Spreng.	Chonakappulu	Herb	W	F
Polygonum aviculare L.	Anjabar	Herb	W	AU
Polygonum glabrum Willd.	Raktha rohitha	Herb	W	FS
Polygonum punctatum Buch. -Ham. ex D. Don	Kangani-machan-pillu	Herb	W	H
Polygonum viviparum L.	Unjwar, Anjbar	Herb	W	F
Polypodium vulgare L.	Bisphaiz	Herb	W	FU
Pongamia pinnata (L.) Pierre [= *P. glabra* Vent.; *Derris indica* (Lam.) Bennett][1]	Honge Beej, Karanji, Pongum seed	Tree	C/W	AF
Portulaca oleracea L.	Lonika	Herb	W	AFSU
Potentilla nepalensis Hook.	Dori Ghas	Herb	W	A
Pothos scandens L.	Bendarli	Climber	W	F
Premna corymbosa Rottl. & Willd., syn. *Premna integrifolia* Linn.	Arni	Shrub	W	FS
Premna latifolia Roxb.	Bakar	Tree	W	A
Premna serratifolia L. [= *P. integrifolia* L.][1]	Arnimool/ Agnimantha	Shrub	W	AFS
Premna tomentosa Willd.	Sonachal	Tree	W	FS
Prosopis cineraria (L.) Druce., syn. *Prosopis spicigera* Linn.	Jhand, Chonkar	Tree	W	AF
Prunus armeniaca L.[2]	Apricot, Chuli	Tree	C	AUFTW
Prunus avium (L.) L.	Wild cherry, Aileya	Tree	W	A

Botanical Name[1 & 2]	**Trade Name**	**Habit**	**Material Source**[3]	**Associate Systems**[4]
Prunus cerasoides D. Don	Padamkasht	Tree	C/W	AFT
Prunus dulcis (Mill.) D.A. Webb.[1], syn. *P. amygdalus* Batsch	Almond, Badam, Magaj Badam	Tree	C	AFH
Prunus mahaleb L.	Mahaleb, Priyangu	Tree	C	AU
Pseudarthria viscida (L.) Wight & Arn.[2]	Moovila	Shrub	W	AFS
Psoralea corylifolia L.[1]	Bawachi, Bavanchi	Herb	W	AFHSTU
Pterocarpus marsupium Roxb.[2]	Damul-akhwain, Vijaysar	Tree	W	AFHSTU
Pterocarpus santalinus L. f., and *P. dalbergioides* Roxb.[1]	Raktachandana, Lal Chandan	Tree	W	AFSTU
Pueraria lobata (Willd.) Ohwi	Mudgaparni, Surpaparni	Climber	C/W	A
Pueraria tuberosa (Roxb. ex Willd.) DC.[2]	Vidhari kand, Indian Kudzu, Vidari, Patal Kumbha	Climber	W	AFTU
Punica granatum L. var. *granatum*	Dadam chal, Anaar	Shrub	C/W	AFHSTU
Quercus infectoria G. Olivier[2]	Majuphal, Oak	Tree	I	AFSUW
Rauvolfia densiflora (Wall.) Benth. *ex* Hook. f., syn. *R. verticillata* (Lour.)	Sarpagandha, Pagal Buti	Shrub	W	F
Rauvolfia micrantha Hook.f., syn. *R. membranifolia* Kerr	Sarpagandha, Pagal Buti	Shrub	W	F
Rauvolfia serpentina (L.) Benth. ex Kurz[1]	Sarpagandha, Pagal Buti	Shrub	W/C	AFHSTUW
Rauvolfia tetraphylla L.	Barachandrika	Shrub	C/W	F
Reinwardtia indica Dumort.[2]	Basanti, Yellow flax	Shrub	W/C	F
Rhaphidophora pertusa (Roxb.) Schott [= *R. laciniata* (Burm. f.) Merr.; *Pothos pertusus* Roxb.]	Ganeshkanda	Climber	C/W	AFS
Rheum australe D. Don [= *R. emodi* Wall. ex Meissn.][2]	Revan chini	Herb	W	A
Rheum moorcroftianum Royle	Revand-chini, Amlaparni	Herb	W	A
Rheum palmatum L.	Revand-chini	Herb	W	ASUW
Rheum spiciforme Royle	Amlaparni	Herb	W	A
Rheum webbianum Royle	Revand-chini	Herb	W	AU
Rhinacanthus nasutus (L.) Kurz	Juyiparni, Vitamallikai	Shrub	C/W	AFS
Rhodiola rosea L.[2]	Sanjeevani, solo	Herb	W	F

Botanical Name[1 & 2]	Trade Name	Habit	Material Source[3]	Associate Systems[4]
Rhododendron anthopogon D. Don[2]	Talisapatra	Shrub	W	A
Rhododendron campanulatum D. Don.	Cherailu	Shrub	W	F
Rhododendron lepidotum Wall.	Talisa, *Abies webbiana* substi.	Shrub	W	F
Rhus succedanea L.	Karkataka shringi	Tree	W	ASTU
Ricinus communis L.[1]	Arand	Shrub	C/W	AFHSTUW
Rosa alba L. (*R. canina* is an equivalent)	Gulseoti (Dog rose)	Shrub	C	AFSU
Rosa centifolia L.	Gulab	Shrub	C	AFSU
Rosa damascena Mill.	Gulab, Rose flowers	Shrub	C	AFHSU
Rosa indica L.[1]	Gulab	Shrub	C	FS
Roscoea alpina Royle	Kshirkakoli	Herb	W	F
Roscoea purpurea Sm.	Kshirkakoli	Herb	W	F
Rosmarinus officinalis L.	Rosemary	Shrub	C/I	FHW
Rotula aquatica Lour.	Pasanbheda	Herb	W	AF
Rourea santaloides Wight & Arn., syn. *R. minor* (Gaertn.) Alston	Varadara	Climber	W	F
Rubia cordifolia L., syn. *R. munjesta* Roxb. [1]	Maddar, Majith, Manjistha	Climber	W	AFSTU
Rumex nepalensis Spreng.	Yellow dock, *R. Crispus* substitute	Herb	W	AHUW
Ruta graveolens L.	Sadab	Herb	C	AFHSUW
Saccharum munja Roxb.	Amaveru	Herb	W	AT
Saccharum officinarum L.	Karumbu	Herb	C	AFHSTU
Saccharum spontaneum L.	Kusha	Herb	W	AFSTU
Salacia chinensis L.	Courondi	Climber	W	AFS
Salacia oblonga Wall. ex Wight & Arn.[2]	Kadalainjil, Saptrangi	Climber	W	FS
Salacia reticulata Wight	Pitila	Shrub	W	AF
Salix tetrasperma Roxb.	Indian willow	Tree	W	AFSU
Salvadora oleoides Decne. [2]	Pilu	Tree	W	AFU
Salvadora persica L.	Goni	Tree	W	AFSU
Salvia aegyptiaca L.	Balangoo	Herb	W	AU
Salvia haematodes L.	Behman	Herb	I	FU
Salvia plebeia R.Br.	Kachora	Herb	W	AS
Salvia sclarea L.	Behman safed	Herb	I/C	U
Santalum album L.[1]	Chandan	Tree	W	AFHSTUW
Sapindus emarginatus Vahl.[2]	Reetha, Soapnut	Tree	W/C	AFU
Sapindus mukorossi Gaertn.[1]	Aretha mota	Tree	C/W	A

Botanical Name[1 & 2]	**Trade Name**	**Habit**	**Material Source[3]**	**Associate Systems[4]**
Saraca asoca (Roxb.) W.J.de Wilde, syn. *S. indica*[1]	Ashoka	Tree	W	AFHS
Sarcostemma viminale (L.) R. Br. [= *S. acidum* (Roxb.) Voigt]	Soma, Somalatha	Climber	W	A
*Saussurea costus (*Falc.) Lipsch. [= *S. lappa* (Decne.) Sch. Bip.][1]	Kuth, Uplet	Herb	C/W	AFSTU
Schizachyrium exile (Hochot.) Stapf	Sprkka	Herb	W	A
Schleichera oleosa (Lour.) Oken	Kusum beeja	Tree	W	AFS
Schrebera swietenoides Roxb.[2]	Ghanti phal	Tree	W	AFST
Scindapsus officinalis (Roxb.) Schott.	Gaj pipal	Climber	W	AFSTU
*Securinega leucopyrus (*Willd.) Mull. -Arg. [= *Flueggea leucopyrus* Willd.], syn. *S. leucopyrus* (Willd.) Mull-Arg	Hartho, Bhuriphali	Shrub	W	AFS
Selinum candollei DC., syn. *S. tenuifolium*	Mura	Herb	W	A
Selinum vaginatum (Edgew.) C.B. Clarke	Butkesh	Herb	W	F
Semecarpus anacardium L. f.[2]	Balave, Bhilavan, Bhilawa, Marking nuts	Tree	W	AFHSTU
Senna italica Mill. [= *Cassia italica* (Mill.) Lam. ex Andr.*; C. obtusus* Roxb.][2]	Nila aavaarai	Herb	W	AFH
Sesamum orientale L. [= *S. indicum* L.]	Til	Herb	C	AS
Sesbania grandiflora (L.) Poir.	Agase, Agatti.	Tree	C	AFSTU
Seseli diffusum (Roxb. ex Sm.) Santapau	Kirmani, Ajwain	Herb	W	A
Setaria italica (L.) Beauv	Kangni	Herb	C	AFSTU
Shorea robusta Gaertn.[1]	Raal	Tree	C/W	AFSTU
Sida acuta Burm. f., syn. *S. carpinifolia* auct. non Linn f.	Bala	Herb	W	AFS
Sida cordifolia L.	Bala	Herb	W	AFST
Sida rhombifolia L.[2]	Bala	Herb	W	AFS
Silybum marianum (L.) Gaertn.[2]	Milk thistle	Herb	C/W	HW
Simmondsia chinensis (Link) CK Schneid.[2]	Jojoba	Shrub	C	W
Sinapis alba L.	Safed Rai, Yellow mustard	Herb	C/W	AHS
Sisymbrium irio L.[2]	Khubkalan	Herb	W	AU

Botanical Name[1 & 2]	**Trade Name**	**Habit**	**Material Source[3]**	**Associate Systems[4]**
Smilax aspera L.	Chopchini (Substi.)	Climber	W	AH
Smilax china L.	Chobchini	Climber	I	ASU
Smilax glabra Roxb.[2]	Chobchini, Lokhandi	Climber	W	A
Smilax zeylanica L.	Chopchini (Subst.)	Climber	W	AFS
Solanum anguivi Lam. [= *S. indicum* auct. non. L.; *S. violaceum* Ortega][2]	Katheli-badhi	Shrub	W	AS
Solanum nigrum L.[1]	Makoi, Kakamachi, Black night shade	Herb	W	AFHSTU
Solanum torvum Sw.	Padarchunda	Shrub	C/W	AFS
Solanum trilobatum L.	Alarka	Climber	W	AFS
Solanum virginianum L. [= *S. surattense* Burm. f.; *S. xanthocarpum* Schrad. & H. Wendl.][2]	Kateli	Herb	W	AH
Soymida febrifuga (Roxb.) A. Juss.[2]	Rohan	Tree	W	AFSTU
Spatholobus parviflorus (DC.) Kuntze, syn. *S. roxburghii* Benth.	Mula	Climber	W	FS
Spermacoce articularis L.f., syn. *S. hispida* Linn.; *Borreria articularis* (Linn. f.) F.N. Williams	Tukah	Herb	W	A
Sphaeranthus africanus L.	Mundi	Herb	W	A
Sphaeranthus amaranthoides Burm.	Cevayam	Herb	W	FS
Sphaeranthus indicus L.[2]	Gorak mundi	Herb	W	AFSTU
Spilanthes oleracea L. [= *S. acmella* Murr. var. *oleracea* C.B. Clarke]	Akarkara, Sarahattika, Vana-mugali	Herb	C/W	A
Spondias pinnata (L. f.) Kurz	Amate	Tree	W	AFSTU
Stephania glabra (Roxb.) Miers	Patha, Rajapatha	Climber	W	A
Sterculia foetida L.	Jangli badam, Java olive	Tree	W/C	AFS
Sterculia urens Roxb.[2]	Karaya, Kateera	Tree	W	AFSW
Stereospermum chelonoides (L. f.) DC.[2] [= *S. suaveolens* (Roxb.) DC.]; *S. personatum* (Hassk.) Chatterjee[2]	Patala	Tree	W	AFS
Stereospermum colais (Buch. -Ham. ex Dillw.) Mabb. [= *S. personatum* (Hassk.) Chatterjee]	Patala, part of Dash Moola	Tree	W	AF
Stevia rebaudiana Bertoni	Stevia	Herb	C	S. America

Botanical Name[1 & 2]	**Trade Name**	**Habit**	**Material Source[3]**	**Associate Systems[4]**
Streblus asper Lour.	Bajradanti	Tree	W	AFS
Strophanthus wightianus Wight	Naithal kizhangu	Climber	W	F
Strychnos nux-vomica L.[1]	Kuchla, Itti Beeja	Tree	W	AFHSTUW
Strychnos potatorum L. f.[2]	Nirmali, Cleaning nuts	Tree	W	AFSTU
Stylocoryna lucens (Hook.f.) Gamble [= *Tarenna alpestris* (Wight) NP Balakr.]	Paphanals	Shrub	W	A
Swertia alata (Royle ex D. Don) CB Clarke	Kiratatikta	Herb	W	A
Swertia angustifolia Ham.	Chiraeta shireen	Herb	W	AU
Swertia chirayita (Roxb. ex Fleming) H Karst.[1], syn. *S. chirata*[1]	Chiraiyata	Herb	W/C	AFH
Symplocos cochinchinensis S. Moore	Lodhra	Tree	W	AFS
Symplocos paniculata (Thunb.) Miq.	Lodhra, Lodhra Pathani	Tree	W	AU
Symplocos racemosa Roxb.[2]	Pathani Lodha	Tree	W	AFSTU
Syzygium aromaticum (L.) Merr. & L.M. Perry[2]	Laung, Cloves	Tree	C	AFSTUW
Syzygium caryophyllatum (L.) Alston [= *Eugenia caryophyllaea* Wight]	Clove, Lavang	Tree	W	AFSU
Syzygium cumini (L.) Skeels[2], syn. S. *jambolanum* Lam.	Jamun	Tree	W/C	AFHSTU
Tabernaemontana divaricata (L.) R.Br. ex M. Roem. & Schult., syn. *T. coronaria* (Jacq.) Willd. [= *Ervatamia coronaria* (Jacq.) Stapf]	Candni	Shrub	C	AF
Tacca aspera Roxb. [= *T. integrifolia* Ker Gawl.]	Varahikand	Herb	W	A
Tagetes erecta L.[2]	African/ Mexican Marygold, Genda, Sandu	Herb	C	AFS
Tamarindus indica L.[2]	Imli	Tree	C/W	AFSTU
Tamarix indica Roxb.	Jhan	Shrub	W	AFU
Tanacetum vulgare L.	Tansy	Herb	I	HW
Taraxacum officinale F.H. Wigg.	Kanphul, Dudhal	Herb	C/W	AFHW
Taxus wallichiana Zucc. [= *T. baccata* L.][2]	Talispatra	Tree	W/C	AFW
Tecomella undulata (Sm.) Seem.	Rohit Aka	Shrub	W	AT

Botanical Name[1 & 2]	**Trade Name**	**Habit**	**Material Source[3]**	**Associate Systems[4]**
Tectona grandis L. f.	Teak	Tree	C/W	AFSTU
Tephrosia purpurea (L.) Pers.[2]	Sarad foka, Sarpankha	Herb	W	AFSU
Teramnus labialis (L. f.) Spreng.	Masparni	Climber	W	AFST
Terminalia arjuna (Roxb. ex DC.) Wight & Arn.[1]	Arjun	Tree	W/C	AFHSTU
Terminalia bellirica (Gaertn.) Roxb.[1]	Behda, Bibhitaki (Part of Triphala)	Tree	W/C	AFSU
Terminalia chebula Retz.[1]	Harda, Haritaki (Part of Triphala)	Tree	W	AFHSTU
Terminalia crenulata Roth.	Asan	Tree	W	AFS
Terminalia paniculata Roth., syn. *T. alata* Herb. Madr. Ex Wall	Vanmaruthu	Tree	W	AFS
Terminalia tomentosa (Roxb. ex DC.) Wight & Arn. [= *T. alata* Heyne ex Roth]; *T. elliptica* Willd	Asan	Tree	W	AST
Thalictrum foliolosum DC.	Mamira	Herb	W	AFS
Thespesia populnea (L.) Sol. ex Correa	Phalisa-Chhal	Tree	C	AFSU
Thymus serpyllum L.	Ban Ajwain, Satar Farsi	Herb	W	AHU
Tiliacora acuminata (Poir.) Miers ex Hook.f. & Thomson	Kappatiga, Bagamushada	Climber	C/W	FS
Tinospora cordifolia (Willd.) Miers ex Hook.f. & Thomson[1]	Giloy, Galo, Amrithaballi	Climber	W	AFHSTU
Tinospora crispa (L.) Hook.f. & Thomson	Giloy, Galo, Amrithaballi	Climber	W	F
Tinospora sinensis (Lour.) Merr. [= *T. malabarica* (Lam.) Hook.f. & Thomson]	Amrata	Climber	W	AFS
Toddalia asiatica (L.) Lam.	Janglikalimirch	Shrub	W	AFS
Toona ciliata M. Roem. [= *Cedrela toona* Roxb.]	Thooniyanoikam	Tree	W	F
Trachyspermum ammi (L.) Sprague[1]	Ajmo, Ajwayan (Essential oil)	Herb	C	AFSTU
Trachyspermum roxburghianum (DC.) Craib	Sath Ajwayan	Herb	W	AFSU
Tragia involucrata L.[2]	Barhanta	Climber	W	AFS
Trapa natans L., syn. *T. bispinosa* Roxb.	Singhada	Herb	C/W	A

Botanical Name[1 & 2]	**Trade Name**	**Habit**	**Material Source[3]**	**Associate Systems[4]**
Trianthema decandra L. [= *Zaleya decandra* (L.) Burm. f.]	Saaranai Ver	Herb	W	AS ? F
Trianthema portulacastrum L.	Lalsabuni	Herb	W	AFS
Tribulus alatus Del.	Gokhru kalan	Herb	W	FU
Tribulus lanuginosus L.	Gokhru	Herb	W	AU
Tribulus rajasthanensis MM Bhandari & VS Sharma	Substitute of Gokhru	Herb	W	AU
Tribulus subramanyamii P Singh, GS Giri & V Singh	Substitute of Gokhru	Herb	W	AU
Tribulus terrestris L.[2]	Gokhru, Gokshura	Herb	W	AFHSTU
Trichodesma zeylanicum (Burm. f.) R. Br.	Dhadhona	Herb	W	AFS
Trichosanthes anguina L., syn. *T. cucumerina* L., *T. cucumerina* L. *anguina* (L.) Haines	Snake gourd	Climber	C	AST
Trichosanthes cordata Roxb.	Vidari	Climber	W	A
Trichosanthes cucumerina L.[2]	Patol panchang	Climber	W	AFS
Trichosanthes dioica Roxb.	Patol (Kadu Parval)	Climber	C	AFHST
Trichosanthes lobata Roxb.	Patola, Tiktapatola	Climber	W	A
Trichosanthes tricuspidata Lour. [= *T. bracteata* (Lam.) Voigt]	Indrayan, Mahakal	Climber	W	AFS
Tridax procumbens L.	Jayanti	Herb	W	AFS
Trigonella corniculata (L.) L.	Kasuri Mathi	Herb	W/C	AUS
Trigonella foenum-graecum L.	Mathi	Herb	C	AFSUW
Tulipa stellata Hook.	Meethi Suranjan	Herb	W	F
Tylophora indica (Burm.f.) Merr. [= *T. asthmatica* (L. f.) Wight & Arn.]	Antamul, Country Ipecacuanha	Climber	W	AFHS
Typha elephantina Grah., non Roxb. [= *T. australis* K. Schum. & Thonn.]	Anaikkorai	Herb	W	AFS
Typhonium trilobatum (L.) Schott	Karunai kizhangu	Herb	W	FS
Uncaria gambier Thwaites	Kath	Climber	I	AS
Uraria lagopodioides (L.) DC.	Kalasi	Herb	W	AFSU
Uraria picta (Jacq.) DC.	Prshniparni, Kalasi	Herb	W	AFST
Uraria rufescens (DC.) Schindl.	Kalasi, *U. picta* substitute	Herb	W	AF
Urginea indica (Roxb) Kunth. [= *Drimia indica* (Roxb.) Jessop]	Janglipyaz, White squills	Herb	W	ASU
Urtica dioica L.	Nettle	Herb	W	AHW
Valeriana hardwickii Wall.	Tagar-ganth, Nihani	Herb	W	AU

Botanical Name[1 & 2]	**Trade Name**	**Habit**	**Material Source[3]**	**Associate Systems[4]**
Valeriana jatamansi Jones., and *V. wallichii* DC. [1], syn. *Nardostachys jatamansi* (Jones) DC.	Mushakbala, Tagar	Herb	W	AF
Valeriana pyrolaefolia Decne. = *V. pyrolaefolia* Decne.	Mushkabala substitute	Herb	W	AF
Vanda tessellata (Roxb.) Hook. ex G. Don	Rasna	Herb	W	AFSU
Vateria indica L.[2]	Mandadhupa, Dupa	Tree	C/W	AFSU
Ventilago madraspatana Gaertn.	Pitti	Climber	W	AFS
Vernonia cinerea (L.) Less. [= *V. conyzoides* DC.]	Dandotpala	Herb	W	AFSTU
Vetiveria zizanioides (L.) Nash, syn. *Andropogon zizanioides* Linn.[1]	Lavancha, Khas, Vetiver	Herb	W/C	AFHSTU
Vigna sublobata (Roxb.) Babu & Sharma [= *V. radiata* (L.) Wilez var. *sublobata* (Roxb.) Verdc.]	Masaparni	Herb	C	AS
Vigna vexillata (L.) A. Rich., syn. *Dolichos* and *Phaseolus vexillatus* L.	Kattupayar	Herb	W	A
Viola canescens Wall.	Banafasha	Herb	W	F
Viola cinerea Boiss.	Banafasha phool	Herb	W	A
Viola odorata L.	Banafasha	Herb	W/C	AFHU
Viola pilosa Bi.[2]	Banafasha	Herb	W	AF
Vitex agnus-castus L.	Chaste tree	Shrub	I	HU
Vitex altissima L. f.	Myrole	Tree	W	FS
Vitex negundo L.[1]	Neergundi	Shrub	W/C	AFSTU
Vitis vinifera L.	Draksh	Climber	C	AFSTU
Walsura trifoliata (A. Juss.) Harms	Valsura	Tree	W	FS
Wedelia chinensis (Osbeck) Merr. [= *W. calendulacea* (L.) Less.]	Bhangra	Herb	C/W	AF
Withania coagulans Dunal[2]	Paneer- Dodi	Shrub	W	AU
Withania somnifera (L.) Dunal[1]	Ashwagandha, Asgandha	Shrub	C/W	AFHSTU
Woodfordia fruticosa (L.) Kurz.[2]	Dhaiphool, Dhavadiphool	Shrub	W	AFSTU
Wrightia arborea (Dennst.) Mabb. syn. *W. tomentosa* Roem. & Schult.	Kutajah	Tree	W	AF

Botanical Name[1 & 2]	**Trade Name**	**Habit**	**Material Source[3]**	**Associate Systems[4]**
Wrightia tinctoria (Roxb.) R.Br.[2]	Indrajau	Tree	C/W	AFHSU
Xylia xylocarpa (Roxb.) Taub.	Trumullu	Tree	W	AF
Zanthoxylum acanthopodium DC.	Kabab Khandan, Tummad	Shrub	W	ASU
Zanthoxylum armatum DC., syn. *Z. alatum* Roxb.	Tejbal	Shrub	W	A
Zanthoxylum rhetsa (Roxb.) DC., syn. *Z. budrunga* DC	Tejbal, Triphala	Tree	W	AFS
Zea mays L.	Maize	Herb	C	AFHSUW
Zehneria umbellata (Klein) Thwaites	Karuvikkilannu	Climber	W	FS
Zingiber officinale Roscoe[2]	Adrak, Ginger, Sonth, (nutraceautical too)	Herb	C	AFSU
Zingiber zerumbet (L.) Roscoe *ex* Sm.	Narkachur	Herb	C/W	AFSU
Ziziphus jujuba (L.) Gaertn. [= *Z. mauritiana* Lam.][2]	Ber	Tree	C/W	AFSTU
Ziziphus sativa Gaertn. [= *Z. jujuba* Mill.; *Z. vulgaris* Lam.]	Unnab	Tree	W	AFU
Ziziphus xylopyrus (Retz.) Willd.[2]	Ghontaphala	Shrub	W	AFS

Based on Checklist of Medicinal Plants in Trade, NMPB www.nmpb.nic.in › content › Ongoing-and-completed-

1. Most demanded and traded commercial MAPs.
2. Other MAPs attracting market demands.
3. C – Cultivated; I – Imported; W – Wild; C/W - Cultivated and Wild; W/C - Wild and cultivated; I/C - Imported and Cultivated; I/W - Imported and Wild
4. A – Ayurveda; F – Folk; H – Homoeopathy; S – Siddha; T – Tibetan; U – Unani; W – Wester

References

Pareek, SK, Gupta, Veena, Bhatt, KC, Negi, KS and Sharma, N. (2005) Medicinal and Aromatic Plants. In: Plant Genetic Resources: Horticulture Crops (Eds. Dhillon, BS, Tyagi, RK, Saxena, S and Randhawa GJ) pp 279-308. Narosa Publishing House Pvt. Ltd., New Delhi, India

5

Inventory of Additional Plant Species Worth Scientific Scrutiny for Medicinal and Aromatic Properties and Potential Use

5.1 Introduction

Considering the richness of floristic diversity of India represented by 47,513 plant species, of which around are 9,500 reported to be of ethnobotanical botanical significance and around 7,500 - 8900 of ethnomedicinal importance (Anonymous, 1994; Pareek *et al.*, 2005). It has been estimated that of these, around 5,000 plant species have been investigated in some detail in search for potential sources of wellness or health or medicinal drugs to cure various human ailments and diseases. As discussed in the previous sections, the Indian literature on medicinal aromatic plants formally documents information on about 2,500 medicinal and aromatic plants under various traditional systems with sufficient details. In addition, indigenous traditional knowledge exists for another 4,000-4,500 medicinal plant species based on long time experience/observations and their use by various tribes and communities and the individual clan or families in cure of specific ailments. However, this knowledge predominantly exists in oral traditions, which are under danger because of the threat to these traditional orthodox societies and loss of the dialect or languages in which they exist. Thus, there is an exceptionally large number of species, whose medicinal properties require validation following appropriate systematic scientific investigations using conventional and available advanced technologies. Further, there are many more little or unknown tribal medicinal plants needing evaluation and chemical and biological screening for their potential in search of promising authentic sources for alleviation various human deficiencies and cure of ailments and diseases, as part of drug formulations for the treatment. Thus, in addition to medicinal and aromatic plants (MAPs) used in various traditional systems, many more less-known or under-utilized MAPs species are worthy of a scientific scrutiny for evaluation of their medicinal potential and possible use as part of authentication of traditional systems and bioprospecting of many more potential plant species for sustainable development and extension of use of herbal medicinal systems with value-addition and/ or new products (Pushpangadan *et al.*, 2018b). Recognizing this fact, an attempt has been made in this section to enlist plants worthy of scientific scrutiny for medicinal properties. It would include plants species falling in following categories:

1. Species known for medicinal use without a scientific detail/rational and need to be subjected to scientific and clinical scrutiny with thorough investigation.
2. Not part of the list of commonly used traditional herbs and those of other species and the ancestral or sister species used in traditional herbal medicinal systems as alternative source.
3. Non-coded/non-listed MAPs species that are part of oral traditions and not recorded in the recognised accessible and scientific literature.
4. Recently explored or reported ancestral or sister species of the established medicinal and aromatic plants/sources of various drug formulations.
5. Lastly, new potential species recorded from different regions of the country in recent times, based on observations and experience of locals in cure of various ailments.

The main source of information about the new prospective/potential/ possible medicinal plants are tribal communities living in and around the forests and the rural communities dependent on herbal solutions for health and treatment of diseases. Living close to nature and by trial and error, empirical reasoning and experimentation, these primitive indigenous societies have developed their own wealth of knowledge pertaining to conservation and sustainable use of plants, animals, and other natural resources of their medicinal and general use. In many cases, the lead for such plants and their products has been provided by the local tribes/communities but it cannot be traced back because the user is not obliged to disclose the origin of source of information, and several other reasons, such as loss of tribal dialect/languages and the loss of economic gains to the holder of knowledge. Further, due to modernization, the previous knowledge systems have been eroding and are corroding fast and at times totally disappearing in the last few decades.

There is no authentic complete list of medicinal plants with their properties, particularly the one including the non-documented plants of oral traditions. Some of these plants have been mentioned with different names in various reports/publications of different indigenous languages from different areas. Therefore, inventorying and creation of a final list of plant species of medicinal importance is required to initiate their effective conservation activities and facilitate use. The realization of this fact has motivated researchers from many countries around the world to undertake ethnobotanical investigations and document the economically important plants with associated traditional knowledge and wisdom of the indigenous people.

The Wealth of India, a dictionary of Indian raw materials and industrial products covering plants published by Council of Scientific and Industrial research (CSIR), has included description of prevalent usage of various plants and other elements of bioresources amongst the local people. This series has substantial documentation but does not refer to area or the people and probably does not cover all the medicinal plant species and available total information.

Recognizing this, Government of India lunched a multidisciplinary, multi-institutional, and action-oriented research project, All India Coordinated Research Project on Ethnobiology (AICRPE) in 1982 (AICRPE, 1982-1998) aiming at inventorying and documenting the multidimensional perspectives of the life, culture, and traditions of the tribes as well as their knowledge system associated with the utilization of the local biological resources.

According to the 2011 census, the tribal people population in India is 8.6 percent of the national population with over 104 million people. The tribes belong to over 550 communities inhabiting in 5000 villages, mostly located in and around forests of the country. The population of the individual tribe is as large as about 1.53 crores (21.1%) of the total population in Madhya Pradesh and as small as 101 like Onge's of Andaman Islands. The tribes in the country occupy about 18.74% of the total land area, mainly in the hilly and forest areas of 19 states and union territories. During the period of AICRPE project (1982-1998), the information on the multidimensional perspectives of the life, culture, tradition, and knowledge systems associated with biotic and abiotic resources of the 550 tribal communities comprising over 83.3 million people belonging to the diverse ethnic groups was recorded (Pushpangadan *et al.*, 2018). The knowledge of these communities on the use of wild plants for food, medicine and for meeting many other material requirements is now considered to be potential information for appropriate scientific and technological intervention for developing value added commercially marketable products. Unfortunately, a significant part of the indigenous traditional knowledge (ITK) exists only in oral tradition and is not qualified for the formal IPR system. The vast amount of information collected by the AICPPE team is locked up as unattended reports for want of proper resources and has not come out in the public/scientific domain permitting their conservation and further use.

The AICPPE efforts revealed that 8,000 wild plants are used by tribal communities for medicinal purpose to support health and cure various ailments/diseases meeting their medicinal requirements, including around 2500 plants recorded under various classical Indian traditional medicinal systems, such as Ayurveda, Amchi, Siddha, Unani, etc. Additionally, it records about 950 more MAPs with new claims of medicinal properties, worthy of scientific scrutiny (Pushpangadan *et al.*, 2018). Also, it records 3,500-3,900 plants for edible purposes, 400-500 for fodder, 550-700 fibres, 300-425 gums, resins, and dyes, 300-325, piscicides or pesticides, and 750-1000 others under tribal use. Thus, this effort created information for a database on total of 10,000 plants, used by Indian tribal and rural communities for meeting their varied requirements (Anonymous, 1994; Pushpangadan *et al.*, 2018; personal communication). However, the list of these plants, particularly of the new one's worthy of scientific scrutiny, is not in public domain and could not be availed from Dr Pushpangadan, the coordinator of the AICPPE project or the Ministry of Environment and Forest to whom the final report of the project was submitted. Therefore, in this section an attempt has been made to enlist some additional/new medicinal plant species falling under four criterions described earlier.

This scenario demonstrates that, in addition to the documented traditional medicinal plant species under various systems, there exists many more plant species with or without associated traditional knowledge, partially explored and/or unexplored/ untapped. They have vast potential and offer opportunities for expansion of herbal use and commercial exploitation of plant-based products, particularly to strengthen and promote the traditional herbal-based healthcare systems/practices to facilitate cost-effective health care of poor local indigenous people and rural masses. It would, therefore, be worthwhile to give brief background/preliminary information regarding these plant species to generate interest among researchers for systematic scientific scrutiny of these, using state of the art technologies dealing with biochemical profiling, molecular characterization, etc. In subsequent section an attempt has been made to list some of the representative new species with preliminary information available about them.

References

AICRPE (All India Co-ordinated Project on Ethnobiology) (1992-1998) Final Technical Report, Ministry of Environment and Forests (MoEF), Government of India.

Anonymous (1994) Ethnobiology in India: A Status Report. All India Coordinated Research Project on Ethnobiology, Ministry of Environment and Forests, New Delhi.

Arora, RK, Nayar, ER and Pandey, A. (2006) Indian centre of floristic and economic plant diversity: A review. In: Hundred Years of Plant Genetic Resources Management in India, (Eds. Singh, Anurudh K, Srinivasan, K, Saxena, S and Dhillon BS) pp 1–28. National Bureau of Plant Genetic Resources (ICAR), New Delhi, India,

Pareek SK, Gupta V, Bhatt KC, Negi KS and Sharma N (2005) Medicinal and Aromatic Plants. In: Plant Genetic Resources: Horticultural Crops (Eds. Dhillon, BS, Tyagi, RK, Saxena, S and Randhawa, GJ) pp 279-308. Narosa Publishing House, New Delhi,

Pushpangadan, P, George Varughese, Ijinu, TP and Chithra, MA. (2018) All India coordinated research project on ethnobiology and genesis of ethnopharmacology research in India including benefit sharing. Annals of Phytomedicine 7(1): 5-12, DOI: 10.21276/ ap.2018.7.1.2

Pushpangadan, P, George Varughese, Ijinu, TP and Chithra, MA. (2018b). Biodiversity, bioprospecting, traditional Knowledge, sustainable development and value-added products: a review. Journal of Traditional Medicine and Clinical Naturopathy 7:256.

5.2 The List of Plant Species Worthy for Scientific Scrutiny for Medicinal and Aromatic Properties and Potential Use

Out of the 8000 wild plant species used by the tribals for medicinal purposes in India, a large number are new claims (Pushpangadan *et al.,* 2014). These species have not been documented as per the international code with full details and worthy of scientific scrutiny to promote conservation and use. The list of such species identified under All India Coordinated Research Project on Ethnobiology (AICRPE) is neither available in public domain nor it was made available by the Chief Coordinator despite several attempt. Therefore, here an attempt has been made to list new/additional species, which has been recorded in different publications based on the information and claims made by local communities and tribes, based on experience and use. This list consists of many phylogenetically and taxonomically close species belonging to the genus of common/important medicinal and aromatic plants recorded in literature. It is possible that many more species belonging to genera with large number species, such as *Aconitum, Andrographis, Asparagus, Berberis, Ceropegia, Chlorophytum, Cymbopogon, Eulophia, Inula, Saussurea, Terminalia*, etc. may also qualify to this list because due to phylogenetic closeness there is the possibility that they may have nearly similar medicinal and aromatic properties as those of known species. They may offer new genetic diversity for principal components and consequently new opportunities.

5.2.1 The List with Description of Worthy Plant Species

***Abutilon ranadei* Woodrow & Stapf. (CR)**

Common name (s): Holly hock, Son Ghanta

Family: Malvaceae

Distribution: Endemic to Northwestern Ghats.

Usable parts and properties: Related to *Abution indicum*. Needs investigation for alternative use.

Source of information (system): Locals of Northwestern Ghats

***Acacia canescens* Grab.**

Common name (s): Ari, Araara, Aadaari, *A. Pennata* wild relative

Family: Leguminosae/ Fabaceae (Mimosaceae)

Distribution: Occur in Bihar and South India.

Usable parts and properties: Bark- Anti-inflammatory, antiseptic, used in skin diseases and hair wash. Flower- Emmenagogue. Plant parts- Locally used in cough, bronchitis, measles, tubercular fistula.

Source of information (system): F

***Acer caesium* Wall. ex Brandis (VU)**

Common name (s): Bluish Grey Maple

Family: Aceraceae

Distribution: Himalayas and southern China.

Usable parts and properties: Bark- Treat muscular swellings, boils, and pimples.

Source of information (system): F

***Achyranthes coynei* Santapau (EN - R)**

Family: Amaranthaceae

Distribution: Endemic to Karnataka and Maharashtra.

Usable parts and properties: Herb- Antimicrobial and antioxidant. Decoction used to easy delivery. Root powder- Kills intestinal worm.

Source of information (system): F

***Aconitum kashmiricum* Stapf ex Coventry (EN)**

Family: Ranunculaceae

Distribution: Wild in northwestern Himalayas.

Parts and properties used: Root- Febrifuge and carminative.

Associated system (s): U

***Aconitum lethale* Griff. (R- Rare)**

Common name (s): Balfour's Monkshood, Gobaree

Family: Ranunculaceae

Distribution: Native range, Himalaya to South Tibet. Wild in Mishmi Hills, Arunachal Pradesh (ArP)

Usable parts and properties: Root- *Vish*/poison, analgesic, anti-inflammatory, anti-rheumatic and vermifuge.

Source of information (system): F

***Aconogonum tortuosum* D. Don, syn. *Polygonum tortuosum* D. Don**

Common name: Nyalo

Family: Polygonaceae

Distribution: Habitat to alpine Himalayas and cold arid Ladakh.

Usable parts and properties: Dried leaves- Anti-bacterial, abdomen pain.

Source of information (system): Local use

***Actinocarya tibetica* Benth.**

Common name: Forget-me-not

Family: Boraginaceae

Distribution: Habitat to alpine Himalayas and cold arid Ladakh.

Usable parts and properties: Whole plant- Source of anti-inflammatory and immunomodulator flavones. Used for fever. Unexplored, medicinal potential species.

Source of information (system): Local use

***Adenia hondala* (Gaertn.) W.J. De Wilde, syn. *A. palmata* (Lam.) Engl.**

Common name (s): Karimuthukku, Modekka, Hondala

Family: Passifloraceae

Distribution: Deciduous forests of Western Ghats and south India.

Usable parts and properties: Tuberous- Treat skin disorder and hernias. Produce Ayurvedic "*vidari*" (aphrodisiac) or "*vidaari* drug.

Source of information (system): AF

***Aegle marmelos* var. *mahurensis* B.R. Zate (R)**

Common name (s): Bael, Bel

Family: Rutaceae

Distribution: Occur all over India, Maharashtra.

Usable parts and properties: Alternative to *Aegle marmelos*.

Source of information (system): F

***Aerva wightii* Hook. f. (EX)**

Family: Amaranthaceae

Distribution: Extinct from Western Ghats. Rediscovered in Tirunelveli District, Tamil Nadu (TN)

Usable parts and properties: Known for antibacterial principle.

Source of information (system): F

***Agave vera-cruz* Mill**

Common name (s): Blue elephant aloe

Family: Asperagaceae

Distribution: Mexican. Naturalized in parts of India.

Usable parts and properties: Leaf juice- Purgative. Used to treat constipation.

Source of information (system): F

***Albizia orissensis* Sahni & Bennet**

Common name (s): New wild relative of Siris, *A. lebbeck*.

Family: Leguminosae/ Fabaceae (Mimosaceae)

Distribution: Habitat to Northern Eastern Ghats.

Usable parts and properties: Unexplored sister species of *A. lebbeck*, needing scrutiny.

Source of information (system): Locals

***Albizia thompsonii* var. *galbana* Haines (R)**

Common name (s): Wild relative of Siris, *A. lebbeck*.

Family: Leguminosae/ Fabaceae (Mimosaceae)

Distribution: Northeastern Ghats Bihar and adjoining areas.

Usable parts and properties: Bark- Skin diseases. Leaves paste- Used to cure ulcers.

Source of information (system): - Locals

***Alhagi maurorum* Medik, syn. *A. pseudalhagi* (M. Bieb.) Desv. ex B. Keller & Shap., *A. camelorum* Fisch. ex DC.**

Sanskrit name: Dhanvayasa

Common name (s): Javasa

Family: Leguminosae/ Fabaceae

Distribution: Western Himalayas.

Usable parts and properties: Leaf oil- Used to treat rheumatism. Flower- Diuretic, laxative, expectorant. Used to treat piles.

Source of information (system): F

***Alpinia malaccensis* (Burm. f.) Roscoe, syn. *A. nutans* var. *sericea* Baker**

Family: Zingiberaceae

Distribution: Occur in northeastern and peninsular India.

Parts and properties used: Rhizome- Cure sores, aromatic source of essential oil. Fruit- Emetic.

Associated system (s): F (traditional medicinal system)

***Alysicarpus vaginalis* (L.) DC., syn. *A. nummularifolius* (L.) DC.**

Common name (s): Alyce clover, Sauri, Chauli

Family: Leguminosae/ Fabaceae

Distribution: Lawn and wastelands of South and Southeast Asia.

Usable parts and properties: Herb- Diuretic, blood purifier. Used in congestion of liver, gallstone, malaria and jaundice, bronchitis, pneumonia, and typhoid.

Source of information (system): AF

***Amanita muscaria* Linn. (mushroom)**

Sanskrit name: Soma

Common name (s): Soma, fly agaric.

Family: Agaricaceae; Amanitaceae

Distribution: Habitat to temperate boreal regions of northern hemisphere, Himalayas & Indian hills in symbiosis of *Pinus* spp.

Usable parts and properties: Fruiting bodies- Medicinal and therapeutic properties are attributed to polysaccharides extract. In Russia used for manufacturing an intoxicating drink.

Source of information (system): F

Amomum fenzlii **Kurz.**

Family: Zingiberaceae

Distribution: Endemic to Nicobar Island.

Usable parts and properties: Insecticidal. Used as bee repellent.

Source of information (system): F

Amorphophallus commutatus **Engl. (VU)**

Common name (s): Dragon Stalk Yam, Jangali Suran

Family: Araceae

Distribution: Occur in Western Ghats.

Usable parts and properties: Tuber paste- Applied externally to cure scabies.

Source of information (system): F

Ampelocissus araneosa **(Dalz.) Gramble (VU), syn.** *Cissus araneosa* **(Dalz.) Planch. ex Gamble;** *Vitis araneosa* **(Dalz.) M. Lawson**

Common name (s): Wild grape

Family: Vitaceae

Distribution: Native, habitat to semi-evergreen forests of peninsular and South India.

Usable parts and properties: Plant- Antioxidant. Root & leaf decoction- Used in venereal diseases. Dried tubers- Astringent and cooling. Medicinal use without scrutiny, needing validation.

Source of information (system): AS

Ampelocissus arnottiana **Planch, syn.** *A. indica* **(L.) Planch (EN)**

Common name (s): Kemballi, Sambarra balli

Family: Vitaceae

Distribution: Endemic to Karnataka.

Usable parts and properties: Roots- Cooling and astringent. Mixed with coconut oil, and employed as a depurative, aperient and diuretic, and to treat eye diseases and ulcers.

Source of information (system): F

Ampelocissus latifolia **(Roxb.) Planch.**

Sanskrit or Ayurvedic name: Amlavetasah, Vanadraksha

Common name (s): Pani-bel

Family: Vitaceae

Distribution: Along stream of forests in South Asia, Himalayas to Western Ghats

Usable parts and properties: Plant- Used for wound healing bleeding nose. Bark- Bone fracture, indigestion, dyspepsia. Needs scientific scrutiny.

Source of information (system): AF

Anisomeles indica **(L.) Kuntze**

Common name (s): Yu-chen-tsao, Indian Catmint, Gobara, Kala Bhranga

Family: Lamiaceae/ Labiatae

Distribution: Deciduous forests and wastelands of Assam, Meghalaya.

Usable parts and properties: Herb- Contain ant inflammatory agents. Used in facial swellings, typhoid, fever, etc.

Source of information (system): F

Anticharis glandulosa **Asch., syn.** ***A. glandulosa*** **var.** ***caerulea*** **Blat& Halbl. (EN-R)**

Common name (s): Gaamesh

Family: Scrophulariaceae

Distribution: Endemic and rare to Thar Desert of western Rajasthan.

Usable parts and properties: The whole plant is dried and made into powder which is used as bath shampoo. Leaves- Purgative and to cure skin diseases. Leaves and flowers boiled in sesame oil and is dropped into ear to cure earache. Needs further study.

Source of information (system): F

Aquilaria malaccensis **Lam. (CR)**

Common name (s): Agarwood (coloured wood), Agru.

Family: Thymelaeaceae

Distribution: Foothills of the Northeast up to 1000 m.

Usable parts and properties: Valued for its fragrant resinous bark- The incense is used against cancer especially of the thyroid gland. Relieve of spasms and lower fever.

Source of information (system): F

Ardisia solanacea **(Poir.) Roxb., syn,** ***A. elliptica*** **Thunb.**

Common name (s): Bisi

Family: Primulaceae

Distribution: Western Ghats and Eastern Ghats.

Usable parts and properties: Roots- Used in fever, dropsy, diarrhoea, and rheumatism. Bark- Treat concussion or bruises.

Source of information (system): F

***Arisaema jacquemontii* Blume, syn. *Arisaema cornutum* Schott**

Common name (s): Chichyda

Family: Araceae

Distribution: Himalayas and South India.

Usable parts and properties: Plant- Known for antioxidant, antifungal, antibacterial and anticancer activities. Treat dermatological disorders.

Source of information (system): F

***Arisaema tortuosum* (Wall.)**

Common name (s): Whipcord Cobra Lily, Bagh Jandhra

Family: Araceae

Distribution: Rhododendron forests of Himalayas.

Usable parts and properties: Herb- Used against diseases caused by inflammation, stress, and Guinea-worm.

Source of information (system): F

***Artemisia indica* Willd. (LC)**

Common name (s): Titepati

Family: Asteraceae/ Compositae

Distribution: Habitat to Garhwal Himalaya, Uttarakhand.

Usable parts and properties: Known for medicinal use and for its essential oil. Needs scientific scrutiny for principal component and action.

Source of information (system): AS

***Artocarpus hirsutus* Lam. (VU)**

Common name (s): Wild jack, Anjili, Hebbalasu

Family: Moraceae

Distribution: Endemic to Western Ghats, Karnataka, Kerala

Usable parts and properties: Plant- Used in traditional medicine to treat skin diseases. Needs further study.

Source of information (system): AFS

***Asparagus rottleri* Baker (EX)**

Common name (s): Satamuli (?)

Family: Asparagaceae

Distribution: South India.

Usable parts and properties: Rhizome- Used to make traditional medicines. Needs investigation as alternative Asparagus.

Source of information (system): F

***Aster tibeticus* Hook.f. (R), syn. *A. flaccidus* var. *flaccidus* Bunge *Jacobaea tibetica* (Hook. f.) B. Nord.**

Common name (s): Lungna, Tatarinow's aster

Family: Asteraceae/ Compositae

Distribution: Trans Himalayan region.

Usable parts and properties: Herb- Anti-bacterial action noted. Leaves, bark, fruits extract- Give medicine.

Source of information (system): F

***Atylosia scarabeoides* (L.) Benth.**

Common name (s): Ban kulatha, *Cajanus cajan* wild relative.

Family: Leguminosae/ Fabaceae

Distribution: Occur in Bihar, Odisha.

Usable parts and properties: Herb- Antidysentery, anticholera, febrifuge. Seeds- Taeniafuge (deworming)

Source of information (system): F

***Balanophora involucrata* Hook. f. (obligate parasite on forest trees), (R - EN)**

Common name (s): Chavya, Mastani Booti

Family: Balanophoraceae

Distribution: Distributed from China to western Himalayas.

Usable parts and properties: Plant- Astringent. Used in piles, rheumatism, menstruation, cough, asthma, haemoptysis, traumatic injury and bleeding, dizziness and gastralgia. Needs further study.

Source of information (system): F

***Baliospermum micranthum* Müll.Arg., syn. *B. calycinum* var. *micranthum* (Müll. Arg.) Chakrab. & N.P. Balakr.**

Sanskrit name: Substitute of Hastidanti

Common name (s): Naagadanti

Family: Euphorbiaceae

Distribution: Native to south China, Meghalaya, Assam, and Myanmar.

Usable parts and properties: Root- Used for kidneys, diuretics. Leaves and seeds-Purgative. Much unknown needs study.

Source of information (system): Garo tribes

***Barleria acanthoides* Vahl.**

Common name (s): Bajardanti, Chapari

Family: Acanthaceae

Distribution: Northwest India, (Rajasthan).

Usable parts and properties: Plant- Used as fodder and medicine associated with specific diseases. Require scientific scrutiny.

Source of information (system): AF

***Barleria gibsonioides* Blatter (R)**

Common name (s): Karav

Family: Acanthaceae

Distribution: Endemic to Western Ghats (Maharashtra).

Usable parts and properties: Root paste- Applied externally for healing wounds.

Source of information (system): F

***Barleria prionitis* ssp. *prionitis* var. *dicantha* Balt. & Hallb (DD)**

Common name (s): Vajradanti, Kat-sareya

Family: Acanthaceae

Distribution: Habitat to rocky hill sides of arid zone (Jodhpur, Barmer).

Usable parts and properties: Sister species of *B. prionitis*. Known for its antidote properties. Needs study as an alternative source.

Source of information (system): F

***Barringtonia asiatica* (L.) Kurz**

Common name (s): Putat Laut, putat, fish poison tree, or sea poison tree

Family: Lecythidaceae

Distribution: Mangrove associated with Indian Ocean sandy rocky shores.

Usable parts and properties: Heated leaves- Used in stomach disorder, rheumatism, fungal infections. Seeds- Removal of intestinal worm. Also used for sores, cough, influenza, sore throat, diarrhoea, swollen spleen.

Source of information (system): FS

***Berberis apiculata* Ahrendt (R), syn. *B. jaeschkeana* var. *apiculata* (Ahrendt) H. B. Naithani and S. N. Biswas**

Family: Berberidaceae

Distribution: New record from Uttarakhand.

Usable parts and properties: Ornamental and medicinal- Being sister species require adequate study for medicinal properties.

Source of information (system): Locals

***Berberis kashmiriana* Ahrendt (R)**

Common name (s): Kashmir Barberry

Family: Berberidaceae

Distribution: Forest slopes of Jammu & Kashmir (J & K).

Usable parts and properties: Fruit edibles. Used in traditional medicine. However, require adequate study.

Source of information (system): F

***Berberis lambertii* Parker (EX - VU)**

Common name (s): Darhaldi, Common Barberry

Family: Berberidaceae

Distribution: Grassy slopes of Himalayas. Rediscovered in Chamoli & Pithoragarh, Uttarakhand.

Usable parts and properties: Plant- Anti-malarial. However, require investigation for medicinal properties.

Source of information (system): F

***Berberis micropetala* C.K. Schneid**

Common name (s): *B. vulgaris* wild relative

Family: Berberidaceae

Distribution: Found wild in Assam, Nagaland.

Usable parts and properties: Roots/other parts- Needs study for berberine and berbamine contents.

Source of information (system): F

***Berberis pseudoumbellata* Parker (DD)**

Family: Berberidaceae

Distribution: Western Himalayas.

Usable parts and properties: Root and stem bark- Source of berberine contents.

Source of information (system): F

Berberis royleana **Ahrendt (DD)**

Common name (s): Common Barberry of the region

Family: Berberidaceae

Distribution: Himalayas and J & K.

Usable parts and properties: Plant and root extract- Medicinal properties needs investigation.

Source of information (system): F

Berberis sublevis **W.W. Smith**

Common name (s): *B. vulgaris* wild relative

Family: Berberidaceae

Distribution: Occur wild in Manipur.

Usable parts and properties: Roots/other parts- Needs study for berberine and berbamine contents.

Source of information (system): F

Berberis wardii **C.K. Schneid (NT)**

Common name (s): Cultigen Barberry wild relative

Family: Berberidaceae

Distribution: Temperate forests slopes of Assam, and Nagaland.

Usable parts and properties: Root- Needs investigation to establish medicinal potential.

Source of information (system): F

Bhesa robusta **(Roxb.) Ding Hou (DD)**

Common name (s): Koliori, Hinguri, Hinuri

Family: Centroplacaceae

Distribution: China to East Himalaya, ArP, Assam, Tripura.

Usable parts and properties: Plant- Ornamental, used in traditional medicines, needing further investigation.

Source of information (system): F

Boehmeria glomerulifera **Miq., syn.** ***B. malabarica*** **Wedd**

Common name (s): Malabar tree nettle

Family: Urticaceae

Distribution: Sikkim to northeast to Western Ghats.

Usable parts and properties: Leaf poultice- Used in snake bite, and inflammation. Root decoction- Used in dysentery.

Source of information (system): F

***Breynia retusa (Dennst.)* Alston., syn. *B. patens Benth., B. turbinata* (Oken) Cordem.**

Sanskrit or Ayurvedic name: Bahupraja, Kaamboji

Common name (s): Kaali Kamboi, Kangi.

Family: Euphorbiaceae

Distribution: Habitat to tropical Indochina, Himalayas, and Deccan Peninsula.

Usable parts and properties: Leaf juice- Used in body pain, skin inflammation, hyperglycaemia, diarrhoea, and diuretic. Fruits- Used in dysentery. Twigs- To treat toothache. Functional Galactagogue (supporting drug in herbal formulations).

Source of information (system): AF

***Calligonum polygonoides* L. (EN)**

Common name (s): Phog

Family: Polygonaceae

Distribution: Western Rajasthan.

Usable parts and properties: Aerial parts- Chewing helps in stomach ailments and toothache.

Source of information (system): F

***Canthium parviflorum* Lam.**

Sanskrit name (s): Kanda Kari

Family: Rubiaceae

Distribution: Found all over India in shrubby forests.

Usable parts and properties: Plant- Used as laxative, to cure gout, diabetes. Can be source of new drug.

Source of information (system): Tribals of south India

***Cayratia pedata* (Lam.) Juss. ex Gagnep. var. *glabra* (CR)**

Common name (s): Kattuppirandai

Family: Vitaceae

Distribution: Western Ghats, Thiashola, Nilgiris, TN 1800-2200 m.

Usable parts and properties: Taken orally to rejuvenate the body and popular health tonic among local tribes without scientific scrutiny.

Source of information (system): AFS

***Cerastium cerastioides* (L.) Britton**

Common name (s): Tyndong

Family: Caryophyllaceae

Distribution: Habitat to alpine Himalayan slopes, 2700 m and Ladakh.

Usable parts and properties: Herb- Promising for relief from headache. Needs scientific scrutiny.

Source of information (system): Locals

***Ceropegia bulbosa* Roxb.**

Common name (s): Hindi: Hedulo, khadula, Kanad: Bittiruka, halike, Marathi: Ankalodya, gayala

Family: Apocynaceae

Distribution: Wild throughout India.

Usable parts and properties: Seeds- Used to cure deafness. Tubers- Used to treat kidney stone, urinary tracts diseases, diarrhoea, dysentery and eaten by ladies to enhance fertility and viability.

Source of information (system): F

***Ceropegia fimbriifera* Bedd. (VU - R)**

Common name (s): Purple-hairy Ceropegia

Family: Apocynaceae

Distribution: Endemic to evergreen forests of peninsular India

Usable parts and properties: Tuber- Edible and have medicinal value, needing validation.

Source of information (system): Local

***Ceropegia pusilla* Wight & Arn. (VU)**

Common name (s): Weak Ceropegia

Family: Apocynaceae

Distribution: Endemic to evergreen forests of southern Western Ghats.

Usable parts and properties: Herb- Antidote to snake bite. Tuber- Nutritive, blood purifier. Has ceropegin used in Ayurvedic drugs- for diarrhoea, dysentery, and syphilis.

Source of information (system): A, Local

***Ceropegia odorata* Nimmo ex J. Graham (EN - CR)**

Common name (s): Fragrant Ceropegia, Jeemikanda

Family: Apocynaceae

Distribution: Rajasthan, Gujarat, Maharashtra.

Usable parts and properties: Plant parts- Chewed in stomach-ache. Tuber- Medicinal, juice used for eye medicine.

Source of information (system): Local

***Chloroxylon swietenia* DC.**

Sanskrit or Ayurvedic name: Bhillotaka, Bhimbilota

Common name (s): Bhirraa, Bharahula

Family: Rutaceae

Distribution: Habitat to dry, deciduous forests, all over peninsular India.

Usable parts and properties: Leaves- Anti-inflammatory, antiseptic. Paste is applied to wounds and skin infection, but without scientific scrutiny, needs validation. Bark- Astringent, decoction used in contusions and rheumatic joint pain.

Source of information (system): AFS

***Chlorophytum malabaricum* Baker (VU)**

Common name (s): Safed Musali wild relative

Family: Asparagaceae

Distribution: Found wild in Baba Budangiri, Karnataka.

Usable parts and properties: Tuber- Must be explored for medicinal properties.

Source of information (system): Locals

***Cinnamomum bejolghota* (Buch. -Ham.) Sweet.**

Common name (s): Another species of Tejpatta, Cinnamon

Family: Lauraceae

Distribution: Nepal to Assam.

Usable parts and properties: Used as a condiment, medicine. Source of essential oil and wood. But needs further study. Traded as condiment.

Source of information (system): Locals

***Cissus repanda* (Wight & Arn.) Vahl., syn. *Vitis repanda* (Vahl.) Wight & Arn.**

Common name (s): Pani bel

Family: Vitaceae

Distribution: Western Ghats.

Usable parts and properties: Root- Possess healing properties. Used in rashes and detoxication.

Source of information (system): F

***Clematis roylei* var. *patens* (Haines) B.M. Kapoor. (EN)**

Common name (s): Royle Clematis

Family: Ranunculaceae

Distribution: Himalayas, HP to Bihar & Chotanagpur

Usable parts and properties: Herb- Used in joint pains (rheumatism) but needs scrutiny.

Source of information (system): Local

***Cleome rutidosperma* DC var. *burmannii* (Wight & Arn.), syn. *C. burmanni* Wight & Arn. (DD)**

Common name (s): Wild relative of Hurhur, *C. viscosa*

Family: Cleomaceae

Distribution: Northeast and peninsular India

Usable parts and properties: Like *C. viscosa*, anthelmintic having cytotoxic potential to cancerous cells, contain aromatic amines, alkynes, etc. Needs further investigation.

Source of information (system): F

***Clerodendrum colebrookianum* Walp. (VU)**

Common name (s): Nefafu, East Indian glory bower

Family: Lamiaceae/ Labiatae (Verbenaceae)

Distribution: Northeast India and ArP.

Usable parts and properties: Medicinal food plant- Antioxidant and antimicrobial, widely used to control high blood pressure.

Source of information (system): F

***Commelina nudiflora* Linn.**

Sanskrit or Ayurvedic name: Kanchata

Common name (s): Kenaa

Family: Commelinaceae

Distribution: Throughout India.

Usable parts and properties: Plant- Antidermatotic, blood purifier. Used in intestinal obstruction, diarrhoea, haemorrhoids, abnormal uterine bleeding, vaginal discharge.

Source of information (system): F

***Convolvulus auricomus* (A. Rich.) Bhandari. = var. *volubilis* (C.B. Clarke); *ferruginosus* Bhandari**

Common name (s): Rata bel, Ratanjot

Family: Convolvulaceae

Distribution: Western Rajasthan, the Indian Desert.

Usable parts and properties: Herb decoction- Cooling drink, used medicinally.

Source of information (system): Bhil's tribe

***Corchorus fascicularis* Lam.**

Sanskrit or Ayurvedic name: Chanchuk, Chanchu

Common name (s): Chanchu shaaka

Family: Tiliaceae

Distribution: Throughout warmer parts of India.

Usable parts and properties: Plant- Astringent, spasmolytic, restorative, mucilaginous. Leaves extract- Antimicrobial.

Source of information (system): AF

***Cordia gharaf* Ehrenb. ex Asch., syn. *C. rothii* Roem. & Schult.**

Common name (s): Lasora, Nani Gundi

Family: Boraginaceae.

Distribution: Dry deciduous forests of Indian desert.

Usable parts and properties: Bark decoction- Traditionally used for gargle. Needs scrutiny to establish medicinal properties.

Source of information (system): F

***Cordia macleodii* Hook. fil. & Thoms (EN), syn. *Gerascanthus macleodii* (Hook. f. & Thoms.) A. Borhidi**

Common name (s): Dahiman, or Dahipalas, Sikari/Phanki

Family: Boraginaceae

Distribution: Peninsular India (Odisha) and Western Ghats.

Usable parts and properties: Plant- Locally known for wound healing, aphrodisiac and hepatoprotective activities.

Source of information (system): F

***Corydalis flabellata* Edgew.**

Common name (s): Lepum

Family: Papaveraceae

Distribution: Habitat to dry rocky slopes of alpine Himalayas 2500-4000 m & HP.

Usable parts and properties: Herb- Locally used in rheumatic pains, stomachache, and menstrual disorders. Root, tuber- Used in skin disorder.

Source of information (system): F

Croton malabaricus **Bedd.**

Common name (s): Yettimara, Thavatta palavu

Family: Euphorbiaceae

Distribution: Habitat to evergreen forests of southern Western Ghats.

Usable parts and properties: Recorded as medicinal plant with properties for cleansing the stomach and intestine. Needs to be explored.

Source of information (system): F

Croton zeylanicus **Mull. Arg., syn.** *C. hypoleucus* **Dalzell, nom.**

Common name (s): Porivatta

Family: Euphorbiaceae

Distribution: Forest undergrowth in peninsular India.

Usable parts and properties: Aqueous leaf extracts- Anti-bacterial. Seed- Used as feed.

Source of information (system): F

Ctenolepis cerasiformis **Naud**

Sanskrit name: Shankhini

Common name (s): Aankha-phuutaa-mani

Family: Cucurbitaceae

Distribution: Wild in wastelands of Gujarat.

Usable parts and properties: Plant- Emetic, drastic purgative. Used for internal tumours, venereal diseases, and abscesses.

Source of information (system): F

Cyclea peltata **Miers (EN), syn.** *C. fissicalyx* **Dunn.**

Common name (s): Patha or Indian moonseed

Family: Menispermaceae

Distribution: Aquatic in evergreen forests of Western Ghats.

Usable parts and properties: Herb- Known wound healer, antidote to poisons, and for various digestive, skin, and inflammatory disorders.

Source of information (system): F

***Cynoglossum wallichii* G. Don.**

Common name(s): Prickly Hound's Tongue

Family: Boraginaceae

Distribution: Native range is Afghanistan to China, including Himalayas (Ladakh).

Parts and properties used: Root juice- Locally used in vomiting. Have medicinal potential.

Source of information (system): F

***Daemonorops draco* Blume., syn. *Calamus draco* Willd**

Sanskrit or Ayurvedic name: Raktaniryaas

Common name (s): Damm-ul-Akhwain, East Indian dragon

Family: Palmae, Aracaceae

Distribution: Indo-Malayan. Resin is imported to India from Indonesia.

Usable parts and properties: Plant resin- Astringent and anti-tumor. Used for diarrhoea, dysentery, and against malignant tumours but require scientific scrutiny.

Source of information (system): AF

***Dalbergia horrida* (Dennst.) Mabb. (= *D. sympathetica* Nimmo) (DD)**

Common name (s): Pedgul

Family: Leguminosae/ Fabaceae

Distribution: Karnataka India to peninsula Malaysia.

Usable parts and properties: Root bark- For treatment of diarrhoea. Root extract- Applied on bites.

Source of information (system): F

***Decaschistia cuddapahensis* Paul and Nayar**

Common name (s): Magasiri

Family: Malvaceae

Distribution: Endemic to Seshachalam Hills, Andhra Pradesh (AP).

Usable parts and properties: Root- Aphrodisiac. Requires scientific scrutiny.

Source of information (system): Tribals of Nallamalai, Eastern Ghats

***Diploknema butyracea* (Roxb.) H.J. Lam., syn. *Aisandra butyracea* (Roxb.) Baehni**

Common name (s): Chiuri

Family: Sapotaceae

Distribution: Southeast Tibet, Uttarakhand to Assam to Andaman.

Usable parts and properties: Plant fat (Chyuri ghee)- Used to treat rheumatism, asthma, ulcer and for cooking and lighting.

Source of information (system): Locals

Dischidia bengalensis **Colebr. (R)**

Common name (s): Thao wan duan, thao hua duan, o lop.

Family: Apocynaceae

Distribution: Assam.

Usable parts and properties: Creeping vine- Ornamental, toxic, medicinal usage needs investigation.

Source of information (system): Ethnic people

Draba tibetica **Hook f. & Thomson**

Common name (s): Chin (Whitlow-grasses)

Family: Brassicaceae/ Cruciferae

Distribution: Habitat to alpine Himalayas.

Usable parts and properties: Herb- Source of tonic. Has medicinal potential.

Source of information (system): F

Drosera indica **L. (NT)**

Common name (s): Kandulessa.

Family: Droseraceae

Distribution: Paleotropical to moist deciduous forests and plains of India.

Usable parts and properties: Plant- Insectivorous. Roots, flowers, fruit capsules- Used to prepare medicine '*Swarna Bhasma*' (Golden ash) for asthma, coughs, lung infections, and stomach ulcers.

Source of information (system): SH Tripuri tribe

Drynaria quercifolia **(L.) J. Smith, syn:** ***Polipodium quercifolium*** **L. (fern)**

Sanskrit name: Ashvakatri

Common name (s): Baandar-Baashing

Family: Polypodiaceae

Distribution: Allover India, in plains and low mountains of Tripura.

Usable parts and properties: Plant- Anthelmintic, expectorant, tonic. Used for loss of appetite, in treatment of chest and skin diseases. Rhizome paste- Used in treatment of diarrhoea, typhoid, cholera, chronic jaundice, fever, headache, skin diseases and syphilis.

Source of information (system): F

***Elaeagnus conferta* Roxb. (LC)**

Common name (s): Marathi Ambgul, Hulige, Palga, Kulangai

Family: Elaeagnaceae

Distribution: Indo-Malay deciduous to evergreen forests of Western Ghats.

Usable parts and properties: Plant- Treat indigestion and diabetes. Fruit- Antimicrobial, antioxidant, contains some minerals like Ca, Mg, K, Mn, P, etc.

Source of information (system): F

***Enhydra fluctuans* Lour.**

Sanskrit name: Hil-mochikaa

Common name (s): Helencha, Harkuch

Family: Asteraceae/ Compositae

Distribution: Semi-aquatic in hills of Bihar, Bengal, and Assam.

Usable parts and properties: Plant- Antioxidative and analgesic. Leaf- Nutritious, antibilious, laxative, demulcent, anti-dermatotic. Used in dyspepsia, nervous disorder, and cutaneous affections.

Source of information (system): F

***Ephedra foliata* Boiss. ex C.A. Mey. (NT)**

Common name (s): Shrubby horsetail

Family: Ephedraceae

Distribution: Punjab, India.

Usable parts and properties: Plant- Produce ephedrine and pseudoephedrine. Tea- is used in colds, coughs, bronchitis, asthma, and arthritis.

Source of information (system): F

***Eriolaena hookeriana* Wight & Arn.**

Common name (s): Bhondia dhaman

Family: Sterculiaceae

Distribution: Degraded forests of peninsular India.

Usable parts and properties: Bark mucilage with water- Cure for stomachaches.

Source of information (system): FS

***Ermania albiflora* (T. Anderson) O.E. Schulz., syn. *Phaeonychium albiflorum* (T. Anderson) Jafri**

Family: Brassicaceae/ Cruciferae

Distribution: Habitat to China and alpine northwest Himalaya, Ladakh.

Usable parts and properties: Plant- Locally used in fever, skin, mouth wounds, toothache. Medicinal potential.

Source of information (system): F

Eryngium foetidum **L., syn.** ***E. antihyctericum*** **Rottb.**

Common name (s): Brahma memedhu

Family: Apiaceae/ Umbelliferae

Distribution: Native of America. Grown in Assam, Odisha, and Meghalaya.

Usable parts and properties: Herb decoction- Used in malaria hemorrhages, fevers, chills, vomiting, burns, haemorrhages, headache, earache, etc.

Source of information (system): F

Erythrina herbacea **L. (EN)**

Common name (s): Coral bean

Family: Leguminosae/ Fabaceae

Distribution: Introduced from America to coral reef of Indian Ocean.

Usable parts and properties: Root infusion- Treat bowel pain. Decoction of roots/ berries- Used to treat nausea, constipation. Seeds or roots- Used to treat flu.

Source of information (system): Local

Erythrina variegata **L. forma** ***mysorensis*****,** ***E. indica*** **(DD)**

Sanskrit or Ayurvedic name: Pārijāta

Common name (s): Pangara

Family: Leguminosae/ Fabaceae

Distribution: Indo-Malayan, southern India peninsula. Grown in fencing.

Usable parts and properties: Leaf and seed- Reported anti-inflammatory and analgesic activity. Not subjected to scientific scrutiny.

Source of information (system): AFSL

Eulophia cullenii **(Wight) Blume (CR)**

Common name (s): Orchid relative of Amarkand

Family: Orchidaceae

Distribution: Endemic and exquisite orchid of southern Western Ghats.

Usable parts and properties: Extract- Terrestrial orchids have been used in traditional medicine. Used in neurological disorders.

Source of information (system): AF

***Eulophia mackinnonii* Duthie**

Common name (s): Amarkand

Family: Orchidaceae

Distribution: Himalayas, ArP, Assam.

Usable parts and properties: Tubers have been used for diarrhoea, stomach pain, rheumatoid arthritis, cancer, asthma, bronchitis, sexual impotency, tuberculosis. Needs more investigation.

Source of information (system): Local

***Eulophia ramentacea* Wt (CR), and many more *Eulophia* spp.**

Common name (s): Amrkand (terrestrial orchid)

Family: Orchidaceae

Distribution: Peninsular India Karnataka, Khandesh, Maharashtra.

Usable parts and properties: Herb- Rich in bioactive substances. Extract and herbal formulations used by indigenous Banjara Community.

Source of information (system): F, Banjara tribe

***Ficus asperrima* Roxb.**

Sanskrit or Ayurvedic name: Kharapatra

Common name (s): Kaala-umar, Karvant

Family: Moraceae

Distribution: Evergreen forests of Madhya Pradesh (MP), Indian western peninsula.

Usable parts and properties: Juice of bark- Given in case enlargement of liver and spleen.

Source of information (system): AF

***Ficus dalhousiae* Miq.**

Sanskrit name: Soma-valka

Common name (s): Kal Aal

Family: Moraceae

Distribution: Dry and moist deciduous forest of South India.

Usable parts and properties: Fruit- Cardiotonic, used during fever. Leaves and bark- Used in affections of the liver and skin diseases.

Source of information (system): F

***Fimbristylis ovata* (Burm,) Kern., syn. *F. monostachya* (L.) Hassk**

Common name (s): Ibha-muulaka

Family: Cyperaceae

Distribution: Occur as weed all over warm India.

Usable parts and properties: Herb- Used in adenitis, scrofula, syphilis; also, in cough, bronchitis and asthma.

Source of information (system): F

***Garcinia rubro-echinata* Kosterm., syn. *G. echinocarpa*; or var. *monticola* Mahes (VU)**

Common name (s): Mangosteen Pura, Para

Family: Clusiaceae/ Guttiferae

Distribution: Evergreen forests of southern Western Ghats.

Usable parts and properties: Tree- Potential cytotoxic. Used as medicine, and in cosmetics, pesticides, paints, and dyes.

Source of information (system): F

***Garcinia talbotii* Raizada Ex Santapau (VU)**

Common name (s): Haldi, Ont, Limboti, Pansara

Family: Clusiaceae/ Guttiferae

Distribution: The native range is southern Western Ghats.

Usable parts and properties: The fruits yield an inferior quality of yellow gum. Dried fruits are used like tamarind in Currie.

Source of information (system): F

***Garcinia travancorica* Bedd. (CR)**

Common name (s): Kolivala, Pulimaranga, Attukaruka

Family: Clusiaceae/ Guttiferae

Distribution: Endemic to the Western Ghats (Agasthyamalai Hills).

Usable parts and properties: Tree- Source of 'gamboge', a gum-resin, used as an ointment.

Source of information (system): F

***Gentiana decumbens* L. f.**

Common name (s): Kuth

Family: Gentianaceae

Distribution: Habitat to alpine slopes of trans Himalayas 3300-4500 m.

Usable parts and properties: Plant tincture- Used as stomachic.

Source of information (system): F

***Gentiana quadrifaria* Blume (var. *zeylanica* (Griseb.) Kusnezov), syn. *G. laxicaulis* Zoll. & Moritzi. = *G. pedicellata* Wall. (VU)**

Common name (s): Silvery Gentian

Family: Gentianaceae

Distribution: Northwest to East hills, Ooty, TN Kerala.

Usable parts and properties: Root- Used to fight inflammation swelling. Antidote to animal poison. Needs further investigation.

Source of information (system): F

***Glycosmis macrocarpa* Wt. (NT-VU)**

Common name (s): Ashvashakota relative, Ban nimbu

Family: Rutaceae

Distribution: Wet evergreen forests of Western Ghats 200 -1100 m. and northeast.

Usable parts and properties: Locally used in treatment of snakebite. May be source of large variety of lead compounds, needing investigation.

Source of information (system): F

***Goodyera repens* (L.) R. Br.**

Common name (s): Creeping Lady's Tresses

Family: Orchidaceae

Distribution: Found wild in Himalayas, Ladakh, Assam, Nagaland.

Usable parts and properties: Leaf infusion- Used to improve the appetite, cold, kidney problems; poultice applied on burns, blood purifier. Treat syphilis.

Source of information (system): F

***Grangea maderaspatana* Poir., syn. *Artemisia maderaspatana* Linn.**

Sanskrit name: Aakaarakarabha

Common name (s): Mastaru

Family: Asteraceae/ Compositae

Distribution: Found throughout greater part of India.

Usable parts and properties: Leaf - Stomachic, sedative, antispasmodic, emmenagogue, de-obstruent, antiseptic. Used in amenorrhea studied for estrogenicity, antifertility, analgesic, anti-inflammatory, antiarthritic, cytotoxic, antioxidant, hepatoprotective, diuretic and antimicrobial activities.

Source of information (system): F

Guaiacum officinale **Linn**

Sanskrit name: Jivadaaru

Common name (s): Chob-hayaat

Family: Zygophyllaceae

Distribution: Introduced from West Indies. Grown as ornamental.

Usable parts and properties: Aqueous extract- Used for antifertility. Wood- Antirheumatic, anti-inflammatory, mild laxative, diuretic, diaphoretic, fungistatic (In 16th century used to cure syphilis).

Source of information (system): F

Guazuma ulmifolia **Lam., syn.** ***G. tomentosa*** **H. B. & K**

Sanskrit name: Pundraaksha

Common name (s): Rudraksham

Family: Sterculiaceae

Distribution: Tropical American. Grown as roadside shade tree in warmer parts of India.

Usable parts and properties: Fruit- Ant catarrhal (in bronchitis). Bark- Demulcent, sudorific. Used in skin diseases. Seed- Astringent, carminative, and antidiarrheal. Needs scientific & clinical evaluation.

Source of information (system): F

Gymnema khandalense **Santapau (EN-CR)**

Common name (s): Genus *Gymnema* is called gur mar, means "destroyer of sugar."

Family: Apocynaceae

Distribution: Endemic to Western Ghats (Goa).

Usable parts and properties: Reported as medicinal plant from Nilackal/Nilakkal in Sabarimala Forest. Leaves have medicinal properties. Extract be investigated for diabetes, etc.

Source of information (system): F

Gymnema montanum **(Roxb.) Hook. (EN)**

Common name (s): *Gymnema* is called gur-mar, meaning "destroyer of sugar."

Family: Apocynaceae

Distribution: Evergreen forests of Kerala.

Usable parts and properties: Used for general health and curing diseases like diabetes by the tribals.

Source of information (system): F

Hackelia uncinata **Fischer.**

Common name (s): Hooked Stickseed

Family: Boraginaceae

Distribution: Habitat alpine Himalayas, and Ladakh.

Usable parts and properties: Arial parts- Used in cold and cough. Medicinal potential.

Source of information (system): F

Hedychium gratum **Wall. ex. Baker (EN - R)**

Common name (s): Spiked ginger lily

Family: Zingiberaceae

Distribution: Endemic to Eastern India (Assam).

Usable parts and properties: Rhizome decoction/powder- have medicinal potential. Essential Oils- Used in food preservation, flavour, and safety.

Source of information (system): -

Hedyotis barberi **(Gamble) A.N. Henry & Subr. (VU - R)**

Family: Rubiaceae

Distribution: Evergreen Forests of Western Ghats.

Usable parts and properties: Locally used in health benefits and disease prevention.

Source of information (system): Locals

Heliotropium keralense **Siv. & Manl (EN)**

Ayurvedic or Common name (s): Thelkkada

Family: Boraginaceae

Distribution: Moist areas of southern Western Ghats.

Usable parts and properties: Traditionally used to treat asthma and skin related diseases.

Source of information (system): A

Helminthostachys zeylanicus **(L.) Hook (EN)**

Common name (s): Kamraj and Tunjuk-langit

Family: Ophioglossaceae

Distribution: Himalayan foothills, Kumaun to coconut groves/ forests of Kerala.

Usable parts and properties: Fern-like plant- Used in neurological disorders, inflammation, antidiabetic.

Source of information (system): F

Hydnocarpus alpina **Wt (EN)**

Common name (s): Maravetti

Family: Achariaceae

Distribution: Endemic to evergreen forests of southern Western Ghats.

Usable parts and properties: Plant- Hypoglycaemic. Traditionally used in the treatment of leprosy.

Source of information (system): FS

Hydnocarpus macrocarpa **(Bedd.) Warb. (VU)**

Common name (s): Chalmogra, Jangli badam, Malamkummatti, Vellananku

Family: Achariaceae

Distribution: Endemic to Western Ghats.

Usable parts and properties: Plant- Contain alkaloids, flavonoids, carbohydrates, etc., Needs research for active principles and action.

Source of information (system): F

Hydnocarpus pentandra **(Buch. -Ham.) Oken (VU)**

Sanskrit name: Garudaphala

Common name (s): Thamana, Koti, Kadu Kawath, Kastel

Family: Achariaceae

Distribution: Endemic to the Western Ghats.

Usable parts and properties: Leaf extract- Antibacterial, antioxidant. Seeds/ seed oil- Used for treating leprosy and other skin disorders.

Source of information (system): F

Ichnocarpus frutescens **R. Br**

Sanskrit name: Gopavalli

Common name (s): Kaalisar

Family: Apocynaceae

Distribution: Found in Uttar Pradesh (UP), MP, Bihar, Assam, Sundarbans of West Bengal (WB).

Usable parts and properties: Root- Demulcent, diuretic, alterative, diaphoretic; used in fevers, dyspepsia, and cutaneous affections. Substitute/adulterant for Indian sarsaparilla, *Hemidesmus indicus* (Needs therapeutic studies).

Source of information (system): F

Ilex khasiana **Purk. (EN - CR)**

Common name (s): Holly tree

Family: Aquifoliaceae

Distribution: Endemic to Meghalaya and northeast India.

Usable parts and properties: Leaves- Locally used in coronary heart diseases, hypertension, hyperlipemia and hepatitis. High commercial value, however, relatively unknown, needing study.

Source of information (system): Local

Indigofera barberi **Gamble (R)**

Common name (s): Adavineelimanadu mokka (Tirumala Hills)

Family: Leguminosae/ Fabaceae

Distribution: Lower hills of South India.

Usable parts and properties: Herb- Antibacterial, DNA gyrase B inhibitors. Used to treat various skin, renal and liver diseases.

Source of information (system): F

Indigofera constricta **(Thw) Trimen (VU - R)**

Common name (s): Wild relative of Neel

Family: Leguminosae/ Fabaceae

Distribution: Semi-evergreen forests of Indian peninsula.

Usable parts and properties: Being related to *I. tinctoria*, which is used for wide range of disorders, therefore it needs investigation.

Source of information (system): Locals

Indigofera coerulea **var.** ***monosperma*** **(Santapau) Santapau (R)**

Common name (s): Wild relative of Neel

Family: Leguminosae/ Fabaceae

Distribution: Dry Forest of north Gujarat, Saurashtra and Kachchh.

Usable parts and properties: Being related to *I. tinctoria*, needs investigation for medicinal properties.

Source of information (system): Locals

Inula falconeri **Hook. f.**

Common name (s): Relative of Pushkarmul

Family: Asteraceae/ Compositae

Distribution: Habitat wild to J & K between 2300-2500 m, Ladakh and western Tibet.

Usable parts and properties: Extracts- Used in allelopathy. Have antifungal and anti-inflammatory activities. Root- Used in sprains. Medicinally potential

Source of information (system): F

***Inula obtusifolia* A. Kerner**

Common name (s): Relative of Pushkarmul

Family: Asteraceae/ Compositae

Distribution: Habitat to rock crevices, dry cliffs, slopes of west Himalayas at 2500-4500 m.

Usable parts and properties: Plant- Used in tuberculosis, chest problems, cough, and antiseptic. Root- Used to cure internal wounds without scientific scrutiny.

Source of information (system): F

***Ipomoea aquatica* Forsk, syn. *I. reptans* Poir**

Sanskrit name: Kalambi

Common name (s): Vellaikeerai

Family: Convolvulaceae

Distribution: Throughout the greater part of India.

Usable parts and properties: Plant- Emetic and purgative. Antidote to arsenical or opium poisoning. Plant juice- Used for liver complaints. Buds- Used for deworming ringworm.

Source of information (system): F

***Iris germanica* Linn**

Sanskrit name (s): Paarseeka Vachaa

Family: Iridaceae

Distribution: Native of Italy, Morocco. Wild on graves in Kashmir. Cultivated.

Usable parts and properties: Root- Diuretic, emetic, cathartic demulcent, antidiarrhoeal, expectorant. Leaf extract- Treat frozen feet.

Source of information (system): F

***Janakia arayalpathra* Joseph & Chandrasekaran (CR)**

Common name (s): Amruthapala

Family: Asclepiadaceae, Periplocaceae

Distribution: Endemic to forests of southern Western Ghat.

Usable parts and properties: Plant- Used by the local 'Kani' tribe as an effective remedy for peptic ulcer.

Source of information (system): Kani tribe

Jasminum rottlerianum* Wall. ex DC., syn. *J. coarctatum* Roxb. var. *coarctatum

Sanskrit name (s): Vana-mallikaa

Family: Oleaceae

Distribution: Western Peninsula, from Konkan to Kerala.

Usable parts and properties: Leaves- Used to treat eczema.

Source of information (system): F

***Juniperus excelsa* Webb ex Parl.**

Common name (s): Padma

Family: Cupressaceae

Distribution: Gilgit-Baltistan (POK), introduced to Chikhli, Maharashtra.

Usable parts and properties: Plant- Antihypertensive, diuretic, appetizer, carminative, stimulant, anticonvulsant. Used in diarrhoea, abdominal spasm, asthma, fever, gonorrhoea, headache, and leucorrhoea.

Source of information (system): F

***Knema attenuata* (Wall.) Warb., syn. *Myristica corticosa* Bedd. (NT)**

Common name: Raktamara, Kaatu, Choora pathiri (Wild Nutmeg).

Family: Myristicaceae

Distribution: Endemic to the Western Ghats.

Usable parts and properties: Stem bark- Used for treating inflammatory conditions. Reported to contains a lignan 'attenuol' with diverse pharmacological activities. But need phytochemical research.

Source of information (system): FS

***Lepidagathis cristata* Willd.**

Common name: Nakka pidi, Lankapindi

Family: Acanthaceae

Distribution: Seshachalam Hills, AP.

Usable parts and properties: Tuber ash + coconut oil- Traditionally applied on burns, wounds.

Source of information (system): Locals

***Leucas lanata* var. *nagpurensis* C.B. Clarke ex Haines (EN)**

Common name: Doron, Durun-phul

Family: Lamiaceae/ Labiatae

Distribution: Wet areas of peninsular India.

Usable parts and properties: *L. lanata* is used in the treatment of bleeding, septic wounds, and many other diseases. It can be an alternative source.

Source of information (system): F

***Lilium mackliniae* Sealy (EN - R)**

Common name: Shirui lily

Family: Liliaceae

Distribution: Shirui Hill ranges, Manipur.

Usable parts and properties: Lily- Used in treating skin and stomach problems.

Source of information (system): F

***Lobelia nicotianifolia* Roth.**

Common name: Wild tobacco, Dhawal.

Family: Campanulaceae.

Distribution: East Himalaya to Western Ghats.

Usable parts and properties: Plant- Poisonous, used to treat bronchitis, asthma, and insect and scorpion bites.

Source of information (system): F

***Luffa umbellata* (Klein) Roem. (EN), syn. *L. acutangula* (cultigen)**

Sanskrit name: Mridangaphalika

Common name: Kurvitori, wild form of *L. acutangula*

Family: Cucurbitaceae

Distribution: Wild along bushes and hedges in Western Ghats.

Usable parts and properties: Matured dried fruit boiled fibre- Used as medicine without scientific scrutiny.

Source of information (system): F

***Lycopodium pseudoclavatum* Ching, syn. *L. japonicum* Thunb. ex Murray (VU)**

Common name(s): Lycopidium

Family: Lycopodiaceae

Distribution: Grasslands of high-altitude Himalayas, and Kerala.

Parts and properties used: Herb- Locally used in treatment of urinary disorder. Assessed medicinally potential.

Source of information (system): Locals

***Madhuca diplostemon* (Clarke) Royen, syn. *Diospyros obovata* Wight (EN)**

Common name: Kavilippa in Malayalam

Family: Sapotaceae

Distribution: Endemic to Western Ghats. Rediscovered in a sacred grove.

Usable parts and properties: Alternative to *Madhuca indica* and *M. longifolia*. Used medicinally. against diabetes, rheumatism, ulcers, bleeding, and tonsillitis.

Source of information (system): F

***Madhuca insignis* (Radlk) Lam. (EX)**

Common name: Kattu Kuligam, Kattu Iluppa

Family: Sapotaceae

Distribution: Endemic to southern Western Ghats. Rediscovered in Karnataka.

Usable parts and properties: Fruits- Edible and used for medicinal purposes? Phytocompounds with known biological functions were discovered. Needs study.

Source of information (system): F

***Madhuca neriifolia* (Moon) Lam (VU)**

Common name: Naanilu mara

Family: Sapotaceae

Distribution: Endemic to Western Ghats and Sri Lanka.

Usable parts and properties: Fruit- Used in rheumatism, cough, asthma, and consumption. Seed oil- Used in rheumatism. Soaked flower- Used in kidney disorder.

Source of information (system): Local

***Mahonia leschenaultia* (Wall. Ex Wight & Arn.) Takeda ex Gamble; *Berberis napaulensis* (DC) Spreng.**

Common name: Daruharidra, Mullukadambu, Toda plant.

Family: Berberidaceae

Distribution: Found in Western Ghats, TN.

Usable parts and properties: Plant/ Berries- Diuretic, demulcent anticancer. Root- Improve liver and renal conditions. Source of isoquinoline. Unexplored and needs investigation.

Source of information (system): Local

Mattiastrum thomsonii **(C.B. Clarke) Kazmi. L.**

Family: Boraginaceae

Distribution: Habitat to alpine Himalaya, including Ladakh.

Parts and properties used: Plant- Diuretic.

Associated system(s): AF

Michelia nilagirica **Zenk. (VU)**

Common name: Pola champa, Shempangan

Family: Magnoliaceae

Distribution: Higher mountains of Western peninsula.

Usable parts and properties: Bark decoction and infusion- Used as a febrifuge.

Source of information (system): FS

Microula tibetica **L.**

Family: Boraginaceae

Distribution: Habitat to Himalayas, including Ladakh.

Parts and properties used: Plant- Pulmonary disorder. Fruit- Heart disorder and ulcers.

Associated system(s): Amchi

Minuartia kashmirica **(Edgew.) Mattf.,** ***Arenaria foliosa*** **Royle ex Edgew. & Hook. Fil.**

Common name: Kashmir Sandwort

Family: Caryophyllaceae

Distribution: Habitat to alpine Himalayas, 1800-5100 m and Ladakh.

Usable parts and properties: Plant- Reported to yield liver tonic. Medicinally potential.

Source of information (system): Locals

Morinda angustifolia **Roxb. (R)**

Common name: Narrow leaf Noni, Indian Mulberry

Family: Rubiaceae

Distribution: Northeast and coastal region of India.

Usable parts and properties: Roots- Febrifuge, tonic and antiseptic and stomachic. Bark- Contains morin din glycoside. Fruits- Diuretic, laxative, emollient and emmenagogue. Used in asthma, arthritic inflammations, and high BP.

Source of information (system): Locals

***Morinda pubescens* J.E. Smith**

Common name: Mulugu chettu, Togaru chettu

Family: Rubiaceae

Distribution: Seshachalam Hills, AP.

Usable parts and properties: Plant- Reported with anticancer, antimicrobial, antifungal, wound healing properties. Bark decoction- Used in rheumatism.

Source of information (system): Locals

***Myristica elliptica* Wall. ex Hook.f. & Thoms., syn. *M. hackenbergii* Diels**

Common name: Nutmeg

Family: Myristicaceae.

Distribution: Habitat to swamps of Western Ghats. Cultivated too.

Usable parts and properties: Used as alternative dried seed- For food flavoring and for medicinal properties to cure stomach problem.

Source of information (system): Locals

***Myrsine capitellata* Wall.**

Common name: Vidanga, *Embelia ribes*, substitute

Family: Myrsinaceae

Distribution: Habitat to Himalayas, and moist hills of north India, 900-1800 m.

Usable parts and properties: Fruit- Edible, anti-depressant. Medicinal properties to be explored.

Source of information (system): Nepal, Northeast tribes

***Nepenthes khasiana* Hook. f. (EN-CR)**

Common name: Kset phare, the pitcher plant

Family: Nepenthaceae

Distribution: Endemic to Meghalaya.

Usable parts and properties: Plant- Used locally in urinary and stomach troubles, diabetes, etc.

Source of information (system): F

***Nepeta leucolaena* Benth. ex Hook. f.**

Common name: Azoomal

Family: Lamiaceae/ Labiatae

Distribution: Wild in trans-Himalayan/alpine region and Ladakh.

Usable parts and properties: Plant- Used in skin disease. Leaves/flower decoction- Gastric ulcer.

Source of information (system): F

***Nicandra physalodes* (L.) Gaertne**

Family: Solanaceae

Distribution: South American, naturalized to peninsular India.

Parts and properties used: Plant- Analgesic, anthelmintic, antibacterial, anti-inflammatory, febrifuge, and tonic. In South America, used to treat unspecified medical disorders.

Source of information (system): F (South America)

***Ochreinauclea missionis* (Wall. ex G. Don) Ridsdale (VU)**

Common name: Attuvanchi Neervanchi Phuga (Malyalam), Attu-vanji (Tamil)

Family: Rubiaceae

Distribution: Native/endemic to Western Ghats.

Usable parts and properties: Stem bark- Antileprotic, antirheumatic. Used in constipation, ulcers. Root- Antirheumatic. Used in skin, eye diseases, dropsy, piles, and haemophilic disorders.

Source of information (system): Local tribe

***Ophioglossum pendulum* L. (fern) (R- Rare)**

Common name: Old-world adder's-tongue

Family: Ophioglossaceae

Distribution: Known, rediscovered in Northeast India.

Usable parts and properties: Epiphytic herb fond- Locally used to improve the hair.

Source of information (system): F

***Panax arunachalensis* sp. nov.**

Common name: Close to *P. sokpaiyensis* and *P. assamicus*

Family: Araliaceae

Distribution: Lower Subansiri district, ArP.

Usable parts and properties: Rhizome- Morphologically distinct from other *Panax* spp. Needs detailed study for an alternative source to strengthen immune system.

Source of information (system): F

Panax assamicus **Banerjee (EN)**

Common name(s): Another Ginseng species

Family: Araliaceae

Distribution: Native to east Himalayas to Assam.

Parts and properties used: Rhizome- Old-age revitalizer. Needs further study. Can be of significant trade value.

Source of information (system): Locals

Panax sokpaiyensis **Shiva K. Sharma & Pandit (EN)**

Common name: A new species

Family: Araliaceae

Distribution: Narrowly endemic to Sikkim and ArP.

Usable parts and properties: Rhizome- High medicinal values with market demand, because of being ginseng *Panax* sp.

Source of information (system): F

Paphiopedilum druryi **(Bedd.) Stein. (CR)**

Common name: Drury's Slipper Orchid

Family: Orchidaceae

Distribution: Endemic to the Agasthyamalai Hills, Western Ghats, TN

Usable parts and properties: Herb- Have medicinal and aromatic properties, needing further investigation.

Source of information (system): F

Parnassia laxmannii **Pall. ex Schult., syn.** ***P. subacaulis*** **Kar. & Kir.**

Common name: (?)

Family: Saxifragaceae

Distribution: Habitat to alpine Himalayas and Ladakh.

Usable parts and properties: Plant- Diuretic. Has medicinal potential.

Source of information (system): F

Pentanema indicum **(L.) Ling**

Common name: Aggikoora chettu

Family: Asteraceae/ Compositae

Distribution: Seshachalam Hills, AP.

Usable parts and properties: Plant- Yields female antifertility drug. Leaf juice- Used in insect sting.

Source of information (system): Tribes of Bihar

Persea macrantha **(Nees) Kost., syn.** ***Machilus macrantha*** **Nees (EN)**

Common name: Kulamavu, Kolamavu

Family: Lauraceae

Distribution: Habitat to evergreen forests above peninsular India 1100-1900m.

Usable parts and properties: Leaf/bark- Used to treat various ailments, like rheumatism, asthma, ulcer, bruise, mental upset, fractures, swellings, weakness, and debility.

Source of information (system): F

Phyllanthus narayanswami **Gamble (EN)**

Common name: Wild Bhumiaonla

Family: Euphorbiaceae

Distribution: Endemic to south India.

Usable parts and properties: Herb- Antibacterial. An alternative medicinal plant needing investigation.

Source of information (system): Locals

Phyllanthus talbotii **Sedgw. (CR-Rare)**

Common name: Wild Bhumiaonla

Family: Euphorbiaceae

Distribution: Endemic to Western Ghats, Karnataka, and Goa.

Usable parts and properties: Plant extract- Used traditionally. Ethno-medicinal studies are needed as alternative *Phyllanthus* spp. properties.

Source of information (system): Tribes of Palamau Tiger Reserve.

Piper barberi **Gamble. (CR)**

Common name: Kattukurumulagu, wild pepper.

Family: Piperaceae

Distribution: Evergreen forests of southern Western Ghats.

Usable parts and properties: Essential oil- Possess antifungal and cytotoxic activities.

Source of information (system): F, Tribal healers *Plathis.*

Piper mullesua **D. Don (VU)**

Common name: Kattukurumulagu, wild pepper.

Family: Piperaceae

Distribution: Endemic to peninsular and northeast India.

Usable parts and properties: Roots- Useful in asthma, bronchitis, dyspepsia, and anorexia. Twenty-eight known compounds isolated from the aerial parts. Needs pharmacological study.

Source of information (system): Local

Plectranthus nilgherricus **Benth., syn.** ***Isodon nilgherricus*** **(VU)**

Family: Lamiaceae/ Labiatae

Distribution: Western Ghats and south India.

Usable parts and properties: Plant- Aromatic. Essential oil- Source of natural bioactive compounds with analgesic and antioxidant agents.

Source of information (system): -

Polygonum plebejum **var.** ***indica*** **(Roth.) Hook. fil., syn.** ***P. prostratum*** **Roxb. ex D. Don. (R)**

Common name: Pani jaluk

Family: Polygonaceae

Distribution: Assam.

Usable parts and properties: Boiled root mixed with butter- Stimulate mammary gland, and sooth alimentary canal.

Source of information (system): Locals

Potentilla cuneata **Wall., syn.** ***P. ambigua*** **Cambess.**

Common name: - Cuneate cinquefoil or five finger cinquefoils

Family: Rosaceae

Distribution: Habitat to Afghanistan, China, alpine Himalayas and Ladakh. Grown.

Usable parts and properties: Flower and leaf- Yields tonic. Medicinally potential.

Source of information (system): F

Przewalskia tangutica **Maxim. (CR)**

Common name: Ai Langdang

Family: Solanaceae

Distribution: Native range central China to Tibet to Sikkim.

Usable parts and properties: Plant- Used in gastrointestinal diseases, analgesic and heal wound and inflammation. Rhizomes- Have amidoamine and scopolamine alkaloid.

Source of information (system): Tibetan

***Pterocarpus dalbergioides* Roxb., syn. = *P. indicus* R. Vig. (EN)**

Common name: Narra, Andaman redwood

Family: Leguminosae/ Fabaceae

Distribution: Native/ Endemic to Andaman and Nicobar Islands.

Usable parts and properties: Bark- Contains tannins, which is astringent, with medicinal potential.

Source of information (system): Local

***Pterospermum xylocarpum* (Gaerth.) Samt. & Wagh. (LC)**

Common name: Tada

Family: Malvaceae

Distribution: TN.

Usable parts and properties: Herbal formulations- Known for use in rheumatism and diabetes.

Source of information (system): FS

***Pupalia lappacea* (L.) Juss.**

Common name: Antudu chettu

Family: Amaranthaceae

Distribution: Seshachalam Hills, AP.

Usable parts and properties: Stem- Locally used as a toothbrush, toothache.

Source of information (system): Locals

***Pycnocycla glauca* Lindl. (EN), syn. *Chamaesciadium glaucum* (Lindl.) M. Hiroe**

Family: Apiaceae/ Umbelliferae

Distribution: Chotanagpur montane forest.

Usable parts and properties: Roots- Locally used in dysentery. Essential oil-containing.

Source of information (system): F

***Ranunculus muricatus* L.**

Common name: Rough-fruit-buttercup

Family: Ranunculaceae

Distribution: Evergreen forests of Western Ghats.

Usable parts and properties: Fruit- Treat intermittent fevers, gout, and asthma. Contain saponins, tannins, phenols, flavonoids, and alkaloids.

Source of information (system): F

***Rheum nobile* Hook. F. & Thoms (EN)**

Common name: Sikkim noble rhubarb, Padamchal

Family: Polygonaceae

Distribution: Himalayan alpine region 4000–4800 m.

Usable parts and properties: Plant- Antioxidant and phenolic constituents. No report on biological activities, needing investigation.

Source of information (system): Tibetan medicine

***Rhododendron concinnoides* Hutch. & Kingdon-Ward (EN)**

Family: Ericaceae

Distribution: Arunachal Pradesh.

Usable parts and properties: Plant- Used for heart, constipation, dysentery, diarrhoea, detoxification, inflammation, bronchitis, asthma, and fever. Needs investigation.

Source of information (system): Local tribes

Rhododendron formosum* Wall., syn. *R. gibsonii* Paxt. and *R. formosum* var. *formosum

Family: Ericaceae

Distribution: Northeast India, Assam, Meghalaya. Cultivated flowering plant.

Usable parts and properties: Plant- Antioxidant, found with anti-inflammatory and hepatoprotective activities in animals.

Source of information (system): Local tribes

***Rhododendron santapaui* Sastry et al. (EN)**

Family: Ericaceae

Distribution: Arunachal Pradesh, temperate rainforests of northeast India.

Usable parts and properties: Leaves- Used to make traditional medicine. But medicinal properties to be assessed.

Source of information (system): Local tribes

***Rosa canina* L.**

Common name: Dog rose = *R. alba*

Family: Rosaceae

Distribution: European, distribution extended to J & K.

Usable parts and properties: Rose hips (part below Patel)/ essential oil- Antioxidant, anti-inflammatory, improve joint health.

Source of information (system): Local

***Rubia sikkimensis* Kurz.**

Common name: Naga madder, Moyum. Relative of *R. tinctorum*

Family: Rubiaceae

Distribution: Habitat to eastern Himalayas and northeast hills.

Usable parts and properties: Root- Give red dye. Medicinal properties need investigation.

Source of information (system): Local tribes

***Sagittaria trifolia* Linn., syn. *S. sagittifolia* Hook. f. (non-L.)**

Common name: Chhotaa Kuuta, Chinese arrowroot

Family: Alismataceae

Distribution: Chinese, found in northeast and Kerala.

Usable parts and properties: Plant- Used as vegetable in Manipur. Discutient, ant galactagogue astringent, anti-inflammatory. Tuber- Used to cure skin diseases. Needs investigation for nutritive and medicinal value.

Source of information (system): F

***Saussurea auriculata* (DC.) Sch.-Bip., *S. hypoleuca* Spreng. ex DC., syn. *Aplotaxis auriculata* DC.**

Sanskrit or Ayurvedic name: Kushtha

Common name: Uplet

Family: Asteraceae/ Compositae

Distribution: Occur in the Himalayas, from Kashmir to Sikkim.

Usable parts and properties: Leaves- Purgative and anti-syphilis. Paste used in venereal diseases. Powder- Used for improvement of semen. Substitute to Kuth, *S. costus*.

Source of information (system): AF

***Saussurea clarkei* Hook. f. (NT-Rare)**

Family: Asteraceae/ Compositae

Distribution: Distributed in western Himalayas.

Usable parts and properties: Plant- Anti-cancer. Needs investigation.

Source of information (system): Local tribes

***Securinega virosa* (Roxb. ex Willd.) Baillon, syn. *Fluggea virosa* (Roxb. ex Willd.) Baillon.**

Family: Euphorbiaceae

Distribution: Found wild all over India.

Usable parts and properties: Bark- Yields Securinine (inhibits cancer cell metastasis) and other alkaloids. Root bark- Induce sleep, fever. Needs investigation.

Source of information (system): (?)

***Semecarpus travancorica* Bedd., syn. *Cassuvium travancoricum* (Bedd.) Kunt (EN)**

Common name: Avukaram, Kattu

Family: Anacardiaceae

Distribution: Endemic to southern Western Ghats.

Usable parts and properties: Plant- Used to extract raw drugs. Needs further study.

Source of information (system): FS

***Smilax lanceifolia* Roxb.**

Common name: Chobachini (Hindi), Shukchin

Family: Liliaceae

Distribution: Himalayas, Sikkim Assam, and Manipur.

Usable parts and properties: Roots- Used for rheumatic affections. Not much information is available about medicinal uses.

Source of information (system): F

***Smilax wightii* L. (DD - R)**

Common name: Jangli aushbah

Family: Liliaceae/ Smilacaceae

Distribution: Endemic to shola forests of Nilgiris, Western Ghats, India

Usable parts and properties: Plant- Reported anti-inflammatory. Used in dysentery, amoebiasis, venereal diseases and urinary complaints, fever, anaemia. Scientifically unexplored.

Source of information (system): F

***Solidago virga-aurea* Linn.**

Common name: Himalayan goldenrod

Family: Asteraceae/ Compositae

Distribution: Distributed from Kashmir to Khasi Hills.

Usable parts and properties: Plant- Ant catarrhal, diaphoretic, anti-inflammatory, antiseptic to mucous membranes. Watering of inflammatory diseases of urinary tract, calculi, and kidney gravel. Medicinal potential.

Source of information (system): Local tribes

***Sphaeranthus amaranthoides* Burm. f.**

Common name: Sivakaranthai (Tamil), Shiva Karandhi

Family: Asteraceae/ Compositae

Distribution: Wet sandy areas in dry deciduous forests of peninsular India.

Usable parts and properties: Herb- Rejuvenator, contains steroids, flavonoids, alkaloids, terpenoids, etc. Used in eczema, blood disorders, stomach worms, filaria, fever, skin diseases, antihelminth and jaundice.

Source of information (system): SF

***Statice macrorrhabdos* Boiss., syn. *Dictyolimon macrorrhabdos* (Boiss.) Rech. f.**

Common name: (?)

Family: Plumbaginaceae

Distribution: Afghanistan to northwest Himalayas.

Usable parts and properties: Flower (oil)- Used in stomach disorder. Medicinally promising.

Source of information (system): F

***Sterculia khasiana* King ex Debb. (EX)**

Family: Malvaceae

Distribution: Endemic to Assam and Khasi Hills of Meghalaya.

Usable parts and properties: Listed and locally used for medicinal purposes.

Source of information (system): Local tribes

***Strychnos aenea* A.W. Hill. (EN)**

Common name: Wild relative *S. nux-vomica*

Family: Strychnaceae

Distribution: Native range is from Assam southwest India.

Usable parts and properties: Root bark- Locally used in cholera. Wood- Used to treat dysentery, fevers, and dyspepsia. Seed- Used to treat colic. Leaf poultice- Applied to wounds/ulcers. Can be alternative to *S. nux vomica.*

Source of information (system): F

Strychnos nux-blanda **A.W. Hill.**

Common name: Substitute of Itti beeja, Kuchla. *S. nux-vomica* relative

Family: Loganiaceae

Distribution: Distributed from Indochina to Assam.

Usable parts and properties: Substitute of *S. nux-vomica.* Seed- Used in migraines, nervousness, and depression.

Source of information (system): F

Syzygium travancoricum **Gamble (CR)**

Common name: Vathamkollimaram, Poriyal

Family: Myrtaceae

Distribution: Endemic to the Western Ghats.

Usable parts and properties: Plant- Astringent, antimicrobial, hypoglycemic, and neuro psychopharmacological. Cure diabetes and arthritis. Source of essential oil.

Source of information (system): F

Terminalia bialata **Steud., syn.** ***T. calamansanay*** **(Blanco) Rolfe.**

Common name: Indian Silver Greywood

Family: Combretaceae

Distribution: Distributed on the east coast, and Andamans.

Usable parts and properties: Bark- Contains tannins. Used as cardiac stimulant. Adulterant of cutch, *Acacia catechu.*

Source of information (system): F

Thymus serpyllum **L.**

Common name: Breck land thyme

Family: Lamiaceae/ Labiatae

Distribution: Native to Europe. Grown in Himalayas (Kashmir-Kumaon) and Nilgiris.

Usable parts and properties: Herb- Gives essential oil. Leaves- Aromatic. Known for pharmacological properties but needs investigation. Used as tonic, herbal tea and flavouring agent.

Source of information (system): Europe

Tragia bicolor **Miq., syn.** ***T. miqueliana*** **Müll. Arg. (VU)**

Family: Euphorbiaceae

Distribution: Endemic weed of evergreen forests of southern Western Ghats.

Usable parts and properties: Plant- Antibacterial, analgesic, antidiabetic, antioxidant. stimulates or inhibits a common biochemical pathway in all the body systems. Used in medicinal formulations but needs scientific evaluation.

Source of information (system): A

Trichosanthes anamalaiensis **Beddome (CR)**

Family: Cucurbitaceae

Distribution: Endemic to Tripura, ArP, AP, Karnataka, Maharashtra, Kerala, TN Andamans.

Usable parts and properties: Fruit- Used in headache and tetanus. Being wild relatives of snake gourd needs investigation for medicinal properties and use.

Source of information (system): F

Utleria salicifolia **Bedd. (CR), syn.** ***Decalepis salicifolia*****, (Lithophyte)**

Common name: Mahalikizhangu

Family: Apocynaceae

Distribution: Endemic to southern Western Ghats.

Usable parts and properties: Tubers- Used in treatment of skin diseases, asthma, and debility due to tuberculosis.

Source of information (system): F

Uvaria hookeri **King (DD)**

Family: Annonaceae

Distribution: Endemic to south India.

Usable parts and properties: Root bark extract- Observed to be antimicrobial & anthelmintic. Attributable to Acer-ogenins.

Source of information (system): F

Valeriana leschenaultia **DC. (CR)**

Sanskrit name: Sugandhabaalaa, Baalaka, Tapaswani.

Common name: Jatamansi related wild spp.

Family: Valerianaceae

Distribution: Endemic to high altitudes southern Western Ghats (reported from 2 localities of moist grassy & shady slopes).

Usable parts and properties: Can be substituted for valerian/ Jatamansi (Tapaswani).

Source of information (system): F

***Valeriana leschenaultii* DC. var. *brunoniana* (Wight & Arn.) C. B. Clarke. (VU - exceedingly rare), syn. *V. brunoniana* Wight & Arn.**

Sanskrit name: Sugandhabalaa, Tagara

Common name: Taggar

Family: Valerianaceae

Distribution: Endemic to Nilgiris hills at an altitude of 1830-2134 m.

Usable parts and properties: Substitute to *V. officinalis* (roots sedative).

Source of information (system): F

***Vateria macrocarpa* B.L. Gupta (CR)**

Common name: Vellappayin

Family: Dipterocarpaceae

Distribution: Endemic to evergreen forest of southern Western Ghats.

Usable parts and properties: Plant resin- Used by traditional healers of Indian tribes. Needs further study.

Source of information (system): F

***Xyris pauciflora* Willd.**

Common name(s): Yellow-eyed-grass or satin flower

Family: Xyridaceae

Distribution: Tropical, Subtropical Asia in moist deciduous forests, marshy areas of Assam Bengal to TN.

Parts and properties used: Plant- Locally used as sedative for insomnia, ring worm.

Source of information (system): F

***Youngia tenuifolia* (Willd.) Babc. & Stebbins.**

Family: Asteraceae/ Compositae

Distribution: Habitat wild to alpine regions of Himalayas.

Usable parts and properties: Whole plant- Locally used as health tonic. Medicinally potential.

Source of information (system): Locals

Zantedeschia aethiopica **(L.) Spreng.**

Common name: Arum lily, Calla lily

Family: Araceae

Distribution: Habitat to cooler Bihar, Odisha, and Western Ghats.

Usable parts and properties: Leaves- Locally used as poultice on sores, injury, gout, and rheumatism.

Source of information (system): Locals

Zosima orientalis **Hoffm., syn.** ***Z. absinthifolia*** **Link.**

Common name(s): ayı eli' or 'peynir out (Turkish)

Family: Apiaceae/ Umbelliferae

Distribution: West Asian, recorded in Maharashtra.

Usable parts and properties: Herb/fruit- Sedative and digestive. Used in cough and bowel disorders. Root- Umbelliferon isolated. Has medicinal potential and needs critical scrutiny.

Source of information (system): A, Locals

1. IUCN threat category in parenthesis after species Latin botanical name
2. Source of information, (system): Ayurveda = A; Folk = F; Homoeopathy = H; Sowa-Rigpa/Amchi/Tibetan = L; Sidha = S; Unani = U

Some More Sources of Information

Bhadrecha, P, Kumar, V and Kumar, M. (2017) Medicinal Plant Growing under Sub-optimal Conditions in trans-Himalaya Region at High Altitude. Defence Life Science Journal 2(1):37-45. DOI: 10.14429/dlsj.2.11107

Botanical Survey of India reports (status of threat)

Chaurasia, OP, Khatoon, N and Singh, SB. (2008) Field Guide: Floral Diversity of Ladakh pp 198. Defense Institute of High-Altitude Research (DIHAR), Defense Research and Development Organization (DRDO), Leh, Ladakh, India.

Jain, BJ, Kumane, SC and Bhattacharya, S. (2006) Medicinal flora of Madhya Pradesh and Chhattisgarh- A review. Indian Journal of Traditional Knowledge 5(2): 237-242.

Pushpangadan, P, George Varughese, Ijinu, TP and Chithra, MA. (2018) All India coordinated research project on ethnobiology and genesis of ethnopharmacology research in India including benefit sharing. Annals of Phytomedicine 7(1): 5-12, DOI: 10.21276/ap.2018.7.1.2

Singh, Anurudh K. (2015) Agricultural Biodiversity Heritage Sites and Systems in India pp 467. Asian Agri-History Research Foundation, Secunderabad, Telangana, India. ISBN 81-903963-4-X

6

Inventory of Medicinal and Aromatic Plants (MAPs) under Various Levels of Threat, Needing Protection

6.1 Introduction

The World Health Organization (WHO) compiled a list of 20,000 medicinal plants used in different parts of the world. An estimated 15,000 medicinal plant species, forming about 21% of the total plant species used for medicinal purposes in the world, fall in the endangered category (Schippmann *et al.,* 2006). In India, medicinal plant species used under different traditional systems to a large extent are still gathered and collected from the wild in from of roots, bark, wood, stem and even the whole plant causing severe destruction of their habitat, resulting in loss of species/populations and erosion of genetic diversity. The remaining few are cultivated as crop plants, but marginally because of lack of assured market and can be safely termed as commercially under-utilized and are also under threat. Therefore, despite India being rich in plants species with medicinal properties, genetic diversity within the species and the associated traditional knowledge regarding their use, the lack of awareness regarding the loss of species and genetic diversity due to destructive harvesting from the wild and many other factors, such as infrastructure development, iintroduction of exotic species is causing depletion of genetic diversity or loss of indigenous MAPs. For example, introduction of *Cassia* and *Acacia* has put in pressure on many Indian tree species, such as neem. The primary impact of these factors and from habitat loss of species results in loss of diversity of species; degradation of habitat that can bring change in species composition and fragmentation of habitat, effecting ecosystem's balance and reduce rate of regeneration. These changes can also lead to added social costs to society because they result in disruption of ecosystem and loss of diversity of natural species and their products that are valuable for human health. These factors coupled with lack of efforts on development of cultivation practices, research on taxonomy, reproductive biology, use, standardization of quality parameters for assessment and conservation have put species existence under threat. In the Indian context, 265 medicinal plant species have already been assessed as threatened in one or more states [The *Foundation for Revitalization of Local Health Traditions* (FRLHT), Database]. International Union for Conservation of Nature (IUCN) updated the Red List in June 2015 and added 44 Indian medicinal plants in the list. In the updated list, 18 plants are categorized as vulnerable, 16 as endangered and 10 as critically endangered species. Some medicinal plants of high commercial value, facing high threat because of non-cultivation and over harvesting from nature due to dependence of industries on natural population(s). Therefore, most MAPs need protection (*in situ*) and or conservation (*ex situ*) following diverse or complementary comprehensive strategies for conservation of their entire gene pool to facilitate their genetic improvement (enhancement) and long-term sustainable use. An attempt has been made to produce a comprehensive list of these species below, which can be prioritized based on their market demand. Of these some may be under threat at global level, while other at the Indian nation or region or local level effecting specific population(s).

6.2 List of Medicinal and Aromatic Plants under Threat

S.No.	Botanical name	IUCN Category
1	*Abutilon ranadei* Woodrow & Stapf.	Critically endangered
2	*Acer caesium* Wall. ex Brandis	Vulnerable
3	*Achyranthes coynei* Santapau	Endangered - Rare
4	*Aconitum balfourii* Stapf., syn. *A. atrox* (Bruchl) Mukherjee	Critically endangered, Rare
5	*Aconitum chasmanthum* Stapf.	Critically endangered
6	*Aconitum deinorrhizum* Stapf., syn. *A. heterophylloides* (Brühl) Stapf.	Critically endangered
7	*Aconitum falconeri* Stapf	Critically endangered
8	*Aconitum ferox* Wall. ex Ser.	Endangered
9	*Aconitum heterophyllum* Wall. ex Royle	Critically endangered
10	*Aconitum kashmiricum* Stapf ex Coventry	Endangered
11	*Aconitum lethale* Griff.	Rare
12	*Aconitum violaceum* Jacq. ex Stapf	Critically endangered
13	*Acorus calamus* L.	Vulnerable
14	*Adenia hondala* (Gaertn.) W.J. De Wilde, syn. *A. palmata* (Lam.) Engl.	Vulnerable
15	*Adhatoda beddomei* C.B. Clarke	Critically Endangered
16	Adina cordifolia (Roxb.) Brandis.	Endangered
17	*Aegle marmelos* (L.) Correa Ex. Schultz	Vulnerable
18	*Aegle marmelos* var. *mahurensis* B.R. Zate	Rare
19	*Aerva wightii* Hook. f.	Extinct
20	*Albizia thompsoni* Brandis	Near Threatened Rare (Nayar & Sastry,1990)
21	*Albizia thompsonii* var. *galbana* Haines	Rare (Red Data Book)
22	*Alpinia galanga* (L.) Willd.	Data Deficient
23	*Amorphophallus commutatus* Engl.	Vulnerable
24	*Amorphophallus paeoniifolius* (Dennst.) Nicolson	Vulnerable
25	*Ampelocissus araneosa* (Dalz. & Cribbs.) Planch.	Vulnerable
26	*Ampelocissus arnottiana* Planch., syn. *A. indica* (L.) Planch	Endangered

S.No.	Botanical name	IUCN Category
27	*Andrographis nallamalayana* J. L. Ellis	Endangered - Rare
28	*Andrographis paniculata* (Burm. F.) Wallich Ex Nees.	Least Concerned
29	*Angelica glauca* Edgew.	Critically Endangered
30	*Anticharis glandulosa* Asch., syn. A. glandulosa var. caerulea Blatt. & Halbl.	Endangered - Rare
31	*Aphanamixis polystachya* (Wall.) Parker	Vulnerable
32	*Aquilaria agallocha* Roxb., syn. *A. malacaccensis* Lam	Critical Endangered
33	*Aquilaria malaccensis* Lam., syn. *A. agallocha* Roxb.	Vulnerable
34	*Aquilqria khasiana* Hallier f.	Eliminated; recollected from Meghalaya (Critical Endangered)
35	*Aristolochia bracteolata* Lam., (= *A. bracteata* Retz.)	Least Concerned
36	*Aristolochia tagala* Cham.	Vulnerable
37	*Arnebia benthamii* (Wall. *ex* G.Don) I.M. Johnst.	Critically Endangered
38	*Artemisia indica* Willd.	Least Concerned
39	*Artocarpus hirsutus* Lam.	Vulnerable
40	*Asparagus rottleri* Baker.	Extinct
41	*Aster tibeticus* Hook. f., syn. *Jacobaea tibetica* (Hook. f.) B. Nord.	Rare
42	*Atropa acuminata* Royle Ex Lindl.	Critically Endangered
43	*Bacopa monnieri* (L.) Pennel	Least Concerned
44	*Balanites aegyptiaca* (L.) Delile	Least Concerned
45	*Balanophora involucrata* Hook. f.	Rare and endangered
46	*Baliospermum montanum* (Willd.) Muell. -Arg.	Vulnerable
47	*Barleria gibsonioides* Blatter	Rare (possibly extinct)
48	*Barleria prionitis* L.	Data deficient
49	*Barleria prionitis* ssp. *dicantha* var. *dicantha* Baltt. & Hallb.	Data deficient
50	*Barleria stocksii* T Anders	Extinct, Rediscovered
51	*Bauhinia malabarica* Roxb.	Endangered
52	*Bauhinia variegata* L.	Data deficient
53	*Berberis affinis* G. Don	Rare
54	*Berberis apiculata* Ahrendt	Rare
55	*Berberis asiatica* Roxb. (Sub sp.); syn. *B. aristata* DC.	Endangered
56	*Berberis chitria* Lindl	Vulnerable
57	*Berberis kashmirian* Ahrendt	Rare
58	*Berberis lambertii* Parker	Vulnerable
59	*Berberis lycium* Royle	Endangered
60	*Berberis pseudoumbellata*	Data deficient
61	*Berberis royleana Ahrendt*	Data deficient
62	*Berberis wardii* C.K. Schneid	Near threat (not evaluated)

S.No.	Botanical name	IUCN Category
63	*Bergenia ciliata* (Haw.) Sternb., syn. *B. ligulata* (Wall.) Engl.	Vulnerable
64	*Bhesa robusta* (Roxb.) Ding Hou	Data deficient
65	*Boswellia ovalifoliolata* Balakr,	Endangered
66	*Boswellia serrata* Roxb.	Eroded
67	*Buchanania lanzan* Spreng.	Low risk
68	*Bunium persicum* (Boiss.) Fedt.	Endangered
69	*Butea monosperma* (Lam.) Taub., *B. frondosa* Willd.	Endangered
70	*Butea monosperma* var. *lutea* (Witt.) Maheswari	Endangered
71	Calligonum polygonoides L.	Endangered
72	*Calophyllum apetalum* Willd.	Vulnerable
73	*Canarium strictum* Roxb.	Vulnerable
74	*Capparis zeylanica* L.	Rare
75	*Cayratia pedata* (Lam.) Juss. ex Gagnep. var. *glabra* Gamble	Critically Endangered
76	*Celastrus paniculatus* Willd.	Vulnerable
77	*Ceropegia fimbriifera* Bedd.	Vulnerable - Rare
78	*Ceropegia pusilla* Wight & Arn.	Vulnerable - Rare
79	*Ceropegia odorata* Nimmo ex J. Graham	Endangered - Critically Endangered
80	*Ceropegia spiralis* Wight.	Vulnerable
81	*Chlorophytum borivilianum* Santapau & R.R. Fern.	Critically endangered
82	*Chlorophytum malabaricum* Baker	Vulnerable
83	*Chonemorpha fragrans* (Moon) Alston	Endangered
84	*Cinnamomum macrocarpum* Hook.	Vulnerable
85	*Cinnamomum sulphuratum* Nees.	Vulnerable
86	*Cinnamomum tamala* Nees & Eberm.	Under pressure, threatened
87	*Cinnamomum wightii* C.F.W. Meissn	Vulnerable
88	*Cinnamomum zeylanicum* Bl.	Data deficient
89	*Clematis roylei* var. *patens* (Haines) B.M. Kapoor.	Endangered
90	*Cleome burmanni* Wight & Arn.	Data deficient
91	*Clerodendrum colebrookianum* Walp.	Vulnerable
92	*Coleus vettiveroides* K.C. Jacob	Data deficient
93	*Commiphora wightii* (Arn.) Bhandari, syn. *C. mukul* (Hook. ex Stocks) Engl.; *Balsamodendron mukul* Hook. ex Stocks	Vulnerable - Critically Endangered
94	*Coptis teeta* Wall.	Vulnerable - Endangered
95	*Cordia macleodii* (*Gerascanthus macleodii* (Hook. fil. & Thoms.) A. Borhidi	Endangered
96	*Cordia rothii* Roem. & Schult., syn. *C sinensis* Lam.	Near Threatened
97	*Cordyceps sinensis* (Berk.) Saac (mushroom)	Endangered
98	*Coscinium fenestratum* (Gaertn.) Coleb.	Critically Endangered

S.No.	Botanical name	IUCN Category
99	*Crinum eleonorae* Blatt. & McCann	Possibly Extinct
100	*Croton scabiosus* Bedd.	Vulnerable
101	*Curculigo orchioides* Gaertn.	Endangered
102	*Curcuma angustifolia* Roxb.	Vulnerable
103	*Curcuma caesia* Roxb.	Critically Endangered
104	*Curcuma pseudomontana* Graham	Vulnerable
105	*Cycas beddomei* Dyer,	Endangered
106	*Cycas circinalis* L..	Critically Endangered
107	*Cyclea fissicalyx* Dunn	Endangered
108	*Dactylorhiza hatagirea* (D. Don.) Soo (orchid)	Populations declining, Critically Endangered
109	*Dalbergia horrida* (Dennst.) Mabb. (= *D. sympathetica* Nimmo)	Data deficient
110	*Decalepis hamiltonii* Wight & Arn.	Endangered
111	*Delphinium denudatum* Wall. Ex Hook. F. & Thoms.	Critically Endangered
112	*Dendrobium ovatum* (L.) Kraenzl.	Least Concerned
113	*Dioscorea deltoidea* Wall. Ex Kunth.	Critically Endangered
114	*Dioscorea wightii* Hook. f.	Rare
115	*Diospyros candolleana* Wight	Vulnerable
116	*Diospyros cordifolia* Roxb., syn. *D. montana* Roxb.; *D. montana* var. *cordifolia* Hiem.	Endangered
117	*Diospyros paniculata* Dalz.	Vulnerable
118	*Dipterocarpus indicus* Bedd.	Endangered
119	*Dioscorea deltoidea* Wall. ex Griseb.	Endangered - Critically Endangered
120	*Drosera indica* L.	Near Threatened
121	*Drosera peltata* Sm. Willd.	Vulnerable
122	*Dysoxylum malabaricum* Bedd.	Endangered
123	*Elaeagnus conferta* Roxb.	Least Concerned
124	*Elaeocarpus serratus* L.	Least Concerned
125	*Embelia ribes* Burm. f.	Endangered
126	*Embelia tsjeriam-cottam* A. Dc.	Vulnerable
127	*Ephedra gerardiana* Wall. ex JA Mey.	Endangered - Critically Endangered
128	*Ephedra foliata* Boiss. ex C. Mey.	Near Threatened
129	*Ephedra sinica* Stapf.	Endangered
130	*Erythrina herbacea* L.	Endangered
131	*Erythrina variegata* forma *mysorensis* (Gamble)	Data Deficient
132	*Eulophia cullenii* (Wight) Blume	Critically Endangered
133	*Eulophia ramentacea* Wt.	Critically Endangered
134	*Flickingeria fugax* (Rchb.f.) Seidenf., syn. *Dendrobium fugax* Rchb. f.	Endangered
135	*Fritillaria roylei* Hook.	Critically Endangered

S.No.	Botanical name	IUCN Category
136	*Garcinia gummi-gutta* (L.) Robs.	Vulnerable
137	*Garcinia indica* (Dup.) Choisy	Vulnerable
138	*Garcinia morella* (Gaertn.) Desr.	Vulnerable
139	*Garcinia pedunculata* Roxb. ex Buch. -Ham.	Endangered
140	*Garcinia rubro-echinata* Kosterm.	Vulnerable
141	*Garcinia talbotii* Raizada ex Santapau	Vulnerable
142	*Garcinia travancorica* Beddome	Critically Endangered
143	*Gardenia gummifera* L. f.	Threatened
144	*Gentiana kurroo* Royle	Critically endangered
145	*Gentiana quadrifaria* Blume, syn. *G. laxicaulis* Zoll. & Moritzi. = *G. pedicellata* Wall.	Vulnerable
146	*Gloriosa superba* L.	Threatened
147	*Glycosmis macrocarpa* Wt.	Near Threatened - Vulnerable
148	*Gymnema khandalense* Santapau	Endangered - Critically endangered
149	*Gymnema montanum* (Roxb.) Hook	Endangered
150	*Gymnocladus assamicus* Kanjilal ex P.C. Kanjilal	Critically endangered
151	*Hedychium coronarium* Koenig.	Near Threatened
152	*Hedychium gratum* Wall. ex. Baker	Endangered - Rare
153	*Hedyotis barberi* (Gamble) A.N. Henry & Subr.	Vulnerable - Rare
154	*Hedychium spicatum* Buch. -Ham.	Vulnerable
155	*Heliotropium keralense* Siv. & Manl	Endangered
156	*Helminthostachys zeylanicus* (L.) Hook.	Endangered
157	*Heracleum candolleanum* (Wt. & Arn.) Gamble	Vulnerable
158	*Heracleum lanatum* Michx., syn. *H. candicans* Wall.	Endangered
159	*Heracleum rigens* Wall.	Data Deficient
160	*Holostemma ada-kodien* Schult.	Vulnerable
161	*Homalomena aromatic* (Roxb.). Schott	Vulnerable
162	*Humboldtia vahliana* Wight	Vulnerable - Endangered
163	*Hydnocarpus alpina* Wt.	Endangered
164	*Hydnocarpus kurzii* (King) Warb.	Endangered
165	*Hydnocarpus macrocarpa* (Bedd.) Warb.	Vulnerable
166	*Hydnocarpus pentandrus* (Buch. -Ham.) Oken	Vulnerable
167	*Ilex khasiana* Purk.	Endangered - Critically Endangered (Rare)
168	*Idigofera barberi* Gamble	Rare
169	*Indigofera constricta* (Thw) Trimen	Vulnerable - Rare
170	*Indigofera coerulea* var. *monosperma* (Santapau) Santapau	Rare
171	*Inula racemosa* Hook. F.	Critically Endangered

S.No.	Botanical name	IUCN Category
172	*Ipomoea turpethum* (L.) R. Br. (= *Operculina turpethum* (L.) Silva Manso	Vulnerable
173	*Janakia arayalpathra* Joseph & Chandrasekaran	Critically Endangered
174	*Jurinea macrocephala* DC., syn. *J. dolomiaea* Boiss	Near Threatened
175	*Kaempferia galanga* L.	Critically Endangered
176	*Kingiodendron pinnatum* (Roxb. Ex Dc.) Harms	Endangered
177	*Knema attenuata* (Wall.) Warb. syn. *Myristica corticosa* Bedd.	Near Threatened
178	*Lamprachaenium microcephalum* Benth.	Endangered
179	*Lemanea australis* Alkins	Critical Endangered
180	*Leucas lanata* var. *nagpurensis* C.B. Clarke ex Haines	Endangered
181	*Lilium mackinliae* Sealy	Endangered - Rare
182	*Lilium polyphyllum* D. Don	Critically endangered
183	*Luffa umbellata* (Klein) Roem. (wild), syn. L *acutangula* (L.) Roxb. (Cultigen)	Endangered
184	*Luvunga scandens* (Blume) Kurz.	Critically Endangered
185	*Lycopodium pseudoclavatum* Ching, syn, *L. japonicum* Thunb. ex Murray	Vulnerable
186	*Madhuca diplostemon* (Clarke) Royen.	Endangered
187	*Madhuca insignis* (Radlk) Lam.	Considered possibly extinct
188	*Madhuca longifolia* (Koen.) Macler	Endangered
189	*Madhuca neriifolia* (Moon) Lam	Vulnerable
190	*Malaxis muscifera* (Lindl.) Kuntze, syn. *Microstylis musifera* Ridley	Critical Endangered
191	*Mallotus philippensis* (Lam.) Muell. -Arg.	Rare
192	*Mappia foetida* (Wight) Miers.	Vulnerable
193	*Meconopsis aculeata* Royle	Critically Endangered
194	*Michelia champaca* L.	Vulnerable
195	*Michelia nilagirica* Zenk.	Vulnerable
196	Microstylis wallichii Lindl.	(Near threat, not assessed)
197	*Morinda angustifolia* Roxb.	Rare
198	*Moringa concanensis* Nimmo ex Dalz. & Gibson	Vulnerable
199	*Moringa oleifera* Lam., syn. *M. pterygosperma* Gaertn.	Near Threat
200	*Myristica dactyloides* Gaertn.	Vulnerable
201	*Myristica malabarica* Lam.	Vulnerable
202	*Nardostachys grandiflora* Wall. ex DC., syn. *N. jatamansi* (D. Don).	Endangered - Critically Endangered
203	*Nepenthes khasiana* Hook. f.	Endangered- Critically Endangered
204	*Nervilia aragoana* Gaud.	Endangered
205	*Nilgirianthus ciliatus* (Nees) Bremek.	Endangered
206	*Ochreinauclea missionis* (Wall. ex G. Don) Ridsdale	Vulnerable

S.No.	Botanical name	IUCN Category
207	*Operculina turpethum* (L.) S. Manso	Near Threat
208	*Oroxylum indicum* (L.) Vent.	Vulnerable
209	*Paeonia emodi* Wall. ex. Royle	Vulnerable
210	*Panax assamicus* Banerjee	Endangered
211	*Panax ginseng* C.A. Mey	Critically Endangered
212	*Panax pseudoginseng* Wall.	Vulnerable - Critically Endangered
213	*Panax sokpaiyensis* Shiva K. Sharma & Pandit (Sikkim)	Endangered, due to limited location & population size
214	*Paphiopedilum druryi* (Bedd.) Stein.	Critically Endangered
215	*Paris polyphylla* Sm.	Vulnerable
216	*Persea macrantha* (Nees) Kost. syn. *Machilus macrantha* Nees	Endangered
217	*Phoenix pusilla* Gaertn.	Least Concerned
218	*Phyllanthus narayanswami* Gamble	Endangered
219	*Phyllanthus talbotii* Sedgw.	Critically endangered - Rare
220	*Physochlaina praealta* (Decne.) Miers, syn. *P. dubia* Pascher (Not recognized)	Vulnerable
221	*Picrorhiza kurroa* Royle ex Benth.	Endangered
222	*Pimpinella tirupatiensis* Balakr. and Subram.	Vulnerable - Endangered
223	*Pinus gerardiana* Wall.	Rare
224	*Piper barberi* Gamble.	Critically Endangered
225	*Piper longum* L	Near threatened – recoded in wild
226	*Piper mullesua* Buch. -Ham. ex D. Don	Vulnerable
227	*Piper nigrum* L.	Vulnerable
228	*Plectranthus barbatus* Andrews [= *Coleus forskohlii* (Poir.) Briq.]	Endangered
229	*Plectranthus nilgherricus* Benth., syn. Isodon nilgherricus (Benth.) H. Hara	Vulnerable
230	*Podophyllum hexandrum* Royle, syn. *P. emodi* Wall. ex Honig.	Endangered-Critically Endangered
231	*Polygonatum cirrhifolium* (Wall.) Royle	Referred Endangered
232	*Polygonatum verticillatum* (L.) All.	Near Threat - Endangered
233	*Polygonum plebejum* var. *sindica* (Roth.) Hook. fil., syn. *P. prostratum* Roxb. ex D. Don	Rare
234	*Przewalskia tangutica* Maxim.	Critically Endangered
235	*Pseudarthria viscida* Wight & Arn.	Near Threatened
236	*Pterocarpus marsupium* Roxb.	Endangered
237	*Pterocarpus santalinus* L. f.	Endangered
238	*Pterocarpus dalbergioides* Roxb. (= *P. indicus* R. Vig.)	Endangered
239	*Pterospermum xylocarpum* (Gaerth.) Samt. & Wagh.	Least concerned (?)

S.No.	Botanical name	IUCN Category
240	*Pueraria tuberosa* (Roxb. ex Willd.) DC.	Near threatened - Vulnerable
241	*Pycnocycla glauca* Lindl., syn. *Chamaesciadium glaucum* (Lindl.) M. Hiroe	Endangered
242	*Rhaphidophora pertusa* Schott.	Vulnerable
243	*Rauvolfia beddomei* Hook. f.	Threatened
244	*Rauvolfia micrantha* Hook. f	Referred Endangered - Critically Endangered
245	*Rauvolfia serpentina* (L.) Benth. ex Kurz.	Rare - Critically Endangered
246	*Rhaphidophora pertusa* (Roxb.) Schott, syn. *R. laciniata* (Burm. f.) Merr.; Pothos pertusus Roxb.	Vulnerable
247	*Rheum australe* D. Don	Vulnerable
248	*Rheum emodi* Wall. ex Meissn.	Under threat overharvesting disturbing habitat
249	*Rheum nobile* Hook. F. & Thoms	Endangered
250	*Rhododendron anthopogon* D. Don	Vulnerable
251	*Rhododendron concinnoides* Hutch. & Kingdon-Ward	Endangered
252	*Rhododendron santapaui* Sastry *et al.*	Endangered
253	*Rhus semialata* Murr.	Vulnerable
254	*Rhynchosia beddomei* Baker	Vulnerable
255	*Roscoea purpurea* Sm., *R. procera* Wall.	Near Threat
226	*Rubia manjith* Roxb. ex Fleming	Vulnerable
257	*Rumex crispus* Linn.	Rare
258	*Salacia oblonga* Wall.	Endangered
259	*Salacia reticulata* Wight	Endangered
260	Salvadora persica L.	Near Threat
261	*Santalum album* L.	Endangered
262	*Sapindus emarginatus* Vahl	Population structure variable
263	*Saraca indica* L., syn. *S. asoca* (Roxb.) W.J.de Wilde	Endangered
264	*Saussurea bracteata* Decne.	Rare
265	*Saussurea clarkei* Hook. f.	Near Threatened - Rare
266	*Saussurea costus* (Falc.) Lipsch., syn. *S. lappa* (Decne.) Sch. Bip.]	Critically Endangered
267	*Saussurea gossypiphora* Don	Endangered
268	*Saussurea involucrata* Kar et Kir.	Endangered
269	*Saussurea obvallata* Wall.	Vulnerable due to over harvesting
270	*Saussurea simpsoniana* (Field & Gard.) Lipsch	Endangered
271	*Schrebera swietenioides* Roxb.	Vulnerable
272	*Selinum candollei* DC., syn. *S. tenuifolium* Sailsb.	Near Threatened

S.No.	Botanical name	IUCN Category
273	*Semecarpus travancorica* Bedd., syn. *Cassuvium travancoricum* (Bedd.) Kunt.	Endangered
274	*Sesbania speciosa* Taub., syn. *S. pubescens* DC. var. *grandiflora* Vatke	Vulnerable
275	*Shorea tumbuggaia* Roxb.	Critically Endangered
276	*Smilax glabra* Roxb.	Critical Endangered
277	*Smilax wightii* L.	Data deficient - Rare
278	*Smilax zeylanica* L.	Vulnerable
279	*Sterculia khasiana* King ex Debb.	Extinct due to habitat loss
280	*Sterculia urens* Roxb.	Endangered
281	*Stereospermum chelonoides* (L. f.) DC., syn. *S. personatum* (Hassk.) Chatterjee; *S. suaveolens* (Roxb.) DC.	Critically Endangered
282	*Strychnos aenea* A.W. Hill.	Endangered
283	*Strychnos minor* Dennst.	Data deficient
284	*Swertia angustifolia* Ham.	Endangered
285	*Swertia chirayita* (Roxb. ex Fleming) H. Karst., syn. *S. chirata* (Wall.) C. B. Clarke.	Vulnerable
286	*Swertia corymbosa* (Griseb.) Wt. ex Clarke Anti-diabetic Raipur	Vulnerable
287	*Swertia lawii* Burkill	Endangered
288	*Symplocos cochinchinensis* S. Moore	Nearly Threat
289	*Symplocos racemosa* Roxb.	Vulnerable
290	*Syzygium alternifolium* (Wight) Walp.	Endangered
291	*Syzygium travancoricum* Gamble	Critically Endangered
292	*Taxus baccata* L., syn. *T. wallichiana* Zucc.	Endangered - Critically Endangered
293	*Tecomella undulata* (Sm.) Seem.	Referred Endangered
294	*Terminalia arjuna* (Roxb. Ex DC.) Wight & Arn.	Near Threatened due to pressure
295	*Terminalia pallida* Brandis	Endangered
294	*Thalictrum foliolosum* DC.	Vulnerable
295	*Tinospora sinensis* (Lour.) Merr., syn. *T. malabarica* (Lam.) Hook.f. & Thomson	Vulnerable
296	*Tragia bicolor* Miq., syn. *T. miqueliana* Müll. Arg.	Vulnerable
297	*Tribulus rajasthanensis* M.M. Bhandari & V.S. Sharma	Critically endangered
298	*Trichopus zeylanicus* Gaertn.	Critically Endangered
299	*Trichosanthes anaimalaienis* Beddome	Critically Endangered
300	*Trichosanthes cucumerina* L.	Data Deficient
301	*Urginea indica* (Roxb) Kunth.	Vulnerable
302	*Utleria salicifolia* Bedd. Syn. *Decalepis salicifolia* Bedd. ex Hook,f. (Venter)	Critically Endangered
303	*Uvaria hookeri* King	Data Deficient

S.No.	Botanical name	IUCN Category
304	*Valeriana hardwickii* Wall.	Vulnerable
305	*Valeriana jatamansi* Jones, syn. *V. wallichii* DC.	Critically endangered
306	*Valeriana leschenaultia* DC.	Critically endangered (reported from two localities of south India moist grassy and shady slopes)
307	*Valeriana leschenaultia* DC. var. *brunoniana* C. B. Clarke.	Vulnerable
308	*Vateria indica* L.	Near Threatened
309	*Vateria macrocarpa* B.L. Gupta	Critically Endangered
310	*Vernonia anthelmintica* (L.) Willd.	Least concern
311	*Vitex trifolia* L. f.	Least concern
312	*Withania coagulans* Dunal.	Near threatened in western Rajasthan due to overharvest
313	*Woodfordia fruticosa* (L.) Kurz.	Least concern

Source: envis.frlht.org/junclist.php? txtbname; Singh (2015)

This list predominantly includes those species of high trade value, which are harvested from the nature due to a huge gap between the supply and demand of medicinal plants to manufacture Ayurvedic drugs and medicines in India. According to the 'All India Trade Survey of Prioritized Medicinal Plants, 2019', demand for high-value medicinal plants increased by 50%, while the availability declined by 26%. This led to increased habitat degradation and levels of over-exploitation by pharmaceutical industries. This also resulted in 65 species (i.e., 10% of the total species) falling into the critically endangered, endangered, vulnerable, and nearly threatened categories. Some of these species' populations are depleting at global level, while others, which are endemic to India, are under general threat because of over harvesting from the nature. Government of India has drawn a list of these species, which continue to be depleted because of indiscriminate harvesting of wild populations affecting export. In addition, there are other medicinal plant species, which are presently traded in lower volumes, but they have high trade potential. Therefore, there is a need for focused management interventions for their conservation and sustainable use.

Many other medicinal plant species are also under threat for various reasons other than over exploitation, such as deforestation, over grazing, human settlement, infrastructure development, climatic disasters etc. causing loss of specific populations and larger habitat destruction. Therefore, there is an urgent need to initiate interim measures for management of high-risk medicinal plant species and to take up re-assessment of wild populations of these species at the regional, national, and global level to evolve comprehensive strategies for *in situ* and *ex situ* conservation, and to facilitate species recovery and sustainable commercial use. As discussed in section 3, in most cases the best strategies can be to domesticate these species and develop best cultivation practices and programs, respectively, for their large-scale propagation to ensure availability of the raw material.

An analysis of medicinal and aromatic plants of trade value has revealed that many species of high trade value, such as *Aconitum heterophyllum, Coscinium fenestratum, Decalepis hamiltonii, Nardostachys grandiflora, Oroxylum indicum, Picrorhiza kurroa, Saraca asoca, Swertia chirayita, Vateria indica* are under threat. The concerns about global depletion of populations of some of these species have prompted their inclusion in the list of The Convention on International Trade in Endangered Species of Wild Fauna and Flora (CITES), an international agreement to which states and regional economic integration organizations adhere voluntarily. Despite efforts of Government of India, wild populations of many of these threatened species continue to be depleted in the face of indiscriminate harvesting leading to their inclusion in the banned/ negative list of exports (Chapter 4).

In addition, there are many more medicinal plants species under threat and need attention from conservation and sustainable use point of view to facilitate greater exploitation. As e-screening and the search of alternative medicines continues to combat new diseases, one never knows how many of them will become important in future and will need to be saved before depletion of their populations and/or extinction. Therefore, an assessment of the population status of medicinal plant species collected in the wild is needed for two purposes. Firstly, to formulate strategies for management of wild populations in respect of species under threat of extinction and secondly, to promote their cultivation for large scale multiplication.

Guidelines developed by IUCN (The World Conservation Union) are available for rapid assessment of the threatened status of wild plant taxa. Foundation for Revitalization of Local Health Traditions (FRLHT), which is a registered Public Trust and Charitable Society, started its activities in 1993 under the guidance of Sam Pitroda and Anant Darshan Shankar. The Indian Ministry of Science and Technology recognizes FRLHT as a scientific and research organization. It has applied these guidelines to rapidly assess the status of prioritized medicinal plants of India and has, over the past ten years, coordinated assessment of threat status of more than four hundred medicinal plants species across the country. However, this initiative marks only the pilot assessment exercises. The effort needs to be carried forward by creating a long-term institutional mechanism with nodal and regional centres to undertake initially rapid threat assessment and subsequent population studies. Supporting such institutional efforts is necessary in view of the large diversity of medicinal plant species in use and the need to ensure survival of those which are threatened with extinction.

The rapid threat assessment exercises would need to be followed up with species-specific field surveys in respect of those species that are assessed as threatened. Urgent field surveys are needed in the case of species that are already assessed as 'critically endangered' to assess the status of their wild populations and to enable planning of appropriate action for their restoration and conservation. The species-specific field surveys would also help in assessing the authenticity of the material being traded in the name of some of the threatened species. Additionally, they may also raise many critical issues needing attention and for initiating appropriate action plan, particularly in relation to threat to wild population, linkages with trade, production of forest species, development, and implementation of effective management systems to meet the high market demand, and promotion of commercial cultivation. Also, to support research and development related with genetic improvement of target species for developing improved varieties for commercial cultivation with higher productivity.

Based on threat, size of sample, nature, and breeding behaviour of species and the objectives of conservation, the conservation strategies can be either *in situ* or *ex situ* or both. *In situ* conservation would mean protection or supportive propagation of species/ genetic diversity within their ecosystem and natural habitat. For many populations, because of their adaptation to a very narrow ecological niche, *in situ* conservation may be the only answer. Similarly, for conservation of specific types associated with a geographical region and/or to include evolutionary dynamics responding to climate change, *in situ* or on farm conservation may be a complementary option in addition to their *ex-situ* conservation in seed gene bank.

For most development agro technique of cultivation of under agricultural ecosystem may be best option. Cultivation of many species like Isabgol (*Plantago ovata*), Amla (*Emblica officinalis*), Ashwagandha (*Withania somnifera*), Kalihari (*Gloriosa superba*), Kuth (*Saussurea costus*), Pushkarmool (*Inula racemosa*), Mentha (*Mentha* spp.), Safed Musli (*Chlorophytum* sp.), Senna (*Cassia angustifolia*), Tulasi (*Ocimum tenuiflorum*), etc. has stabilized and integrated into the prevailing agriculture systems of the area because of these interventions. The focus of promoting commercial cultivation now needs to shift from these species to the threatened species, which are in short supply. Some such species namely *Aconitum heterophyllum, Swertia chirayita, Picrorhiza kurroa* and *Nardostachys grandiflora,* already included in the NMPB's priority list, need to be given more attention. Other species of conservation concern like *Coscinium fenestratum, Gymnema sylvestre, Embelia tsjerium-cottam, Nilgirianthus ciliatus, Decalepis hamiltonii, Onosma hispidum*, etc. would need concerted research inputs before their commercial cultivation could be promoted. Further, there is also a strong case for promoting the domestication and commercial cultivation of high value imported species like *Anacyclus pyrethrum, Cinchona* spp. and high volume imported species like *Piper chaba.*

To meet the growing demand of medicinal and aromatic plants of significance and trade value, the Department of Indian System of Medicine and Homeopathy has identified 165 plant species for developing agro-technique for their cultivation. This besides providing raw material as per the demand of drug industry shall help conservation of important species for long-term sustainable use.

6.3 Medicinal and Aromatic Plants Needing Development of Cultivation Practices for Protection/Conservation and to Support Commercial use

S.No.	Botanical name	Trade name
1	*Abroma augusta* L.	Ulatkambal, Devil's cotton
2	*Aconitum balfourii* (Benth.) Muk., syn. *A. atrox* (Bruhl) Muk.	Vatsnabha
3	*Aconitum chasmanthum* Stapf.	Vatsnabha
4	*Aconitum ferox* Wall. ex Ser.	Vachhnag, Atis
5	*Aconitum heterophyllum* Wall. ex Royle	Atis, Aconite
6	*Aconitum palmatum* D. Don, syn. *A. bisma* (Buch.-Ham.) Rap.	Bikhma
7	*Alstonia scholaris* (L.) R.Br.	Saptaparnachal, Chitvan
8	*Anacyclus pyrethrum* (L.) Cass.	Akarkara
9	*Andrographis paniculate* (Burm. f.) Wall. *ex* Nees	Kalmegh
10	*Aquilaria agallocha* Roxb. syn. *A. malacaccensis*	Agar kala, Akil
11	*Aristolochia indica* L.	Ishwar mool
12	*Asparagus adscendens* Roxb.	Pili Satawar
13	*Asparagus racemosus* Willd.	Shatavari, Satawar
14	*Bacopa monnieri* (L.) Wettst., syn. *Herpestis monnieria*	Jal Brahmi
15	*Baliospermum montanum* (Willd.) Mull. Arg., syn. *B. solanifolium* (Geiseler) Suresh	Dantimool, Nagadanti
16	*Bergenia ciliata* (Haw.) Sternb., syn. *B. ligulata*	Pashnabheda
17	*Blepharis edulis* auct. non (Forssk.) Pers., syn. *B. persica* (Burm. f.) O.Kuntze.	Uttangan
18	*Boswellia serrata* Roxb.	Kandur, Salai Guggul, Dupa
19	*Callicarpa macrophylla* Vahl.	Priyangu
20	*Celastrus paniculatus* Willd.	Malkangani, Jyothismathi
21	*Centella asiatica* (L.) Urb.	Brahmi
22	*Chlorophytum* spp.	Safed Musali
23	*Chrysanthemum indicum* L.	Pyrethrum
24	*Cinchona* spp.	Quinine
25	*Clerodendrum serratum* (L.) Moon	Bharangi
26	*Colchicum luteum* Baker	Hirantotiya, Suranjan
27	*Coleus zeylanicus* (Benth.) Cramer, and *C. vettiveroides*, syn. *Plectranthus vettiveroides*	Valakah, Kuruver
28	*Commiphora wightii (*Arn.) Bhandari, syn. *C. mukul; Balsamodendron mukul* Hook. ex Stocks	Guggul
29	*Convolvulus microphyllus* Sieb. ex Spreng., syn. *C. prostratus*; *C. pluricaulis*	Shankapushpi
30	*Coscinium fenestratum* (Gaertn.) Colebr.	Maramanjal, Daruharidra
31	*Crataegus oxyacantha* L.	Hawthorn
32	*Cryptolepis buchananii* R.Br. ex Roem. & Schult.	Medaksinghi, Dudhi

S.No.	Botanical name	Trade name
33	*Curculigo orchioides* Gaertn.	Kali musali
34	*Datura* spp.	Datura
35	*Decalepis hamiltonii* Wight & Arn.	Magali
36	*Desmodium gangeticum* (L.) DC.	Salparni, Prasniparni
37	*Desmostachya bipinnata* (L.) Stapf.	Hamsapapdi
38	*Dioscorea bulbifera* L.	Varahi kand
39	*Elaeocarpus sphaericus* (Gaertn.) K.Schum, syn. *E. granitrus* Roxb.	Rudraksha
40	*Embelia ribes* Brum. f.	Vaividang
41	*Embelia tsjerium-cottam* (Roem. & Schult.) DC.	Vaividang
42	*Evolvulus alsinoides* (L.) L.	Shankhavali, Shankhpush
43	*Fumaria indica* (Hauskn.) Pugsley, syn. *F. parviflora*	Shahtara
44	*Garcinia pedunculata* Roxb. ex Buch. -Ham.	Thaikala
45	*Gmelina arborea* Roxb.	Ghambar chal
46	*Gymnema sylvestre* R.Br. ex Schult.	Gudmar, Sirukurinjan
47	*Habenaria edgeworthii* Hook. f. ex Collett and *H. intermedia*	Riddhi
48	*Hedychium spicatum* Sm., syn. *H. acuminatum*	Kapoor kachri, spiked Ginger Lily
49	*Hemidesmus indicus* (L.) R.Br.	Anatmool, Sariwa
50	*Holostemma ada-kodien* Schult., syn. *H. annularis, H. rheedii* Wall.	Jeevanti
51	*Hydnocarpus* spp.	Chaulmoogra
52	*Inula racemosa* Hook. f.	Pushkarmool
53	*Ipomoea mauritiana* Jacq., syn. *I. digitata* A. Meeuse	Palmudhukkan kizhangu
54	*Ipomoea petaloidea* Choisy	Vrddhadaru
55	*Juniperus communis* L., syn. *J. effuses*	Hauber Sukpa
56	*Leonotis nepetifolia* (L.) R.Br.	Hajurchei
57	*Lilium polyphyllum* D. Don	Kakoli
58	*Lycopodium clavatum* L.	Club Moss, Bendarali
59	*Mallotus philippensis* (Lam.) Mull. -Arg.	Kameela
60	*Marsdenia tenacissima* (Roxb.) Moon	Nishod sufed
61	*Mesua ferrea* L., syn. *M. nagassarium*	Nagakesari, Sirunagappo
62	*Microstylis muscifera* (Lindl.) Ridl., syn. *Malaxis muscifera*	Jivaka
63	*Microstylis wallichii* Lindl.	Jeevak
64	*Mucuna pruriens* (L.) DC., syn. *M. prurita M. pruriens* var. *utilis*	Kavach
65	*Nardostachys jatamansi* (D. Don) DC. *N. grandiflora*	Balchad, Jatamansi
66	*Nilgirianthus ciliates* (Nees) Bremek.	Kurinji, Karvi
67	*Onosma bracteatum* Wall.	Gule-Gazbaan
68	*Onosma hispidum* Wall. ex G.Don	Ratanjot

S.No.	Botanical name	Trade name
69	*Operculina turpethum* (L.) J. Silva Manso, syn. *Merremia turpethum* (L.) Shah & Bhatt.	Nishoth
70	*Orchis latifolia* L.	Salam misri
71	*Oroxylum indicum* (L.) Benth. ex Kurz	Tetuchaal
72	*Paederia foetida* L.	Prasaarani
73	*Pelargonium graveolens* L. Heritt.	Rose geranium (introduction)
74	*Phaseolus trilobus* (L.) Aiton, syn. *Vigna trilobata*	Jangali Moong
75	*Phyllanthus amarus* Schumach. & Thonn.	Bhumiamla
76	*Pimpinella anisum* L.	Ajamoda, Anise
77	*Picrorhiza kurroa* Royle ex Benth.	Buinowla, Kutki
78	*Piper chaba* Hunter	Kabab chini, Chavya, Gajpippal
79	*Pistacia integerrima* Stew. ex Brand., syn. *P. chinensis* var. *integerrima*	Kakarsinghi
80	*Pluchea lanceolata* (DC.) Oliv. & Hiern	Rasna
81	*Plumbago zeylanica* L.	Chitrak,
82	*Podophyllum hexandrum* Royle, syn. *P. emodi*	Bankakri, Himalayan mayapple
83	*Polygonatum cirrhifolium* (Wall.) Royle	Meda-m-meda
84	*Polygonatum verticillatum* (L.) All.	Meda-m-meda
85	*Prosopis cineraria* (L.) Druce, syn. *P. spicigera*	Jhand, Chonkar
86	*Prunus cerasoides* D. Don	Himalayan cherry
87	*Psoralea corylifolia* L.	Baabchi
88	*Pterocarpus marsupium* Roxb.	Damul- akhwain, Vijaysar
89	*Pterocarpus santalinus* L. f., and *P. dalbergioides*	Lal Chandan, Sandal Surkh
90	*Pueraria tuberosa* (Roxb. ex Willd.) DC.	Vidhari kand, Vidari
91	*Rheum australe* D. Don., *R. emodi* wall. ex Meissn	Revan chini
92	*Rubia cordifolia* L.	Maddar, Manjitti, Majith, b-Tsod
93	*Salvadora persica* L.	Goni, Miswaak
94	*Saraca indica* L., syn. *S. asoca*	Ashoka
95	*Sarcostemma viminale* (L.) R.Br., syn. *S. acidum S. brevistigma*	Soma
96	*Scindapsus officinalis* (Roxb.) Schott.	Gaj pipal
97	*Semecarpus anacardium* L. f.	Balave, Bhilavan
98	*Sisymbrium irio* L.	Khubkalan
99	*Solanum indicum* L.	Barahantaa, Katkari with bigger fruit
100	*Solanum nigrum* L.	Makoi, Kakamachi
101	*Solanum virginianum* L., syn. *surattense* Burm,f.; *S. xanthocarpum* Schard. & Wendl.	Kateli
102	*Sphaeranthus indicus* L.	Gorak-mundi

S.No.	Botanical name	Trade name
103	*Stereospermum chelonoides* (L. f.) DC., syn. *personatum H., S. suaveolens* DC. (species name in italics)	Patala, Padal
104	*Streblus asper* Lour.	Bajradanti
105	*Strychnos nux-vomica* L.	Kuchla, Itti beeja, Nux Vomica
106	*Strychnos potatorum* L. f.	Nirmali, Nux Vomica, Cleaning nuts
107	*Swertia chirayita* (Roxb. ex Fleming) H. Karst., syn. *S. chirata* Buch.-Ham.	Chiraiyata
108	*Symplocos racemosa* Roxb.	Pathani lodh
109	*Tecomella undulata* (Sm.) Seem.	Rohitaka
110	*Terminalia arjuna* (Roxb. Ex DC.) Wight & Arn.	Arjun
111	*Terminalia chebula* Retz.	Harda, Himaj
112	*Tinospora cordifolia* (Willd.) Miers ex Hook. f. & Thomson	Giloy
113	*Tinospora sinensis* (Lour.) Merr.	Amrta, Gilai (Giloy substitute)
114	*Tribulus terrestris* L., *T. lanuginosus* Linnaeus	Gokhru
115	*Tylophora indica* (Burm. f.) Merr., syn. *T. asthmatica*	Antamul, Ipecacuahna
116	*Uraria picta* (Jacq.) DC.	Oripai
117	*Vateria indica* L.	Mandadhupa, Dupa
118	*Viola* spp.	Banafasha
119	*Zanthoxylum armatum* DC., syn. *Z. alatum* Roxb.	Tejbal, Tejyovathi

Based on National Medicinal Plant Board, ISM & H, New Delhi, 2000 report and Pareek *et al.*, 2005

References

Anonymous (2000) National Medicinal Plant Board, [Indian systems of medicine and homoeopathy (ISM&H)] Report of the Task Force on Conservation and Sustainable Use of Medicinal Plants. Planning Commission, Government of India.

FRLHT (Foundation for Revitalisation of Local Health Traditions) Encyclopaedic Database on Indian Medicinal Plants.

Nayar, MP and Sastry, ARK. (1987). Red data Book of Indian Plants, Botanical Survey of India, Calcutta.

Nayar, MP and Sastry, ARK (1990). Red Data Book of Indian Plants pp 271. Volume 3. Calcutta: Botanical Survey of India.

Pareek, SK, Gupta, Veena, Bhatt, KC, Negi, KS and Sharma, N. (2005) Medicinal and Aromatic Plants. In: Plant Genetic Resources: Horticulture Crops (Eds. Dhillon, BS, Tyagi, RK, Saxena, S and Randhawa, GJ) pp 279-308. Narosa Publishing House Pvt. Ltd., New Delhi, India

Schippmann, Uwe, Leaman Danna, and Cunningham, AB. (2006) A Comparison of Cultivation and Wild Collection of Medicinal and Aromatic Plants Under Sustainability Aspects. In: Medicinal and Aromatic Plants, (Eds. Bogers RJ, Craker, LE and Lange, D) pp 75-95. Springer. Printed in the Netherlands.

Singh, Anurudh K. (2015) Agricultural Biodiversity Heritage Sites and Systems in India pp 467. Asian Agri-History Research Foundation, Secunderabad, Telangana, India, ISBN 81-903963-4-X

Ved DK, and Goraya GS. (2007) Demand and Supply of Medicinal Plants in India. NMPB, New Delhi and FRLHT, Bangalore, India.

7

Inventory of Important Medicinal and Aromatic Plants (MAPs) Species Habitat to Different Biogeographic Regions of India

7.1 Introduction

The occurrence of different medicinal and aromatic plant species is often associated with a specific ecology and thereby, to a biogeographical or phytogeographical region of the Indian Subcontinent. The Indian Subcontinent is a confluence of at least two biogeographical realms, i.e., Palearctic or Palaearctic (Himalayas) and Oriental (Sino-Indian- mainland India to much of insular the Southeast Asia). The most acknowledged classification of biogeographical zones was proposed by Udvardy (1975), which recognized 12 biogeographical provinces in India falling in these two realms. Based on varied climate, terrain, and faunistic information, Roger and Panwar (1988) tried to distinguish India into 10 distinct bio-geographical regions. Their classification did account for distribution of the plant-communities because they provided information about key components to identify the probable habitat for the presence of animals. An attempt for phytogeographical zonation, based on type of vegetation by Gadgil and Meher Homji (1990), distinguished 16 phytogeographical zones. Rodgers *et al.* (2002) further revised their biogeographical maps using geographical information system (GIS) techniques. They distinguished 10 biogeographical zones and 26 biotic provinces. This classification included diversity in various factors such as altitude, moisture, topography, rainfall, etc.

Since most medicinal and aromatic plants are procured/harvested from nature, the description of their natural distribution/adaptation in various biogeographical region is important from source of raw material point of view. This is predominantly important to understand the habitat of these plant species in relation to the commonly agreed 10 biogeographic zones of India to provide information about their natural adaption to promote conservation through cultivation and use. Considering the variation existing in altitude in the Eastern Ghats, impacting the forest and vegetation types as per altitude, Singh (2017) has proposed further classification of the Deccan Peninsula, which comprises of Central highlands and Deccan Plateau (Peninsula) (Rodgers and Panwar, 1988) into Peninsular India and Eastern Ghats (Fig. 7.1).

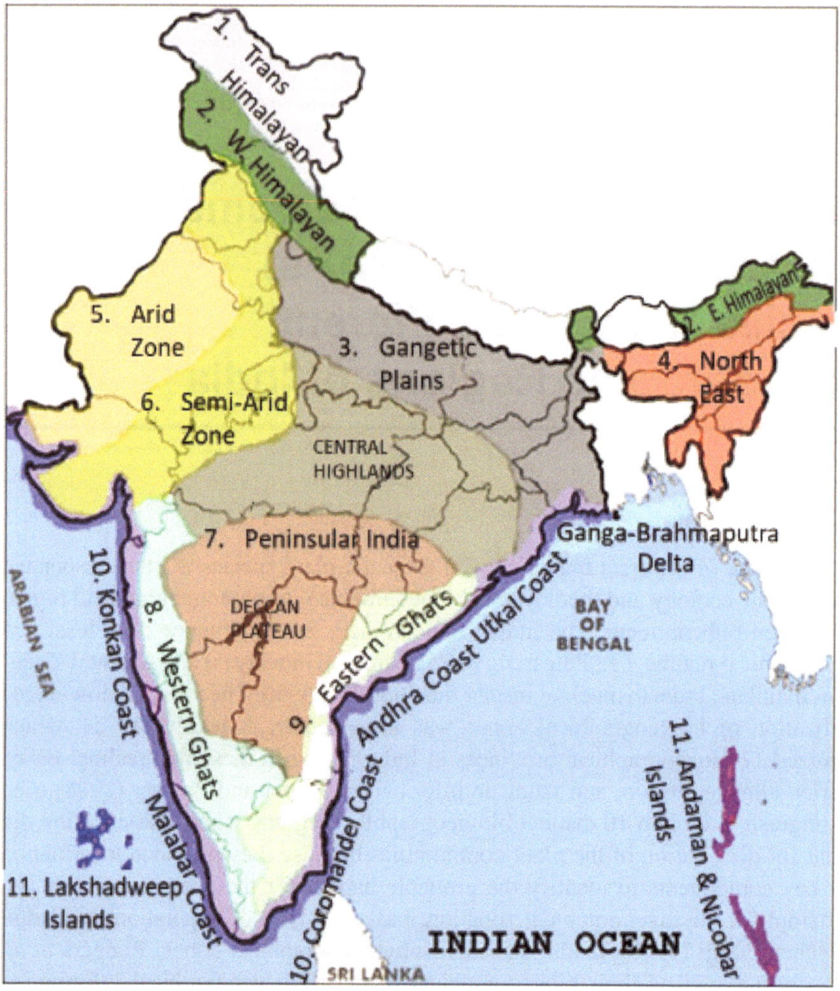

Fig. 7.1: Biogeographical regions of India as impacted by the biodiversity, including plant diversity (improved up on Rodgers and Panwar, 1988)

Endemism is the state of a species being native to a single defined geographic location or defined zone of a state or nation, i.e., organisms that are indigenous. As per Nayar (1996), 5,725 species are endemic to India. These have risen to 6,000 as per the recent survey (Goyal and Arora, 2009). Of these 3,471 species are distributed in the Himalayas, 2,051 species confined to the peninsular region, and around 239 species to the Andaman and Nicobar Islands (Nayar, 1996). Nayar (1996) identified three mega centres and 25 micro centres of endemism in India. The three mega centres of endemism in India are the Western and the Eastern Himalayas, the Northeastern Indo-Myanmar region, and the Western Ghats. The Central Himalaya not only includes the fringe endemic species of Eastern and Western Himalayas but is a crucible of speciation.

Trans Himalayan region, because of being the part of extreme ecology with severe stresses, presents a unique floristic diversity, which has been the basis of Tibetan medicinal system, Amchi or Sowa Rigpa. Sowa Rigpa is derived from Bhoti or Bodhi language, which means 'Knowledge of Healing'. In addition to known medicinal plant species, the Himalayan region (including trans Himalayan) in-houses many medicinal potential plant species used by locals, which need scientific scrutiny for wider and regular use. The Himalayan region floristic diversity has been further enriched with extended distribution of many plant species including medicinal and aromatic plants from Southern Europe, Mediterranean and Central Asia.

The Northeast region, particularly the hill region, one of the biodiversity Hotspots of the Globe (Mittermeier *et al.*, 2004; Upadhaya *et al.*, 2012), occupies 7.7 per cent of India's total geographic area and supports about 50 per cent of the biodiversity in the country with high concentration of endemism (31.58 per cent). It represents the transitional zone between India, Indo-Myanmar-Malayan, and Indo-China regions. It inhabits around 225 tribes using plant species for various purposes including medicine. Chatterjee (1940) estimated around 3,196 species, which are endemic to the region. Over exploitation of medicinal plants from wild in this region has depleted the populations of many plant species.

Peninsular India is part of the Gondwanaland landmass with geological lineage of great antiquity. It has a high degree of endemism and is the second richest endemic centre after Himalayas. Most of the endemic species are paleoendemic having found favourable ecological niches in hill ranges and the Western and Eastern Ghats. The moist evergreen forests of the Western Ghats, due to varied topography and microclimatic regimes, have some areas of active speciation. The Southern Western Ghats and Northern Western Ghats have been considered mega centres of endemism. Nayar (1996) estimated around 2015 endemic flowering species in Peninsular India representing around 12% of the Indian flora.

The Indian Arid zone (Thar desert of Rajasthan and parts of Gujarat) is one of the 12 biogeographical provinces of India (Rodgers and Panwar, 1988), and one of 16 phytogeographical zones classified by Gadgil and Meher-Homji (1990). Most of the region consists of dry undulating plain of hardened sand and the remaining area is largely a rolling plain of loose sand that form shifting sand dunes. The region presents extreme arid harsh environmental conditions and climate, because of which the region is referred as *maroosthali* in Hindi (meaning region of death). Species diversity is not extraordinarily rich in the region with only around 10 per cent of endemic species. Besides, the cosmopolitan and tropical species, it has predominance of western elements, which goes up to 65.8% and eastern elements that are up to 34.2%. There is an overall dominance of the African element, 37.1%, as compared to the Oriental elements, which are only 20.6%. Dominant numbers of species occurring in the region are from west (Arabian, African), followed by Indian subcontinent (Hindustani) and east (Indo-Malayan). Singh (2004) identified 84 taxa from the region with economic potential, including medicine.

A perusal of distribution of medicinal and aromatic plants reflects that many species are widely adapted and are distributed throughout the subcontinent under specific habitat with greater diversity overlapping over various biogeographic regions. There are species habitat to specific ecologies and are endemic to a specific biogeographical region, and many more to a very narrow habitat, confining to specific biotic province.

References

Chatterjee, D. (1940) Studies on endemic flora of India and Burma. Journal of Asiatic Society of Bengal Science 5:19-97.

Gadgil, M and Meher-Homji, VM. (1990) "Ecological diversity" In: Conservation in Developing Countries: Problems and Prospects (Eds: Daniel JC, Serrao JS) pp. 175–198. Bombay Natural History Society. Oxford University Press, Delhi.

Goyal, AK and Arora, S. (2009) Chapter 1. Overview of biodiversity status, trends and threats. In: India's Fourth National Report to the Convention on Biological Diversity pp 15-53, Ministry of Environment and Forest, Government of India, New Delhi.

Mittermeier, RA, Robles, Gil P, Hoffmann, M, Pilgrim, J, Brooks, T, Mittermeier, CG, Lamoreux, J, da Fonseca, GAB (2004) Hotspots revisited: Earth's biologically richest and most endangered ecoregions. CEMEX, Mexico City, Mexico.

Nayar, MP. (1996) Hot spots of endemic plants of India, Nepal and Bhutan pp 253. Tropical Botanic Garden and Research Institute, Palode, Thiruvananthapuram, Kerala, India.

Rodgers, WA, Panwar, HS and Mathur VB. (2002) Executive summary. In: Wildlife Protected Area Network in India: A Review p 44, Wildlife Institute of India, Dehra Dun, India.

Rodgers, WA and Panwar, HS. (1988) Planning a Wildlife Protected Area Network in India. Vol. 1 and 2. A report prepared for the Department of Environment, Forests and Wildlife, Government of India at the Wildlife Institute of India, Dehra Dun, India

Singh, AK. (2004) Endangered Economic Species of Indian Desert. Genetic Resources and Crop Evolution 51(4): 371-380.

Singh, Anurudh K. (2017) Revisiting the Status of Cultivated Plant Species Agrobiodiversity in India: An Overview. Proc Indian Natn Sci Acad 83 (1):151-174. (DOI: 10.16943/ptinsa/2016/v82/48406)

Udvardy, MDF. (1975) A Classification of the Biogeographical Provinces of the World. ICUN Occasional Paper No. 18, p 49, Morges, Switerzerland.

Upadhaya, K, Choudhary, H and Odyuo, N. (2012) Florestic diversity of northeast India and its conservation In: International Environmental Economics: Biodiversity and Ecology (Ed. Mandal, RK) pp 61-71. Discovery Publishing House Pvt. Ltd., New Delhi. ISBN: 978-93-5056-152-2

7.2 List of Important Medicinal and Aromatic (MAPs) Found in Different Biogeographical Regions of India

An inventory of important medicinal and aromatic plant species is required to facilitate collection and conservation activities on priority species, promoting use. Botanical Survey of India is doing this work but needs coordination and involvement of other stakeholders to produce a effective list. An analysis regarding the distribution of medicinal plants shows they are distributed across diverse habitats and landscapes across the country in different ecosystems, predominantly, Himalayas, East and Northeast of India, Gangetic plains, Indian arid zone stretching from Thar Desert to Kutch, Semi-arid zone extending from Punjab to Gujarat, Indian Peninsula, and the

Western Ghats. In the list presented below, an attempt has been made to list some important species. Nevertheless, the distribution of medicinal and aromatic plant species needs regular update through survey involving all stakeholders to understand the changing pattern, which have been occurring at a very fast pace, because of various developmental activities occurring all around, causing shift in species distribution and composition. Most of the medicinal plants in the list belong to higher flowering angiosperms, but several lower plant groups like algae, fungi, lichens, bryophytes, pteridophytes etc., needs attention and inclusion.

Important MAPs Found in Different Biogeographical Regions of India

Bio-geographic Zones	Important representative medicinal plant species	Total species identified[1]
1. Trans Himalayan region	*Aconitum deinorrhizum, Allium carolnianum, Aquilegia fragrans, Arnebia euchroma, A. guttata, Artemissia* spp., *Aster tibeticus, Astragalus* spp., *Berberis ilicinia, Bergenia stracheyi, Biebersteinia odora, Capparis spinosa, Carum carvi, Clematis tibetana, Cirsium wallichii, Corydalis govaniana, Cousinia falconeri, Cremanthodium reniforme, Delphinium* spp., *Echinops cornigerus, Ephedra gerardiana, Ferula jaeschkeana; Gentianella paludosa, Hippophae rhamnoides, Hyoscyamus niger, Inula* spp.- *I. obtusifolia, I. royleana, I. racemosa; Juniperus macropoda, J. excelsa, Jurine macrocephala, Lactuca lessertiana, Mentha longifolia, Myricaria squamosa, Nepeta floccose, Orchis latifolia; Picrorhiza kurroa, Podophyllum hexandrum, Physochlania praalta, Plantago depressa, Potentilla anserina, Saussurea* spp.- *S. bracteate, S. lappa, S. sacra, Viola biflora,*	700
2. Himalayan region (Western, Central and Eastern Himalayas)	*Achillea mllefolium, Aconitum* spp.- *A. chasmanthus, A. ferox, A. heterophyllum, A. lethale, A. violaceum,* etc., *Allium* spp.- *A. stracheyi, Alpinia galanga, Angelica glauca, A. euchroma , Arnebia benthamii, Asparagus adscendens, Atropa acuminata, Baliospermum montanum, Berberis aristata, B. asiatica, B. tinctoria, Bunium persicum, Centella asiatica, Chlorophytum arundinaceum, Cinchona officinalis, Cinnamomum* spp.- *C. camphora, C. cassia, C. tamala, Coptis teeta, Costus speciosus, Curculigo orchioides, Curcuma amada, Dactylorhiza hatagirea, Delphinium denudatum, Dioscorea deltoidea, D. prazeri, Ephedra gerardiana, Eurya arunachalensis, Gentiana kurroo, Holarrhena antidysenterica, Inula racemosa, Ocimum* spp.- *O. basilicum, O. canum, O. gratissimum, O. sanctum, Onosma bracteatum, Panax sikkimensis, Picrohiza kurroa, Piper longum, Pistacia chinensis,*	2900

Bio-geographic Zones	**Important representative medicinal plant species**	**Total species identified[1]**
	Podophyllum hexandrum, P sikkimense, Polygonatum cirrhifolium, Potentilla fulgens, Nardostachys grandiflora (N. jatamansii), Rauvolfia serpentina, Rheum spp.- *Rheum australe, R. palmatum, R. spiciforme, Rhododendron* spp.- *R. anthopogon, R. formosum, R. santapaui, Rubia manjith, R. sikkimensis, Saussurea* spp.- *S. costus (S. lappa), Senna sophera, Swertia chirayata, Terminalia chebula, Thymus serpyllum, Taxus baccata, T. baccata* ssp. *wallichiana, Thymus serphyllum, Valeriana jatamansi*	
3. Northeast (Assam and Northeast Hills)	*Achyranthes aspera, Acorus calamus, Aegle marmelos, Alpinia galanga, Ambroma augusta, Aquilaria agallocha (Aquilaria malaccensis), Arisaema jacquemontii, Aristolochia bracteata, A. tagala, Asparagus racemosus, Baliospermum micranthum, Berberis* spp.- *B. lycium, B. micropetala, B. wardii, Butea monosperma, Canarium strictum, Catharanthus roseus, Centella asiatica, Cinnamomun bejolghota, C. tomala, Cissus quadrangularis (Vitis quadrangularis), Clerodendrum colebrookianum, C. serratum, Costus speciosus, Croton tiglium, Curculigo orchioides, Curcuma* spp.- *C. amada, C. aromatica, C. caesia, C. longa, C. zeodoaria, Cuscuta reflexa, Cymbopogon* spp.- C. *jwarancusa, Dioscorea alata, Emblica officinalis, Eryngium foetidum, Eugenia jambolana (Syzygium cumini), Fagopyrum dibotrys, Garcina* spp.-, *Gloriosa superba, Gynocardia odorata, Hedychium coronarium, H. spicatum, Hodgsonia macrocarpa (Trichosanthes macrocarpa), Holarrhena antidysenterica, Houttuynia cordata, Hydnocarpus kurzii, Litsea cubeba, Lycopodium clavatum, Mesua ferrea, Morinda angustifolia, Mucuna pruriens, Myrica esculenta, Nepenthes khasiana, Ocimum* spp.- *O. sanctum, Panax ginseng, P. pseudoginseng, Paederia foetida, Phlogacanthus thyrsiflorus*, Piper spp.- *P. mullesua (P. brachystachyum), P. cubeba, P. longum, P. pepuloides, P. nigrum, Plantago ovata, Plumbago zeylanica, Rauvolfia serpentina, Rhus javanica (R. semialata), Saraca indica, Scroporia dulcis, Smilax chinensis, S. glabra, S. lanceifolia, Solanum* spp.- *S. khasianum, Strychnos nux-vomica, Swertia* spp.- *S. charayita, S. paniculata, S. macrosperma, Taxus baccata, (T. baccata)*, L. var. *wallichiana, Terminalia* spp. T. *chebula, Tinospora cordifolia, Vetivaria zizaniodes, Viburnum foetidum, Wedelia calandulacea, Zanthoxylum armatum, Z. rhetsa* etc.	2000

Bio-geographic Zones	Important representative medicinal plant species	Total species identified[1]
4. Gangetic plain	*Acanthus ilicifolius*, *Acorus calamus*, *Adhatoda zeylanica*, *Aegle marmelos, Alangium salviifolium*, *Aloe barbadensis*, *Andrographis paniculata*, *Aristolochia indica, Asparagus curillus*, *A. racemosus*, *A. sarmentosus*, *Cassia fistula, Catharanthus roseus*, *Celastrus paniculatus*, *Chlorophytum borivillianum*, *Clausena excavata, Crateaeva nurvala, Curcuma* spp- *C. amada*, *C. angustifolia*, *C. aromatica C. zedoaria*, *Cymbopogon* spp- *C. flexuosus*, *C. martini*, *C. nardus*, *C. winterianus*, *Cyprus scariosus*, *Datura stramonium*, *Desmodium gangeticum*, *Digera muricata*, *Enhydra fluctuans*, *Eupatorium triplinerve* (*ayapana*), *Feronia elephantum, Gloriosa superba*, *Hemidesmus indicus*, *Holarrhenaq pubescens*, *Lactuca intybacea* syn. *L. runcinata* (*L. remotiflora*), *Mallotus philippensis, Mentha arvensis*, *M. piperita*, *Mucuna pruriens* (*M. deeringiana*), *Murraya koenighii, Nardostachys grandiflora*, *Nelumbo nucifera, Ocimum sanctum*, *Oroxylum indicum*, *Pandanus fascicularis*, *Papaver somniferum*, *Peganum harmala*, *Phyllanthus fraternus*, *Piper longum, Pluchea lanceolata, Plumbago zeylanica*, *Pogostemon cablin*, *Psoralea corylifolia* (*Cullen corylifolium*), *Pterocarpus marsupium*, *Rauvolfia serpentina*, *Ricinus communis*, *Rosa damascene*, *Rubia cordifolia* (manjishtha), *Sesbania cannabina*, *Soymida febrifuga*, *Strychnos nux-vomica*, *Swertia chirata*, *Terminalia arjuna*, *T. bellirica*, *Tinospora cordifolia, Trichosanthes tricuspidate*, *Trigonella corniculata*, *Tylophora indica*, *Urginea indica*, *Withania somnifera*, *Ziziphus jujuba.*	1000
5. Arid Zone (Thar Desert, Parts of Gujarat)	*Abutilon indicum*, *Acacia nilotica*, *Achyranthes aspera*, *Alhagi camelorum, A. maurorum, Aloe barbadensis*, *Anticharis glandulosa* var. *caerulea*, *Argemone mexicana*, *Balanites* spp.- *B. aegyptiaca, Barleria acanthoides*, *B. prionitis* var. *dicantha*, *Blepharis sindica*, *Boerhavia diffusa*, *Boswellia serrata*, *Cadaba fruticose*, *Calligonum polygonoides*, *Calotropis* spp.- *C. procera*, *C. gigantca*, *Capparis decidua*, *Cassia angustifolia*, *C. auriculata*, *Cissus quadrangularis*, *Citrullus colocynthis, Cleome vahliana*, *Clitoria ternatea*, *Coleus amboinicus*, *Commiphora wightii, C. caudata*, *Convolvulus auricomus* var. *volubilis* and var. *ferruginosus*, *C. microphyllus*, *Cordia gharaf*, *Cressa cretica*, *Crotolaria burhia*, *Diospyros* spp., *Eclipta prostrata*, *Fagonia cretica*, *Gloriosa superba*, *Grewia tenax*,	500

Bio-geographic Zones	Important representative medicinal plant species	Total species identified[1]
	Guaiacum officinale, Adhatoda zeylanica (Justicia adhatoda), Leonotis nepatefolia, Leptadenia pyrotechnica, Pedalium murex, Peganum harmala, Plantago ovata, Prosopis cineraria, Psoralea corylifolia, Salvadora oleoides, S. persica, Senna auriculata, Tecomella undulata, Tribulus terrestris, T. rajasthanensis, Withania coagulens, W. somnifera, Ziziphus nummularia, Zygophyllum simplex, etc.	
6. Semi-arid Zone	*Acacia nilotica, Acorus calamus, Adhatoda vasica, Aloe barbadensis, Alstonia scholaris, Andrographis peniculata, Asparagus racemose, Azadirachta indica (Melia azadirachta),, Balanites aegyptiaca, Baliospermum montanum, Boerhavia diffusa, Bombax ceiba, Boswellia serrata, Bryophyllum pinnatum (Kalanchoe pinnata), Caesalpinia bonduc, Calotropis procera, Cannabis sativa, Canscora decussata, Capparis decidua, Cassia italica, Celastrus paniculatus, Centella asiatica, Chlorophytum borivilianum, Cissus quadrangularis, Clitoria ternatea* var. *pilosula, Coleus amboinicus, Commiphora wightii, Crataeva religiosa (C. nurvala), Curculigo orchiodes, Cymbopogon citratus, C. martini, Embelia tsjeriam-cottam, Emblica officinalis, Ficus religiosa, Garuga pinnata, Helicteres isora, Holarrhena antidysenterica, Leonotis nepetifolia, Martynia annua, Myristica fragrans, Pedalium murex, Phyllanthus reticulatus (Kirganelia reticulata), Plumbago zeylanica, Pongamia pinnata, Psoralea corylifolia, Strychnos nux-vomica, Terminalia* spp.- *T. arjuna, T. bellirica, T. chebula, Tinospora cordifolia, Tribulus alatus, Tribulus terrestris, Vetiveria zizanoides, Vitex negundo, Withania somnifera, Woodfordia fruticosa*	1000

Bio-geographic Zones	Important representative medicinal plant species	Total species identified[1]
7. Eastern Ghats	*Abrus precatorius*, *Aegle marmelos*, *Aganosma dichotoma*, *Alpinia galanga*, *Alstonia scholaris*, *Andrographis paniculata*, *Anthocephalus cadamba*, *Argyreia speciosa*, *Azadirachta indica*, *Azima tetracantha*, *Bauhinia variegata*, *Bixa orellana*, *Bridelia montana*, *Bryonopsis laciniosa*, *Buchanania lanzan*, *Butea monosperma*, *Caesalpinia bonduc* (*C. bonducella*), *Cassia fistula*, *Catharanthus roseus*, *Celastrus paniculatus*, *Centella asiatica*, *Cissampelos pareira*, *Cissus quadrangularis*, *Clerodendrum serratum*, *Clitoria ternatea*, *Cocculus hirsutus*, *Costus speciosus*, *Costus igneus* (introduced), *Curculigo orchioides*, *Cymbopogon flexuosus*, *C. martini*, *Decalepis hamiltonii*, *Elaeocarpus sphaericus*, *Emblica officinalis* (*Phyllanthus emblica*), *Garuga pinnata*, *Gloriosa superba*, *Gymnema sylvestris*, *Helicteres isora, Hemidesmus indicus*, *Hypericum* spp., *Lawsonia inermis*, *Leptadenia reticulata*, *Madhuca indica*, *Mallotus philippensis*, *Morinda citrifolia*, *M. pubescens*, *Mucuna monosperma*, *M. pruriens*, *Ocimum tenuiflorum* (*O. sanctum*), *Oroxylum indicum*, *Phyllanthus amarus*, *Piper longum*, *Plumbago zeylanica*, *Pogostemon* spp., *Pterocarpus marsupium*, *P. santalinus*, *Pueraria tuberosa*, *Rauvolfia serpentina*, *Rhynchosia beddomei*, *Rubia cordifolia*, *Santalum album*, *Sarcostemma viminale* (*S. acidum*), *Solanum* spp., *Stephania hernandiifolia*, *Strychnos nux-vomica*, *Tacca pinnatifida* (*T. leontopetaloides*), *Terminalia* spp.- *T. alata*, *T. arjuna*, *T. bellirica*, *T. chebula*, *Tinospora cordifolia*, *Trichosanthes tricuspidata*, *Tylophora indica*, *Vanda* spp., *Vernonia cinerea*, *Vitex negundo*, *Withania somnifera*, *Woodfordia fruticosa*, *Wrightia tinctoria*, *Xyris indica*,	828 (Natrajan *et al.*, 2020)

Bio-geographic Zones	**Important representative medicinal plant species**	**Total species identified**[1]
8. Peninsular India (Central Highlands and Deccan Plateau/ Peninsula)	*Acorus calamus, Aegle marmelos, Adhatoda vasica, Agave vera-cruz, Aloe vera, Achyranthes aspera, Allium porrum, Alstonia scholaris, Andrographis paniculata, Aristolochia indica, Asparagus racemosus, Bacopa monnieri, Balanites aegyptiaca, Baliospermum montanum, Blumea eriantha, Boerhaavia diffusa, Bryonia laciniosa, Butea monosperma, Caesalpinia digyna, Canthium parviflorum, Cassia tora, Celastrus paniculatus, Chlorophytum* spp.- *C. borivillianum C. tuberosum, Cissus quadrangularis, Clerodendron serratum, Corchorus depressus, Costus speciosus, Cruculigo orchioides, Cucurma amada, Cymbopogon martini, Cynodon dactylon, Decalepis hamiltonii, Eclipta alba, Embelia tsjeriam-cottam, Emblica officinalis, Eriolaena hookeriana, Evolvulus alsinoides, Gloriosa superba, Gymnema sylvestre, Helicteres isora, Hemidesmus indicus, Hibiscus cannabinus, H. sabdariffa, Imperata cylindrica, Laptadenia reticulata, Leucas aspersa, Phaseolus trilobus, Phyllanthus amara, P. niruri, Piper longum, Plumbago zeylanica, Pterocarpus marsupium, P. santalinus. Psoralea corylifolia, Rauvolfia serpentina, Santalum album, Schrebera sweitenoides, Semecarpus anacardium, Smilax ovalifolia (S. macrophylla), Spilanthes acmella, Strychnos nux-vomica, Terminalia* spp.- *T. arjuna, T. bellirica, T. chebula, Tinospora cordifolia, Tylophora indica, Ventilago madersuspatana, Vitex negundo, Withania somnifera, Woodfordia fruticose, Wrightia tinctoria*	3000
9. Western Ghats	*Achyranthes aspera, Aloe barbadense (A. vera), Amomum aromaticum, Andrographis echioide, A. paniculata, , Anisochilus carnosus, Asparagus racemosus var. javanicus, Bacopa monnieri, Bauhinia variegata, Boehmeria malabarica, Centella asiatica, Ceropegia hirsuta, Chlorophytum spp.- C. tuberosum, C. borivilianum, Cinnamomum tamala, Cissus quadrangularis , C. repanda, Citrus aurantifolia, Coleus amboinicus, Coscinium fenestratum, Curcuma spp., C. amada, Cymbopogon spp.- C. citratus, C. flexuosus var. coimbatorensis, C. martini, C. nardus, Desmodium gangeticum, Elettaria cardamomum, Garcinia indica , Gymnema sylvestre, Helminthostachys zeylanica, Hemidesmus indicus, Hydnocarpus pentandrus, Kaempferia galanga, Lobelia nicotianifolia, Myristica malabarcia, Nilgirianthus ciliatus, Ocimum tenuiflorum*	2000

Bio-geographic Zones	Important representative medicinal plant species	Total species identified[1]
9. Western Ghats	*(O. sanctum), Pergularia daemia, Piper spp.- P. niger, P. longum, P. betel, Rauvolfia serpentina, Ruta graveolens, Saraca asoca, Terminalia spp.- T.arjuna, T. bellirica, T. catappa, T. chebula, Tinospora cordifolia, Tylophora indica, Strychnos nux-vomica, Vateria indica, Vetiveria zizanioides (Andropogon zizanioides), Vitex negundo, Zingiber spp.- (Z. officinale)*	2000
10. Coastal region	*Acanthus ilicifolius, Aglaia roxburghiana, Albizia lebbeck, Alstonia scholaris, Ardisia solanacea, Areca catechu, Avicennia marina, Cassia fistula, Casuarina equisetifolia, Cocos nucifera, Dischidia bengalensis, Excoecaria agallocha, Gloriosa superba, Glycosmis pentaphylla, Ichnocarpus frutescens, Mucuna pruriens, Myristica elliptica, Pandanus fascicularis, (P. odoratissimus, P. odorifer), Phoenix paludosa, Piper longum, Rhizophora mucronata, Sansevieria hyacinthoides, S. roxburghiana, Saraca indica (S. asoca), Sonneratia caseolaris. Terminalia bialata, Vitex trifolia, Vetiveria zizanioides, Wedelia biflora, Zanthoxylum armatum*	500
11. Island region (Andaman and Nicobar, Lakshadweep Islands)	*Acanthus ilicifolius, Adenanthera pavonina, Aisandra butyrace (Diploknema butyracea). Amomum fenzlii, Aphanamixis polystachya, Asparagus racemosus, Barringtonia asiatica, Caesalpinia digyna, Claophyllum inophyllum, Delima scandens, Dioscorea esculenta, Dipterocarpus turbinatus, Dryobalanops aromatica, Geophila reniformis, Morinda citrifolia, Mucuna monosperma, Mussaenda frondosa, Ophiorrhiza mungos, Pandanus fascicularis, Phoenix paludosa, Premna corymbosa, Randia dumetorum, Sandoricum koetjape, Terminalia bialata, Zanonia indica,*	1000

References/ Source of Information

Lakshman, CD. (2016) Biodiversity and conservation of medicinal and aromatic plants. Advances in Plants & Agriculture Research 5(4):561-566. DOI: 10.15406/apar. 2016.05.00186

Natrajan, S, Venkateswaran, Kamla, Rao, PS, Reddy, TM and Rajasekharan, PE. (2020) Threatened Medicinal Plants of Eastern Ghats and Their Conservation In: Conservation and Utilization of Threatened Medicinal Plants. Publisher: Springer Nature Switzerland. AG DOI: 10.1007/978-3-030-39793-7_2

Singh, Anurudh K. (2015) Agricultural Biodiversity Heritage Sites and Systems in India pp 467. Asian Agri-History Research Foundation, Secunderabad, Telangana, India. ISBN 81-903963-4-X

8

Future Perspective

8.1 Introduction

Considering increased awareness and interest in using herbal products because of their safer nature, the world market of traditional medicinal and aromatic plants is growing at a very fast pace. The global herbal industry is projected to be worth 5 trillion US$ by 2050 (the world bank report, 2000). These resources not only provide primary health care, but also nutraceutical support for wellness. Also, it has been experienced that they may help in curing some of the deadly and painful diseases such as cancer, HIV, AIDS, rheumatism, arthritis etc. India, due to its rich heritage in use of medicinal and aromatic plants under various traditional medicinal systems, which are important source of information about plant properties and use, can become a major player in the global context and can increase its market share substantially. To achieve this, an effective, efficient, and vibrant program is required to cover various aspects of management, research, and development.

8.2 Management, Conservation, and Search for New and Alternative of Resources

To safeguard our MAPs wealth for long-term sustainable use in Indian health care system, to compete globally for their wider use worldwide, and to support wellness and health care programs, the future emphasis is required on:

A holistic approach for management and conservation of the rich heritage of medicinal aromatic plants available in India. It may include a systematic survey, identification and creation of an inventory and collection, characterization and evaluation of medicinal properties followed by conservation to facilitate utilization of these resources with sustainable production of plant-based products.

To provide due credit to contribution of local tribes and communities, documentation of associated indigenous knowledge must be one of the main components of these efforts. Documentation of indigenous knowledge regarding the use of plants species as medicine and validation of the claims through laboratory tests for handling of issues related with Intellectual Property Rights (IPR) shall help empowerment of tribes or communities for a fair and equitable benefit sharing of profits accrued from use and commercialization of such knowledge and material.

A large number of medicinal aromatic plants are used in the Indian Traditional Medicinal Systems and are still being collected from nature, threatening their existence. Therefore, *in situ* protection and development of agro techniques for their commercial cultivation should be another priority effort to ensure the availability of raw material and to reduce threat concerns.

Important MAPs may be incorporated in the cropping systems of agroforestry as an under crop, intercrop, cash crop, companion crop, plantation crop, orchards, or as a part of forestry program, etc. Considering concerns of the effect of edaphic factors on the raw material, emphasis must be on organic cultivation of MAPs as per the present global trends.

New and non-investigated species should be studied in detail for phenology, reproductive biology, and for their phylogenetic relationship with related and sister species through biosystematics investigations to identify alternative sources and genetic resources for genetic enhancement or genetic improvement of MAPs used as crops.

To facilitate *ex situ* conservation of genetic diversity of important MAPs, study on the seed storage behaviour of more and more medicinal plant species, for cost effective seed storage should be an area of attention. Development of protocols for *ex situ* conservation of explants in tissue culture, genomic DNA or DNA sequences as an alternative method of conservation should be other areas of research.

Development of essential basic DNA profile of base species for finger printing of valuable resources having importance from IPR point of view and use of biotechnology to address the unsolved problems related to the area of taxonomy and phylogeny may be another priority.

In case of traditionally cultivated, important MAPs, such as Amla (*Emblica officinalis*), Aswagandha (*Withania Somnifera*), Isabgol/Psyllium (*Plantago ovata*), Atees (*Aconitum heterophyllum*), Angurshafa (*Atropa belladonna*), Kutki (*Picrorhiza kurroa*), Kuth (*Saussurea lappa*), Trepatra (*Trifolium pratense*), Tagar (*Valeriana wallichii*), etc. the efforts can be initiated for genetic enhancement and improvement to increase the productivity of principal components. These efforts may include use of both conventional and molecular breeding.

In case of vegetatively propagated MAPs, efforts may also be initiated to use biotechnological approaches for controlled *in vitro* propagation to produce and harvest of the principal component/ compounds *in vitro*, ensuring the consistency and quality aspects.

8.3 Search for Hidden or Non-coded Knowledge about the use of Herbal Medicines Locally

In India, majority of the tribal and rural population is predominantly dependent on herbal medicine for wellness, health care and cure of various ailments and diseases provided by the local practitioners. In many cases, the information about the plants or crude drug or formulation used is a family secret and is not divulged because of economic reasons. The practitioners' do not keep any record, and the information is mainly passed on verbally from generation to generations. There is no national legislation to ensure sharing of this knowledge. There is a need to search for such knowledge and bring it into the public domain for larger human welfare with assurance to the knowledge holder on sharing of commercial benefits accrued from their use through available transparent legislations. In this regard, there have been some pioneering efforts at some organizations to collect and document such knowledge in partnership with provisions of fair and equitable benefit sharing accrued from commercialization of developed product. For example, Jawaharlal Nehru Tropical Botanic Garden and Research Institute (JNTBGRI) developed a benefit-sharing model through All India Coordinated Research Project on Ethnobiology (AICRPE) with Kani tribe of Kerala, on the use of the drug, *Jeevani*, extracted from plant, *Trichopus zeylanicus* Gaertn. ssp. *travancoricus* (Bedd Burkill *ex* Narayanan). This is a pioneering model, wherein the benefits accrued from the development of a product based on an ethnobotanical lead were shared with the holders of that traditional knowledge. Considering the significant outcome of this model in community empowerment, income generation and poverty eradication/alleviation of a tribal community, Pushpangadan was awarded with the UN-Equator Initiative Prize (under individual category) at the World Summit on Sustainable Development held in Johannesburg in August 2002. Patanjali Yog Peeth, Haridwar has also taken such initiative (personal communication) searching for such associated knowledge regarding the use of some herbs for certain ailments. The Kani model of equitable benefit sharing involving a medicinal plant and associated traditional knowledge could be adopted widely (George *et al.*, 2016). However, there is a need for a national mechanism that would encourage search for such hidden knowledge, facilitating wider use of traditional herbs/drugs and empowerment of knowledge holding communities.

World Health Organization (WHO) has shown great interest in documenting the use of medicinal plants with associated knowledge used by tribal communities in different parts of the world. Many countries have intensified their efforts in documenting the ethnomedicinal data on medicinal plants. Such efforts need to be extended with systematic scientific scrutiny for the claims of tribal or rural healers on use of Indian herbs. Once validated, these ethnomedical preparations can be used with or without improvisation and clinical trials, with due approval of the Central Drugs Standard Control Organization (CDSCO) under Directorate General of Health Services, Ministry of Health and Family Welfare, Government of India, the national regulatory body for pharmaceuticals and medical devices.

8.4 Market Development in Herbal Medicine Industry with Reference to Trade

There is great demand for herbal medicine in the developed as well as developing countries, because of their wider biological activities, higher margins of safety than the synthetic drugs, and lesser costs (Chaudhri, 1996). India is sitting on a gold mine of well-recorded and well-practiced knowledge of traditional herbal medicine. The basic requirements for entering global, market, particularly that of developed countries will include well-documented traditional use, standards followed, medicinal plants formulations free from pesticides, heavy metals, etc. Standardization must be based on chemical and activity profile, and safety and stability of byproduct/drug. Effective herbal drug development is possible only through the development of scientifically standardized herbal products. The health care systems are going to become more and more expensive, therefore, efforts should be made to develop technologies with an objective of essentially introducing and integrating herbal medicine system in our health care. There is an enormous scope/opportunity for this in India, to emerge as a major player in the global herbal-based products and medicines. Let us hope that drug manufactured in accordance with principles of Ayurveda, Amchi, Siddha and Unani will reach new horizons to make them the best in the world by maintaining the quality of the herbal drugs, which would ensure maintenance of efficacy and efficiency. If this is achieved, then, there is nothing to stop them from competing with the modern medicine and other traditional systems (Chinese) with added advantages of fewer side effects and lower costs. However, for this to happen, the following aspects will need attention.

- The country needs to develop an institution at the national level to keep watch on the new developments in the field of plant-based products at global level to facilitate development of appropriate strategies to gain the leadership in selective products having high demand.
- Development and promotion of marketing system involving farmers or cultivators, entrepreneurs, and industries for establishment of a regular market for medicinal plants. This would require a strong linkage among growers, buyers, researchers, and institutions associated with development and promotion. to complement the efforts of each other.
- The policies, plans and research and development activities should be demand driven and market oriented, which can address the present and future needs.

8.5 Research on Issues and Concerns

There have been limited clinical trials in case of traditional herbal medicines to determine their efficacy and safety like that of modern medicines, because they are often administered and recommended based on long-term experiences and standing cultural traditions. Being used under unorganized systems till recently, finding appropriate ways to conduct this type of research and trials have been an ongoing challenge. However, now with the establishment of the national system under the Ministry of health, these trials shall be a regular essential component parallel to the modern medicine system. Nevertheless, this lack of research has not impeded the use and popularity, at least at the national level. However, to promote wider global use, research, and trials to establish efficacy, efficiency and safety would be essential under a recognized and legalized system.

Although recently India has been successful in promoting its therapies with more research and science-based approach, but it still needs more extensive and evidence-based research to fulfil the market and IPR requirements of uniqueness, essential characteristics, and uniformity. In addition to MAPs, numerous nutraceutical combinations included in Indian practices have entered the international market through exploration of ethnopharmacological claims made by different traditional practitioners. But this needs further research in established modern medicinal science directions.

Herbal medicines that are part of the various Indian Traditional Medicinal Systems are locally well known and used by the majority population of rural and poor folks. However, considering their effectiveness in cure of some chronic ailments and to widen their usage by entering in the international market because of renewed interest in herbal medicines, there is a need to further research to popularize Indian MAPs at global level by bringing their medicinal properties into public and scientific domain with various levels of research and publications.

To address IPR issues, including that for patenting of the herbal products and drugs at national and global level, and market concerns of quality, uniformity of results, etc., there are several priority areas for research needing immediate attention, such as –

Development of information on essential characteristic through biochemical profile of commonly used important MAPs. Identification of principal components responsible for the medicinal properties claimed, molecular characterization of these component(s) and mode of action, efficacy, side effects, etc. starting from commonly used MAPs.

Market statistical research on Ayurvedic medicines purchased via the Internet shows that quality and contamination of Indian herbal products have been the main market concerns. Researchers at Boston University School of Medicine (BUSM) have found that one-fifth of the Indian-manufactured Ayurvedic medicines purchased via the Internet contain lead, mercury, or arsenic. In addition, the herbal remedies were also discovered to be contaminated with a total of four fungi (Saper *et al.*, 2004; Ernst, 2002). These concerns need redressal.

Although herbal medicines have been used in India for thousands of years, however, to enter the global market requires quality assurance, ensuring the efficiency and effectivity of the by-products and/or the drugs. Therefore, basic research programs need to be strengthened with focus on quality and consistency assurance, biosafety measures at the manufacturer and the national level following accepted standards.

To overcome contaminations from pesticide residues and heavy metals, there should be control measures to implement necessary standard operating procedure (SOP) starting from source of raw material from mature (wild) or cultivation. A SOP is a procedure specific to operations that describe the activities necessary to complete tasks in accordance with industry regulations, provincial laws or even just your own standards for running the business. It sets step-by-step instructions compiled by the concerned organization to help workers carry out routine operations. SOPs aim to achieve efficiency, quality output and uniformity in performance of the product, while reducing miscommunication and failure to comply with industry regulations.

Consistency in composition and in biological activity are essential requirements for the safe and effective use of therapeutic agents, affecting the trade in the market. Several factors can affect these aspects, starting from collection of raw material from natural habitat or from cultivation. It requires an effective protocol for collection of plant or plant part material ensuring appropriate physiological age and freedom from any infection of pest and pathogen. Similarly, the cultivated MAPs, either as monocrops or as part of a cropping system, should be preferably under organic farming system to ensure zero residue of pesticides and fungicides. The demand for organic medicinal plants has increased in last decades.

Harvest of raw material must be followed by a hygienic post-harvest handling protocol to ensure appropriate conditions, particularly those of temperature and relative humidity to maintain required moisture level and restrict further deterioration of the raw material because of high temperature inviting post-harvest pests and pathogens contaminating the material.

Good laboratory practices (GLP) and good manufacturing practices (GMPs) are required in order to conform to the guidelines recommended by agencies that control the authorization and licensing to manufacturers and sale of food and beverages, cosmetics, pharmaceutical products, dietary supplements, and medical devices. GMPs are also needed to produce good quality medicinal products. Without all these measures, it is impossible to realize the dream of having a major share in the herbal drug industry, despite having a gold mine of well-documented and well-practiced knowledge of traditional herbal medicines. Therefore, the traditional Indian medicine systems require monitoring at various levels, including policies, quality standards, integration of practices, research models and the complementary integration keeping the public health at the central position. Most Indian traditional medicine systems have great traditions with strong philosophical and experience (in not clinical) base that could play an important role in development of new therapies, drug discovery and development processes.

Keeping the above concerns about abiotic and biotic contaminations of the Ayurvedic herbal medicines in mind, there is an urgent need for systematic scientific studies to quantify the frequency and potential risk of heavy metal poisoning from Indian traditional medicinal system medicines in the different age group of patients through clinical trials. Microbial load can be reduced under permissible level of irradiation, which needs to be investigated for respective pathogens. Unfortunately, such moderation approaches and regulatory mechanisms may not be sufficient to ensure safe use of traditional medicines. Therefore, side by side culturally, there is a need to provide appropriate education that can inform the public about the potential toxicity associated with the different products or the practice (Sharma, 2000; Anonymous, 2004).

While promoting Indian therapies and drugs with scientific evidence, an approach of integrative medicine by selective incorporation of elements of the modern methods, particularly those of diagnosis can be of great help to support therapies. India needs a clear policy for such integration without compromise on the strategies that are science-based. Efforts are needed to establish and validate pharmacoepidemiologic evidence regarding safety and practice of Ayurvedic medicines (Vaidya *et al.*, 2003).

To address all the issues related with research and usage concerns, a team of trained manpower is required in all aspects of MAPs management and use. It may need quality education, with improved curriculum, which shall also help integration of modern technologies (diagnostic) in an attempt to produce hybrid curricula, providing and producing adequately trained graduates for either modern and/or traditional systems. Further, the younger generation should be exposed to the world class facilities available in India or abroad. The policies, plans and research and development activities should be demand driven and market oriented, which can address the present and future needs.

In recent times, significant progress has been made around genomics and proteomics, which has opened new gateways in therapeutics and drug discovery. These developments must be applied to traditional herbal medicines as well. To date, only a small fraction of the vast diversity of plant metabolism has been explored to produce new medicines and other products. The emergence of new herbal genomics research, medicinal plant genomics consortium, together with advances in other omics information may help for the speedy discovery of previously unknown metabolic pathways and enzymes (Chakraborty, 2018). Therefore, for the speedy discovery of unknown metabolic pathways/enzymes, interdisciplinary research on herbal genomics together with large-scale sequencing tools and metabolomics and proteomics is necessary.

Genomics can be used to screen the target molecules of the action and to promote the action of traditional medicines, to identify the new effective components of traditional medicines, and to explore the mechanisms of the effects. It will help to overcome the shortcomings of conventional methods. Application of the theories and techniques of genomics in further studies of traditional medicines would be of great significance in integration of traditional medicines with modern biological sciences and technology and in promoting globalization of traditional medicines. China has already started using genomics and proteomics technologies on Chinese traditional medicine. It is high time for Indian researchers to apply these tools to speed up the process of drug discovery and details of their action.

All the programs related to researchable issues and market concerns call for a mission mode program in the field of MAPs to address the emerging issues. If all the related issues are addressed in right earnest, then India can become a major world player in the field of plant-based products, particularly for MAPs.

8.6 Need of Standardization

As herbal medicinal products are complex mixtures, which originate from biological sources, great efforts are necessary to guarantee consistency and adequate quality. By carefully selecting the plant material and a standardized manufacturing process, the pattern and concentration of constituents should be kept at constant as possible, as this is a prerequisite for reproducible therapeutic results.

A projection is being made that after information technology, herbal technology will be India's biggest revenue earner (Chaudhri, 1996; Pushpangadan *et al.*, 2018). India has capability of playing great role, as supplier of herbal products not only to meet the domestic needs, but also to take advantage of the tremendous export potential. To be a global supplier of herbal medicines conforming to international specification the country must prepare itself to address the issues related with quality evaluation, maintaining the internationally expectable quality standards to produce, and managing the low limit of contamination like microbial counts, pesticides residue, heavy metals counts, radio activity etc. It would be demanding the attention on the following aspects:

- Proper botanical identification of all medicinal plants that are part of one or the other Indian System of Medicine using an authenticated taxonomic key during collection of raw material from nature.
- Considering the contamination of Ayurvedic medicines obtained from the market with heavy metals as one of the major constraints, cultivation of MAPs should be practiced with selection of the fields devoid of heavy metals through soil-testing. Whereas irrigation should be ensured with water devoid of such contamination.
- All herbal ingredients involved in the formulations to be specified by their botanical names besides their popular/common names.
- Processing of medicinal plant raw material in a scientific, economic, and safe way using similar ones as used for modern drugs.
- Isolation and chemical (molecular) characterization of principal components including inorganic constituents, wherever possible.
- Pharmacological and clinical studies to ascertain their efficacy and safety of their dosage in the different age group of patients.
- Standardization to ensure uniformity. The use of medicinal plants in combination to be limited to facilitate analysis and to apply quality control and standardization parameters to herbal drug preparations.
- Establishment of quality testing laboratories for chemical analysis and identification of the active ingredient and potency of the drugs to be used.
- In case of adverse effect, the possible/recommended emergency ant dotes to be utilized to mediate the effect.
- Documentation and information dissemination of all investigation and research.

Such scientific details will project herbal medicines in a proper perspective and help in their sustained growth in global market (Handa and Kapoor, 1995). The World Health Organization currently recommends and encourages traditional herbal remedies in natural health care programs because these drugs are easily available at low cost and are comparatively safe. People's faith in such remedies is reflected in a whooping annual turnover of herbal market besides a healthy annual growth rate and the increasing export potential. This has attracted several large and medium scale pharmaceutical industries and even multinationals to jump on the band wagon herbal medicines.

8.7 International Cooperation

Singh (2019) emphasized that using the transparent international mechanisms of fair and equitable benefit sharing mechanism such as Nagoya Protocol under Convention on Biological Diversity be used to facilitate greater international cooperation in development and use of plant-based phytonutrients, Phyto therapeutics, and Phyto cosmetics. Their use shall ensure community rights, IPR and the right for fair and equitable benefit sharing arising from commercialization benefiting both biodiversity rich countries like India and technology rich countries of the developed world from Europe and Americas.

India needs to focus on research and development in biochemical profiling, identification of active principles in the sources, followed by molecular characterization, and developing agrotechnology and processing technology meeting the requirements of global and national IPR standards for patenting and quality control. In this regard, the international cooperation under the provisions of bilateral system of Nagoya Protocol, which is a supplementary agreement of the CBD, may speed up achieving the above objectives.

References

Anonymous (2004) Indian Herbal Segment Booming. By Darlington Jose Hector, 14 January 2004, Times News Network, News Capsule.

Chakraborty, P. (2018) Herbal genomics as tools for dissecting new metabolic pathways of unexplored medicinal plants and drug discovery. Biochimie Open 16: 9-16.

Chaudhri, RD. (1996) Herbal Drugs Industry. Eastern Publisher, New Delhi, 1st Edn.

Ernst, E. (2002). Heavy metals in traditional Indian remedies. European Journal of Clinical Pharmacology 57:891-96.

George, V, Ijinu, TP, Chithra, MA and Pushpangadan, P. (2016) Can local health traditions and tribal medicines strengthen Ayurveda? Case study 1. Trichopus zeylanicus ssp. travancoricus Burkill ex Narayanan, Journal of Traditional and Folk Practices 4(2):3-16

Handa, SS and Kapoor, VK. (1995) Textbook of Pharmacognosy. Vallabh Prakshan, Delhi, 2nd Edn.

Secretariat of the Convention on Biodiversity (2011) Nagoya Protocol on Access to Genetic Resources and the Fair and Equitable Sharing of Benefits Arising from their Utilization to the Convention on Biological Diversity: text and annex pp 15. Montreal, Canada, Secretariat of the Convention on Biodiversity. DOI: ISBN 92-9225-306-9

Pushpangadan, P, George Varughese, Ijinu, TP and Chithra, MA. (2018) All India coordinated research project on ethnobiology and genesis of ethnopharmacology research in India including benefit sharing. Annals of Phytomedicine 7(1): 5-12. DOI: 10.21276/ap.2018.7.1.2

Saper, RB, Kales, SN, Paquin, J, Burns, MJ, Eisenberg, DM, Davis RB, et al. (2004) Heavy metal content of ayurvedic herbal medicine products. Journal of the American Medical Association 292:2868-73.

Sharma, DC. (2000) India raises standards for traditional drugs. Lancet 356:231.

Singh, Anurudh K. (2019) Nagoya protocol of CBD, mechanism to facilitate international collaborative development of plant-based products: India a case study. International Journal of Phytocosmetics and Natural Ingredients 6:1. doi:10.15171/ijpni.2019.01.

Vaidya, RA, Vaidya, ADB, Patwardhan, B, Tillu, G, Rao, Y. (2003) Ayurvedic pharmacoepidemiology: a proposed new discipline. Journal of the Association of Physicians of India 51:528.

Glossary of Technical Terms Used

Abhyanga: Full body massage with warm oil practiced by two medics who massage to let the warm oil into the tissues of the entire body. This helps loosen and facilitate the removal of accumulated ama (toxins) and the doshas (*vata*, *pitta* and *kapha*) from the body.

Achara or **Aachaar Rasayana** (Behavioural therapy)**:** A unique Ayurvedic concept of mind rejuvenation and to have a calm mind by disciplined training with equal consideration to physical, psychological, food and nutritional and behavioural patterns of an individual (advocated by *Acharya* Charaka).

Agada Tantra: Toxicology, a branch of Ayurveda dealing with poisons.

Agni: Digestive fire; universal principle, which refers to both metabolic activity and human vitality.

Agrotechnology: Agricultural technologies or crop cultivations practices applied for efficient production.

Akasa (Vacuum-Ether): Free open space, ether, sky, or atmosphere.

Akhlat (Humours): They are the liquid part of the human body. The four humours are *Dam* (blood), *Balgham* (phlegm), *Safra* (yellow bile), *Sauda* (black bile). An equilibrium is required between them for a healthy body. Any disbalance in equilibrium results in diseases.

Allopathy: It is an archaic term used to define science based modern medicine system.

Ama: Metabolic waste products and toxins deposited in cells and circulated in the body. Retention of toxins in the blood results in toxaemia. Almost every disease is a result of toxicity omits crisis. Toxins are vital for *prana* (vital life energy), *ojas* (immunity), and *tejas* (cell metabolic energy)

Anuvasan Basti: Enema given with an oily substance.

Asthapana or **Niruha Basti**: Decoction Enema Therapy

Asthi: It is one of the seven components and the bone tissues that supports the body by giving it protection, shape, nourishment, and longevity.

Ayana: Path of action of essence of, the *Rasa*.

Ayurveda: The sacred knowledge of life. Constituted of two words, *Ayur* meaning life and *Veda* meaning knowledge. Ayurveda encompasses the secrets of why Man needs to cooperate with Nature completely in order to insure his wellbeing.

Bhootavidya: The science dealing with microorganisms and evil spirits, demonology (Psychological disorders).

Biodiversity: It refers to the variation of all living organisms, their genetic material, and the ecosystems of which they are a part. It is described at three levels: genetic, species, and ecosystem diversity.

Biodiversity Hotspots: A biodiversity hotspot is a biogeographic region with significant levels of biodiversity and has many endemic species and faces high risk for destruction. Around the world, 36 areas have qualified as hotspots.

Biogeographical Regions: These are geographical areas that are defined based on the biological (plant and animal) species found in them impacting the ecology and habitat, which provides invaluable information to ecologists and natural resources managers for understanding large scale processes that affect species and ecosystems.

Biological Species: It is the most widely accepted species concept. It defines species in terms of interbreeding. Ernst Mayr defined a species as "the groups of interbreeding natural populations that are reproductively isolated from other such groups."

Biosystematics: The study of living organisms based on observational and experimental data on the breeding system for classification of biological units into taxa, making taxonomic decisions, based on relationships, variability, and dynamic of interrelationships.

Brihat Trayi or **Brihattrayi:** "The Great Triad", refers to three early Sanskrit ncyclopaedias of medicine, which are the core texts of the indigenous Indian medical system of Ayurveda.

Convention on Biological Diversity: The Convention on Biological Diversity (CBD) is an international treaty that was adopted at the Earth Summit in Rio de Janeiro in 1992 and entered into force on 29 December 1993. The Convention has three main goals, (i) conservation of biological diversity, (ii) sustainable use of its components and (iii) fair and equitable sharing of benefits arising from use of genetic resources.

Dashamula or **Dash Moola**: Potent magical concoction of ten dried roots of ten different plants, which have been widely used in Ayurveda for ages due to its amazing health benefits.

Dhanvantari or Dhanwantari: The physician of the gods, as per Hindu mythology (Presently, a designation of honour)

Dhatu (Body Tissue): The basic structural and nutritional body factor that supports or nourishes the seven body tissues, the *rasa*, *rakta*, *mamsa*, *meda*, *asthi*, *majja* and *shukra*.

Dosha: Bioenergetic regulation system that governs the functions of the human body and determines people 's individual characteristics, both mental and physical. There are three doshas: *vata*, *pitta* and *kapha*.

Endemism: Refers to a species or a taxon associated with a region. By extension, this term is used to refer to species which are found only in that region.

Ethnobiology: The scientific study of dynamic relationships among peoples, biota, and environments i.e., the way living things are treated or used in different human cultures.

Ethnobotany: A multidisciplinary science that deals with the traditional and natural relationship between human societies and plants.

Ethnopharmacology: Natural sciences research on medicinal, aromatic, and toxic plants linked with socio-cultural studies.

***Ex-Situ* Conservation**: Means conservation of components of genetic material of biological diversity outside their natural habitat.

Experimental taxonomy: The classification of organisms based on experimental facts drawn from different lines of scientific investigations has been termed "experimental taxonomy."

Floristic Diversity: Floral diversity refers to the diversity of plants occurring in a specific region during particular era. It generally refers to the diversity of naturally occurring indigenous or native plants.

Garshan: Full body dry massage with raw silk gloves.

Genetic Diversity: The genetic variation (genes or alleles) within a population and among the populations of a species is generally referred to as genetic diversity.

Genetic Resources: The genetic variability available in gene pool of a species useful for enhancing/improving genetic potential of a cultivated species about agronomic features, resilience against stresses, nutritional traits, etc. over present levels.

Genomics: A branch of biotechnology concerned with applying the techniques of genetics and molecular biology to the genetic mapping and DNA sequencing of sets of genes or the complete genome of selected organisms, with organizing the results in databases, and with applications of the data to identify differences and similarity focusing on the structure, function, evolution, mapping, and editing of genome.

Ghee: Heated and melted butter fat; considered to be the elixir of life.

Granthas: These are the transcriptions (manuscripts/books) of knowledge. In Sanskrit, grantha is literally 'a knot', because of the practice of binding inscribed scripts on palm leaves using a length of thread held by knots.

Intellectual Property Rights: Intellectual property rights (IPR) are legal rights that are conferred to the owner of an intellectual creation. The IPR are granted by means of protection through appropriate legislation, based on the type of creation, that generally include patents, copyright, trademark, industrial designs, geographical indications, trade secrets, protection of layout design of integrated circuits and protection of new plant varieties. The intellectual property right protection entitles the owner of the intellectual property or his assignee the exclusive right to fully utilize the invention/creation for commercial gain generally for a fixed period of time.

International Treaty on Plant Genetic Resources for Food and Agriculture (ITPGRFA): The International Treaty on Plant Genetic Resources for Food and Agriculture, popularly known as the International Seed Treaty, is a comprehensive international agreement in harmony with Convention on Biological Diversity, which aims at guaranteeing food security through the conservation, exchange and sustainable use of the world's plant genetic resources for food and agriculture, as well as the fair and equitable benefit sharing arising from its use. It will implement a *Multilateral System* of access and benefit sharing, among those countries that ratify the treaty, for a list of 64 food and forage crops (the genera and species are listed in Annex 1 to the Treaty).

Jal: Water, fluid

Kamya Rasayan: Kamya Rasayanas are promoters of normal health, used to fulfil a wish or desire or to serve a special purpose (*kama* – desire). It is of four types, Prana Kamya, Medha Kamya, Ayush Kamya and Chakshu Kamya

Kapha: It is one of the three doshas, the dosha derived from the elements of water and earth; responsible for body structure and fluid balance.

Kaumar Bhritya or **Kaumarbhritya:** is one of the Most Important Branch of Ayurveda dealing with the care of the child from the Conception to till the Maturity It is one of the Eight Branches (*Astang*) of Ayurveda.

Kaya Chikitsa or **Kayachikitsa:** It is one of the important branches of Ayurveda. This branch deals with various physical and psychological diseases like Fever, *Rakta-pitta*, *Shotha*, *Prameha*, etc.

Ksiti or **Kshiti** (Earth): In Ayurveda, it refers to "earth" and is mentioned in a list of 53 synonyms, covering land and soil.

Majja: It means the bone marrow. Being one of the seven dhatus it is soft and has the main function of filling up the *asthi* and nourishing the *shukra*.

Mansa: Refers to mind or mental (diseases). One of the seven basic tissues.

Marma: Connection point between matter and consciousness; on the skin there are 107 *marma* points that can be stimulated by touch.

Meda: It is the fat tissue supported by *mamsa* dhatu. The main function of these one of the seven bodily tissues is to support the human body and lubricate it. *Meda's* presence in excess can cause obesity and physical weakness.

Medhya Rasayan: They are group of medicinal plants described in Ayurveda with multi-fold benefits, specifically to improve memory and intellect by Prabhava (specific action).

Mizaj (Temperament): As per this concept, all the three natural sources of drugs have four qualities in different proportion, i. e. heat, cold, dryness and moisture and new *Mizaj* is formed by their interaction. The dominant quality, which prevails forms the temperament of drug.

Nagoya Protocol: It is a supplementary agreement to the Convention on Biological Diversity (CBD) on Access to Genetic Resources and the Fair and Equitable Sharing of Benefits Arising from their commercial utilization. It was adopted on 29 October 2010 in Nagoya, Japan and entered into force on 12 October 2014, 90 days after the deposit of the fiftieth instrument of ratification.

Naimittic or **Naimittika Rasayan:** They can be defined as Rasayana specific to a disease which will improve the vitality of individual towards the specific disease.

Nasya: One of the treatments of panchakarma. consists of a series of oil massages, compresses and nasal irrigation, herbal steam bath for the head.

Nutraceuticals: Food substance that provides food, nutrition and medical or health benefits, including prevention and treatment.

Ojas: Fine material metabolic end product that is produced with proper digestion.

Ojas-vridhias or **Ojas Vriddhi (Increase of Ojas):** It does not cause any disease; it provides satisfaction, strength, nutrition to whole body.

Paleoendemic: Refers to those species that were formerly widespread but are now restricted to a smaller area.

Pancha Mahabhutha (Jung-Wa-Nga): In Ayurvedic medicine, the five elements of which all matter is composed: air, earth, ether, water, and fire, which are physical basis of body impregnated with Atma or soul (life element).

Panchakarma: Ayurvedic cleansing treatment that rids the body of harmful deposits and restores the balance of the doshas.

Panchakarma Therapy: Five-fold therapy, involving five activities, namely, Vomiting, Purgation, *Niruham*, *Anuvaasan*, and *Nasyam* to help eliminate the impurities from the human body.

Pashchat: After

Patent: It is a type of intellectual property right given to innovator/owner for a product or a process and exclude others from making, using, or selling an invention for a limited period. To get a patent, technical information about the invention must be disclosed to the public in a patent application.

Pesticides: Any substance used to kill, repel, or control certain forms of plant or animal life that are pests. It includes herbicide, insecticides nematicide, molluscicide, piscicide, avicide, rodenticide, bactericide, insect repellent, animal repellent, antimicrobial, fungicide, and lampricide.

Pharmacoepidemiological: It is the study of the utilization and effects of drugs in large numbers of people to provides an estimate of the probability of beneficial effects in a population and the probability of adverse effects.

Phylogenetics: A study that relates to the evolutionary development and diversification of a species or group of organisms.

Phytocosmetics: It is a part of cosmetology, which consists of using plants in cosmetics.

Phytogeographical: It is the branch of botany that deals with the geographical distribution of plants, and their influence on the earth's surface.

Phytonutraceutics: The foods containing the Phyto-chemicals are termed as nutraceuticals include various nutrients, dietary supplements, specially designed diets or herbal products of both plant and animal origin.

Pinda Sweda: Full body massage with warm rice and cereal decoction.

Piscicides: Chemical substance which is poisonous to fish. The primary use for *piscicides* is to eliminate a dominant species of fish in a body of water.

Pitta: Dosha derived from the element of fire, responsible for digestive and metabolic processes and the heat balance.

Plant genetic resources: Genetic material of plants, including modern cultivars, landraces, and wild relatives of crop plants, of value as a resource for present and future generations of people.

Prakriti: Primordial matter; the cosmic mother, the divine feminine energy behind creation- the feminine potential from which all form emerges.

Prana: Life-giving respiratory and nervous energy.

Proteomics: It is the large-scale study of proteins. Proteomics enables the identification and adding of new ever-increasing numbers of proteins.

Purva: Before or prior

Rajas: One of the three *maha-gunas*, rajas are the principle that ignites energy, movement, passion, and the ability to act.

Rajasic: A substance, experience, or mental state infused with the qualities of rajas: kinetic energy, movement, passion, and action.

Rakta (Blood)**:** It is one of the seven dhatus (*Rasa*, *Rakta*, *Mamsa*, *Meda*, *Asthi*, *Majja* and *Sukhra*), classified in Ayurveda. The. The word '*Rakta*' is derived from Sanskrit word 'Raj Ranjane' which indicates red colour.

Rasa: Plasma, it is derived from the digested food and is circulated to the entire body by channels. The main function of this first of the seven dhatus is to provide nutrition to all cells of the body and the plasma dhatu.

Rasayana Therapy: Literally means rejuvenation i.e., the therapy that rejuvenates or regenerates body- mind, prevents decay and postpones aging.

Rasayana: A substance that nourishes and tones the entire body; and is believed to enhance immunity, stamina, and longevity.

Reproductive Biology: It refers to both sexual and asexual reproduction. In plants, it covers general plant biology, pollination, pollen-pistil interaction, post-fertilization changes and seed dormancy. Its understanding helps in genetic improvement of plants using conventional or biotechnological methods.

Samhita: It refers to compendium or treatise or Sanskrit text of Ayurveda and the ancient medical science of India. describing ancient theories on human body, etiology, symptomology, and therapeutics for a wide range of diseases.

Sapta Dhatu or **Saptadhatus (Luszung-Dun)**: According to Ayurveda the human body is primarily made up of *saptadhatus,* that are also part of the body's protective mechanism. The seven Dhatus are plasma, blood, muscle, fat, bone, bone marrow and reproductive fluid. In Ayurveda, they are called *Rasa*, *Rakta*, *Mamsa*, *Meda*, *Asthi*, *Majja* and *Sukhra* respectively.

Sattva: One of the three *maha gunas*, *sattva* is the principle that gives rise to equilibrium, clarity, light, intelligence, compassion, insight, and wisdom.

Sattvic: A substance, experience, or mental state infused with the qualities of sattva: light, clarity, intelligence, compassion, and wisdom.

Shalakya: The word Shalakya in Sanskrit, means a probe. Therefore, Shalakya tantra, refers to treatment done using appropriate tools for ingesting medication to the affected parts of the body, commonly, above the clavicle disorders of Ear, Nose, Throat, Eye, Dental, Head and Neck. The branch is also described as Urdhwangchikitsa.

Shalya: *Shalya* means that which creates pain or agony. Shalya tantra is a branch of Ayurveda which deals with surgery and provides a basic understanding of the principles of modern surgery. Area of strength in Ayurvedic surgery is management of *vrana* (wound).

Shashtra Pramaana: It is the scriptural evidence, the direct, ultimate authority on the truth as it is, for it is directly from Sadashiva, the Adi Guru (original Guru), the source of all that is, Veda-Agamas.

Siddhas (Siddharth purusha): It means "one who is accomplished." It refers to perfected masters who have achieved a high degree of physical as well as spiritual perfection or enlightenment.

Similia similibus curantur: Homeopathic axiom expressing the law of similar or the doctrine that any drug capable of producing detrimental symptoms in healthy individuals will relieve similar symptoms occurring as an expression of disease. Symptoms like those of the diseased person.

Shirovirechana (Nasal insufflations therapy): It is the therapeutic cleansing of the head and neck region, which purges excess mucous and toxins from the sinus cavities and surrounding tissues.

Sister Species: Ancestral or taxonomically closely related species.

Shodhana Therapy: It is a purification therapy. The aggravated Doshas from the body are expelled out in this procedure, thereby eliminating the internal causative factors of the disease. The metabolic process causes generation of large quantities of toxic byproducts in the body.

Soma: Refers both to a mysterious sacred plant, and to a drink made from the juice of that plant; the drink is said to be an elixir of life, giving immortality to anyone who drinks it.

Sub-Dosha: Each of the doshas (*vata, pitta, kapha*) has five sub-doshas (or sub-types), that have specific actions, overseeing the mind and emotions as well as the functions of specific organs.

Sukra (Shukra)**:** It is the last of the seven dhatus, the semen or the male reproductive tissue and has the major functions of producing sensation of ejaculation, fondness and strengthens the body.

Tamas: It is the principle responsible for inertia, darkness, heaviness, slowness, sleep, and decay.

Tamasic: A substance, experience, or mental state infused with the qualities of tamas: inertia, darkness, heaviness, slowness, sleepiness, and decay.

Taxonomy: A branch of biology engaged in the classification of organisms, especially according to their natural relationships. It covers the laws of and principles of such classification.

Tejas (Fire): Solar energy; the positive subtle essence of *agni* and of *pitta* that governs intelligence, discernment, enthusiasm, and all types of digestion and transformation; it shares a subtle functional integrity with *ojas* and *prana*.

Trade Related Aspects of Intellectual Property Rights: The Agreement on Trade Related Aspects of Intellectual Property Rights (TRIPS) is a treaty administered by the World Trade Organization (WTO) which sets down minimum standards for forms of intellectual property (IP) regulation. It was negotiated at the end of the Uruguay Round of the General Agreement on Tariffs and Trade (GATT) treaty in 1994. It introduced intellectual property law into the international trading system for the first time and remains the most comprehensive international agreement on intellectual property to date.

Tridosha (Nespa-Sum): In Ayurveda theory of Tridosha originated from the theory of the three elements of the universe. They are *vata* (wind), *pitta* (bile), and *Kapha* (phlegm), corresponding to the three elements of the universe: air, fire, and water.

Triphla: Powders of three dried fruits: Amla, Bahera and Harada, which has unique ability to gently cleanse and detoxify the digestive tract, support regularity, and simultaneously offer deep nourishment to the tissues.

Udvarthana: Full body rub massage with a pulp made from oils and cereals, which is performed simultaneously by two therapists.

Upanishads: Refers to the Sanskrit texts thought to be the direct teachings received from ancient Indian sages or Rishis. These texts were written in a passionate poetic verse describing mystical states and spiritual concepts or in descriptive short stories and dialogues between historical figures. There are hundreds of Upanishads, but only eleven "principal" Upanishads, commented on by the ancient sage Shankara, discussing the concept of Brahman, Atman, Rebirth, and accumulation of Karma.

Vajikarana: One of the eight branches of ayurvedic medicine, dealing with all types of sexual function and dysfunction. The therapy provides strength, potency, virility, sexual excitement, erection of sexual organ and pleasure in intercourse.

Vamana: Refers to therapeutic vomiting which is a medicated emesis.

Vasti (Enemas): It's a therapeutic procedure, in which medicines are administered into the body through lower pathways (anus). It cleanses the accumulated toxins from all over the body.

Vata: One of the three dosha derived from the elements of air and space responsible for the motion sequences in the body. Regulates the activity of mind and body and controls the other two doshas.

Vayu (Air): Refers to the "airy" vital forces of the body. *Vayu* is classified as *Udana Vayu*. Ether; head region; the sense organs, consciousness.

Vedic: Pertaining to the Vedic period of ancient India, from approximately 1750–500 BCE, when the Vedas, a large body of religious texts were written.

Vegetal: Plant based food products.

Vikriti: Refers to an individual's current state of health or body; the specific ratio of three doshas, opposed to the natural ratio.

Virechana or Virechana Karma: It is a purification therapy and one among the five sacred, healing therapies of the Panchakarma. Synonym to purgation, the cleansing the body of toxins by taking laxatives.

Vyadhi-Kshamatva or **Vyadhikshamatva:** The word *vyadhi* meaning is to harm, to injure, to damage, or to hurt. The word *kshamatva* means to composed, to suppress anger or to keep quiet or to resist. So, the word means to be patient towards resisting the disease.